# Static and Dynamic Analysis of Engineering Structures

# Static and Dynamic Analysis of Engineering Structures

## Incorporating the Boundary Element Method

Levon G. Petrosian

Washington District Department of Transportation, USA

Vladimir A. Ambartsumian

Structural Engineering and Structural Mechanics
Armenian National University of Architecture & Construction, Armenia

WILEY

*Registered Offices*
John Wiley & Sons, Inc., 111 River Street, Hoboken, NJ 07030, USA
John Wiley & Sons Ltd, The Atrium, Southern Gate, Chichester, West Sussex, PO19 8SQ, UK

*Editorial Office*
The Atrium, Southern Gate, Chichester, West Sussex, PO19 8SQ, UK

For details of our global editorial offices, customer services, and more information about Wiley products visit us at www.wiley.com.

Wiley also publishes its books in a variety of electronic formats and by print-on-demand. Some content that appears in standard print versions of this book may not be available in other formats.

*Library of Congress Cataloging-in-Publication Data*

Names: Petrosian, Levon G. (Levon Gregory), 1- author. |
    Ambartsumian, V. A. (Vladimir Alexander), author.
Title: Static and dynamic analysis of engineering structures :
    incorporating the boundary element method / Levon G. Petrosian,
    Professor of Structural Engineering and Structural Mechanics,
    Washington District Department of Transportation, USA
    Vladimir A. Ambartsumian, Professor of Structural Engineering and Structural Mechanics,
    Armenian National University of Architecture & Construction, Armenia
Description: First edition. | Hoboken, NJ : John Wiley & Sons, Inc., 2020.
    | Includes bibliographical references and index.
Identifiers: LCCN 2019034987 (print) | LCCN 2019034988 (ebook) | ISBN
    9781119592839 (hardback) | ISBN 9781119592884 (adobe pdf) | ISBN
    9781119592938 (epub)
Subjects: LCSH: Structural analysis (Engineering)–Mathematics. | Boundary
    element methods.
Classification: LCC TA640 .P48 2020 (print) | LCC TA640 (ebook) | DDC
    624.1/7–dc23
LC record available at https://lccn.loc.gov/2019034987
LC ebook record available at https://lccn.loc.gov/2019034988

Cover Design: Wiley
Cover Image: © aaaimages /Getty Images

Set in 11.5/14pt STIXTwoText by SPi Global, Chennai, India

Printed and bound by CPI Group (UK) Ltd, Croydon, CR0 4YY

10  9  8  7  6  5  4  3  2  1

In Memory of
my Teacher, Mentor, and Friend

ALEXANDER IZRAELEVICH TSEITLIN
(1933 – 2011)
Doctor of Technical Science, Professor,
Member of the Russian & International Engineering
Academy of Science

# Contents

# About the Authors

Levon G. Petrosian is a Professor of Structural Engineering and Structural Mechanics. He received his B.S. and M.S. degrees from the Armenian National University of Architecture & Construction. He received a Ph.D. from the Moscow State University of Civil Engineering (MGSU) and the degree of Doctor of Technical Science from the Moscow Research Center of Construction, Research Institute of Building Structures (TSNIISK). Dr. Petrosian is the author of more than 60 scientific works and publications, including monographs and text books. He has been associated with transportation and structural engineering industries for over 40 years as a research engineer and scientist. Prior to his move to the United States, Dr. Petrosian was the Chairman of Department of Structural Mechanics at the Armenian National University of Architecture & Construction as well as the Executive Director and the Head of the Armenian Earthquake Engineering Research Institute. Currently, Dr. Petrosian serves as the Chief of the Plan Review Division at the District of Columbia Department of Transportation.

Vladimir A. Ambartsumian, Professor of Structural Engineering and Structural Mechanics received B.S., M.S., Ph.D., and the degree of Doctor of Technical Science from the Armenian National University of Architecture & Construction. He was a Professor of Structural Mechanics at the Armenian National University of Architecture & Construction. He authored and co-authored numerous scientific publications.

# Preface

This book presents both methodological and practical purposes of static and dynamic analysis of engineering structures. Therefore, accounting of all methods is accompanied by the solution of the specific problems, which not only are illustrative material, but also may represent independent interest in the study of various technical issues. It provides an overview and applications of all modern as well as well-known classic methods of calculation of various structure mechanics problems in one place.

The book generalizes all existing classical and modern methods of calculations of engineering problems and structural mechanics problems over the span of the last 50 years. Through comprehensive analysis, the book shows analytical and mechanical relationships between classical and modern methods of solving boundary value problems.

The book features extensive use of the generalized functions for describing the impacts on structures, and substantiations of the methods of the apparatus of the generalized functions.

The book illustrates the modern methods of static and dynamic analysis of structures and the methods for solving boundary value problems of structural mechanics and soil mechanics.

The book includes examples of solving different problems of static and dynamic calculation of beams, plates, shells, multi-body systems, regular structures, bridge structures, and underground constructions.

The book provides a wide spectrum of applications of modern technics and methods of calculation of static, dynamic, and seismic problems of engineering design.

The book shows which methods and techniques should be used for specific and complex problems. These methods are most effective for solving deferent problems of static and dynamic analysis of structures.

Finally, the book presents a new model of a cohesive elastic base covering a wide range of properties of real soils. The limiting cases of the model are the Winkler base and the isotropic elastic half-space.

The book provides numerous solutions of various static and dynamic problems of the theory of elasticity.

Levon G. Petrosian
Professor of Structural Engineering and Structural Mechanics
Department of Transportation
Plan Review Division Chief
Washington DC, USA

# Introduction

The classical methods of structural mechanics widely used for static and dynamic design of structures, in particular the methods of force and displacement/deformation, are based on very transparent mechanical representations associated with the replacement of a given structure to some other, more convenient to calculate "primary system," that is loaded in such a way that the entire primary system or one of its parts is turned into a given. In the theory of elasticity and in mathematical physics a similar method of solving the boundary-value problems is used and considered classical; the method of potential. When applying this method, the assigned domain is actually substituted by some unlimited domain, loaded with some additional load, such that the solution in the given domain and chosen unlimited domain (analog of primary system) are identical. The reason for choosing the primary system in a form of unlimited domain, for which solutions of the corresponding boundary-value problems are constructed, has proven to be very successful and has found development in such engineering methods as the method of compensating loads, extended domains, and boundary elements, as well as in purely mathematical methods of generalized solutions (GSMs), boundary integral equations, and delta-transforms. From the listed methods, which have progressed in the recent years, the greatest development was obtained in the method of boundary integral equations with the discreteness of the boundary, also called the method of boundary element by the analogy with the known finite element method (FEM). Today FEM is the main tool of calculations and analysis of construction in design and research practice. Methods of the Boundary Element Method (BEM) type have proved to be very effective to successfully compete in plane (two-dimensional) and spatial (three–dimensional) problems with FEM and Finite Difference Method (FDM), since they do not require discreteness of the entire domain, but only its boundary. This leads to per-unit reduction of the dimensionality of a problem with more effective computational algorithms. Another advantage of the BEM is the ability to solve problems in unlimited domains, which is paramount for the dynamics of constructions, soil mechanics, and other areas of the theory of elasticity and structural mechanics.

In this book, along with the general approach to the construction of boundary equations of different types, two new methods are applied; Delta Transformation Method and Spectral Method of Boundary Elements (SMBEs) which make obtaining effective solutions of various boundary-value problems of structural mechanics possible. Application of the methods is illustrated by solving a number of problems in the calculation of structures on static and dynamic impacts.

Contemporary structural mechanics and theories of construction investigate stresses on many objects, calculation of which is associated with the solution of boundary-value problems for differential operators. In this case, in contrast to the classical problems of mathematical physics, the boundary-value problems of structural mechanics have higher order and more complex structures. Therefore, both general methods of solving differential and integral equations, and specific engineering methods purely based on mechanical performances are widely used. In this book we will examine the methods of solving boundary-value problems; typical for structural mechanics and related fields. All of these methods, while different in form, can be united by one basic procedure – the Integral Transform, which is either an integral part of the method itself, or is used for its substantiation or conclusion. In the modern technical and physical literature, the Method of Integral Transforms is presented in essence as apparatus connected with the use of classical Fourier transform, Laplace, Melina, Hankel, as well as the transforms of Kantorovich-Lebedev, Meler-Fock, and others. Each of the transforms is obtained within a comparatively clearly outlined field of application.

At the same time, the variety of problems which are encountered in the various sections of science cannot be solved only with the aid of traditional integral transforms. Therefore, further development of the method of integral transforms should proceed in the direction of expanding the quantity of practical suitable kernels, and generalization and necessary application of the methods of constructing the kernels of transforms. This book presents a general approach to the method of integral transforms based on the spectral theory of the linear differential operators, and provides new transforms which will aid us in solving various problems relevant to bars, beams, plates, and shells, in particular.

Another equally important task is to show the close relationship of integral transforms with different modern methods and with types of potentials; in particular, with the method of delta-transform (A.I. Tseitlin [370]), which may be the basis for a mathematical foundation of many methods of structural mechanics.

The book pursues both methodological and practical purposes, and the accounting of all methods is accompanied by solutions of the specific problems, which are not merely illustrative in nature, but may represent independent interest in study of various technical issues. Two special features of the book are an extensive use of the generalized functions for describing the impacts on structures and substantiations of the methods of the apparatus of the generalized functions. In structural mechanics the theory of the generalized functions does not yet apply as thoroughly as in other sections of mechanics and physics. In this book it is shown that the theory of generalized functions allows us to obtain broad generalizations of the classical methods of structural mechanics, to simplify solutions of many problems, and identify the relationships among the methods.

This book illustrates the modern methods of static and dynamic analysis of structures and the methods for solving boundary-value problems of structural mechanics and soil mechanics based on the application of boundary equations. The systems with non-linear and variable elastic characteristics are examined. The fundamentals of the general theory of oscillations are considered. Examples of solving different problems of static and dynamic calculation of beams, plates, shells, multi-body systems, regular structures, bridge structures, underground constructions, and structures on an elastic foundation are given according to the methods presented in the book. The book also analyzes the impact of seismic influences on regular structures.

Further, the book offers a method of physical realization of the dynamic systems based on the examples of an elastic-viscous foundation with internal friction. The internal friction in the foundation is described by the models of Kelvin-Voigt and Sorokin.

Chapter 1 shows the main methods used in the dynamic analysis of bars, plates, and shells. Historically, one of the first methods of solving equations of mathematical physics is the method of separation of variables. The application of this method is described in the examples of solving problems of free oscillation of rectangular and circular membranes.

The range of problems which can be solved accurately in a closed form is limited, and therefore, there are often variational methods used in practice. The application of two variational methods – the Rayleigh-Ritz method and the Galerkin method are shown in the examples of the transverse oscillations of bars. The use of integral equations, in particular the reduction of problems with oscillation of one-dimensional structures to the integral equations of Volterra of the second kind and Fredholm of the second kind, are demonstrated. The methods used for numerical solutions of problems of dynamics are described. These are the FEM and FDM. In either the FEM or FDM the whole domain of the partial differential equations (PDEs) requires discretization. The FEM is presented in an example of a plane problem of the theory of elasticity. A diagram of solving a problem of non-stationary oscillations of a cantilever plate is shown. The oscillations of a compressed bar are solved using the FDM.

The BEM originated from the works carried out by several research groups during the 1960s in the application of boundary integral equations for the solution of engineering problems. These researchers were looking for a different solution from the FEM which was starting to become more widely established for computational analysis of structural mechanics problems.

The Boundary Integral Equations Method (BIEM) is generally found in the methods of Theory of Potential and Boundary Integral Equations, but the basic features and idea of the method of boundary equations can be also found in the classical methods of structural mechanics that have developed considerably earlier than the corresponding methods of mathematical physics. BIEM in structural mechanics were known in the western countries through the work of former Soviet Union researchers and scientists such as N.I. Muskelishvili, S.G. Miklin, V.D. Kupradze, V.Z. Parton, P.I. Perlin, G.J. Popov, A.I. Tseitlin, and Y.V. Veryuzhsky. At that time these methods were considered to be difficult to implement numerically.

Successful approaches of structural mechanics to the calculation of complex systems by replacing them with simpler systems and making them more accessible for calculations carried all signs of the Methods of Boundary Equations. Since structural mechanics initially developed mainly as a science of bar systems, the application of such classical methods as methods of force, displacement, or mixed method led to the boundary equations in a discrete set of joints.

Thus, as a rule, auxiliary tasks for which one-dimensional analogues of Green's function – a unit reaction and unit displacement, were determined and corresponded to elements of the system (e.g. the horizontal and vertical elements of the frame), so that the construction of auxiliary conditions was associated with the narrowing of the domain to several sub-domains. The first one-dimensional analogues of the BEM appeared in structural mechanics with an introduction of infinite systems, which initiated the Compensating Loads Method (CLM), Extended-Domain Method (EDM), and BEM, as well as in purely mathematical methods of GSM, BIEM, and the Delta-Transform Method (DTM). Among these methods developed in recent decades, the most developed is the method of Boundary Integral Equations with the discretization of the boundary, also known as the BEM. BEM is similar to the method of finite elements, which is now the major instrument for actual calculations and analysis of structures in the design and research practice. The BEM is a technique for solving a range of engineering and physical problems. The heart of the BEM technique lies in the integral equation formulation for a given boundary value problem. The mathematical basis of this approach, of course, is classical Green's function, and as mentioned earlier it corresponds with the influence function of structural mechanics. The BEM has the distinction in and advantage of the fact that only the surfaces of the domain need to be meshed. Methods such as boundary elements have proved to be very effective, and they successfully compete in two and three-dimensional problems with the (FEM) and the (FDM), because unlike these methods, BEM does not require sampling (discreteness) of the entire area, but only its border, which decreases dimension of the problem by one unit and leads to more efficient computing algorithms. The advantages in the BEM arise from the fact that only the boundary or boundaries of the domain of the PDEs require sub-division to produce a surface or boundary mesh.

Another advantage of BEM is a possibility of solving problems in an unbounded domain, which is very important for the structural dynamics, soil mechanics, and other fields of the theory of elasticity and structural mechanics.

This book describes both a general approach to the construction of various types of boundary equations and a spectral BEM, which allows us to obtain effective solutions of various boundary value problems of structural mechanics. The application of SMBE is illustrated by solving a number of problems on the structural analysis of the design of construction on the static and dynamic impacts.

The issue of the equivalence of boundary and initial value problems with the inhomogeneity in the boundary (initial) conditions and the right side of the equation in respect to the design of structures on the static and dynamic loads are considered. This book shows a "standard" form of the problem, obtained by using the delta transform and allowing replacement of the boundary or initial conditions by a certain corresponding load on the structures in the form of a delta-function and its derivatives. The construction of the "standard" form is illustrated on the simplest examples of the boundary value problem for beams and the Cauchy problem for the system with one degree of freedom.

There is a general scheme for solving the problems of the theory of elasticity by the CLM in the first chapter.

Chapter 2 examines the application of BEM in the multidimensional problems. The integral equations of BEM are considered; construction of the boundary equations using the method of delta-transform is given; and three basic schemes for constructing boundary integral equations using the DTM are analyzed. Simple proof of the complete equivalence of two basic versions of BEM – direct and indirect – is given. The SMBE, based on expansion of differential operators considering in the extended domain is proposed. In particular, for this purpose the Fourier and Hankel transforms, as well as expansion of beam functions widely used in structural mechanics can be used. Problems described by the integro-differential system of equations are examined. As an example, a contact problem for a plane structure with stress-strained conditions in a certain domain is described by a linear differential operator, some given boundary conditions, and contact conditions with a linear-deformed medium.

Chapter 3 outlines the issues of vibrations of bars, arches, and combined systems. The free nonlinear vibration systems with one and many degrees of freedom, and the systems with distributed mass at various dependencies of the restoring force from moving are examined. The values of the periods of free vibration are obtained by various methods, such as direct integration of the equations of motion and the use of asymptotic methods.

A study of free vibrations of a beam with an arbitrary law of variation of the cross section is also discussed in the third chapter. Classical results for the oscillation problems of a cantilever beam of minimum mass are shown. The outlined solutions have specific applications in the dynamic structural analysis. In particular, the chapter shows the application of the solution of the problem of forced oscillations of the shifted flexural-cantilever beam under the seismic analysis of the structures.

Along with the study of the oscillations of elements with a straight axis, the oscillations of curvilinear elements, in particular circular arches and rings, are also examined. These include the equation of oscillations of S.P. Timoshenko for circular arches and rings, as well as some simple results on values of natural frequencies of these elements.

The third chapter further describes free vibrations of combined systems such as "rigid beam-flexible arch" considering the dynamic thrust when the axis of the arch is outlined by a square parabola and a circular arch.

The chapter shows an experimentally-theoretical method for calculating the combined systems taking into account the joint dynamic work of the basic span to adjacent bridge trestles using the factor "increasing" frequency.

We show a method for seismic analysis of combined systems with regard to their length using real seismograms of powerful earthquakes. There are general equations of motion for a system with a finite number of degrees of freedom in a spatial deformation structures.

Chapter 4 is devoted to the vibrations of plates and shells. We show the optimum design of the shear cantilever plate. The minimum of the total mass of the plate is considered as the optimality criterion. The problem is solved considering the limitation of the thickness of the plate.

This chapter gives the experimental results of free vibration of cantilever plates with rectangular holes. We also show a satisfactory match between the results of theoretical calculations and the experiment, and present the asymmetric vibrations of thin elastic spherical shells and equations for determining the frequency of a hemisphere with free ends.

The application of the SMBE to the oscillation of the plates on an elastic foundation is examined. In particular a spectral BEM to the oscillations or free vibration of plate of arbitrary shape lying on Winklerian foundation and being under a load is provided.

Chapter 5 examines issues of elastic waves spread and their interaction with engineering structures. The chapter examines the problem of propagation of elastic waves in inhomogeneous layered medium and analyzes cases of change of shear modulus and the density of the exponential and power law. This chapter solves the problem of interactions of elastic waves with a semi-cylindrical base structures. The issues of interaction of seismic waves on the tunnel lining are examined. An approach to the study of stress around a circular lining, located near the surface of the half-space, is outlined. The chapter proposes a method for studying the dynamic stresses affecting the cross-section lining. Numerical results are stated.

Chapter 6 describes features of solving dynamic methods of boundary equations. It is known that most building materials have damping properties, characterized by the frequency's independent loss. However, there is a certain dependency of damping parameters (coefficients loss, decrement) on the frequency of the deformation for soils in the experiments. Frequency-dependent losses are typical for a number of modern materials, such as certain plastics. It is therefore, a very important task to describe the dynamic behavior of soil and structural materials possessing both frequency-independent and frequency-dependent internal friction. We discuss the types of problems in relation to the construction of the Green function for the corresponding environment. From the theory of dynamic systems, it follows that between the real and imaginary parts of the transmission function of the casual system, certain integral dependencies which define the formulas of Hilbert transform, must exist. This chapter presents a new way of calculation of Hilbert transform through Fourier transformation of corresponding functions, which simplifies the problem, due to extensive Fourier transforms tables. The presented method of calculating the Hilbert transform allows to comparatively easily analyze the aprioristic models of the dynamic systems based on the creation of complex rigidity or complex pliability of system. This method makes it possible to analytically determine one of the components of the complex rigidity in accordance with the experimentally defined other component. A few examples of the calculation of the Hilbert transform of delta-functions, Heaviside function, and a power function are examined.

The above-described method of calculating the Hilbert transform is applied to the analysis of dynamic models bases of internal friction, described by the hypothesis of Voigt, Sorokin, and Schlippe-Boc. Further, construction of Green's function for the unlimited plate, which lies on Winklerian type elasto-viscous base with the arbitrary dependence of the parameters of the complex coefficient of bed on the frequency is examined. The importance, from a practical point of view, of the task on oscillations of the system with one degree of freedom, whose elastic-viscous properties (parameters) depend on frequency of oscillations, is considered.

The question of construction of the Green's function for one-dimensional dynamical systems with damping described by different model frequency-independent and frequency-dependent internal friction is examined. We examine stationary one-dimensional wave tasks, based on an example of the longitudinal free vibrations of the isotropic uniform unlimited bar with unit harmonic load in the central section or the lateral oscillations of string. The following basic models of medium are investigated:

1. common linear model;
2. elastic-viscous model of Voigt; and
3. the model of the frequency-independent elastic-viscous resistance.

The construction of the Green's function for the non-stationary transverse free vibration of a bar and plate taking into account the frequency-independent elastic-viscous resistance is considered. A numerical realization of a direct BEM for the solution of plane stationary problems of the linear theory of elasticity, taking into account the internal friction, is presented.

An urgent problem of structural mechanics, connected in particular, with the development and improvement of foundation engineering, a major trend in the construction business, is the theory of analysis of structures on elastic foundation. This theory provides for reliability and efficiency of some of the most popular and important structures, such as; foundations of buildings, road and airport paving, coating slopes of hydraulic structures, dams, base layers of floors of industrial buildings, power floors of test cases, and more.

Improving the methods of analysis of structures based on actual soil properties is one of the major hurdles in creating a more efficient and reliable design solution, and preventing undesirable consequences in construction on weak and subsiding soils, and in seismically active areas near mining and blasting. This problem has become very significant after the devastating earthquakes in Armenia, China, Chili, Italy, Japan, Haiti, Pakistan, and Philippines in connection with large-scale reconstruction work and the revaluation intensities of a seismically active zone.

Chapter 7 is devoted to the static and dynamic analysis of structures on elastic foundation. Description of a new model of cohesive elastic foundation, covering a wide range of properties of real soils, is examined. The limiting cases for this model are the Winkler foundation and isotropic elastic half-space. The proposed model, along with the versatility, has several advantages over the earth foundation models currently applied in practical calculations. In particular, under a concentrated load the model gives the final displacements and stresses, allows for jumps of displacements on the surface of the base, and does not result in endless movement solutions of plane problems. The proposed model has solved a number of problems in the analysis of structures on the elastic foundation, such as; membranes, hard foundation, bare beams, and slabs; and free oscillations of rectangular foundation slabs are also examined. This chapter shows techniques for determining the model parameters under the results of experimental studies of soils.

The application of SMBE to the calculation of structures on the elastic foundation is further examined. The equilibrium of the membrane, bend of the non-isolated beams and plates with the free ends, lying on the basis with the proposed base kernel, are examined.

The numerical solution of contact task for the plane and axisymmetrical (axisymmetric) rigid die, which rests on the basis with the proposed kernel and a new iteration technique for solving contact problems, are given. The proposed generalized model of elastic foundation due to the regularity of its kernel paves the way for the application of effective methods of solving contact problems and problems of design structures on elastic foundation, associated with different ways of directly reducing the formulated problem to the solution of the integral equation of Fredholm of the second kind. The effectiveness of this method is determined by the ability to apply a simple iterative procedure for obtaining the most important characteristic (variable) – contact pressure. Then the calculation of structures can be produced with the combined action of external load and obtained contact pressure. To illustrate the method of calculation a solution to the problem of the bending of the beam of finite length, free lying on the elastic foundation, described by the generalized model is given.

The authors offer methods and applications (not always a simple task considering boundary-value problems), in a sufficiently simple and clear way in order to make it easily accessible for readers with mathematical and technical background at university level. However, for comprehension of the book, initial knowledge in such areas of mathematics as the theory of the generalized functions, transforms, and theory of linear integral equations, is necessary. The specified mathematical apparatus in recent decades has widely penetrated the technical sciences, especially structural mechanics, and knowledgeable and sophisticated readers should be familiar with it. Within the framework of the limited size of this publication, dedication was made to emphasize BEM calculations of constructions, however it is not possible to illustrate all aspects of the rapidly developing areas of BEM. One of the purposes of this publication is to familiarize the readers with the different methods of building the boundary equations, including those based on the orthogonal expansions that lead to the boundary algebraic equations.

The techniques of solving boundary integral equations, the choice of boundary elements, approximation of functions, etc. remain beyond the scope of this work. The readers who are interested in these issues should pursue monographs and numerous articles by M.H. Aliabadi, P.K. Banerjee, R. Butterfield, C.A. Brebbia, S. Walker, T.A. Cruse, F.J. Rizzo, P. Fedelinski, S. Hirose, S. Mellings, V.Z. Parton, P.I. Perlin, U.V. Veryuzhsky, J.O. Watson, and others.

The book is intended mainly for professionals and specialists in the field of structural mechanics and related areas. In preparing the book, the authors tried to make it accessible not only for scientists, researchers, and graduate students in the field of structural mechanics and related areas, but also for engineers working in design centers and organizations.

In the appendix to the book printouts are given of the programs of solution of contact problem for the base with the regular kernel, and tables of the contact stresses obtained with solution of plane and axisymmetrical (axisymmetric) contact problems.

Additionally, the book can serve as a guide for students of technical colleges in the study of the relevant sections for the course of structural mechanics.

This book was written in collaboration with Dr. Ambartsumian, who passed away several years prior to its completion and publication. I would like to acknowledge his contribution to this monograph and his dedication to science and research.

I wish to take this opportunity to express my appreciation to my former teachers and colleagues who over the years have so graciously advised and encouraged discussions that led to preparation of this text. In particular, I gratefully acknowledge indebtedness to professors A.I. Tseitlin, A.R. Rzhanitsin, B.G. Korenev, N.N. Leontiev, V.I. Travush, D.N. Sobolev, A.A. Babloyan, V.A. Ilichev, E.E. Khachiyan, M.A. Dashevsky, V.A. Smirnov, and A.W.Taylor.

I would especially like to thank my wife, Natalie, and my daughters, Galina and Irina, for their support, patience and love.

Finally, I want to thank John Wiley and Sons Inc. and the staff for their effective cooperation and their great care in preparing this edition of the book.

The work was spread between authors the following way:

Paragraphs 1.1, 1.2, 1.3, 1.4, and 1.5 of chapter 1; and paragraph 5.1 of chapter 5 are written together. Paragraphs 1.6, 1.7, 1.8, 1.9, 1.10, 1.11, and 1.12 of chapter 1; paragraphs 2.1, 2.2, 2.3, 2.4, and 2.5 of chapter 2, paragraphs 3.8, 3.9, 3.10, and 3.11 of chapter 3; paragraph 4.4 of chapter 4, chapter 6, and chapter 7 of the book are written by L.G. Petrosian:

Paragraphs 3.1, 3.2, 3.3, 3.4, 3.5, 3.6, and 3.7 of chapter 3; paragraphs 4.1, 4.2, and 4.3 of chapter 4; paragraphs 5.2, 5.3, 5.4, and 5.5 of chapter 5 of the book are written by V.A. Ambartsumian.

I am deeply grateful to those users of this book who have been kind enough to write me their impressions and criticisms. Any further comment and suggestion for improvement of the book will be gratefully received.

# Chapter 1
# Methods of Dynamic Design of Structural Elements

## 1.1  The Method of Separation Variables

Let us briefly present the substance of the Fourier Method or a method of separation of variables. The *separation of variables* is the process in which solutions are found for separable differential equations. The method of separation of variables relies upon the assumption that a function of the form $u(x, t) = X(x)T(t)$ will be a solution to linear homogeneous partial differential equations (PDEs) in $x$ and $t$. This is called product solution, and provided the boundary conditions are also linear and homogeneous, this will also satisfy the boundary conditions. This technique works because if the product of functions of independent variables is a constant, each function must separately be a constant as well. Success requires the choice of an appropriate coordinate system and may not be attainable at all depending on the equation. Consider the problem of oscillations if the string is fixed at the ends. The problem is reduced to the solution of the equations of hyperbolic type:

$$\frac{\partial^2 u}{\partial t^2}(x, t) = a^2 \frac{\partial^2 u}{\partial x^2}(x, t). \tag{1.1.1}$$

Let $u(x, t)$ denote the vertical displacement of the string from the $x$ axis at position $x$ and time $t$. In the derivation of (1.1.1) it was assumed that the extension of individual sections of the string during the oscillation does not

*Static and Dynamic Analysis of Engineering Structures: Incorporating the Boundary Element Method,*
First Edition. Levon G. Petrosian and Vladimir A. Ambartsumian.
© 2020 John Wiley & Sons Ltd. Published 2020 by John Wiley & Sons Ltd.

occur and, therefore, according to the Hooke's law, the tension $T_0 = |T|$ does not depend on time or on $x$. The axis $x$ coincides with the direction of a string in equilibrium position. The string is a thin string that does not resist bending, and is not associated with a change in its length.

The string has length $l$: $(0 < x < l)$. The function $u(x,t)$ satisfies the initial conditions.

The conditions that specify the initial state of the system are:

$$u(x, 0) = u_0(x)$$

$$\frac{\partial u}{\partial t}(x, 0) = u_1(x). \tag{1.1.2}$$

The conditions at the boundary of the problem are:

$$u(0, t) = 0 \quad u(l, t) = 0. \tag{1.1.3}$$

The idea of separation of variables is simple. The idea is to assume the solution to the PDE (1.1.1) has a special form of solution, namely $(x, t) = X(x)T(t)$. We attempt to convert the given PDE into several ordinary differential equations. When a problem is posed, such as our problem for $u(x, t)$, one can look for a product solution in the form of $u(x, t) = X(x)T(t)$. The solution can be done by inserting $X(x)T(t)$ into the PDE for the variable $u$, and then separating the variables so that each side of the equation depends on only one variable. Once the equation has been broken up into separate equations of one variable, the problem can be solved like an ordinary differential equation. We will first seek particular solutions of Eq. (1.1.1) that are not equal to zero and satisfying the boundary conditions (1.1.3), in the form of:

$$u(x, t) = X(x)T(t). \tag{1.1.4}$$

Substituting (1.1.4) in (1.1.1), we come to equations:

$$T''(t) + a^2 \lambda T(t) = 0; \tag{1.1.5}$$

$$X''(x) + \lambda X(x) = 0 \tag{1.1.6}$$

where $\lambda$ is constant. The value of $\lambda$ is not specified in the equation, finding the values of $\lambda$ for which there exists a non-trivial solution of (1.1.1) satisfying the boundary conditions. For obtaining the non-trivial solutions of (1.1.4) it is necessary to find non-trivial solutions satisfying the conditions:

$$X(0) = 0, \quad X(l) = 0. \tag{1.1.7}$$

This brings us to the problem of Sturm-Liouville or eigenvalue problem. In mathematics, a certain class of PDEs are subject to extra constraints, known as boundary values, on the solution.

The eigenvalues of this problem are the numbers:

$$\lambda_\kappa = \left(\frac{\pi\kappa}{l}\right)^2 \quad (k = 1, 2, \dots)$$

These eigenvalues correspond to the normalized unique eigenfunctions:

$$X_k(x) = \sqrt{\frac{2}{l}} \ \sin \frac{\pi\kappa x}{l}$$

when $\lambda = \lambda_\kappa$, Eq. (1.1.5) has the general solution

$$T_k(t) = a_k \cos \frac{k\pi at}{l} + b_k \sin \frac{k\pi at.}{l}..$$

Therefore the function

$$u_k(x, t) = X_k(x)T_k(t) = \left(a_k \cos \frac{k\pi at}{l} + b_k \sin \frac{k\pi at}{l}\right)\sin \frac{k\pi x}{l}$$

satisfies Eq. (1.1.1) and the boundary conditions (1.1.3) for any $a_k$ and $b_k$. The solution of Eq. (1.1.1), satisfying the conditions (1.1.2)–(1.1.3), is found in the form of a series

$$u(x, t) = X_k(x)T_k(t) = \sum_{k=1}^{\infty} \left(a_k \cos \frac{k\pi at}{l} + b_k \sin \frac{k\pi at}{l}\right)\sin \frac{k\pi x}{l} \quad (1.1.8)$$

If this series converges uniformly and can be twice differentiated term by term, the sum of the series will satisfy Eq. (1.1.1) and the boundary (1.1.3). Determining constants $a_k$ and $b_k$, such that the sum of the series (1.1.8) satisfies the initial condition (1.1.2), the following equations are obtained:

$$u_0(x) = \sum_{k=1}^{\infty} a_k \sin \frac{k\pi x}{l}, \quad (1.1.9)$$

$$u_1(x) = \sum_{k=1}^{\infty} b_k \frac{k\pi a}{l} \sin \frac{k\pi x}{l}. \quad (1.1.10)$$

Equations (1.1.9) and (1.1.10) give a decomposition of functions $u_0(x)$ and $u_1(x)$ in the Fourier sine series in the interval $(0, l)$.

The coefficients of these decompositions are calculated from the known formulas:

$$a_k = \frac{2}{l} \int_0^l u_0(x) \sin \frac{k\pi x}{l}\, dx, \qquad b_k = \frac{2}{k\pi a} \int_0^l u_1(x) \sin \frac{k\pi x}{l}\, dx.$$

So, while the sum (1.1.8) can be very complicated, each term is quite simple. For each fixed $t$ and each fixed $x$ the term $\sin \frac{k\pi x}{l} \left( a_k \cos \frac{k\pi a t}{l} + b_k \sin \frac{k\pi a t}{l} \right)$ is just a constant time $\sin \frac{k\pi x}{l}$, and constant time $\cos \frac{k\pi a t}{l}$ plus a constant time $\sin \frac{k\pi a t}{l}$ for each fixed $x$. As $x$ runs from 0 to $l$, the argument of $\sin \frac{k\pi x}{l}$ runs from 0 to $k\pi$, which is $k$ half-periods of sin. If the string has density $\rho$ and tension $T$, then $a = \sqrt{\dfrac{T}{\rho}}$ .

The method of separation of variables is one of the most widespread methods of the solution of tasks on oscillations of bars, plates, shells, membranes, and other constructive elements.

We will illustrate this method on an example of a problem of free oscillations of a rectangular membrane. In the Cartesian coordinates, oscillations of a membrane are described by the wave equation of the type:

$$\frac{\partial^2 w}{\partial x^2} + \frac{\partial^2 w}{\partial y^2} - \frac{1}{\alpha^2} \frac{\partial^2 w}{\partial t^2} = 0 \tag{1.1.11}$$

$$\alpha^2 = \frac{T}{\rho}; 0 \le x \le a; \ 0 \le y \le b$$

where $W(x, y, t)$ – displacement of the membrane; $T$ – tension of the membrane; $-\rho$ – density per unit area; and $a$ and $b$ – dimensions of the membrane in the plan.

Boundary and initial conditions are as follows:

$$W(0, y, t) = W(a, y, t) = W(x, 0, t) = W(x, b, t) = 0$$

$$W(x, y, 0) = \varphi(x, y)$$

$$\frac{dw}{dt}(x, y, 0) = \psi(x, y) \tag{1.1.12}$$

The particular solutions of Eq. (1.1.11) we are looking for are in the form of:

$$W(x, y, t) = T(t) \times X(x) \times Y(y) \tag{1.1.13}$$

Substituting (1.1.13) into (1.1.11) and dividing $T(t) \times X(x) \times Y(y)$, we will obtain:

$$\frac{X''}{X} + \frac{Y''}{Y} - \frac{1}{\alpha^2}\frac{T''}{T} = 0 \qquad (1.1.14)$$

Designating (denoting):

$$\frac{X''}{X} = -k_1^2; \qquad \frac{Y''}{Y} = -k_2^2; \qquad \frac{T''}{\alpha^2 T} = -\omega^2 \qquad (1.1.15)$$

we obtain the following differential equations:

$$T'' + (\alpha\omega)^2 T = 0$$

$$X'' + k_1^2 X = 0 \qquad (1.1.16)$$

$$Y'' + k_2^2 Y = 0$$

With a view to (1.1.15) from (1.1.14) we will obtain the interrelation

$$\omega^2 = k_1^2 + k_2^2 \qquad (1.1.17)$$

The solutions of (1.1.16) will be

$$T(t) = A\cos\alpha\omega t + B\sin\alpha\omega t$$

$$X(x) = C\cos k_1 x + D\sin k_1 x \qquad (1.1.18)$$

$$Y(y) = E\cos k_2 y + F\sin k_2 y$$

Using boundary conditions, we find:

$$C = E = 0$$

$$\sin k_1 a = 0 \quad \sin k_2 b = 0$$

$$k_{1m} = \frac{m\pi}{a}; \quad k_{2n} = \frac{n\pi}{b};$$

$$(m = 1, 2, ...) \quad (n = 1, 2, ...)$$

From (1.1.17) follows that:

$$\omega_{mn}^2 = k_{1m}^2 + k_{2n}^2 = \pi^2\left(\frac{m^2}{a^2} + \frac{n^2}{b^2}\right) \qquad (1.1.19)$$

The general solution of the task bearing in the mind (1.1.13) is represented in the form of:

$$W(x,y,t) = \sum_{m=1}^{\infty} \sum_{n=1}^{\infty} (A_{mn} \cos \alpha\omega_{mn} t + B_{mn} \sin \alpha\omega_{mn} t) \sin \frac{m\pi}{a}x \sin \frac{n\pi}{b}y$$

$$(1.1.20)$$

The coefficients $A_{mn}$; $B_{mn}$ are determined from the initial conditions (1.1.12), using the orthogonality of the eigenfunctions of $X_{mn}(x)$ and $Y_{mn}(y)$ in the intervals $0 \leq x \leq a$; $0 \leq y \leq b$.

Thus, the use of the method of separation of variables made it possible to present a solution to the problem in the form of a series in which each term is the product of three functions depending on $x$, $y$, and $t$.

We will consider the free oscillations of a circular membrane. The differential equation of oscillations of a circular membrane in polar coordinates has the form:

$$\frac{\partial^2 w}{\partial r^2} + \frac{1}{r}\frac{\partial w}{\partial r} + \frac{1}{r^2}\frac{\partial^2 w}{\partial \theta^2} - \frac{1}{\alpha^2}\frac{\partial^2 w}{\partial t^2} = 0 \qquad (1.1.21)$$

where $r$, $\theta$–polar coordinates.

Using the method of separation of variables, we search for the solution in the form of:

$$W(r,\theta,t) = T(t)\,V(r)\Phi(\theta) \qquad (1.1.22)$$

Substituting (1.1.22) into (1.1.21) and dividing by $T(t)\,V(r)\Phi(\theta)$, we will obtain:

$$\frac{V'' + \frac{1}{r}V'}{V} + \frac{1}{r^2}\frac{\Phi''}{\Phi} - \frac{1}{\alpha^2}\frac{T''}{T} = 0 \qquad (1.1.23)$$

Denoting:

$$\frac{\Phi''}{\Phi} = -n^2; \quad \frac{T''}{\alpha^2 T} = -\omega^2$$

And then we can write:

$$\Phi'' + n^2\Phi = 0$$

$$T'' + \alpha^2\omega^2 T = 0$$

$$V'' + \frac{1}{r}V' + \left(\frac{\omega^2}{\alpha^2} - \frac{n^2}{r^2}\right)V = 0 \qquad (1.1.24)$$

Solutions of these equations are:

$$\Phi(\theta) = A\cos n\theta + B\sin n\theta$$

$$T(t) = C \sin \alpha \omega t + D \cos \omega \alpha t \tag{1.1.25}$$

$$V(r) = EI_n \left( \frac{\omega r}{\alpha} \right) + FY_n \left( \frac{\omega r}{\alpha} \right)$$

where $I_n \left( \frac{\omega r}{\alpha} \right)$; $Y_n \left( \frac{\omega r}{\alpha} \right)$ – respectively, Bessel functions of $n$th order of the first and second kind. If a round membrane is considered that of the condition of periodicity on $\theta$, it follows that coefficient $n$ is an integer ($n = 0, 1, 2, \ldots$).

If we have a membrane with a sector shape in plan, the value of the coefficient is defined from the boundary conditions given on the edges $\theta = 0, \theta_1$

So, for example, if it is given that:

$$W(r, 0, t) = W(r, \theta_1, t) = 0 \tag{1.1.26}$$

using (1.1.22), (1.1.25), and (1.1.26) we find that

$$n = \frac{k \pi}{\theta_1}; \qquad (k = 0, 1, 2, 3 \ldots) \tag{1.1.27}$$

Since $n$ in this case is not a whole number, then the solution of the third Eq. (1.1.24) is represented as:

$$V(r) = EI_n \left( \frac{\omega r}{\alpha} \right) + FI_{-n} \left( \frac{\omega r}{\alpha} \right) \tag{1.1.28}$$

Using the obtained particular solutions (1.1.25) as in the case of a rectangular membrane, it is possible to compose the general solution that can satisfy the boundary and initial conditions.

## 1.2    The Variational Methods

Let us apply the Hamilton's principle to the elastic oscillation body, according to which:

$$\delta \Gamma = \delta \int_{t_0}^{t_1} (\Pi - K) dt = 0 \tag{1.2.1}$$

where $\Pi$, $K$ – are potential and kinetic energy of the system, respectively. According to this principle, in the time interval between $t = t_0$ and $t = t_1$, the system displacements are such that a variation of $\delta$ integral is zero. In terms of static loads from condition (1.2.1) the following condition of the minimum of potential energy has the form:

$$\delta\Pi = 0 \quad \text{or} \quad \Pi = \min \tag{1.2.2}$$

The variation method provides a general prescription for improving on any wave function with a parameter by minimizing that function with respect to the parameter.

The Hamilton Functional has the form for a bar transverse oscillation:

$$\Gamma = \frac{1}{2}\int_{t_0}^{t_1}\int_0^l\left[EI\left(\frac{\partial^2 y}{dx^2}\right)^2 - \rho(x)\left(\frac{\partial y}{\partial t}\right)^2\right]dxdt \tag{1.2.3}$$

where $E$, $\rho$ – the module of elasticity and the linear density of the bar, respectively; $I$–inertia moment of transverse section of the bar; $l$ – length of the bar; and $y$ – transverse displacement of the bar.

Consider the problem of free oscillations, taking:

$$y(x, t) = Y(y)\sin\omega t, \tag{1.2.4}$$

substituting in (1.2.3) and integrating within one period, we obtain:

$$\Gamma = \frac{\pi}{2\omega}\int_0^l[EI(Y'')^2 - \rho(x)\omega^2 Y^2]dx \tag{1.2.5}$$

One of the commonly used variational methods is the Ritz method. The application of this method uses Hamilton's functional of the form (1.2.5). The variational problem in applying the Ritz method is reduced to search the extremum of function of many variables. The scheme for the solution of a problem of this method consists of the following: solution of the problem is sought in the form of;

$$y(x) = \sum_{i=1}^n a_i\varphi_i(x) \tag{1.2.6}$$

where $\varphi_i(x)$ – fundamental functions, are satisfying the boundary conditions of the problem; and $a_i$ – parameters, whose values are determined in the process of solving the problem. After substituting (1.2.6) in (1.2.5), the Hamilton functional becomes a function of the variables: $a_i$ $(i = 1, 2, ... , n)$. The function's extremum conditions $L(a_i)$:

$$\frac{dL}{da_i} = 0, \quad i = 1, 2, ... n \tag{1.2.7}$$

give $n$ equations, from which can be defined the values $a_i$.

In the case of transverse oscillations of a beam, substituting (1.2.6) in (1.2.5), the functional (1.2.5) gives the following expression:

$$\Gamma(a_i) = \sum_{i=1}^{n}\sum_{k=1}^{n} U_{ik}a_i a_k - \omega^2 \sum_{i=1}^{n}\sum_{k=1}^{n} T_{ik}a_i a_k,$$

where

$$U_{ik} = \int_0^l EI\varphi_i'' \varphi_k'' dx,$$

$$T_{ik} = \int_0^l \rho\varphi_i\varphi_k dx. \tag{1.2.8}$$

Conditions (1.2.7) take the form:

$$\sum_{k=1}^{n}[U_{ik} - \omega^2 T_{ik}]a_k = 0; \quad i = 1, 2, \ldots n \tag{1.2.9}$$

The homogeneous system of equations (1.2.9) has a non-zero solution if the determinant comprised of the coefficients of unknowns is equal to zero. That brings us to the equation of frequencies:

$$\|U_{ik} - \omega^2 T_{ik}\| = 0, \quad i, k = 1, 2, \ldots \; n. \tag{1.2.10}$$

The special case of a method of Ritz is the method of a Rayleigh by which the frequency of the main tone can be determined. Assuming that $i = 1, k = 1$, from (1.2.10) we will obtain:

$$\omega_1^2 = \frac{U_{11}}{T_{11}} = \frac{\displaystyle\int_0^l EI\varphi_1''^2 dx}{\displaystyle\int_0^l \rho\varphi_1^2 dx}. \tag{1.2.11}$$

The expression (1.2.11) presents Rayleigh formula for transverse oscillations of a beam. In the case of longitudinal oscillations of a bar, the frequency equation has the same form (1.2.6), except that the value of $U_{ik}$ will be determined by the formula:

$$U_{ik} = \int_0^l EF\varphi_i' \varphi_k' \, dx \tag{1.2.12}$$

At torsional or shift oscillations of the bar in expression (1.2.12), modulus of rigidity at twisting and shifting, $GI_r$ and $kFG$, should be used instead of rigidity of longitudinal deformation *EF*. Here *G* – modules of shear; $I_r$ – polar moment of inertia; *F*–cross-sectional area; and *k* – coefficient, considering uneven distribution of shearing stresses over the cross section of the bar [73, 358].

One of the most often applied variational methods is the method of Galerkin; let us express the essence of this method. Assume that differential equation of one-dimensional oscillations is given in the form of:

$$L(Y) = 0 \qquad\qquad (1.2.13)$$

The solution to the problem we search for in the form of (1.2.6). The essence of the method is the requirement that the function $L(Y)$ was orthogonally with all functions $\varphi_i(x)$; $i = 1, 2, \ldots, n$, that the condition was satisfied:

$$\int_0^l L(Y)\varphi_i(x)dx = 0, \qquad i = 1, 2, \ldots .n \qquad (1.2.14)$$

If $Y(x)$ would be the exact solution of the original Eq. (1.2.13), then the conditions (1.2.14) would be satisfied identically. At approximate problem solving we make a decision as (1.2.6) and substituting (1.2.14) we will obtain a system of *n* equations regarding functions $a_i$. Defining $a_i$ we will obtain the desired displacement of the system. In the case of transverse oscillations of the bar, the system (1.2.14) will have an appearance:

$$\sum_{k=1}^n [V_{ik} - \omega^2 T_{ik}]a_k = 0, \qquad i = 1, 2, \ldots n; \qquad (1.2.15)$$

where $T_{ik}$ are defined by expression (1.2.8)

$$V_{ik} = \int_0^l (EI\varphi_i'')\varphi_k dx. \qquad (1.2.16)$$

The respecting equation of frequencies has the form:

$$\|V_{ik} - \omega^2 T_{ik}\| = 0, \qquad i = 1, 2, \ldots n. \qquad (1.2.17)$$

If we compare the Ritz and Galerkin methods, it should be noted that the range of the application of Galerkin method is wider, since its application is not connected to the existence of Hamilton's type functional that is applied with the use of Ritz's method.

## 1.3 Integral Equations and Integral Transforms Methods

One of the most effective ways of solving boundary value problems are methods based on the use of integral equations. The problem of oscillations of bars, plates, and shells can be reduced to integral equations of Volterra or Fredholm. We will show the methodology of reducing an ordinary differential equation to the integral equation of Volterra of the second kind [233]. Let us consider a differential equation of $n$ order:

$$y^n(x) + a_1(x)y^{(n-1)}(x) + \ldots + a_n(x)y(x) = f(x) \tag{1.3.1}$$

where: $y^n(x) = \dfrac{d^n y}{dx^n}$; $a_1(x), \ldots a_n(x)$, $f(x)$ – given functions.

The equation of the type (1.3.1) reduces the consideration of different tasks, in particular, the problem about established oscillations of the beam of variable section. This equation has the following form:

$$EI(x)y^{IV}(x) + 2EI^I(x)y^{III}(x) + EI^{II}(x)y^{II}(x) - mF(x)\omega^2 y(x) = q(x) \tag{1.3.2}$$

where $E$ – module of elasticity; $I(x)$ – moment of inertia, variable along the length; $F(x)$ – cross-sectional area; $m$ – mass of uniform length; $\omega$ – circular frequency; and $q(x)$ – distributed dynamic load.

Dividing all members of (1.3.2) on $EI(x)$ we find that Eq. (1.3.2) is a special case of (1.3.1). In this case:

$$n = IV; \qquad a_1(x) = 2\frac{I^I(x)}{I(x)}; \qquad a_2(x) = \frac{I^{II}(x)}{I(x)};$$

$$a_3(x) = 0; \qquad a_4(x) = -m\omega^2\frac{F(x)}{EI(x)}; \qquad f(x) = \frac{q(x)}{EI(x)}.$$

Assuming that the initial conditions of equations (1.3.1) are given as:

$$y(b) = y_0; \qquad y^I(b) = y_0{}^I; \qquad y^{n-1}(b) = y_0{}^{n-1}.$$

Denote:

$$y^n(x) = \psi(x) \tag{1.3.3}$$

Integrating (1.3.3) we find

$$y^{(n-1)}(x) = \int_b^x \psi(\xi)d\xi + y^{(n-1)}(b) \tag{1.3.4}$$

Integrating (1.3.4) we obtain:

$$y^{(n-2)}(x) = \int_b^x \int_b^x \psi(\xi)d\xi^2 + y^{(n-1)}(b)(x-b) + y^{(n-2)}(b); \qquad (1.3.5)$$

$$y^{(n-m)}(x) = \int_b^x \int_b^x \cdots \int_b^x \psi(\xi)d\xi^m + \sum_{k=1}^m y^{(n-k)}(b)\frac{(x-b)^{m-k}}{(m-k)!}; \qquad (1.3.6)$$

$$y(x) = \int_b^x \int_b^x \cdots \int_b^x \psi(\xi)d\xi^n + \sum_{k=1}^n y^{(n-k)}(b)\frac{(x-b)^{n-k}}{(n-k)!}. \qquad (1.3.7)$$

Substituting the expressions (1.3.4–1.3.7) for $y(x)$ into (1.3.1), and using the formula of multiple integrals to the sample

$$\int_b^x \int_b^x \cdots \int_b^x \psi(\xi)d\xi^m = \int_b^x \frac{(x-\xi)^{m-1}}{(m-1)!} \int_b^x \psi(\xi)d\xi, \qquad (1.3.8)$$

we will obtain:

$$\psi(x) + \int_b^x \sum_{m=1}^n a_m(x)\frac{(x-\xi)^{m-1}}{(m-1)!} \psi(\xi)d\xi = F(x) \qquad (1.3.9)$$

where

$$F(x) = f(x) - \sum_{m=1}^n \sum_{k=1}^m y^{(n-k)}(b)a_m(x)\frac{(x-b)^{(m-k)}}{(m-k)!}.$$

Designating

$$\sum_{m=1}^n a_m(x)\frac{(x-\xi)^{m-1}}{(m-1)!} = K(x,\xi)$$

Eq. (1.3.9) reduces to the integral equation of Volterra of the second kind.

$$\psi(x) + \int_b^x K(x,\xi)\psi(\xi)d\xi = F(x) \qquad (1.3.10)$$

The solution of (1.3.10) can be found as an infinite series, using known methods of solving the integral equations of Volterra of the second kind.

Let us consider oscillation of the beam with a constant cross-section. The equation of oscillations has the form (1.3.2), where the following is: $I^I = I^{II} = 0$. The motion of this beam can be described by a Fredholm integral equation of the second kind in the following form:

$$y(x) = \int_0^l [mF\omega^2 y(\xi) + q(\xi)]G(x,\xi)d\xi, \qquad (1.3.11)$$

where: $G(x, \xi)$ – Green's function or influence function, representing displacement in the cross-section $x$ from static unit single concentrated load applied transversely at section $\xi$. Equation (1.3.11) can be obtained from the following considerations. To determine the displacement in cross section $x$, Green's function is multiplied by the values of the inertia forces and external loads of those acting on the elementary length and summarized – integrated over the entire length of the beam. The Green's function $G(x, \xi)$ for a beam simply supported at each end has the form:

$$G(x, \xi) = \begin{cases} -\dfrac{(l - \xi)x}{6lEI}(x^2 - 2\xi l + \xi^2), 0 \le x \le \xi \le l \\[4mm] -\dfrac{(l - x)\xi}{6lEI}(\xi^2 - 2xl + x^2); 0 \le \xi \le x \le l \end{cases} \tag{1.3.12}$$

From expression (1.3.12) it follows that $G(x, \xi) = G(\xi, x)$, which is the consequence of the reciprocity theorem of the displacement (Betty Theorem). For a cantilever beam fixed in the cross section $x = 0$, the Green's function has the form:

$$G(x, \xi) = \begin{cases} -\dfrac{x^2(3\xi - x)}{6EI}; 0 \le x \le \xi \le l \\[4mm] -\dfrac{\xi^2(3x - \xi)}{6EI} \quad 0 \le \xi \le x \le l \end{cases} \tag{1.3.13}$$

The equations of type (1.3.11) can be composed for plates and shells. In these cases the motion will be described by the two-dimensional equations of Fredholm of the second kind.

In applied mathematics the use of integral decomposition for the solution of various differential and integral equations is called the method of integral transform. Fourier's integral is the simplest and the most widely used integral decomposition:

$$f(x) = \frac{1}{2\pi} \int_{-\infty}^{\infty} e^{-ix\xi} \int_{-\infty}^{\infty} e^{i\xi\lambda} f(\lambda) d\lambda d\xi \tag{1.3.14}$$

If function $f(x)$ is integrated in the interval $(-\infty, \infty)$ and in any finite interval has finite variation, then the right part of formula (1.3.14) is equal to $f(x)$ at each point of continuity of this function, and is equal to $1/2[f(x + 0) + f(x - 0)]$ at each point of discontinuity. We will designate

$$\frac{1}{2\pi} \int_{-\infty}^{\infty} e^{i\xi x} f(x) dx = F(\xi). \tag{1.3.15}$$

We can write the integral expression (1.3.14) of function $f(x)$ in the following form:

$$f(x) = \frac{1}{2\pi} \int_{-\infty}^{\infty} e^{-i\xi x} F(\xi) d\xi. \qquad (1.3.16)$$

The function $F(\xi)$ is called the integral Fourier transform or transform of function $f(x)$, and expression (1.3.16) by the inversion formula for the integral Fourier transform [108]. The formulas (1.3.15), (1.3.16) are called dual, or considering that $f(x)$ and $F(\xi)$ are Fourier's transform of each other, and also are formulas of inversion. If $f(x)$ is an even function, then formulas (1.3.15), (1.3.16) will take the form:

$$F(\xi) = \sqrt{\frac{2}{\pi}} \int_0^{\infty} \cos \xi x f(x) dx;$$

$$f(x) = \sqrt{\frac{2}{\pi}} \int_0^{\infty} \cos \xi x F(\xi) d\xi. \qquad (1.3.17)$$

Formulas (1.3.17) define a Fourier cosine-transform. Another transformation is Fourier sine-transform obtained from (1.3.15), (1.3.16) with the odd $f(x)$;

$$F(\xi) = \sqrt{\frac{2}{\pi}} \int_0^{\infty} f(x) \sin \xi x dx,$$

$$f(x) = \sqrt{\frac{2}{\pi}} \int_0^{\infty} F(\xi) \sin \xi x d\xi. \qquad (1.3.18)$$

Let us examine the application of a method of integral transform to the solution of PDEs based on the example problem about the lateral oscillations of infinite bar. The equation of oscillations of a bar can be written as:

$$EI\frac{\partial^4 w}{\partial x^4} + m\frac{\partial^2 w}{\partial t^2} = q(x, t), \qquad (1.3.19)$$

where $EI$ – stiffness of the bar; $m$ – mass per unit length; $q$ – external load; and $w$ – deflection.

Boundary and initial conditions are:

$$w(\pm\infty, t) = \frac{\partial w}{\partial x}(\pm\infty, t) = \ldots = 0;$$

$$w(x, 0) = \varphi(x), \quad \frac{\partial w}{\partial t}(x, 0) = \psi(x). \qquad (1.3.20)$$

Let us multiply the left and right parts of Eq. (1.3.19) to $(2\pi)^{-1/2}e^{i\xi x}$ and then integrate over $x$ from $-\infty$ to $+\infty$, we will obtain

$$\frac{EI}{\sqrt{2\pi}}\int_{-\infty}^{\infty}\frac{\partial^4 w}{\partial x^4}e^{i\xi x}dx + \frac{m}{\sqrt{2\pi}}\int_{-\infty}^{\infty}\frac{\partial^2 w}{\partial x^2}e^{i\xi x}dx = \frac{1}{\sqrt{2\pi}}\int_{-\infty}^{\infty}q(x,t)e^{i\xi x}dx.$$

$$(1.3.21)$$

We will designate the transforms

$$W(\xi,t) = (2\pi)^{-1/2}\int_{-\infty}^{\infty}w(x,t)e^{i\xi x}dx;$$

$$Q(\xi,t) = (2\pi)^{-1/2}\int_{-\infty}^{\infty}q(x,t)e^{i\xi x}dx.$$

The expression (1.3.21) integrates by (in) parts, and in the result we will find

$$\frac{EIe^{i\xi x}}{\sqrt{2\pi}}\left[\frac{\partial^3 w}{\partial x^3} - i\xi\frac{\partial^2 w}{\partial x^2} - \xi^2\frac{\partial w}{\partial x} + i\xi^3 w\right]_{-\infty}^{\infty} + EI\xi^4 W(\xi,t) + m\frac{d^2 W}{dt^2} = Q(\xi,t).$$

$$(1.3.22)$$

Using boundary conditions (1.3.20) we have

$$\frac{\partial^2 W}{\partial t^2} + \frac{EI\xi^4}{m}W = \frac{Q}{m}.$$

$$(1.3.23)$$

Thus, with an aid of the Fourier transformation, the equation in the partial derivatives (1.3.19) is reduced to the ordinary differential equation. After determination from Eq. (1.3.23) of transform $W(\xi,t)$, we will obtain the solution to the problem according to the inversion formula (1.3.16)

$$w(x,t) = \frac{1}{\sqrt{2\pi}}\int_{-\infty}^{\infty}W(\xi,t)e^{-i\xi x}dx.$$

$$(1.3.24)$$

Note that for solving Eq. (1.6.23), it is necessary to enter Fourier's transforms for the functions, entering in the initial conditions:

$$W(\xi,0) = \frac{1}{\sqrt{2\pi}}\int_{-\infty}^{\infty}\varphi(x)e^{i\xi x}dx = \Phi(\xi);$$

$$\frac{dW}{dt}(\xi,0) = \frac{1}{\sqrt{2\pi}}\int_{-\infty}^{\infty}\psi(x)e^{i\xi x}dx = \Psi(\xi).$$

This is the basic scheme for solving differential equations by the method of integral transformations. The multiplication of the left and right parts of the equation in a certain function (kernel of integral transform) and further integration of domain by changing the variable, which is excluded from the equation as a result of the application of transformation, is characteristic for this scheme. Sometimes under the integral transformations, the classical method of Fourier based on finding solutions to a problem in a form of the integral of Fourier's type, with unknown density, determined as a result of the substitution of integral into the initial differential equation is described. As we see, both methods are equivalent, however, the first one, which is the common method of integral transformations, is more convenient, especially for solutions of nonhomogeneous equations with inhomogeneous boundary conditions.

Thus, the method of integral transformations is based on the decomposition of the solution into an integral by a certain set of functions, chosen based on a form of differential equation, variable range within domain, and boundary conditions. In addition to the above described Fourier transforms, applied mathematics uses other integral transformation; in particular the transforms of Laplace, Mellin, Hankel, Melera-Foka, Meyer, Kantorovich-Lebedev, and Weber. All these transforms are introduced on basis of the corresponding integral decompositions. Since the transforms for them are obtained as a result of the integration in the infinite intervals, they are called transforms with infinite limits. On the other hand, it is possible to introduce the integral transforms based on the decomposition in series. If the functions $f(x)$ of a certain class are presented in the form of series of orthogonal system $\{u_n(x)\}$, i.e.

$$f(x) = \sum_n F_n u_n(x) \qquad (a \leq x \leq b), \qquad (1.3.25)$$

where the coefficients of decomposition $F_n$ are determined from formula

$$F_n = \int_a^b f(x) u_n(x) dx. \qquad (1.3.26)$$

Of course such interpretation does not add anything to the possibilities of common method of series or method of separation of variables, however, the procedure itself of determination of transforms, i.e. the coefficients of the series, is typical of the conventional method of integral transforms and is more convenient. These advantages are especially evident in the solutions of problems with the inhomogeneous boundary conditions. The specified transformations are called transformations with final limits I. Sneddon [343], or final transformations, since in (1.6.26) integration extends to the finite interval. Division into the transformations with finite and infinite limits conditionally and does not reflect the principle of the matter. The domain of change of the variable in terms of which transformation is achieved is not essential here, and consequently neither is the interval of integration; but the nature of inversion formulas, determined by the form of the assumed as basis method of decomposition discrete or continuous is of importance. Exceptionally continuous integral decompositions were used in the classical method of integral transformation. However, the tendency to expand the circle solved by this method's problems led to the start in the integral transformations of discrete and discrete-continuous decompositions (in series). The general theory of integral transformations, which covers from unified point of view an entire diversity of the questions, connected with the substantiation, by application, and by the development of method, is not yet completed. The most general approach in this sense is associated with the use of a spectral theory of linear operators. If necessary, this approach is especially effective in the building of new transformations and corresponding integral decompositions.

## 1.4   The Finite Element Method

The Finite Element Method is a method of analysis in which a structure is discretized into elements connected at nodes; the shape of the element displacement field is assumed, partial or complete compatibility is maintained among the element interfaces, and nodal displacements are determined by using energy variational principles or equilibrium methods.

During application of the method of finite elements, a continuous medium is broken into subdomains called finite elements. In the case of a two-dimensional problem, the finite elements take the form of a triangle or rectangle; in a three-dimensional case the type is that of a tetrahedron or rectangular prism. Depending on the geometry of the problem, finite elements can have a more complicated (complex geometric) form. The breakdown of the medium into the finite elements substitutes the system with an infinite number of degrees of freedom with a system with a finite number of degrees of freedom. The displacements of the nodes of the final elements are taken as the unknown.

The function, which determines displacements inside the finite element, expressed through the displacements of nodal points is selected. Knowing the nodes of displacements we can determine both the strains and stresses within the limits of final element and on its boundaries. To determine the nodes of displacements, either the condition of the equilibrium of the nodes of finite elements or condition of the minimum of the functional of the corresponding variational problem must be adapted.

It should be noted that historically one of the first prototypes of the application of a method of finite elements is the method of deformations (displacements) of the calculation of plane frame structures to static load, and subsequently to stability and oscillations. The finite elements in this case are beams and columns of the frame structure. The angular and linear displacements of the frame nodes are the unknowns. In this particular case, the function describing deformations of a finite element is defined precisely (within the theory of bars), and therefore this method gives the exact solution of the problem.

Let us proceed to the derivation of the equations of finite elements method using the example of the plane problem of the theory of elasticity [1, 102]. Displacement at any point of the finite element $|f|$ is a function of coordinates $x, y$, and is determined as a vector:

$$|f| = \begin{vmatrix} U(x,y) \\ V(x,y) \end{vmatrix} \qquad (1.4.1)$$

$U(x, y)$, $V(x, y)$ – horizontal and vertical displacements of the point with coordinates $x, y$. The displacement at any point inside the triangular finite element is defined as:

$$|f| = |N|\,|\delta|^e = |N_i, N_j, N_m,| \begin{vmatrix} \delta_i \\ \delta_j \\ \delta_m \end{vmatrix}, \qquad (1.4.2)$$

where:

- $|\delta|^e$ – matrix of the displacements of nodal points at any point inside the triangular finite element with the index "e";
- $\delta_i$ – displacement of node "i" of finite element,

$$\delta_i = \begin{vmatrix} U_i \\ V_i \end{vmatrix};$$

- $U_i$, $V_i$ – projections of the displacement of node "i" in the directions of $x$ and $y$; and

- $|N|$ – matrix, which characterizes the distribution of displacements in the finite element.

In order to obtain the concrete type of this matrix, we assume a certain distribution of displacements within the finite element. The linear polynomials are the simplest representation of displacements [415].

$$U = \alpha_1 + \alpha_2 x + \alpha_3 y$$
$$V = \alpha_4 + \alpha_5 x + \alpha_6 y \tag{1.4.3}$$

If we substitute into the first equation of the coordinates of the points $i$, $j$, and $m$, then we will obtain:

$$U_i = \alpha_1 + \alpha_2 x_i + \alpha_3 y_i$$
$$U_j = \alpha_1 + \alpha_2 x_j + \alpha_3 y_j \tag{1.4.4}$$
$$U_m = \alpha_1 + \alpha_2 x_m + \alpha_3 y_m$$

Determining $\alpha_1, \alpha_2,$ and $\alpha_3$ from this system and substituting in the first Eq. (1.4.3), we will obtain:

$$U = \frac{1}{2\Delta}\{(a_i + b_i x + c_i y)U_i + (\alpha_j + b_j x + c_j y)U_j + (\alpha_m + b_m x + c_m y)U_m\} \tag{1.4.5}$$

where

$$a_i = x_j y_m - x_m y_j,$$
$$b_i = y_j - y_m, \tag{1.4.6}$$
$$c_i = x_m - x_j$$

$$2\Delta = \begin{vmatrix} 1 & X_i & Y_i \\ 1 & X_j & Y_j \\ 1 & X_m & Y_m \end{vmatrix} \tag{1.4.7}$$

Similarly, it is possible to obtain the dependence of the vertical displacement $V$ using the displacements of the nodes of the finite element.

$$V = \frac{1}{2\Delta}\{(a_i + b_i x + c_i y)V_i + (\alpha_j + b_j x + c_j y)V_j + (\alpha_m + b_m x + c_m y)V_m\} \tag{1.4.8}$$

Relationships (1.4.5) and (1.4.8) can be represented in the matrix form:

$$|f| = \begin{vmatrix} U \\ V \end{vmatrix} = |N||\delta|^e = [IN_i^I, IN_j^I, IN_m^I]|\delta|^e$$

$$N_i^I = \frac{a_i + b_i x + c_i y}{2\Delta} \tag{1.4.9}$$

The values $N_j^I$, $N_m^I$ are obtained by replacement of index $i$ with $j$ and $m$; $I$ – the unit matrix of the dimensionality of $2 \times 2$.

Using the known displacements (1.4.9) it is possible to determine the values of the deformation in dependence on the displacement of nodes. This dependence can be presented in the matrix form:

$$|\varepsilon| = |B| |\delta|^e = |B_i \, B_j \, B_m| |\delta|^e. \tag{1.4.10}$$

For definition of a specific type of the matrix $|B|$ we use known relationships [415]:

$$|\varepsilon| = \begin{vmatrix} \varepsilon_x \\ \varepsilon_y \\ \gamma_{xy} \end{vmatrix} = \begin{vmatrix} \dfrac{\partial U}{\partial x} \\[2mm] \dfrac{\partial V}{\partial y} \\[2mm] \left(\dfrac{\partial U}{\partial y} + \dfrac{\partial V}{\partial x}\right) \end{vmatrix} \tag{1.4.11}$$

Substituting (1.4.9) to (1.4.11), we will obtain:

$$|\varepsilon| = \frac{1}{2\Delta} \begin{vmatrix} b_i & 0 & b_j & 0 & b_m & 0 \\ 0 & c_i & 0 & c_j & 0 & c_m \\ c_i & b_i & c_j & b_j & c_m & b_m \end{vmatrix} |\delta|^e = \frac{1}{2\Delta} |B_i \, B_j \, B_m| |\delta|^e = |B| |\delta|^e, \tag{1.4.12}$$

$$b_i = \frac{\partial N_i^I}{\partial x}; \qquad c_i = \frac{\partial N_i^I}{\partial y}$$

$b_j$, $c_j$, $b_m$, and $c_m$ are obtained by the replacement of index $i$ with the index $j$ and $m$.

Under the plane strain conditions the relationship between the strains and stresses is following [415]:

$$\varepsilon_x = \frac{1}{E}(\sigma_x - \nu\sigma_y - \nu\sigma_z), \qquad \varepsilon_y = \frac{1}{E}(-\nu\sigma_x + \sigma_y - \nu\sigma_z), \tag{1.4.13}$$

$$\gamma_{xy} = \frac{2(1 + \nu)}{E} \sigma_{xy}, \qquad \varepsilon_z = \frac{1}{E}(-\nu\sigma_x - \nu\sigma_y + \sigma_z), \tag{1.4.14}$$

Let us show the relationship between the stresses and strains in a matrix form. Excluding $\sigma_z$, we will determine the dependence of stresses from the deformations:

$$|\sigma| = |D| |\varepsilon|$$

$$\begin{vmatrix} \sigma_x \\ \sigma_y \\ \sigma_z \end{vmatrix} = |D| \begin{vmatrix} \varepsilon_x \\ \varepsilon_y \\ \gamma_{xy} \end{vmatrix} \tag{1.4.15}$$

where:

$$|D| = \frac{E(1-v)}{(1+v)(1-2v)} \begin{vmatrix} 1 & \dfrac{v}{1-v} & 0 \\ \dfrac{v}{1-v} & 1 & 0 \\ 0 & 0 & \dfrac{1-2v}{2(1-v)} \end{vmatrix}$$

The simplest and most efficient way to compose the equations of the method of finite elements is to use the virtual work principle. According to the virtual work principal, with the arbitrary nodal displacement, the sum of the external and internal work accomplished by different forces and stresses on this displacement is equal to zero. Suppose that $d|\delta|^e$ – is a virtual displacement of node, then (1.4.2) and (1.4.10) can be presented in the form:

$$d|f| = |N|d|\delta|^e, \quad d|\varepsilon| = |B|d|\delta|^e. \tag{1.4.16}$$

The work of nodal forces is equal to the product of each force component on the corresponding displacement, that is:

$$(d|\delta|^e)^T |F|^e \tag{1.4.17}$$

Here through $(d|\delta|^e)^T$ – the corresponding transposed matrix is designated.

The work of internal forces is equal to the work of stresses and distributed forces $|P|$ per unit volume:

$$d|\varepsilon|^T |\sigma| - d|f|^T |P| \tag{1.4.18}$$

Having (1.4.10) and (1.4.2) and using a formula of the transposition of the matrix product

$$(|A| \, |B|)^T = |B|^T \, |A|^T \tag{1.4.19}$$

we can present (1.4.18) in the form of:

$$(d|\delta|^e)^T \, (|B|^T |\sigma| - |N|^T |P|) \tag{1.4.20}$$

Equating the work of external forces (1.4.17) to the work of internal forces, which is obtained by integrating (1.4.20) by the volume, we obtain:

$$(d|\delta|^e)^T |F|^e = (d|\delta|^e)^T \left( \int |B|^T |\sigma| dV - \int |N|^T |P| dV \right) \tag{1.4.21}$$

Since this equation is correct at any arbitrary displacement, we can equate the expressions before $d \ |\delta|^e$ of the left and right part of the Eq. (1.4.21). Then substituting (1.4.10) and (1.4.14), we obtain:

$$|F|^e = \left( \int |B|^T |D| |B| dV \right) |\delta|^e - \int |N|^T |P| dV \qquad (1.4.22)$$

Eq. (1.4.22) is the condition of equality of the external and internal forces acting on the finite element with the number "$e$". Equation (1.4.22) can be represented as:

$$|F|^e = |K|^e \ |\delta|^e + |F|_p^e \qquad (1.4.23)$$

where $|K|^e$ – stiffness matrix. The types of matrix $|K|^e$ and $|F|_p^e$ are the same, comparing (1.4.23) with (1.4.22). In the expression of the works of internal forces sometimes the influence of initial deformations and stresses is considered as well. In that case (1.4.23) is written in the form of:

$$|F|^e = |K|^e \ |\delta|^e + |F|_p^e + |F|_{\varepsilon_0}^e + |F|_{\sigma_0}^e \qquad (1.4.24)$$

where:

$$|K|^e = \int |B|^T |D| |B| dV,$$

$$|F|_p^e = - \int |N|^T |P| dV,$$

$$|F|_{\varepsilon_0}^e = - \int |B|^T |D| |\varepsilon_0| dV,$$

$$|F|_{\sigma_0}^e = \int |B|^T |\sigma_0| dV. \qquad (1.4.25)$$

$|\varepsilon_0|$, $|\sigma_0|$–matrices, which characterize initial strains and initial stresses [415].

Thus, the solution of the static problem of the theory of elasticity by the method of finite elements is reduced to the composition of the equations of equilibrium for each finite element of the following form:

$$|F|^e = 0; \qquad\qquad e = 1, 2, ... \ . \qquad (1.4.26)$$

After the determination of the nodal displacements $|\delta|^e$, $e = 1, 2, ...$, displacements and stresses can be determined by the formulas given above.

For obtaining the equations of dynamic equilibrium in the expression (1.4.22) instead of $|P|$, accept the following:

$$\overline{|P|} - \rho \frac{\partial^2}{\partial t^2} |f| - \mu \frac{\partial}{\partial t} |f| \tag{1.4.27}$$

where: $\rho$ – mass of unit volume, and $\mu$–coefficient accounting for damping according to the hypothesis of a linear damping (Voigt model). Substituting (1.4.27) in (1.4.22) and taking into account (1.4.26) we obtain:

$$|K|^e |\delta|^e + |C|^e \frac{\partial}{\partial t} |\delta|^e + |M|^e \frac{\partial^2}{\partial t^2} |\delta|^e + \overline{|F|}^e = 0 \tag{1.4.28}$$

where:

$$|C|^e = \mu \int |N|^T |N| dV; \quad |M|^e = \rho \int |N|^T |N| dV; \quad \overline{|F|}^e = - \int |N|^T \overline{|P|} dV.$$

$|C|^e$ and $|M|^e$ – the matrices of damping and masses of the element, respectively.

According to (1.4.10) we have:

$$|N| = [IN_i^I, IN_j^I, IN_m^I] \tag{1.4.29}$$

where:

$$I = \begin{vmatrix} 1 & 0 \\ 0 & 1 \end{vmatrix}$$

The matrix $|M|^e$ can be represented in the form:

$$|M|^e = \rho t \, I \iint |N|^T |N| dx dy \tag{1.4.30}$$

or

$$|M|^e = \frac{\rho t \Delta}{3} \begin{vmatrix} c & d & d \\ d & c & d \\ d & d & c \end{vmatrix},$$

where $c, d$ – are sub matrices expressed in the following forms:

$$c = \begin{vmatrix} 1/2 & 0 \\ 0 & 1/2 \end{vmatrix} \; ; \qquad d = \begin{vmatrix} 1/4 & 0 \\ 0 & 1/4 \end{vmatrix},$$

$t$ – thickness of a plate; and $\Delta$ – area of a triangle.

The method of finite elements is successfully applied in non-stationary problems as well.

Let us consider the non-stationary vibrations of cantilever plate, which is either a plane stress or plane strain conditions. If the law of vibration accelerations of the ground $y_0^{II}(t)$ is known, then vibration of the system can be represented as:

$$|M|\frac{\partial^2}{\partial t^2}|\delta| + |C|\frac{\partial}{\partial t}|\delta| + |K||\delta| = - |M|y_0^{II}(t). \tag{1.4.31}$$

Next the problem of free oscillations of the system is considered, that is, the system of the following type is solved:

$$(|K| - \omega^2|M|)|\delta_0| = 0 \tag{1.4.32}$$

and the natural frequencies of the system $\omega_i$, $i = 1, 2 \ldots$, and their modes of oscillation $|\delta_0|_i$, $i = 1, 2 \ldots$ are determined.

Any non-stationary motion of the system can be sought in the form:

$$|\delta| = [|\delta_0|_1, |\delta_0|_2, \quad \cdots \quad |\delta_0|_n] |q(t)| = |\Delta_0| |q|, \tag{1.4.33}$$

wherein $|\Delta_0|$ includes the normalized eigenfunctions, and $|q|$ – the generalized coordinates to be determined. Substituting (1.4.33) in (1.4.31) and multiplying all members of the equation by $|\Delta_0|^T$, we will obtain:

$$|\Delta_0|^T|M||\Delta_0|\frac{\partial^2}{\partial t^2}|q| + |\Delta_0|^T|C||\Delta_0|\frac{\partial}{\partial t}|q| + |\Delta_0|^T|K||\Delta_0||q| = -|\Delta_0|^T|M|y_0^{II}$$
$$\tag{1.4.34}$$

For eigenfunctions $|\delta_0|_i$ the property of orthogonality takes place [415, 416]:

$$|\delta_0|_i^T|M||\delta_0|_j = \begin{Bmatrix} 0 & i \neq j \\ 1 & i = j \end{Bmatrix} \tag{1.4.35}$$

From (1.4.32) follows that

$$|K||\delta_0|_i = \omega_i^2 |M||\delta_0|_i$$

Hence formula (1.4.35) can be presented also in the form:

$$|\delta_0|_i^T|K||\delta_0|_j = \begin{Bmatrix} 0 & i \neq j \\ \omega_i^2 & i = j \end{Bmatrix} \tag{1.4.36}$$

Since the matrix $|C|$ is proportional to $|M|$, we have orthogonality of eigenfunctions of the following form:

$$|\delta_0|_i^T|C||\delta_0|_j = \begin{Bmatrix} 0 & i \neq j \\ 2n_i\omega_i^2 & i = j \end{Bmatrix} \tag{1.4.37}$$

Using conditions (1.4.35), (1.4.36), and (1.4.37) leads the system of ordinary differential equations to the following form:

$$\frac{\partial^2}{\partial t^2}q_i + 2n_i\omega_i\frac{d}{dt}q_i + \omega_i^2\,q_i = -\,\left|\delta_0\right|_i{}^T\left|M\right|y_0^{II} \tag{1.4.38}$$

The solution is:

$$q_i = -\,\left|\delta_0\right|_i{}^T\left|M\right|\int_0^t y_0^{II}(\xi)\;e^{-n_i\omega_i(t-\xi)}\sin\omega_i(t-\xi)d\xi \tag{1.4.39}$$

Then the displacement of any nodal point of the structure can be determined using (1.4.33).

## 1.5   The Finite Difference Method

The Finite Difference Method is a method of analysis in which the governing differential equation is satisfied at discrete points of a structure. The finite difference approximations for derivatives are one of the simplest and oldest methods of solving differential equations. The essence of the finite difference method consists in the replacement of the derivatives in the differential equation with finite differences, using differential quotients, as a result of which these equations are reduced to the system of linear algebraic equations.

We will consider application of the finite difference method on the example of transverse oscillation of a compressed beam of a constant cross section. The differential equation of free oscillation of such beam is given by:

$$\frac{\partial^4 y}{\partial x^4} + \lambda\frac{d^2 y}{dx^2} - \alpha^2 y = 0 \tag{1.5.1}$$

$$\lambda = \frac{N}{EI};\quad \alpha^2 = \frac{\omega^2\rho}{EI}$$

where $y$ – displacement, and $N$ – normal compressive force.

We will divide the beam into $n$ equal parts.

$$l = n\cdot\Delta x;\qquad \Delta x = \frac{l}{n} = l\Delta\xi;\qquad \Delta\xi = \frac{1}{n}\ .$$

Derivatives for point $k$ will replace with the ratio of differences:

$$\frac{\partial^4 y}{\partial\xi^4}\ \rightarrow\ \frac{\Delta^4 y}{\Delta\xi^4}\ \bigg|_k = n^4(y_{k-2} - 4y_{k-1} + 6y_k - 4y_{k+1} + y_{k+2})$$

$$\frac{\partial^2 y}{\partial \xi^2} \rightarrow \left.\frac{\Delta^2 y}{\Delta \xi^2}\right|_k = n^2(y_{k-1} - 2y_k + y_{k+1}); \qquad \Delta \xi = \frac{x}{l} \qquad (1.5.2)$$

Substitution (1.5.2) in (1.5.3) gives:

$$y_{k-2} - y_{k-1}\left(4 - \frac{\lambda}{n^2}\right) + y_k\left(6 - \frac{2\lambda}{n^2} - \frac{\alpha^2}{n^4}\right)$$

$$- y_{k+1}\left(4 - \frac{\lambda}{n^2}\right) + y_{k+2} = 0; \quad k = 1, 2, \dots, n-1 \qquad (1.5.3)$$

For the hinge supported beam, boundary conditions take the form:

$$y(0) = Y(l) = 0; \qquad \frac{\partial^2 y}{\partial \xi^2}(0) = \frac{\partial^2 y}{\partial \xi^2}(l) = 0. \qquad (1.5.4)$$

Dividing the beam into $n$ parts, the point $n = 0$ corresponds to the left support, and the point $n$ corresponds to the right support. The boundary conditions (1.5.4), considering (1.5.2), will take the form:

$$y_0 = 0; \quad y_n = 0; \quad y_{-1} - 2y_0 + y_1 = 0; \quad y_{n-1} - 2y_n + y_{n+1} = 0. \qquad (1.5.5)$$

Assuming that $n = 5$, then $n = 1, 2, 3, 4$. Now, let us compose Eqs. (1.5.3), considering (1.5.5):

$$\left(5 - \frac{2\lambda}{n^2} - \frac{\alpha^2}{n^4}\right)y_1 - \left(4 - \frac{\lambda}{n^2}\right)y_2 + y_3 = 0$$

$$-\left(4 - \frac{\lambda}{n^2}\right)y_1 + \left(6 - \frac{2\lambda}{n^2} - \frac{\alpha^2}{n^4}\right)y_2 - \left(4 - \frac{\lambda}{n^2}\right)y_3 + y_4 = 0$$

$$y_1 - \left(4 - \frac{\lambda}{n^2}\right)y_2 + \left(6 - \frac{2\lambda}{n^2} - \frac{\alpha^2}{n^4}\right)y_3 - \left(4 - \frac{\lambda}{n^2}\right)y_4 = 0$$

$$y_2 - \left(4 - \frac{\lambda}{n^2}\right)y_3 + \left(5 - \frac{2\lambda}{n^2} - \frac{\alpha^2}{n^4}\right)y_4 = 0 \qquad (1.5.6)$$

Equating the determinant of the coefficients $y_1$, $y_2$, $y_3$, and $y_4$ to zero, we obtain the equation of the fourth degree with respect to $\alpha^2$. The solution of the equation gives four values of the frequencies of free oscillations of a beam. In this case an accurate solution of Eq. (1.5.3) can be obtained in the form of:

$$y_k = A \sin \beta k, \qquad k = 1, 2, \dots n - 1. \qquad (1.5.7)$$

This satisfies the boundary conditions at the left end of the beam, i.e. the following conditions:

$$y_0 = 0, \qquad y_1 = -y_{-1}.$$

Now, let us satisfy the boundary conditions at the right end of the beam:

$$y_n = 0, \qquad y_{n+1} = -y_{n-1}.$$

We will obtain:

$$\sin \beta n = 0 \; ; \qquad \beta = \frac{\pi i}{n}; \qquad i = 1, 2, \dots, n$$

and therefore:

$$y_k = A \sin \frac{\pi i}{n} k. \tag{1.5.8}$$

Substituting (1.5.8) in (1.5.3), we will obtain:

$$\omega_i = \frac{n^2}{l^2} \sqrt{\frac{EI}{\rho}} \cdot \sqrt{4\left(1 + \cos \frac{\pi i}{n}\right)^2 - \frac{2\lambda}{n^2}\left(1 + \cos \frac{\pi i}{n}\right)} \tag{1.5.9}$$

$$i = 1, 2, 3 \dots \qquad\qquad n = 1, 2 \dots$$

Using this expression we can determine the frequency of the $i$th tone (mode) of the beam oscillations, which is divided into $n$ parts. When $n \to \infty$, the oscillation frequency $\omega_i$ tends to an exact number. Alternatively, with $\lambda = 0$, the smaller the $n$ the less accurate is the value of the oscillation frequency $\omega_i$. With $\lambda = 0$, expression (1.5.9) for the beam divided into five parts $n = 5$ provides a 5% error in determination of the frequency of fundamental tone $\omega_1$.

## 1.6   The Generalized Method of Integral Transformation

The rapidly developed apparatus of boundary integral equations, used for solving different boundary value problems and design constructions, is a highly effective way of obtaining numerical results, competing with such widespread methods as finite element and finite difference methods. The wide spectrum of questions, which can be united under the name of boundary equations (methods of potentials, integral equations, compensating loads, extended domain, delta transform, etc.), should include the classical methods of structural mechanics, also based on composition of boundary equations.

The classical integral transforms of Fourier, Hankel, Laplace, Mellin, and others can be considered as decomposition of eigenfunctions of one-dimensional self-conjugate and non-self-conjugate differential operators. For a given boundary-value problem the appropriate transformation is dictated by the differential operator, the fundamental domain, and the boundary conditions prescribed. Based on the foregoing theory, all known integral transformations can be built.

The development of methods of integral transformation has to follow two pathways. The first pathway is followed through on expansion of the range of kernels, used for solving boundary-value problems of mathematical physics, and mechanics. The second pathway is through the line of the application of the method to the boundary-value problem, which is different from generating problems under conditions of domain definition, boundary conditions, presence of internal conditions, and so on. Examples of such use of integral transformations in the extended domain can be found in [215]. The integral transformations in the extended domain and similar methods were applied to areas with internal cuts [186, 316, 321], more complex internal conditions [360–364], and in problems with mixed boundary conditions [120, 340]. G.J. Popov for the first time combined and generalized these approaches, extending them to the multi-coherent domains and heterogeneous media [291, 292]. Another trend in the development of integral transformations is application of non-orthogonal kernel, A.I. Tseitlin, L.G. Petrosian [371, 377, 378].

Let us first consider integral transformations as decomposition of eigenfunctions of linear differential operators. We present some general knowledge from the theory of these operators below [240]. The linear differential operator, known as the operator $L = L(y)$, is originated by the differential expression:

$$l(y) = p_0(x)y^{(n)} + p_1(x)y^{(n-1)} + ... + p_n(x)y \tag{1.6.1}$$

and edge (boundary) conditions

$$U_s(y) = 0, \qquad s = 1, 2, ..., m \leq 2n, \tag{1.6.2}$$

where function $p(x)$ – coefficients of differential expression; $n$ – order of the differential equation; $U_s(y)$ – linearly independent forms relative to the values of functions $y$, and its first $n-1$ derivatives at the boundary points $a$, $b$ of the interval $[a, b]$, where differential expression (1.6.1) operates. The differential expression conjugates to $l(y)$ and is named:

$$l^*(y) = (-1)^n (p_0 y)^{(n)} + (-1)^{(n-1)} (p_1 y)^{(n-1)} + \dots + p_n y \qquad (1.6.3)$$

Differential expression is called self-conjugate if $l(y) = l^*(y)$. Any self-conjugate differential expression with real coefficients can be reduced to the following form:

$$l(y) = (p_0 y^{(n)})^n + (p_1 y^{(n-1)})^{(n-1)} + \dots + p_n y \qquad (1.6.4)$$

The differential expression can be conveniently represented by so-called quasi-
derivatives $l(y) = y^{[2n]}$, which are defined by the formulas:

$$y^{[0]} = y, \quad y^{[k]} = \frac{d^k y}{dx^k}, \quad y^{[n]} = p_0 \frac{d^n y}{dx^n} \quad (k = 1, 2, \dots, n-1);$$

$$y^{[n+k]} = p_k \frac{d^{n-k} y}{dx^{n-k}} - \frac{d}{dx}(y^{[n+k-1]}) \qquad (k = 1, 2, \dots, n) \qquad (1.6.5)$$

The boundary conditions expressed through quasi-derivatives according to the formulas of the type $y^{[k]}(x) = 0$, is called canonical.

If $y$ and $z$ are two arbitrary functions, having continuous derivatives up to the $n$th order inclusive on $(a, b)$ interval, for such arbitrary functions $y$ and $z$, then the Lagrange's identity takes place

$$\int_a^b l(y)\bar{z}dx - \int_a^b y\, l^*(\bar{z})dx = \left[ \sum_{s=1}^{2n} U_s(y) V_{2n-s+1}(\bar{z}) \right]_a^b = [y, \bar{z}]_a^b, \qquad (1.6.6)$$

where $V$ – linearly independent forms relative to the values of functions $z$, and its first $n-1$ derivatives; and $\bar{z}$ – designation of the conjugate value.

Boundary (edging) conditions of $V_{2n-s+1}(\bar{z}) = 0$ are called conjugate to the conditions (1.6.2), and if they are equivalent to the boundary conditions (1.6.2), then they are called self-conjugate.

A differential operator generated by self-conjugate differential expression and self-conjugate boundary (edging) conditions is also called self-conjugate.

For a self-conjugate operator Lagrange's identity takes the following form:

$$\int_a^b l(y)z dx - \int_a^b y l(z)dx = [y, z]_a^b. \qquad (1.6.7)$$

A differential operator is called regular if the interval $[a, b]$ is finite and the function $p_0^{-1}(x)$ is summable on it, otherwise the operator is call singular.

The non-trivial solution of the equation is called the eigenfunction of operator $L$:

$$Ly(x, \lambda) = \lambda y(x, \lambda) \tag{1.6.8}$$

Parameter values of $\lambda$ with which eigenfunctions are in the domain of definition of operator $L$ are called eigenvalues. One eigenvalue may correspond to one or more (up to $n$) of the eigenfunctions; their number is called the multiplicity of the eigenvalue. The eigenvalues of the conjugate operators are the conjugate numbers. If the eigenfunctions $y(x, \lambda)$, and $z(x, \mu)$ of conjugate operators $L$ and $L^*$ are orthogonal, i.e. $\lambda \neq \bar{\mu}$

$$\int_a^b y(x, \lambda)\bar{z}(x, \bar{\mu})dx = 0$$

If we assume with $\lambda = \bar{\mu} \int_a^b y(x, \lambda)\bar{z}(x, \bar{\mu})dx = 1$ (condition of standardizing /normalization the eigenfunctions), then

$$\int_a^b y(x, \lambda)\bar{z}(x, \bar{\mu})dx = \delta_{\lambda\bar{\mu}} \tag{1.6.9}$$

where

$\delta_{\lambda\bar{\mu}}$ −Kronecker's symbol: $\delta_{\lambda\bar{\mu}} = 1(\lambda - \bar{\mu})$;    $\delta_{\lambda\bar{\mu}} = 0$   $(\lambda \neq \bar{\mu})$.
In case of the self-conjugate operator:

$$\int_a^b y(x, \lambda)y(x, \mu)dx = \delta_{\lambda\mu}, \tag{1.6.10}$$

where $y$ – normalized eigenfunctions.

The differential expression (1.6.1) considered on the interval $(a, b)$, can originate various operators.

The operator $L$ with a maximum domain of definition in $L_2(a, b)$, i.e. a square summable on $(a, b)$, is an operator which operates according to the expression (1.6.1) to all functions $y(x)$ from $L_2(a, b)$, having absolutely continuous quasi-derivatives up to $2n - 1$th order inclusive, where $y^{[2n]}$ belongs to $L_2(a, b)$. This space of functions is denoted D in [368, 371, 377].

Another operator $L_0$ is defined on the set of all functions $y(x) \in D$, which satisfy conditions:

$$y^{[k]}|_{x=a} = y^{[k]}|_{x=b} = 0 \; (k = 0, 1, 2, \dots, 2n - 1)$$

in the regular case, and $[y, z]_a^b = 0$ in the singular case for all $z \in D$. In the case of one singular end, the domain of definition of operator $L_0$ consists only of those functions $y(x) \in D$, which satisfy the conditions:

$$y^{[k]}|_{x=a} = 0 \ (k = 0, 1, 2, \dots, 2n - 1)$$

$$[y, z]_a^b = 0 \ z \in D.$$

Operator $L_0$ allows various self-conjugate extensions. The domain of definition of $D'$ of any self-conjugate operator $L_u$ consists of those functions $y(x) \in D$, which satisfy the following conditions:

if $y, \ z \in D'$, then $[y, z]_a^b = 0$;

if $z \in D$ and for all $y(x) \in D'$, $[y, z]_a^b = 0$, then $z \in D'$.

Note that $L_0 = L^*$, i.e. the operators $L_0$ and $L$ are conjugate.

The pair of numbers which corresponds to the number of linearly independent solutions of the equations

$$A^*u = \lambda u, \ A^*u = \bar{\lambda} u \ \ [Im \, \lambda > 0, u \in L_2 \ (a, b)],$$

are called the index of the defect operator of $A$. The index of defect of operator $L_0$ is defined by a pair of identical integers $(m, m)$, moreover since $0 \le m \le 2n$, it is designated by one number, $m$. If the index of operator $L_0$ is equal $2n$, then the spectra of all its self-adjoint extensions are discrete. In a general case, the spectrum can be discrete-continuous, and the continuous part of all self-conjugate extensions is identical (same). If one end of the interval $(a, b)$ is regular, then $n \le m \le 2n$. In particular, if functions $(p_0^{-1})', p_1, p_2, \dots, p_n$ are summarized in $[0, \infty)$, and $\lim_{x \to +\infty} p_0(x) > 0$; then the index of defect (deficiency) of operator $L_0$ is $(n, n)$. The same index has an operator $L_0$, when $(p_0^{-1})', p_1, p_2, \dots, p_{n-1}, p(x)$ are summarized on interval $[0, \infty)$, $\lim_{x \to +\infty} p_0(x) > 0$; $p_n(x) = p(x) + q(x)$, where $q(x)$ – measured and substantially limited in interval $[0, \infty)$, function, for example, when

$$l(y) = (-1)^n \frac{d^{2n}y}{dx^{2n}} + q(x)y.$$

If the index of defect (deficiency) of the operator is $n$, then

$$[y, z]_b = 0$$

for all $y, z \in D$.

Any material $\lambda$, with which the number belonging to $L_2(a, b)$ of linearly independent solutions of equation

$$l(y) = \lambda y$$

is lower than the deficiency number of operator $L_0$, belongs to the discrete or continuous part of the spectrum of any self-adjoint extension of operator $L_0$. If in this case one of the ends is regular, then $\lambda$ belongs to the continuous part of the spectrum. Any self-adjoint operator of $L_u$ can be determined with the aid of a certain operator function $P_\lambda$ as follows: if $y$ belongs to the domain of definition of operator $L_u$, then

$$L_u y = \int_{-\infty}^{\infty} \lambda d[P_\lambda y]$$

Operator function $P_\lambda$ is called the spectral function of the operator $L_u$. Spectral function is related to rezolvent of the self-adjoint operator by

$$R_\lambda = \int_{-\infty}^{\infty} \frac{dP_\mu}{\mu - \lambda},$$

where $\lambda$ is a regular point of the operator.

We note that the resolvent of any self-adjoint operator $L_u$ is an integral operator; the resolvent's kernel $G(x, \xi, \mu)$ builds as Green's function of the operator $L_u - \mu E$, i.e. as the solution of the equation

$$l(y) - \mu y = \delta(x - \xi)$$

In the case of one regular end $(a)$, the kernel of a rezolvent satisfies the conditions:

$$\int_a^b |G(x, \xi, \mu)|^2 d\xi < \infty; \quad \int_a^b |G(x, \xi, \mu)|^2 dx < \infty.$$

Any function $f(x)$ from $L_2(a, b)$ can be decomposed according to the eigenfunctions of the self-adjoint differential operator. Let $L_u$ be the self-adjoint operator, generated by differential equation $l(y)$ in the interval $(a, b)$, and $\{u_k(x, \lambda)\}_{k=1}^{2n}$ be the solution of equation $l(y) = \lambda y$, which satisfy the following conditions:

$$u_j^{[\nu-1]}|_{x=a^*} = \delta_{j\nu}, \quad (a < a^* < b),$$

which leads to the following inversion formulas:

$$F_j(\lambda) = \int_a^b f(x)u_j(x,\lambda)dx;$$

$$f(x) = \int_{-\infty}^{\infty} \sum_{j,k=1}^{2n} F_j(\lambda)u_k(x,\lambda)d\tau_{jk}(\lambda).$$

Here $\|\tau_{jk}(\lambda)\|$ — matrix distribution function.

The decomposition of the kernel of the resolvent of operator $L_u$ takes the following form:

$$G(x,\xi,\mu) = \int_{-\infty}^{\infty} \sum_{j,k=1}^{2n} \frac{u_k(x,\lambda)u_j(\xi,\lambda)d\tau_{jk}(\lambda)}{\lambda - \mu}$$

In the case when one end of the interval (for example, $a$) is regular, the order of the matrix distribution function can be lowered. In particular, if the index of defect of operator $L_0$ is $(n, n)$, then summing in the inversion formulas the above equation is produced only up to $n$.

If functions $(p_0^{-1})', p_1, p_2,..., p_n$ are summarized on interval $[0, \infty)$, and $\lim_{x\to\infty} p_0(x) > 0$, then for the operator $L_u$, acting on the semi axis $[0, \infty)$ and without discrete spectrum, the invasion formulas take the form:

$$F(\lambda) = \int_0^{\infty} f(x)u(x,\lambda)dx;$$

$$f(x) = \int_0^{\infty} F(\lambda)u(x,\lambda)u(x,\lambda)d\lambda,$$

where $u(x, \lambda)$ is the solution of the equation $l(u) = \lambda u$, which satisfies corresponding self-adjoint boundary conditions at the end of $x = 0$.

The spectral function of distribution can be easily determined through the kernel of the resolvent of operator. Directing $x$ to $a - 0$ and $\xi$ to $a + 0$, in the formula of decomposition of a kernel of the operator $L_u$, we obtain:

$$G(a-0, a+0, \mu) = \int_{-\infty}^{\infty} \sum_{j,k=1}^{2n} \frac{d\tau_{jk}(\lambda)}{\lambda - \mu} .$$

We will use further the inversion formula of Stieltjes, according to which the integral equation

$$\varphi(\mu) = \int_{-\infty}^{\infty} \frac{d\omega(t)}{t - \mu} \qquad (\mu = \alpha + i\beta)$$

has a solution

$$\omega(a) - \omega(b) = \frac{\sin\beta}{2\pi i} \lim_{\beta \to 0} \int_a^b [\varphi(\mu) - \varphi(\overline{\mu})]d\alpha.$$

The last formula can be transformed and brought to different form. For example, it is possible to write

$$\omega(a + \delta) - \omega(\delta) = -\frac{\sin\beta}{\pi} \lim_{\beta \to 0} \int_\delta^{a+\delta} Im\varphi(\alpha + i\beta)d\alpha.$$

At the points of the continuity of function $\omega$, the inversion formula takes the form:

$$\omega'(a) = -\frac{\sin\beta}{\pi} \lim_{\beta \to 0} Im\varphi(\alpha + i\beta).$$

Designating $G_{q-1,\,\nu-1}(a-0, a+0, \mu) = M_{\nu q}(\mu)$, where indexes designate quasi-
derivatives of $x$ and $\xi$ respectively, from the formula of spectral functions of distribution and from the formula of transformed solution of integral equation, it is possible to obtain [370, 371, 376].

$$\tau_{jk}(\lambda) = \frac{1}{\pi} \lim_{\delta \to +0} \lim_{\varepsilon \to +0} \int_\delta^{\lambda+\delta} Im M_{jk}(\lambda + i\varepsilon)d\lambda.$$

Matrix $M_{jk}$ is called the characteristic matrix of operator $L_u$.

Thus integral transformation can be built on the basis of eigenfunction decomposition of regular and singular differential equations.

In principle, any expansion can be interpreted as the integral transformation, only if the analytical connection or numerical procedure is known, which makes it possible to establish a correspondence between transforms (coefficients of decomposition) and original (initial) functions.

For example, the system of functions $u(\xi, x)$, is the solution to the equation

$$Lu - \lambda\sigma'u = 0,$$

where $L$ – some linear self-adjoint operator; $\sigma'$ – weighting function is considered in the general case as the generalized function; and $\lambda$ – spectral parameter.

Let the eigenfunctions $u(\xi, x)$ satisfy the conditions of orthogonality and completeness in the form

$$\sigma'(x) \int_{-\infty}^{\infty} u(\lambda, x)u(\lambda, x_1)d\tau(\lambda) = \delta(x - x_1);$$

$$\tau'(\lambda) \int_{-\infty}^{\infty} u(\lambda, x)u(\lambda_1, x)d\sigma(x) = \delta(\lambda - \lambda_1),$$

where $\tau(\lambda)$ – spectral function of distribution.

Then there are integral transformations with symmetrical inversion formulas in the form of Stieltjes integrals:

$$f(x) = \int_{-\infty}^{\infty} F(\lambda)u(\lambda, x)d\tau(\lambda);$$

$$F(\lambda) = \int_{-\infty}^{\infty} f(x)u(\lambda, x)d\sigma(x).$$

If the spectral function $\tau(\lambda)$ and $\sigma'(x)$ function are continuous, then Stieltjes's formulas define standard integral transformation with infinite limits.

With the discrete spectrum of the decomposition, when the spectral function $\tau(\lambda)$ is a function of the jumps, and mean spectral density $\tau'(\lambda)$ is a sequence of delta-functions, the inversion formulas define finite integral transformation, convenient for solving boundary value problems with discrete spectrum. But if in this case the weighting function is a function of the jumps, then transformation becomes discrete, and it can be applied to the solution of discrete boundary-value problems of those described by equations in finite differences.

Below we will consider further generalizations of the method of integral transformations in particular, we will use Lagrange's identity within the classical scheme of integral transformation by extending and narrowing of the domain and adding kernels which are eigenfunctions of close operators and non-orthogonal kernels, A.I. Tseitlin, L.G. Petrosian [371, 376, 377]. Let us present these results for the purpose of the substantiation (validation) of the methods of building the boundary equations.

First, we will consider integral transformations with the orthogonal kernels which are eigenfunctions of some one-dimensional linear differential operator $L$ and satisfies the equation:

$$lu(x, \lambda) = \lambda u(x, \lambda) \qquad (x_1 < x < x_2) \qquad (1.6.11)$$

It is assumed that the self-conjugate operator $L$ created by differential operation of order $2n$

$$lu(x) = \sum_{m=0}^{n} (p_m u^{(n-m)}(x))^{(n-m)} \qquad (1.6.12)$$

with $n$ times differentiable in the domain $D = (x_1, x_2)$ by coefficients $p_m(x)$ and by the decomposed boundary conditions:

$$A_j u(x_1) = 0; \quad A_j u(x_2) = 0 \quad (j = 1, 2, . \ldots , s; \quad n \leq s \leq 2n) \quad (1.6.13)$$

$A_j u$ are understood to be linearly independent actual forms relative to the values of function $u(x)$ and its derivatives up to $2n - 1$ order, inclusive. The boundary conditions assigned at each end of the interval equal to $s$, while different forms of $A_j$ at the both ends of the interval are limited from $n$ to $2n$. When $s < 2n$, let us supplement the set of boundary conditions with some linear independent forms to the complete system of $2n$ forms. We will write down of the dual formulas decomposition of eigenfunctions of operator $L$ with some spectral function $\tau(\lambda)$:

$$\int_{x_1}^{x_2} f(x)u(x, \lambda)dx = (f, u) = F(\lambda) \quad (1.6.14)$$

$$\int_{-\infty}^{\infty} F(\lambda)u(x, \lambda)d\tau(\lambda) = f(x)$$

which determine $u$ – transformation of function $f(x)$ and its conversion. Here $\tau(\lambda)$ – a non-decreasing function, uninterrupted in a continuous spectrum and stepped (with countable number of jumps) at the discrete spectrum of decomposition.

By a jump $j$ of a function $g(x)$ at a point $x_0$ we mean the difference between the right-hand and left-hand limits of $g(x)$ at $x_0$; that is $j = g(x_0 + 0) - g(x_0 - 0)$.

Formulas (1.6.14) cover integral transforms with the infinite limits, usually having a continuous spectrum, and with the finite limits, when spectrum is discrete.

In the classical scheme of integral transformation, the formula (1.6.14) is applied for solving the boundary-value problems, generated by the differential operation (1.6.12) on $(x_1, x_2)$ with the homogeneous or inhomogeneous boundary conditions. Let us extend this method to the more general case of the boundary-value problems, generated operators close to $L$. Assume, that it is assigned the boundary-value problem

$$l_* w(x) = q(x) \quad (a < x < b) \quad (1.6.15)$$

with close operation $l_*$, and boundary conditions:

$$B_j w(a + 0) = g_{ja}; \quad B_j w(b - 0) = g_{jb}; \quad (j = 1, 2, \ldots , s; n \leq s \leq 2n)$$
$$(1.6.16)$$

For the sake of simplicity, we will consider that the order of operations $l$ and $l_*$ is identical. However, a case where operation $l_*$ has order multiple with respect to $l$, could be analogously considered. In accordance with the standard schema, let us multiply the left and right sides of Eq. (1.6.15) by $u(x, \lambda)$, and integrate over the interval $(x_1, x_2)$. If the $(x_1, x_2)$ is wider than $(a, b)$, then $w(x)$ and $q(x)$ must be prolonged somehow out of the given interval $(a, b)$. We have:

$$(u, l_* w) = (u, q) \tag{1.6.17}$$

The method of integral transformations further follows partial integration.

It is more convenient however, to use Lagrange's formula directly, on each interval of continuity $(\alpha, \beta)$, instead of the integration by parts

$$(u, l_* w) = (l_* u, w) + \left[ \sum_{j=1}^{2n} B_j w(x) B_{2n-j+1} u(x, \lambda) \right]_{\alpha}^{\beta},$$

assuming that the function $w(x)$ is prolonged out of $(a, b)$ in such a way that it satisfies the boundary conditions (1.6.13), when $x = x_1$ and $x = x_2$, we obtain:

$$(l_* u, w) = (u, q) + \Gamma(a, \lambda) + \Gamma(b, \lambda) \tag{1.6.18}$$

where

$$\Gamma(a, \lambda) = \sum_{j=1}^{2n} \Delta B_j w(x) B_{2n-j+1} u(x, \lambda) \Big|_{x=a},$$

$$\Delta B_j w(x) = B_j w(x + 0) - B_j w(x - 0).$$

If $l_* = l$, $B_j = A_j$, and $(a, b)$ coincides $(x_1, x_2)$ that of (1.6.18) taking into account (1.6.14) follows standard result for the transformation of the required function

$$W(\lambda) = \lambda^{-1}(u, q) \tag{1.6.19}$$

From where we will obtain according to the conversion formula

$$w(x) = \int_{-\infty}^{\infty} \lambda^{-1}(u, q) u(x, \lambda) d\tau(\lambda). \tag{1.6.20}$$

Let us consider separately three cases of the possible generalizations of the method of the integral transforms, when these conditions are not satisfied:

- transformation in the extended $(x_1, x_2) \supset (a, b)$ and the narrowed $(x_1, x_2) \subset (a, b)$ domain;
- transformation with close kernel $l_* \neq l$;
- transformation with close boundary conditions $A_j \neq B_j$.

1. In the case of the extended domain formula (1.6.18) with $l_* = l$ gives:

$$\lambda W(\lambda) = (u, q) + \Gamma(a, \lambda) + \Gamma(b, \lambda)$$

from where we will obtain according to the conversion formula

$$w(x) == w_q(x) + \sum_{j=1}^{2n} \Delta B_j w(a) R_{2n-j+1}(x, a) + \sum_{j=1}^{2n} \Delta B_j w(b) R_{2n-j+1}(x, b)$$

$$(1.6.21)$$

where

$$w_q(x) = \int_{-\infty}^{\infty} \lambda^{-1} u(x, \lambda)(u, q) d\tau(\lambda)$$

is the solution for the problem about the action of the given load in the extended domain (area);

$$R_{2n-j+1}(x, a) = \int_{-\infty}^{\infty} \lambda^{-1} u(x, \lambda) B_{2n-j+1} u(x_1, \lambda) \Big|_{x_1=a} d\tau(\lambda).$$

Using boundary conditions (1.6.16), we will obtain the system of the boundary algebraic equations

$$B_j w_q(a + 0) + \sum_{j=1}^{2n}$$

$$[\Delta B_j w(a) B_j R_{2n-j+1}(x, a)|_{x=a} + \Delta B_j w(b) B_j R_{2n-j+1}(x, b)|_{x=a}] = g_{ja}$$

$$(1.6.22)$$

$$B_j w_q(b - 0) + \sum_{j=1}^{2n}$$

$$[\Delta B_j w(a) B_j R_{2n-j+1}(x, a)|_{x=b} + \Delta B_j w(b) B_j R_{2n-j+1}(x, b)|_{x=b}] = g_{jb}$$

After solving this system relative to $\Delta B_j w(a)$, $\Delta B_j w(b)$ according to the formula (1.6.21) we will obtain a final solution of the problem. During the narrowing of the domain, which can be connected with the presence of internal conditions or localization of the solutions close to the

zone of indignation, integration can be implemented on invariable intervals [340] or on a few intervals covering the standard for this transforms
domain.

   In multidimensional problems with a complex configuration, it
is possible to use transformation in the narrowed domain, which
is obtained as a result of the partition of the given domain of
boundary-value problems into several standard sub-domains. A
similar method in series was examined, in particular in [325].

2. Transformation with a close kernel is convenient when the solution of a
   boundary-value problem with close differential operations and the similar boundary conditions, as in the given problem, are known. Imple-
   menting transformation on a given interval $(a, b)$, according to (1.6.18)
   we will obtain:

$$(l_* u, w) = (u, q) - G(\lambda), \tag{1.6.23}$$

where $G(\lambda)$ – the known function, obtained from the bilinear
Lagrange form.

   We will present differential expression $l_*$ in the form of $l_* = l + \tilde{l}$,
where $\tilde{l}$ is an additional operator. Then from (1.6.23) according to the
conversion formula (1.6.14), we obtain the integral equation of the second kind relative to $w(x)$

$$w(x) = \int_a^b K(x, x_1) w(x_1) dx_1 + v(x), \tag{1.6.24}$$

where

$$K(x, x_1) = - \int_{-\infty}^{\infty} \lambda^{-1} u(x, \lambda) \tilde{l} u(x_1, \lambda) d\tau(\lambda);$$

$$v(x) = \int_{-\infty}^{\infty} \lambda^{-1} [(u, q) - G(\lambda)] u(x, \lambda) d\tau(\lambda).$$

   If $(a, b)$ is finite interval and spectrum of boundary-value problem is
discrete, then:

$$K(x, x_1) = - \sum_{i=1}^{\infty} \lambda_i^{-1} u(x, \lambda_i) \tilde{l} u(x_1, \lambda_i) \tau_i;$$

$$v(x) = \sum_{i=1}^{\infty} \lambda_i^{-1} [(u, q) - G(\lambda_i)] u(x, \lambda_i) \tau_i, \tag{1.6.25}$$

where $\tau_i$ – normalizing numbers of discrete decomposition.

Being limited to finite number terms of series $m$ in (1.6.25), the solution of Eq. (1.6.24) can be obtained according to the formula

$$w(x) = \sum_{i=1}^{m} C_i \lambda_i^{-1} u(x, \lambda_i) + v(x), \qquad (1.6.26)$$

where $C_i$ – solution of the system of the algebraic equations:

$$C_i = \sum_{j=1}^{m} K_{ij} C_j + f_i;$$

$$K_{ij} = \lambda_i^{-1} \tau_j \int_a^b u(x, \lambda_i) \tilde{l} u(x, \lambda_i) dx;$$

$$f_i = \tau_i \int_a^b v(x) \tilde{l} u(x, \lambda_i) dx$$

From Eq. (1.6.23) with the discrete spectrum, it is also possible to obtain a system of algebraic equations, a similar system to the method of Bubnov-Galerkin

$$\sum_{i=1}^{m} (u_i, w)(\tilde{l} u_j, u_i) \tau_i = (u_j, q) - G(\lambda_j) \qquad (j = 1, 2, \dots, m), \quad (1.6.27)$$

if we substitute in (1.6.23) the decomposition

$$w(x) = \sum_{i=1}^{\infty} (u_i, w) u_i \tau_i, \qquad (1.6.28)$$

being limited in it by first $m$ terms of series. After solving this system relative to $W_i = (u_i, w)$ according to the formula (1.6.28) it is possible to obtain the required solution of the problem.

3. In many cases the difficulties of solving the boundary-value problem are connected to the inconvenient boundary conditions. If a solution of a similar problem with any other boundary conditions is known, then the use of transformation with its own kernel of this problem reduces to the boundary equations, whose solution is simpler, the closer the boundary conditions are to the original and close problems. Let $l_* = l$; $x_1 = a - 0$; $x_2 = b + 0$; $A_j \neq B_j$. Then, assuming for all considering $j$

$$B_j w(x_1) = B_j w(x_2) = 0 \qquad (1.6.29)$$

we will obtain

$$w(x) = w_q(x) + \sum_{j=1}^{2n} B_j w(a+0) R_{2n-j+1}(x,a) - \sum_{j=1}^{2n} B_j w(b-0) R_{2n-j+1}(x,b)$$

(1.6.30)

Using boundary conditions (1.6.16), it is possible to exclude $2n$ boundary expressions on the right side of (1.6.30). The number of remaining unknowns $B_j w(a+0)$; $B_j w(b-0)$, can be reduced, if $u(x,\lambda)$ satisfies some of the boundary conditions (1.6.13). For determining the unknown we are using boundary equations which are obtained by substituting (1.6.30) into the boundary conditions (1.6.16). The number of equations will be equal to the number of incongruent differential $A_j$ and $B_j$.

These generalizations of the method of the integral transforms also known as generalized integral transformation [268, 291], are especially effective in solving the multi-dimensional problems.

When choosing the kernel that satisfies the differential equation of a problem, the application of the generalized integral transform reduces the problem to the solution of boundary equations – integral with the continuous spectrum and algebraic with the discrete. If the kernel of transformation satisfies only boundary conditions, then we come to the integral or algebraic equations, obtained in the entire domain of definition of the operator of considered problem. The given generalizations of the method of integral transforms do not go beyond the ideas, connected with the use of the boundary-value problems of orthogonal or bioorthogonal decompositions with the discrete and continuous spectrum. The generalizations are substantially different in their basis and can be obtained by means of the input into the procedure of the integral transformation of non-orthogonal kernels. As such kernels we will use Green's functions and fundamental solutions of close problems, as well as delta-function. Let us first consider a boundary-value problem (1.6.15, 1.6.16) in the finite interval $(a,b)$.

We will accept as the kernel of transformation the Green's function $G(x_*,x)$ of the closed boundary-value problem, described by differential Eq. (1.6.15) and boundary conditions (1.6.16). We will multiply the left and right sides of Eq. (1.6.15) by $G(x_*,x)$ and will integrate over the extended domain $(x_1,x_2)$. As a result we find:

$$w(x_*) = w_q(x_*) + \sum_{j=1}^{2n} [\Delta B_j w(a) B_{2n-j+1} G(x_*,x)|_{x=a} + \Delta B_j w(b) B_{2n-j+1} G(x_*,x)|_{x=b}],$$

(1.6.31)

where

$$w_q(x_*) = \int_{x_1}^{x_2} q(x)G(x_*,x)dx. \tag{1.6.32}$$

From (1.6.31) we will obtain the system of boundary equations:

$$B_j w_q(a+0) + B_j[\Gamma_1(x_*,a) + \Gamma_1(x_*,b)]_{x_*=a+0} = g_{ja},$$

$$B_j w_q(b-0) + B_j[\Gamma_1(x_*,a) + \Gamma_1(x_*,b)]_{x_*=b-0} = g_{jb}, \tag{1.6.33}$$

$$\Gamma_1(x_*,x) = \sum_{j=1}^{2n} \Delta B_j w(a) B_{2n-j+1} G(x_*,x).$$

In (1.6.31) and (1.6.33) there are $4n$ unknown values of $\Delta B_j w(a)$; $\Delta B_j w(b)$. From them, $2n$ unknowns can be arbitrarily assigned a value, for example, equal to zero. The selection of these values is determined by the substitution of boundary conditions at points $x = a - 0$ and $x = b + 0$ with the extension of domain and the design of the Green function.

Design of the Green function by itself in many cases represents a complex task, therefore, it is desirable to select such a close operator, for which the Green function either can be obtained in closed form or constructed with an aid of a sufficiently simple algorithm. Determining $G(x, x_1)$ in the infinite interval is the simplest solution of boundary-value problems. In this case, as you can see, the method considered is the analog to the classical method of potential. The fundamental solution of Eq. (1.6.15) can serve as another non-orthogonal kernel. Since we have various forms of boundary conditions that can be satisfied with fundamental solutions, the system of boundary equations (1.6.33) can be simplified by only keeping some of forms of boundary conditions. In particular, taking a fundamental solution, which satisfies the system of $2n$ homogeneous boundary conditions at one end of the interval $(a, b)$, as a kernel of transformation we will obtain an analogue of the method of initial values (initial parameters), and in the multidimensional case a method of initial functions. Finally, as the non-orthogonal kernel of integral transformation, delta-function can be accepted (A.I. Tseitlin [370]). The method of delta-transform is examined in detail in the following paragraph, where we will dwell on the various solutions, and then we will trace their connection to the classical methods of structural mechanics.

Before proceeding let us pause to establish the following fundamental definitions and concepts pertaining to the extended domain, boundary-value and fundamental solutions, and the choice of a primary system. Thus, for a solution of a problem, we can choose various intervals of integration [A, B], boundary conditions on [A, B], fundamental solutions, and any arbitrary constant to be determined [371].

*Extended Domain.* An interval of integration [A, B], can coincide with (a, b), or to be infinite. In multidimensional problems there can be a convenient choice of a finite interval of the integration including (a, b). The unlimited extended domain is characteristic for the method of the compensating loads, generalized solutions, and methods of the type of potential.

*Boundary conditions.* In particular for [A, B] these can be taken arbitrarily, so that the building of fundamental solutions would not cause difficulties. The choices of an interval of integration, boundary conditions, and discontinuities of the boundaries on (a, b) determine a problem, similar to the examined, mechanical interpretation of which can be connected to the widely used structural mechanics concept of a primary system. In the case of unlimited domain [A, B], the usual conditions of limitedness are established, or conditions of the type of radiation.

*Fundamental Solutions.* It is more convenient to accept Green's function for the extended domain as a fundamental solution. In an interval of integration that coincides with [a, b], the fundamental solution can be built so that it becomes zero when $x < a$, while the generalization of the method of initial parameters for the arbitrary systems, described by one-dimensional differential operators is obtained.

*Primary System.* The choice of a primary system predetermines the scheme of the solutions and the type of the resolving system of equations. If jumps (discontinuities) remain on the borders of (a, b) and at internal points $x = x_i$ corresponding to the introduction of additional force factors, then the resolving system of equations will be an analogue of the canonical equations of the method of force, and the fundamental solutions, which is analogue of unit displacements. While preserving discontinuities corresponding to the natural conditions, we come to the generalization of the method of displacement. The fundamental solutions in this case will correspond to the unit reactions. With other combinations of jumps (discontinuities, breaks) we will obtain an analogue of the mixed method.

## 1.7 The Method of Delta-Transform

The method of delta-transform arose as the apparatus, justifying the application of the generalized functions for constructing boundary equations of different types. On the bases of this method and the method of the generalized integral transforms it is possible to give a sufficiently general scheme of construction and classification of all existing methods and types of boundary equations. It should be noted that a number of subtle, specific questions, solutions of which this scheme could convert into a strict theory of boundary equations, still requires in-depth analysis and relevant proofs. Nevertheless the method of delta-transform makes it possible to connect virtually all methods of constructing boundary integral equations into a united system. Moreover the method of delta-transform gives a clear mechanical interpretation of the various methods, and highlights the close connections between them, in particular between the methods of potentials and *boundary integral equations method* (BIEM).

Below you will find the results of delta and Green transformation schemas [278, 377]. First, we will discuss the usage of delta transformation for solving general types of self-conjugate boundary-value problems. Secondly, we will show the connection between this method and Green transformation method [291, 292, 344].

First we will consider an application of delta-transform to the solution of a self-conjugate boundary-value problem of general view, and then we will point to a connection of this method to the integral transformation with kernel in a form of the Green's function of a considered or close boundary-value problem.

One-dimensional boundary-value problem for the operator $L$, generated by the linear self-conjugate differential operation

$$lw(x) = \sum_{m=0}^{n} (p_m(x)w^{(n-m)}(x))^{(n-m)}, \qquad (a < x < b), \qquad (1.7.1)$$

and by the boundary conditions

$$A_j w(x)|_{x=a+0} = 0; \quad A_j w(x)|_{x=b-0} = 0, \qquad (j = 1, 2, \ldots, s), \qquad (1.7.2)$$

which we for certainty will also consider self-conjugate, is described by the equation

$$lw(x) = q(x) \qquad (a < x < b), \qquad (1.7.3)$$

and boundary conditions (1.7.1).

Let us apply Eq. (1.7.3) of delta-transform in the maximally extended length, i.e. integral transform with the kernel in a form of a delta function on an infinite interval. For this purpose we will multiply both sides of the Eq. (1.7.3) on $\delta(\xi - x)$ and will integrate along the entire axis$(-\infty < x < \infty)$, continuing $w(x)$ and $q(x)$ out of $(a, b)$. In accordance with (1.6.17) we obtain with $u = \delta(\xi - x)$:

$$
\int_{-\infty}^{\infty} w(x) l\delta(\xi - x)dx = q(\xi) + \left[ \sum_{j=1}^{2n} \Delta B_j w(x) B_{2n-j+1}\delta(\xi - x) \right]_{x=a}
$$

$$
+ \left[ \sum_{j=1}^{2n} \Delta B_j w(x) B_{2n-j+1}\delta(\xi - x) \right]_{x=b} \tag{1.7.4}
$$

The integral on the left side of the formula (1.7.4) on the entire real axis, except for points $x = a$, $x = b$, according to the basic property of delta-function (Gelfaund I.M., Shilov G.E. [115]) equals to $lw(\xi)$. Therefore we will rewrite (1.7.4) in the following more simplified form:

$$
lw(\xi) = q(\xi) + \sum_{j=1}^{2n} \Delta B_j w(x) B_{2n-j+1}\delta(\xi - x) \Big|_{x=a,b.} \qquad (|\xi| < \infty) \tag{1.7.5}
$$

The analysis of Eq. (1.7.5) presents a certain interest. First, along with differential expression this equation contains the operators of boundary conditions of the delta-functions. Here we equalize the boundary forms $B_j$ with differential forms of boundary conditions (1.7.1), i.e. assume $B_j = A_j$. Second, it is set in the extended interval. Third, Eq. (1.7.5) contains $4n$ unknown step (jump) functions $\Delta B_j w(a)$, and $\Delta B_j w(b)$, for determining where there are only $2n$ boundary conditions (1.7.1). Finally, the right-hand side of Eq. (1.7.5) includes differential operators of the delta-functions representing the generalized forces and kinematic actions. Let us examine this question in more detail. It is known that, during the design of beams the force and moment can be described mathematically as a delta-function and its first derivative. Similarly, higher moments can be introduced also. So, the second derivative of the delta-function, that is the first derivative of the moment, describes the kinematic actions corresponding to the unit step of the angle of rotation in the section $x = \xi$ and the third derivative of the delta-function corresponding to the unit step of displacement (with the accuracy to the multiplier of $EI$).

The central issue in the method of delta-transform is the choice of unknown jump functions $\Delta B_j w(a)$, and $\Delta B_j w(b)$. Of the $4n$ step functions, $2n$ can be set to zero in order to define the remaining $2n$ step functions from $2n$ boundary conditions. For this purpose the following mechanical considerations can be used in classical methods of structural mechanics when choosing the primary system. Assuming any jump function $\Delta B_j w$ equals to zero, we fix one or extend the considered system through boundary points. For example, if we accept the first $n-1$ continuous derivatives of the function $w(x)$, corresponding to geometrical factors (shape), then we continue construction through the boundary sections as continuous. In this case, unknowns become the jump functions of the higher derivatives, which correspond to the forces factors. Hence, in this case the primary system is chosen as an infinite continuous construction, described by Eq. (1.7.3), and loaded in the boundary sections by external forces and moments. An alternative approach is to accept continuous force factors in the boundary sections, assuming the jump functions of the corresponding boundary forms equal to zero. Such an approach will correspond to the continuation of construction through the boundary section, where only kinematic impact is set, i.e. in boundary sections jump functions of displacements and angles of rotations are assigned.

These two methods of extension of domain and choosing the basic (main) system are similar to classical methods of forces and displacements in structural mechanics. If we keep different jumps $\Delta B_j w(x)$ in the sections $x = a$, $x = b$, then we will obtain the analogue of the mixed method of structural mechanics. It should be noted that not all jumps can be assumed to be zero, and consequently not all of them can be kept in the solution of a problem. In fact, the choice of force and kinematic factors, determined by the jumps of the differential forms, must provide linear independence of the solution of the homogeneous equation, forming Green function and its derivatives. Therefore, the jumps of adjoint differential forms cannot be saved or omitted if, for example, we assume $\Delta B_j w(x) = 0$ in one of the boundary sections, then for this section it is already impossible to accept $\Delta B_{2j-j+1} w(x)$. Besides the two indicated variants of the choice of the unknown jumps, determining the force and kinematic impacts, it is possible to indicate the following passible loads of boundary sections. One of such loads is connected to preservation of all jumps, i.e. all possible types of force and kinematic impacts; the other, contrary to this, is connected to the complete absence of any compensating impacts in the boundary points. In the latter case, in order to obtain a nontrivial solution to a problem, it is necessary to place the compensating force and kinematic impacts in the extended domain outside (beyond) the boundaries of the interval $(a, b)$. This case is equivalent to the modification of the method of compensating loads proposed by the B.G. Korenev [186].

With the retention of all jumps of differential forms and additional conditions, setting the relationship between the unknown jumps, so that to the number of jumps $4n$ would correspond to the number of boundary conditions $2n$ and the number of additional conditions to $2n$, must be set.

Thus, for systematization of the equations of the method of delta-transform we should examine three basic cases when the compensating actions are located at the boundary points.

1. $2n$ jumps of geometric factors described by differential forms to $n - 1$th order inclusive are set to equal to zero. The external force impacts are unknown. Solving Eq. (1.7.5) and satisfying the boundary conditions (1.7.2), we come to the system of $2n$ boundary equations relative to the jumps of the force factors. This scheme of solution corresponds to the common form of the force method, method of compensating loads, and indirect method of boundary elements.

2. $2n$ jumps of force factors described by differential forms, which contain the derivatives of order $n$ and above, are set to equal to zero. The external kinematic impacts are unknown. After solving Eq. (1.7.5) and satisfying boundary conditions (1.7.2) we come to the system of $2n$ boundary equations relative to the jumps of the kinematic parameters in the boundary sections. In this case the higher derivatives of the delta-function in the right side of Eq. (1.7.5) remain, which leads to irregular equations and, in the case of using the transformations (conversions) leads to divergent decompositions. The obtained equations can be regularized through the isolation (separation) of delta-function and its derivatives, making the equation easily solvable. In its mechanical content this method is the analogue of the method of displacements.

3. All jumps of differential forms remain in the solution. In this case they must be set, as mentioned above, to $2n$ additional conditions connecting these jumps. In particular, choosing any $n$ of the boundary forms, corresponding to any self-adjoint boundary conditions in the section $x = a$ and $n$ of the possible other forms, which also correspond to some self-adjoint boundary conditions in the section $x = b$, has us assume them to be equal to the values of the corresponding forms on the internal boundaries of

interval $(a, b)$. This gives $2n$ additional conditions:

$$\Delta B_j w(a) = B_j w(a + 0), \quad \Delta B_j w(b) = B_j w(b - 0).$$

Then, respectively,

$$B_j w(a - 0) = 0, \quad B_j w(b + 0) = 0$$

Since we can set $q(x) = 0$ with $x < a$, $x > b$ and boundary conditions are homogeneous, then the solution of Eq. (1.7.1) in the closed interval $[a, b]$ must be trivial; that is, identically equal to zero. Therefore, the remaining boundary forms $B_j w(a + 0)$, $B_j w(b - 0)$ are also equal to zero, and therefore the jumps, which we for distinction will designate through $B_i$, are equal

$$\Delta B_i w(a) = B_i w(a + 0), \quad \Delta B_i w(b) = B_i w(b - 0).$$

Thus, all jumps are equal to the values of the corresponding differential forms on the internal boundaries (borders) of the interval $(a, b)$. Therefore, instead of (1.7.5), in this case we can write the following:

$$lw(\xi) = q(\xi) + \sum_{j=1}^{2n} B_j w(x) B_{2n-j+1} \delta(\xi - x) \Big|_{x=a,b.} \tag{1.7.6}$$

Further, it is possible to satisfy $2n$ conditions in Eq. (1.7.2) since $(B_j = A_j)$, and the last equation will contain only $2n$ unknowns, making it possible to solve any correctly stated problem. This method of composition of boundary equations is equivalent to the direct boundary element method (DBEM).

For solving Eq. (1.7.5) we can recommend two methods – Green's function and the method of integral transformation. In the first case it is assumed that Green's function of boundary value problem (1.7.1) in the extended domain is known. Further we will consider generalized Green's functions, which are solutions to the following equations.

$$l_\xi G_k(\xi, x) = B_k^\xi \delta(\xi - x). \tag{1.7.7}$$

Here, the index $\xi$ means, that the differential operations are taken on variable $\xi$. These solutions can be easily built, if the operators $l_\xi^{-1}$ and $B_k^\xi$ are permutable. Applying operation $l_\xi^{-1}$ to the left and right sides of Eq. (1.7.7), we obtain

$$G_k(\xi, x) = B_k^\xi l_\xi^{-1} \delta(\xi - x).$$

Since Green's function satisfied the equation:

$$l_\xi G_k(\xi, x) = \delta(\xi - x),$$

and, consequently,

$$l_\xi^{-1}\delta(\xi - x) = G(\xi, x),$$

then

$$G_k(\xi, x) = B_k^\xi G(\xi, x). \tag{1.7.8}$$

Taking into account (1.7.7), the solution of Eq. (1.7.5) can be written in the form of:

$$w(\xi) = w_q(\xi) + \sum_{j=1}^{2n} \Delta B_j w(x) G_{2n-j+1}(\xi, x) \bigg|_{x=a,b} \tag{1.7.9}$$

where $w_q(\xi)$ is the solution of $lw(\xi) = q(\xi)$, satisfying some boundary conditions at the ends of the extended domain.

Similar equations can be obtained by direct application of Green's function transformation to Eq. (1.7.1). Green's function transformation, as mentioned above, will be called integral transformation with the kernel in a form of Green's function of the considered or close boundary-value problem.

It is not difficult to see that, assuming $u = G(\xi, x)$ in Eq. (1.6.18), we will obtain (1.7.9). The advantage of the delta-transform, in comparison with the transformation of Green's function, is the clear mechanical interpretation of compensating impacts, and the use of other methods, such as the integral transformation for solving Eq. (1.7.5), but not reducing it to Eq. (1.7.9).

It should be noted that type of representation in (1.7.9), analogous to Somigliana's formula in the theory of elasticity, plays an important role in the boundary elements and boundary equations methods. In particular, direct methods of boundary conditions are built on the basis of the use of Green's function for an infinite domain and integral identity of Betty within the limits of the given domain $(a, b)$, which reduces to the type of formula in (1.7.9), where we have values of these forms at the boundary points instead of jumps $\Delta B_j w(x)$. As we see, unlike the DBEM, the application of the method of delta-transform or Green's function transformation, provides more opportunities for constructing boundary equations. Instead of one scheme of direct BEM, the delta-transform and Green's function transformation methods lead to a wide variety of boundary equations and give a unified scheme for composing them.

Let us look at one additional method of composition (building) of boundary equations, deriving from (1.7.6), and being the discrete analogue of the V.D. Kupradze method [199–201]. After solving (1.7.6) using Green's function, we will obtain:

$$w(\xi) = w_q(\xi) + \sum_{j=1}^{2n} B_j w(x) B_{2n-j+1} G(\xi, x) \Bigg|_{x=a,b.}$$

Taking into account that, with $\xi < a$, $\xi > b$, and $w(\xi) \equiv 0$, it is possible to choose $2n$ points of $\xi_i$ out of the interval $(a, b)$, and compose $2n$ equations,

$$w_q(\xi_i) + \sum_{j=1}^{2n} B_j w(x) B_{2n-j+1} G(\xi, x) \Bigg|_{x=a,b} = 0,$$

which ensure the uniqueness of the solution to Eq. (1.7.6).

Consider the application of delta-transform and Green transformation further to solve problems in the minimally extended domain, i.e. extending interval $(a, b)$ to $[a, b]$, by inclusion of boundary points $x_1 = a - 0$, and $x_2 = b + 0$ in the extended domain. After multiplication of both sides of the equation by delta function and integration in the limits of $(a, b)$, similarly (1.7.6) we obtain

$$lw(\xi) = q(\xi) + \sum_{j=1}^{2n} B_j w(a - 0) B_{2n-j+1} \delta(\xi - a)$$

$$+ \sum_{j=1}^{2n} B_j w(b + 0) B_{2n-j+1} \delta(\xi - b), \quad (a \le \xi \le b). \qquad (1.7.10)$$

Using the Green transformation or solving Eq. (1.7.10) by using Green's function we will have:

$$w(\xi) = w_q(\xi) + \sum_{j=1}^{2n} B_j w(a - 0) B_{2n-j+1} G(\xi, a) + \sum_{j=1}^{2n} B_j w(b + 0) B_{2n-j+1} G(\xi, b).$$

$$(1.7.11)$$

Here $w_q(\xi)$ is the particular solution of the equation with the right side $q(x)$, boundary conditions (1.6.13), and $G(\xi, a)$ – Green's function of close boundary-value problem (1.7.3), (1.7.2). Of course, if Green's function of the considered problem (1.7.1), (1.7.2) is known, then there is no necessity in the composition of boundary equations. In (1.7.11) the $2n$ forms $B_j G(\xi, a)$ and $B_j G(\xi, b)$, as a consequence of boundary conditions (1.7.2) satisfying

Green's function, are equal to zero. Substituting (1.7.11) into the boundary conditions (1.7.2), we obtain the necessary system of $2n$ boundary equations. Instead of Green's function for solving Eq. (1.7.11) a decomposition, acting on $[a, b]$ may be used. During the composition of boundary equations using the conditions (1.7.2) it is necessary to calculate expression of the type

$$G_i^j(\xi, a) = A_i^\xi A_j^x G(\xi, x), \tag{1.7.12}$$

where indexes $\xi$ and $x$ indicate variables on which differentiation is produced. The total order $m$ of both differential operators $A_i$ and $A_j$ determines the property of function $G_i^j(\xi, x)$. When $m < 2n$ this function is regular; when $m > 2n$ it belongs to the class of generalized functions with the carrier in the point $\xi = x$. Therefore, in the boundary equations it is desirable that figure operators from Green's function have the total order $m < 2n$.

For the illustration of the applicability of the method of delta-transformation and Green's transformation, let us examine a traditional problem about the bend of a beam on elastic foundation, convenient due to its simplicity and clarity for the analysis of different schemes of the composition of boundary equations [186, 278, 279].

The equation of the equilibrium of beam can be written in the form:

$$EI y^{IV}(x) + ky(x) = q(x), \qquad (a < x < b) \tag{1.7.13}$$

where $EI$ – rigidity of the beam; $k$ – the coefficient of the bed base; $y(x)$ – displacement of the beam; and $q(x)$ – external load. We will consider this problem with boundary conditions corresponding to the hinged ends supported beam, assuming that $a = 0$, $b = l$,

$$y(+0) = y^{II}(+0) = y(l - 0) = y^{II}(l - 0) = 0. \tag{1.7.14}$$

The solution of the problem will be found in the extended interval

$$-\infty < x < \infty.$$

We will define Green's function for infinite domain as a diminishing at infinity solution of the following equation:

$$EI \frac{d^4}{d\xi^4} G(\xi, x) + kG(\xi, x) = \delta(\xi - x), \qquad (-\infty < x < \infty) \tag{1.7.15}$$

In addition, we will introduce the generalized Green's functions which satisfy the equations:

$$EI \frac{d^4}{dx^4} G_j(\xi, x) + kG_j(\xi, x) = \frac{d^j}{d\xi^j} \delta(\xi - x), \qquad (-\infty < x < \infty) \tag{1.7.16}$$

Applying the Green's transformation, we will multiply the left and right sides of (1.7.13) by $G(\xi, x)$ and will integrate along all axes

$$EI \int_{-\infty}^{\infty} y^{IV}(x)G(\xi,x)dx + k \int_{-\infty}^{\infty} y(x)G(\xi,x)dx = y_q(\xi), \qquad (1.7.17)$$

where

$$y_q(\xi) = \int_{-\infty}^{\infty} G(\xi,x)q(x)dx$$

is the displacement of the infinite beam due to the load $q(x)$, continued somehow (in particular, zero) out of the interval $(0, l)$. Applying integration by parts, we obtain:

$$y(\xi) = y_q(\xi) + EI \sum_{j=0}^{3}[\Delta y^{(3-j)}(0)G_j(\xi) + \Delta y^{(3-j)}(l)G_j(\xi - l)] \qquad (1.7.18)$$

Here:

$$G_0(\xi, l) = G_0(\xi - x) = \frac{\lambda}{2k}e^{-\lambda|\xi-x|}[\cos\lambda(\xi - x) + \sin\lambda(\xi - x)];$$

$$G_1(\xi, l) = G_1(\xi - x) = \frac{\lambda^2}{k}e^{-\lambda|\xi-x|}\sin\lambda(\xi - x); \qquad (1.7.19)$$

$$G_2(\xi, l) = G_2(\xi - x) = \frac{\lambda^3}{k}e^{-\lambda|\xi-x|}[\sin\lambda(\xi - x) - \cos\lambda(\xi - x)];$$

$$G_3(\xi, l) = G_3(\xi - x) = -\frac{2\lambda^4}{k}sign(\xi - x)e^{-\lambda|\xi-x|}\cos\lambda(\xi - x); \qquad (1.7.20)$$

where $\lambda = \sqrt[4]{\frac{k}{4EI}}$.

Solution (1.7.18) contains eight arbitrary constants. We will consider the three basic variants indicated above, that are distinguished by the design scheme of the beam in the extended domain. In the beginning let us consider the equations of the indirect method of boundary elements. We will assume that, in sections of the infinite beam $\xi = 0$, and $\xi = l$, deflections and angle of rotation are continuous, i.e.

$$\Delta y(0) = \Delta y^I(0) = \Delta y(l) = \Delta y^I(l) = 0. \qquad (1.7.21)$$

Then in (1.7.18) only the breaks of second and third derivatives of the function $y(\xi)$ (members such as: $\Delta y^{II}G_1$ and $\Delta y^{III}G_0$), which is equivalent to uploading infinite beam forces and moments in the sections $\xi = 0$, and $\xi = l$, remain.

The solution (1.7.18), takes the form:

$$y(\xi) = y_q(\xi) + EI[\Delta y^{II}(0)G_1(\xi) + \Delta y^{III}(0)G_0(\xi) + \Delta y^{II}(l)G_1(\xi - l)$$
$$+ \Delta y^{III}(l)G_0(\xi - l)].$$

After satisfaction of the boundary conditions (1.7.14), we will obtain the system of boundary equations (the dots indicate differentiation with respect to $\xi$):

$$y_q(0) + M_0 G_1(0) - Q_0 G_0(0) - M_l G_1(l) + Q_l G_0(l) = 0,$$

$$\ddot{y}_q(0) + M_0 \ddot{G}_1(0) - Q_0 \ddot{G}_0(0) - M_l \ddot{G}_1(l) + Q_l \ddot{G}_0(l) = 0,$$

$$y_q(l) + M_0 G_1(l) - Q_0 G_0(l) - M_l G_1(0) + Q_l G_0(0) = 0,$$

$$\ddot{y}_q(l) + M_0 \ddot{G}_1(l) - Q_0 \ddot{G}_0(l) - M_l \ddot{G}_1(0) + Q_l \ddot{G}_0(0) = 0. \tag{1.7.22}$$

Here, and below the following designations are used:

$$y_0 = EI\Delta y(0), \quad \varphi_0 = EI\Delta y^I(0), \quad M_0 = -EI\Delta y^{II}(0), \quad Q_0 = -EI\Delta y^{III}(0),$$

$$y_l = EI\Delta y(l), \quad \varphi_l = EI\Delta y^I(l), \quad M_l = -EI\Delta y^{II}(l), \quad Q_l = -EI\Delta y^{III}(l).$$

Note, that the system (1.7.22) is regular, since it does not contain derivatives of Green's function higher than the third order. This can clearly be seen if we consider that:

$$G_n(\xi - x) = (-1)^n \frac{d^n}{d\xi^n} G(\xi - x), \tag{1.7.23}$$

and rewrite the system (1.7.22) in the form:

$$y_q(0) - M_0 \dot{G}(0) - Q_0 G(0) + M_l \dot{G}(l) + Q_l G(l) = 0,$$

$$\ddot{y}_q(0) - M_0 \dddot{G}(+0) - Q_0 \ddot{G}(0) + M_l \dddot{G}(l) + Q_l \ddot{G}(l) = 0,$$

$$y_q(l) - M_0 \dot{G}(l) - Q_0 G(l) + M_l \dot{G}(0) + Q_l G(0) = 0,$$

$$\ddot{y}_q(l) - M_0 \dddot{G}(l) - Q_0 \ddot{G}(l) + M_l \dddot{G}(-0) + Q_l \ddot{G}(0) = 0. \tag{1.7.24}$$

In (1.7.22) and (1.7.24) for simplicity reasons, the signs of the arguments of those functions which are continuous at the point of consideration, are omitted. Therefore, only values $\dddot{G}(+0)$ and $\dddot{G}(-0)$ are distinguished, since third derivatives of Green functions in this problem have a unit jump in zero. Furthermore, the following is taken into account:

$$G_n(\xi - x) = (-1)^n G_n(x - \xi).$$

In the system (1.7.24) the unknowns are force factors – moments and forces – are applied at boundary points and conditions of continuity (1.7.22) belonging to kinematic parameters. It is obvious that this variant of indirect MBE is close to the method of force.

*Force Method.* A method of analysis in which the structure is subdivided into statically determinate components. Compatibility among the components is restored by determining the interface forces.

Let us examine the second case, when in the extended domain, moments and shear forces are set to be continuous in sections $x = 0$ and $x = l$. Then in the solution of the equation of equilibrium the members $\Delta y G_3(\xi, x)$, and $\Delta y^l G_2(\xi, x)$, at $x = 0$ and $x = l$, are saved (kept) corresponding to the upload of an infinite beam angular or linear displacements:

$$y(\xi) = y_q(\xi) + EI[\Delta y^l(0)G_2(\xi) - \Delta y(0)G_3(\xi) - \Delta y^l(l)G_2(\xi - l)$$
$$+ \Delta y(l)G_3(\xi - l)]. \tag{1.7.25}$$

Satisfying the boundary conditions (1.7.14), we should differentiate the function (1.7.25), which contains the breaking solution of $G_3(\xi)$ twice. Therefore, it is necessary to consider the derivatives higher than the third order of the Green's function. Differentiating $n$ times the left and right part of the Eq. (1.7.16) we will obtain:

$$\frac{d^{n+4}}{d\xi^{n+4}}G(\xi,x) = \frac{1}{EI}\left[\delta^{(n)}(\xi - x) - k\frac{d^n}{d\xi^n}G(\xi,x)\right]. \tag{1.7.26}$$

If $\xi \neq x$ the singular addend of $\delta^{(n)}(\xi - x)$ in (1.7.26) can be ignored, since at all points, except $x = \xi$, it will give a trivial solution. It should be noted that the question from a mathematical point of view is quite thin and needs a strict proof, although this physical validity is beyond doubt. Nevertheless, in all known modifications of the methods of the boundary elements, using a hidden form of operation of separation and exclusion of singular components of the solution, this question is usually passed over in silence (ignored). Finally, using (1.7.26) and (1.7.23) we will obtain:

$$y_q(0) + \varphi_0 \ddot{G}(0) + y_0 \ddot{G}(+0) - \varphi_l \ddot{G}(l) + y_l \ddot{G}(l) = 0,$$

$$\ddot{y}_q(0) - \frac{k}{EI} \varphi_0 G(0) - \frac{k}{EI} y_0 \dot{G}(0) + \frac{k}{EI} \varphi_l G(l) - \frac{k}{EI} y_l \dot{G}(l) = 0,$$

$$y_q(l) - \varphi_0 \ddot{G}(l) + y_0 \ddot{G}(l) - \varphi_l \ddot{G}(0) + y_l \ddot{G}(-0) = 0,$$

$$\ddot{y}_q(0) - \frac{k}{EI} \varphi_0 G(l) + \frac{k}{EI} y_0 \dot{G}(l) + \frac{k}{EI} \varphi_l G(0) - \frac{k}{EI} y_l \dot{G}(0) = 0. \qquad (1.7.27)$$

In the boundary equations (1.7.27), unlike in the system (1.7.24), the unknowns are no longer forces and moments, but displacements and angles of rotation, while the conditions of continuity relate to force factors, and not kinematics. This kind of indirect boundary element method is similar to the method of displacement.

*Displacement Method.* A method of analysis in which the structure is subdivided into components whose stiffness can be independently calculated. Equilibrium and compatibility among the components is restored by determining the deformations at the interfaces.

We have indicated that besides the two given methods of corresponding force and displacement/deformation it is possible to use a combination of the jumps of force and kinematic factors, corresponding to the mixed method of structural mechanics. This method is conveniently used when given the boundary conditions containing both the force and kinematic parameters, in particular with boundary conditions (1.7.14). Let us set the equation in (1.7.18) equal to zero jumps of those parameters that are not included in the boundary conditions (1.7.14)

$$\Delta y^I(0) = \Delta y^I(l) = \Delta y^{III}(0) = \Delta y^{III}(l) = 0,$$

as a result we will obtain

$$y(\xi) = y_q(\xi) - EI[\Delta y^{II}(0)G_1(\xi) + \Delta y(0)G_3(\xi) + \Delta y^{II}(l)G_1(\xi - l)]$$
$$+ \Delta y(l)G_3(\xi - l)].$$

Hence, satisfying the boundary conditions (1.7.14), and taking into consideration (1.7.26) we will obtain the following boundary equations:

$$y_q(0) + M_0 \dot{G}(0) - y_0 \ddot{G}(0) + M_l \dot{G}(l) - y_l \ddot{G}(l) = 0,$$

$$\ddot{y}_q(0) + M_0 \ddot{G}(0) + \frac{k}{EI} y_0 \dot{G}(0) + M_l \ddot{G}(l) + \frac{k}{EI} y_l \dot{G}(l) = 0,$$

$$y_q(l) + M_0 \dot{G}(l) - y_0 \ddot{G}(l) + M_l \dot{G}(0) - y_l \ddot{G}(-0) = 0,$$

$$\ddot{y}_q(l) + M_0 \ddot{G}(l) + \frac{k}{EI} y_0 \dot{G}(l) + M_l \ddot{G}(-0) + \frac{k}{EI} y_l \dot{G}(0) = 0. \qquad (1.7.28)$$

Now, let us consider the direct method of boundary elements. In this case, all jumps in the boundary sections remain, but in addition to the $2n$ boundary conditions $2n$ more (in this case four) additional conditions of the following type are added

$$\Delta y^{(n)}(0) = y^{(n)}(+0), \qquad \Delta y^{(m)}(l) = y^{(m)}(l-0) \qquad (1.7.29)$$

to $n$ conditions on each boundary. Moreover, the differential forms in (1.7.29) should correspond to any self-adjoint boundary conditions. With these additional conditions all jumps of force and kinematic parameters are equal to the values of these parameters at the boundary points $x = +0$, $x = l-0$, and function $y(x)$ and all its derivatives at $x < 0$, $x > l$ become identically equal to zero. Actually, it follows from (1.7.29), that

$$y^{(n)}(-0) = 0, \qquad y^{(m)}(l+0) = 0.$$

For an example we will take the main types of self-adjoint boundary conditions at the boundary point $x = 0$

$$y(0) = y^I(0) = 0 \qquad \text{(build − in end, cantilevers)}$$

$$y(0) = y^{II}(0) = 0 \qquad \text{(hinge support, simple support)}$$

$$y^I(0) = y^{III}(0) = 0 \quad \text{(sliding cantilevers)}$$

$$y^{II}(0) = y^{III}(0) = 0 \text{ (free edge)} \qquad (1.7.30)$$

In the absence of loading in extended domain ($q(x) \equiv 0$ with $x < 0$, and $x > l$) and any conditions (1.7.30) the beam with $x < 0$ remains in the undisturbed condition and, therefore, $y(x) = 0$ with $x < 0$.

Thus, instead of (1.7.18) now we have

$$y(\xi) = y_q(\xi) + EI \sum_{j=0}^{3} [y^{(3-j)}(0) G_j(\xi) + y^{(3-j)}(l) G_j(\xi - l)] \qquad (1.7.31)$$

Using in (1.7.31) boundary conditions (1.7.14), we obtain

$$y(\xi) = y_q(\xi) - Q_0 G_0(\xi) + \varphi_0 G_2(\xi) - Q_l G_0(\xi - l) + \varphi_l G_2(\xi - l) \quad (1.7.32)$$

Now we can write

$$\varphi_0 = \dot{y}_q(0) - Q_0 \dot{G}(0) + \varphi_0 \ddot{G}(+0) - Q_l \dot{G}(l) + \varphi_l \ddot{G}(l),$$

$$Q_0 = \dddot{y}_q(0) - Q_0 \dddot{G}(+0) - \frac{k}{EI} \varphi_0 \dot{G}(0) - Q_l \ddot{G}(l) - \frac{k}{EI} \varphi_l \dot{G}(l),$$

$$\varphi_l = \dot{y}_q(l) - Q_0 \dot{G}(l) + \varphi_0 \ddot{G}(l) - Q_l \dot{G}(0) + \varphi_l \ddot{G}(-0),$$

$$Q_0 = \dddot{y}_q(l) - Q_0 \dddot{G}(l) - \frac{k}{EI} \varphi_0 \dot{G}(l) - Q_l \ddot{G}(-0) - \frac{k}{EI} \varphi_l \dot{G}(0), \quad (1.7.33)$$

The considered options are not limited to the opportunities of composing different boundary equations. In particular, we reference a possibility to assume all zero jumps of the force and kinematic parameters at point $x = 0$ or at point $x = l$, and consider points located out of the interval $(0, l)$. Since out of the interval, $y(\xi) \equiv 0$, then substituting in (1.7.32) or in to the expressions for $y^I(\xi), y^{II}(\xi), y^{III}(\xi)$ some values $\xi = \xi_i$ ($i = 1, 2, 3, 4; \xi \in (0, l)$), we obtain a system of boundary equations, containing only regular functions.

Further, we will analyze the connection of the method of the generalized integral transformations to the method of delta-transformation. Let us show that in the case of applying the integral transformations in the finite intervals with kernels – which do not satisfy boundary conditions at the ends of these intervals – the equation is reduced to the same boundary equations, as in the method of delta-transformation [370] and classical methods of structural mechanics, therefore making it a very general method of the boundary-value problems. First we will consider integral transformation with the orthogonal kernels, which are the eigenfunctions of some one-dimensional linear differential operator $L$, and satisfying the equation:

$$lu(x, \lambda) = \lambda \sigma'(x) u(x, \lambda) \qquad (x_1 < x < x_2) \qquad (1.7.34)$$

As before, let us assume that the self-adjoint operator $L$ is generated by the linear differential operation of order $2n$

$$lu(x) = \sum_{m=0}^{n} (p_m u^{(n-m)}(x))^{(n-m)} \qquad (1.7.35)$$

with $n$ times differentiable in the domain $D = (x_1, x_2)$ by coefficients $p_m(x)$ and by the decomposed boundary conditions:

$$A_j u(x_1) = 0; \quad A_j u(x_2) = 0 \quad (j = 1, 2, \ldots, s; n \leq s \leq 2n) \quad (1.7.36)$$

By $A_j u$ we understand the linearly independent actual forms relative to the values of functions $u(x)$ and its derivatives up to $2n - 1$ order, inclusive. At each end of the interval $s$ boundary conditions are assigned, while we can have it at both ends, from $n$ to $2n$ are different forms of $A_j$. When $s < 2n$, let us supplement a set of boundary conditions with some linear independent forms with a complete system of $2n$ forms. The nondecreasing function $\sigma(x)$ is an analogue of the function of the distribution of the masses, and $\lambda$ is the spectral parameter.

As is known [240], any function from $L_2(D)$ can be represented as decomposition of eigenfunctions of operator $L$ with some spectral function $\tau(\lambda)$. This decomposition can be written in the form of dual formulas:

$$(f, \sigma' u) = F(\lambda)$$

$$(F, \tau' u) = f(x) \quad (1.7.37)$$

where the scalar products

$$\int_{x_1}^{x_2} f(x) u(x, \lambda) d\sigma(x) = F(\lambda), \qquad \int_{-\infty}^{\infty} F(\lambda) u(x, \lambda) d\tau(\lambda) = f(x),$$

define $u$ as the transformation of function $f(x)$ and its inversion.

In the classical scheme of integral transformation (1.7.37) is applied for solving the boundary-value problems generated by the differential operation (1.7.35) on $(x_1, x_2)$ with the homogeneous or inhomogeneous boundary conditions. Let us extend this method to a more general case of the boundary-value problems, where generated operators are close to $L$. Assume the boundary-value problems are as follows:

$$l_* w(x) = q(x) \qquad (a < x < b) \quad (1.7.38)$$

with close operation $l_*$, and boundary conditions:

$$B_J w(a + 0) = g_{ja}; \quad B_J w(b - 0) = g_{jb}; \quad (j = 1, 2, \ldots, s; n \leq s \leq 2n). \quad (1.7.39)$$

For simplicity we will consider that the order of operations $l$ and $l_*$ is identical. Analogously, the case where operation $l_*$ has order multiple with respect to $l$ can be considered. In accordance with the standard schema, let us multiply the left and right sides of Eq. (1.7.38) by $u(x, \lambda)$, and integrate by measure $\sigma(x)$ over the interval $(x_1, x_2)$. If $(x_1, x_2)$ is wider than $(a, b)$, then $w(x)$ and $q(x)$ must be prolonged somehow out of the given interval $(a, b)$. We have:

$$(u, l_* w) = (u, q) \tag{1.7.40}$$

The method of integral transformations further follows partial integration.

It is more convenient however, instead of the integration by parts, to use Lagrange's formula [240] directly on each interval of continuity $(a, b)$

$$(u, l_* w) = (l_* u, w) + \left[ \sum_{j=1}^{2n} B_j w(x) B_{2n-j+1} u(x, \lambda) \right]_a^b = (u, q), \tag{1.7.41}$$

assuming that the function $w(x)$ is prolonged out of $(a, b)$ in such a way that it satisfies when $x = x_1$ and $x = x_2$ boundary conditions (1.7.36). We obtain

$$(l_* u, w) = (u, q) + \Gamma(a, \lambda) + \Gamma(b, \lambda),$$

where

$$\Gamma(a, \lambda) = \sum_{j=1}^{2n} \Delta B_j w(x) B_{2n-j+1} u(x, \lambda) \bigg|_{x=a},$$

$$\Delta B_j w(x) = B_j w(x + 0) - B_j w(x - 0)$$

If $l_* = l$, $B_j = A_j$ and $(a, b)$ coincides with $(x_1, x_2)$ in (1.7.41), taking into account (1.7.37), standard results for the transformation of the required function follows:

$$W(\lambda) = \lambda^{-1}(u, q) \tag{1.7.42}$$

From where we obtain using conversion formula

$$w(x) = \int_{-\infty}^{\infty} \lambda^{-1}(u, q) u(x, \lambda) d\tau(\lambda). \tag{1.7.43}$$

or

$$w(x) = (\lambda^{-1}(u, q), \tau' u)$$

Let us examine the application of the generalized method of integral transformations and method of delta-transformation on the example of a cross-bending of the bar. The equation of the equilibrium of the bar is:

$$EIy^{IV}(x) = q(x) \qquad (0 < x < l) \tag{1.7.44}$$

where $EI$ – rigidity of the bar; $y(x)$ – displacement of the bar; and $q(x)$ – external load. We will consider a problem with boundary conditions, corresponding to the hinged end-supported bar, assuming that $a = 0$, $b = l$,

$$y(0) = y''(0) = y(l) = y'(l) = 0. \tag{1.7.45}$$

Either the method of finite integral transformation [262, 268, 343] or, close in form, the method of decomposition of eigenfunctions, can be used for solving Eq. (1.7.44) with boundary conditions (1.7.45). In this case the beam functions of the hinge-supported beam can be chosen as the kernel of transformation. But if we use the generalized transformation, which differs from the conventional methods by the fact that the kernel of the transformation $X(\lambda_n, x)$ does not satisfy the boundary conditions of problem, then after the multiplication of the left and right sides of the Eq. (1.7.44) by this kernel and integration of the interval $(0, l)$, we will obtain:

$$\lambda_n EIY_n = Q_n - EI[y'''(x)X(\lambda_n x) - y''(x)X'(\lambda_n x) +$$
$$+ y'(x)X''(\lambda_n x) - y(x)X'''(\lambda_n x)]_{x=0}^{x=l} \tag{1.7.46}$$

Here it is designated

$$Y_n = \int_0^l y(x)X(\lambda_n x)dx, \qquad Q_n = \int_0^l q(x)X(\lambda_n x)dx, \tag{1.7.47}$$

and it is taken into account, that $X(\lambda_n x)$ satisfies the equation

$$X^{IV}(\lambda_n x) - \lambda_n X(\lambda_n x) = 0. \tag{1.7.48}$$

If $X(\lambda_n x)$ satisfies the boundary conditions (1.7.45), then the expression in parentheses disappears, and we obtain the standard solution by the method of finite integral transformations. We will assume that $X(\lambda_n x)$ satisfies not the condition in (1.7.45), but the boundary conditions for the build-in support (cantilever beam)

$$X(0) = X'(0) = X(\lambda_n l) = X'(\lambda_n l) = 0. \tag{1.7.49}$$

Then from (1.7.45) we will obtain:

$$\lambda_n EIY_n = Q_n - y'(l)X''(\lambda_n l) + y'(0)X''(0). \tag{1.7.50}$$

Let us introduce two unknown parameters – $y'(0)$ and $y'(l)$ – angles of rotations of the end sections to the right part of Eq. (1.7.49). From (1.7.49) using inversion formula we obtain:

$$y(x) = y_q(x) - y'(l)\bar{y}_l(x) + y'(0)\bar{y}_0(x) \tag{1.7.51}$$

where,

$$y_q(x) = (EI)^{-1} \sum_{n=1}^{\infty} \lambda_n^{-1} Q_n X(\lambda_n x) \qquad (1.7.52)$$

is a deflection of the build-in support (cantilever beam/bar) from the external load $q(x)$;

$$\bar{y}_l(x) = (EI)^{-1} \sum_{n=1}^{\infty} \lambda_n^{-1} X_n(\lambda_n x) X''(\lambda_n x)$$

$$\bar{y}_0(x) = (EI)^{-1} \sum_{n=1}^{\infty} \lambda_n^{-1} X_n(\lambda_n x) X''(0) \qquad (1.7.53)$$

If we set $q(x) = \delta''(x)$ or $q(x) = \delta''(x-l)$, where $\delta(x)$ – delta-function, to the formula (1.7.46), i.e. consider that the load in the form of concentrated angular deflection is applied to the end sections of the bar (the unit angles of the rotation of the ends of the beam), we will have respectively:

$$Q_n = X''(0), \text{ and } Q_n = X''(\lambda_n x). \qquad (1.7.54)$$

Comparing (1.7.52) and (1.7.53) and taking into account (1.7.54) we can conclude, that the values $\bar{y}_0(x)$ and $\bar{y}_l(x)$ represent the deflections of beam/bar at the unit turning of the end sections $x = 0$ and $x = l$, respectively. Further satisfying force boundary conditions, we obtain a system of equations for determining the unknown parameters $y'(0)$ and $y'(l)$, completely coinciding with the canonical equations of the method of displacement.

$$y_q''(0) - y'(l)\bar{y}_l''(0) + y'(0)\bar{y}_0''(0) = 0$$

$$y_q''(l) - y'(l)\bar{y}_l''(l) + y'(0)\bar{y}_0''(l) = 0 \qquad (1.7.55)$$

Similarly we will obtain a system of the canonical equations of the method of forces if we use decomposition on eigenfunctions of a free beam (the case of the beam with free ends).

Let us examine now the same problem by delta-transformation method. To do this, we will multiply both parts of Eq. (1.7.44) by $\delta(x_1 - x)$ and will integrate from $+0$ to $l - 0$;

$$EI \int_{+0}^{l-0} \vartheta^{IV}(x)\delta(x_1 - x)dx = \int_{+0}^{l-0} q(x)\delta(x_1 - x)dx = q(x_1) \qquad (1.7.56)$$

Considering that,

$$EI \int_{+0}^{l-0} \vartheta^{IV}(x)\delta(x_1 - x)dx$$

$$= \vartheta^{III}(l-0)\delta(x_1 - l) - \vartheta^{III}(+0)\delta(x_1) - \vartheta^{II}(l-0)\delta^{I}(x_1 - l)$$
$$+ \vartheta^{II}(+0)\delta^{I}(x_1) + \vartheta^{I}(l-0)\delta^{II}(x_1 - x) - \vartheta^{I}(+0)\delta^{II}(x_1)$$
$$- \vartheta(l-0)\delta^{III}(x_1 - l) + \vartheta(+0)\delta^{III}(x_1)$$
$$+ \int_{+0}^{l-0} \vartheta(x)\delta^{IV}(x_1 - x)dx$$

And that after the extraction of all discontinuities (breakages) of a function $\vartheta(x)$

$$\int_{+0}^{l-0} \vartheta(x)\delta^{IV}(x_1 - x)dx = \vartheta^{IV}(x_1),$$

we will obtain

$$EI\vartheta^{IV}(x_1) = q(x_1) - EI[\vartheta^{III}(l-0)\delta(x_1 - l) - \vartheta^{III}(+0)\delta(x_1)$$
$$- \vartheta^{II}(l-0)\delta^{I}(x_1 - l) + \vartheta^{II}(+0)\delta^{I}(x_1)$$
$$+ \vartheta^{I}(l-0)\delta^{II}(x_1 - l) - \vartheta^{I}(+0)\delta^{II}(x_1)$$
$$- \vartheta(l-0)\delta^{III}(x_1 - l) + \vartheta(+0)\delta^{III}(x_1)] \tag{1.7.57}$$

Using boundary conditions (1.7.45), which in this case should be written in the form of

$$y(+0) = y''(+0) = y(l-0) = y''(l-0) = 0, \tag{1.7.58}$$

we will obtain

$$EI\vartheta^{IV}(x_1) = q(x_1) - EI[\vartheta^{III}(l-0)\delta(x_1 - l - 0) - \vartheta^{III}(+0)\delta(x_1)]$$
$$+ \vartheta^{I}(l-0)\delta^{II}(x_1 - l - 0) - \vartheta^{I}(+0)\,\delta^{II}(x_1)]. \tag{1.7.59}$$

It should be noted that the derivatives of delta-function in formulas given above are taken on the carrier $x$, but not on the current coordinate $x_1$.

Let us enlarge the domain $0 < x_1 < l$ to $0 \leq x_1 \leq l$, i.e. consider inclusion of the points $x_1 = -0$ and $x_1 = l + 0$. In the extended domain the solution of problem is known;

$$EIG_n{}^{IV}(x_1, a) = \frac{d^n}{da^n}\delta(x_1 - a) \tag{1.7.60}$$

with the boundary conditions;

$$G_n(-0, a) = G_n{}^I(-0, a) = G_n(l + 0, a) = G_n{}^I(l + 0, a), \qquad (1.7.61)$$

i.e. Green's functions (influence functions) for built-in support (cantilever beam/bar), representing in this case the known primary system. Then the solution of Eq. (1.7.59) can be written as:

$$\vartheta(x_1) = \vartheta_q(x_1) - Q(l - 0)G_0(x_1, l) + Q(+0)G_0(x_1, 0)$$
$$+ EI\,\vartheta^I(l - 0)G_2(x_1, l) - EI\vartheta^I(+0)G_2(x_1, 0) \qquad (1.7.62)$$

In formula (1.7.60) the following loads correspond to the derivatives of delta-function: $n = 0$ – concentrated force; $n = 1$ – moment; $n = 2$ – concentrated angular (strain) deformation (jump in the rotation angle); and $n = 3$ – concentrated linear (strain) deformation (jump in the displacement). Therefore, in formula (1.7.62) displacements from the concentrated force $G_0$ and from the angular (strain) deformation $G_2$ are entered. Using boundary conditions (1.7.45) we will obtain the system of boundary equations:

$$\vartheta_q{}^{II}(0) + EI[\vartheta^I(l - 0)G_2{}^{II}(0, l) - \vartheta^I(+0)G_2{}^{II}(0, 0)] = 0$$

$$\vartheta_q{}^{II}(l) + EI[\vartheta^I(l - 0)G_2{}^{II}(l, l) - \vartheta^I(+0)G_2{}^{II}(l, 0)] = 0 \qquad (1.7.63)$$

It is here we take into account that $G_0(0, l) = G_0(l, 0) = 0$. It is easy to see that Eqs. (1.7.63), just as Eqs. (1.7.55), represent the canonical equations of the method of displacements. Thus, in a case of a one-dimensional problem, the generalized methods of finite integral transformation and delta-transformation coincide with the canonical equations of the classical methods of structural mechanics. However, the greatest interest in the application of the methods is in multidimensional problems because we are able to obtain various types of boundary equations, including the equations of the forces and displacements methods. The results for the beam presented here bear illustrative nature (character).

## 1.8   The Generalized Functions in Structural Mechanics

Known simplifications can be obtained for the solutions of boundary and initial-boundary problems using an apparatus of the theory of the generalized functions.

The most frequently used functions are Dirac and Heaviside. Dirac function or delta-function provides the analytical description to the concentrated parameters (forces, moments, concentrated masses, etc.) and is determined by following its own basic properties.

$$\int_a^b f(x)\delta(x-c)dx = \begin{cases} f(x) & (a < c < b) \\ 0 & (c < a, c > b) \end{cases}$$

The Heaviside step function $\theta(x)$ is equal to zero when $x < 0$, and 1 if $x \geq 0$. Both these functions are connected together by the following differential dependence

$$\frac{d\theta}{dx} = \delta(x)$$

Clearly, the step function can describe a load and other physical functions different from zero in a certain interval. For example, if a certain function $f(x)$ is other than zero in an interval $(a, b)$, then it can be written on the entire $x$ axis as:

$$\bar{f}(x) = [\theta(x-a) - \theta(x-b)]f(x)$$

Delta-function can be used to describe a unit single/concentrated force; $\delta^I(x)$ – will designate a unit moment, $\delta^{II}(x)$ – angular deformation, and $\delta^{III}(x)$ – jump of the angle of rotation, determined with the accuracy to $EI$.

The use of generalized functions allows us to formalize the transition from concentrated impacts, determining the right part of the differential equations, to the boundary or initial conditions and, vice versa, the transition from inhomogeneous boundary and initial conditions to the homogeneous with the transfer of heterogeneity to the right part of the equation.

It should be noted that the theory of the generalized functions in its modern form is still insufficiently incorporated into structural mechanics, although attempts to create a similar apparatus for calculating beams, plates [113], and other constructions were undertaken a long time ago (Gersevanov [117]). In their time, the functional interrupters – breakers (the limits of specially selected functions: step functions) were important achievements, however, after the creation of sufficiently strict theory of the generalized functions the necessity to use them disappeared. The generalized functions unify analytical operations above the continuous and discontinuous functions and give mathematical sense to such purely technical concepts as concentrated force and moment. It is evident that a concentrated load at the point $x_0$ can be written as

$$q(x) = q_0 \delta^{(n)}(x - x_0) \tag{1.8.1}$$

where $q_0$ – intensity of load, with the dimensions of unit of force per unit length; $n = 0$ in the case of force; and $n = 1$ in the case of moment. In structural mechanics, uploading by the moments of higher orders is also considered. For example, in [186], and [363] concentrated deformations load was used, which is equivalent to the dual and triple differentiation of the solutions, obtained by the action of concentrated force. It is easy to see that in this case in formula (1.8.1) it is possible to place $n = 2$ when loading with the angular deformations and $n = 3$ when loading with linear deformations (displacements).

Let us keep in mind, that during the integration of generalized functions the dimensionality of results changes per unit of length, for example when the concentrated force

$$\int_{-\infty}^{\infty} q(x)dx = q_0 \int_{-\infty}^{\infty} \delta(x)dx = q_0,$$

where $q_0$, unlike (1.8.1), already has dimension of force.

In the integration of derivatives of delta-function the dimensionality of results changes, again, because of the differentiation of functions in accordance with the formula

$$\int_{-\infty}^{\infty} \delta^{(q)}(x - x_0)\varphi(x)dx = (-1)^q \varphi^{(q)}(x_0) \tag{1.8.2}$$

In Table 1.1 loads and corresponding generalized functions are given, based on an example of a beam [371]. The intensity of a load for the delta-function and its derivatives is determined as a result of the integration of the equation of the equilibrium of the beam

$$EIy^{IV}(x) = q_0\delta^{(n)}(x - a) \qquad (n = 0, 1\,2, 3) \tag{1.8.3}$$

$n + 1$ times. Equation (1.8.3) may conveniently be written in the so-called quasi-derivatives (see formula (1.6.5))

$$y^{[IV]}(x) = q_0\delta^{[n]}(x - a) \qquad (n = 0, 1, 2, 3), \tag{1.8.4}$$

where,

$$y^{[0]}(x) = y(x), \quad y^{[I]}(x) = D_x y(x) = y'(x);$$

$$y^{[II]}(x) = EIy''(x); \quad y^{[III]}(x) = -EIy'''(x); \quad y^{[IV]}(x) = EIy'''(x). \tag{1.8.5}$$

The solution to Eq. (1.8.5), satisfies certain boundary conditions, and is expressed through the generalized Green's functions corresponding to

**Table 1.1** Loads and corresponding generalized functions.

| Loading diagram | Generalized function | Loading intensity ($M$ – unit length) |
|---|---|---|
| | $q = q_0 \Theta(x-a)$ | $q_0$ |
| | $q = q_0[\Theta(x-a) - \Theta(x-b)]$ | $q_0$ |
| | $q = q_0(x-a)_+$ | $q_0$ |
| | $q = q_0(x-a)^{\alpha}_+$ | $q_0$ |
| | $q = q_0\delta(x-a)$ | $Q_{0/M}$ |
| | $q = q_0\delta^{[']}(x-a)$ | $M_{0/M^2}$ |
| | $q = q_0\delta^{['']}(x-a)$ | $\varphi_0 EI_{/M^3}$ |
| | $q = q_0\delta^{[''']}(x-a)$ | $y_0 EI_{/M^4}$ |

**Table 1.2**   Unit displacements and unit reactions expressed through generalized green function.

| Unit factor | External unit impact | | | |
| --- | --- | --- | --- | --- |
| | Force $\delta$ | Moment $\delta^{[']}$ | Rotation angle $\delta^{['']}$ | Displacement $\delta^{[''']}$ |
| Displacement $y$ | $G_0 = G_Q^y$ | $G_1 = G_M^y$ | $G_2 = G_\varphi^y$ | $G_3 = G_y^y$ |
| Angle of rotation $\varphi$ | $G_0^{[I]} = G_Q^\varphi$ | $G_1^{[I]} = G_M^\varphi$ | $G_2^{[I]} = G_\varphi^\varphi$ | $G_3^{[I]} = G_y^\varphi$ |
| Bending moment $M$ | $G_0^{[II]} = G_Q^M$ | $G_1^{[II]} = G_M^M$ | $G_2^{[II]} = G_\varphi^M$ | $G_3^{[II]} = G_y^M$ |
| Shear force $Q$ | $G_0^{[III]} = G_Q^Q$ | $G_1^{[III]} = G_M^Q$ | $G_2^{[III]} = G_\varphi^Q$ | $G_3^{[III]} = G_y^Q$ |

boundary-value problems. We will designate the solution of Eq. (1.8.5) through $G_n(x, a)$ when $q_0 = 1$. The generalized Green's function $G(x, y)$ represents a single displacement from the force considered above (force, moment) and kinematic (the jump-step function of the angle of rotation, the jump-step function of the displacement) impacts. Quasi-derivatives $G_n^{[m]}(x, a)$ are a unit of angles of rotation and unit of reactions of the beam (bending moment and shear force). Table 1.2 shows the unit of displacements and unit of reactions, expressed through the generalized Green's function. The top index (superscript) designates the parameter, which is described by Green's function, and the bottom index denotes parameter that undergoes a jump due to an applied external impact. As an example $G_3^{[II]}(x, a) = G_y^{[M]}(x, a)$ is a bending moment in the section with the current coordinate $x$, due to unit jump of displacement applied in the section $x = a$. The following formula is given in A.I. Tseitlin [371], and describes the generalized Green's functions:

$$G_n^{[M]}(x, a) = D_a^{[n]} G_0^{[m]}(x, a).$$

## 1.9   General Approaches to Constructing Boundary Equations, and Standardized Form of Boundary Value Problems

The application of the methods of boundary equations and boundary elements to the solution of boundary-value problems is the most effective when considering two-dimensional and three-dimensional problems.

Nevertheless, in the simplest one-dimensional problems the application of these methods is often quite convenient. Therefore, first we will consider the reduction of one-dimensional boundary-value problems to boundary equations of different types, which will establish a link between the different methods of the type of boundary elements and to a certain extent, classify these methods. On the other hand the following examples, along with an illustrative function can represent an independent interest. Let us examine a linear self-adjoint differential expression of order $2n$:

$$lw(x) = \sum_{m=0}^{n} (p_m(x)w^{(n-m)}(x))^{(n-m)}, \quad (a < x < b) \tag{1.9.1}$$

with $n - m$ times differentiable on the interval $(a, b)$ coefficients $p_m(x)$, where $p_0(x) \neq 0$.

We will join the expression (1.9.1) with the decomposed boundary conditions in a form of the linear independent substantial forms, relative to the values of the function $w(x)$ and its derivatives up to $2n - 1$ orders, inclusive in the boundary points $x = a + 0$ and $x = b - 0$:

$$A_j w(x)|_{x=a+0} = 0; \quad A_j w(x)|_{x=b-0} = 0, \quad (j = 1, 2, \dots, s) \tag{1.9.2}$$

The boundary conditions assigned at each end of the interval $(a, b)$ equal to $s$, while different forms of $A_j$ at the both ends of the interval are limited from $n$ to $2n$.

Thus the total of both ends of the interval can meet $s$ various forms $A_j$. If the total number of various forms $s < 2n$, then we will complement the system of boundary conditions with some linearly-independent forms up to the complete system of $2n$ forms.

On each section of continuity $(\alpha, \beta)$ the functions $w(x)$, $v(x)$ and their derivatives up to $2n$ order inclusive for differential operation (1.9.1) are valid per the Lagrange-Green formula (see M.A. Naimark [240])

$$\int_{\alpha}^{\beta} v(x)lw(x)dx = \int_{\alpha}^{\beta} w(x)lv(x)dx + [v, w]_{\alpha}^{\beta}, \tag{1.9.3}$$

where the Lagrange's brackets $[v, w]$ have the form:

$$[v, w]_{\alpha}^{\beta} = \sum_{j=1}^{2n} B_j v(x) B_{2n-j+1}^* w(x) \Bigg|_{x=\alpha}^{x=\beta}. \tag{1.9.4}$$

Here $B_j$ – the arbitrary system of the linearly-independent boundary forms; $B^*_{2n-j+1}$ – conjugate system.

A one-dimensional boundary-value problem for an operator $L$, generated by the linear self-conjugate differential operation (1.9.1) and by the boundary conditions (1.9.2), which we certainty will also consider as a self-conjugate, is described by the equation:

$$lw(x) = q(x) \qquad\qquad (a < x < b), \qquad\qquad (1.9.5)$$

and boundary conditions (1.9.2).

The general solution of Eq. (1.9.5) is equal to the sum of any particular solution of the nonhomogeneous equation (1.9.5) and the general solution of the homogeneous equation $lw(x) = 0$, satisfying the boundary conditions (1.9.2). Let us choose as the particular solution of the inhomogeneous equation any solution of the Eq. (1.9.5) in any domain $\Omega$, including $(a, b)$.

At the maximum such domain (henceforth an extended domain) is the entire real axis $(-\infty, \infty)$; at the maximum is a section $[a, b]$, which is obtained from $(a, b)$ by inclusion of the boundary points $x = a - 0$ and $x = b + 0$. We will designate this particular solution through $w_q(x)$. Then the general solution of Eq. (1.9.5) can be represented in the form:

$$w(x) = w_q(x) + \sum_{j=1}^{2n} C_j \varphi_j(x), \qquad\qquad (1.9.6)$$

where $\varphi_j(x)$ – are a linearly-independent solution of the homogeneous equation

$$l\varphi(x) = 0 \qquad\qquad (1.9.7)$$

$C_j$ – arbitrary constants.

Determining arbitrary constants from the boundary conditions in (1.9.2) and completing building the solutions for problem presents a system of equations, which can be named boundary, since these questions are built for the boundary points. This is, of course, an elementary way to find solutions. The methods of boundary equations use essentially the same procedure, but they operate in this case with the solutions of problems only in the extended domain. Any system of $2n$ linearly-independent solutions of Eq. (1.9.7) is called a fundamental system, and the fundamental solution of Eq. (1.9.5) is called any solution of equation

$$lE(x, \xi) = \delta(x - \xi) \qquad\qquad (1.9.8)$$

If it satisfies some homogeneous boundary conditions, then it is the function of Green $G(x, \xi)$ of the corresponding boundary-value problem. Green's function for extended unlimited domain represents the greatest interests, since in this case it can easily be built, and usually has a simple structure. In this case $G(x, \xi)$ along with Eq. (1.9.8) satisfies conditions of limitedness at infinity. Green's function is the convenient tool for analyzing boundary-value problems and is the basis for most methods of this type of boundary equations. In calculations of structures, Green's function is usually the displacement of the system at any section $x$ due to unit concentrated force at point $\xi$. Let us consider the equation:

$$lG_k(x, \xi) = \frac{d^k}{d\xi^k}\delta(x - \xi), \tag{1.9.9}$$

then $G_k(x, \xi)$ will represent the displacement of the system under the action of unit moments and kinematic perturbation of the type of unit displacement and unit angles of rotation (with the accuracy to functional coefficients).

The Green's function of self-conjugate operators, which we examine here, they are symmetrical, i.e.:

$$G(x, \xi) = G(\xi, x)$$

As it is known from [291], the Green's function can be constructed from the homogeneous solution,

$$G(x, \xi) = \begin{cases} \sum\limits_{ij=1}^{2n} \alpha_{ij}\varphi_i(x)\varphi_j(\xi) & (x \leq \xi) \\[2em] \sum\limits_{ij=1}^{2n} \alpha_{ij}\varphi_j(x)\varphi_i(\xi) & (\xi \leq x) \end{cases} \tag{1.9.10}$$

where $\alpha_{ij}$ – some constant coefficients, and $\varphi_j(\xi)$ – linearly-independent solutions of Eq. (1.9.7).

If Green's function of problems (2.1.5) and (2.1.2) is known then the solution can be obtained by the formula

$$w(x) = \int_a^b G(x, \xi)q(\xi)d\xi. \tag{1.9.11}$$

Thus, the solution of nonhomogeneous equation (1.9.5) can be obtained, if any particular solution and fundamental system of the solutions of the homogeneous equation is known. For obtaining a particular solution, corresponding to the right side $q(x)$, it is convenient to use Green's function of the operator $l$ in any extended domain. In particular, the Green's function for differential operators in unlimited domain is known. Therefore, using the Green's function for the extended domain of $\Omega$ we can obtain a particular solution of the Eq. (1.9.5) in the form of:

$$w_q(x) = \int_\Omega G(x, \xi) q(\xi) d\xi. \tag{1.9.12}$$

The idea behind the method of boundary equations allows us to obtain the general solution (1.9.6) with the same Green's function used in (1.9.2), without determining the fundamental system. Alternatively, in addition to the Green's function, the solution in (1.9.5) in the extended domain can be used. The simplest mechanical representation of the stress – strain conditions of the construction, which has undergone the action of static or dynamic loads, can be used. The foregoing indicates the opportunity of a dual determination of fastening and boundary conditions of a corresponding boundary-value problem, by counter-proposing some value of energy conjugate factor to each given geometrical or force factor, i.e.: force – displacement; moment – angle of rotation; and vice versa. For example, if a beam is given a boundary condition corresponding to a build-in at one end (cantilever beam), then this condition can be considered as equal to zero displacement and angle of rotation in the end section, or as action in this section such as force and moment, which together with the given load become zero displacement and angle of rotation. In the latter case the problem is solved relative to the boundary factors, force and moment for the beam, having the arbitrary conditions of fastening of end section.

This fact has quite obvious mathematical basis. Let us examine as an illustration a beam with the build-in ends, which lies on the elastic Winkler foundation. The equation of the equilibrium of beam can be written as:

$$EIy^{IV}(x) + ky(x) = q(x), \tag{1.9.13}$$

where $EI$, $l$ – rigidity and span of beam; $k$ – coefficient of bed; $y(x)$ – deflection; and $q(x)$ – external load. The boundary conditions will be set on the internal borders of the interval $(0, l)$.

$$y(+0) = y'(+0) = 0, \qquad y(l-0) = y'(l-0) = 0. \tag{1.9.14}$$

The general solution of Eq. (1.9.13) can be obtained by adding the general solution of the corresponding homogeneous equation $y_0(x)$, and the particular solution of nonhomogeneous equation $y_n(x)$. We will accept as a particular solution Eq. (1.9.13) for any domain $\Omega$, including $(0, l)$ with any boundary conditions on the border of $\Omega$. The domain $\Omega$, or the extended domain, can represent the entire real axis, or any limited domain up to $[0, l]$, with the attached points of external border of this interval $x = -0$ and $x = l + 0$. Assume that in the extended domain, are assigned the boundary conditions, corresponding to the free ends of the beam are assigned, for example:

$$y_n^{II}(-0) = y_n^{III}(-0) = y_n^{II}(l+0) = y_n^{III}(l+0) = 0. \qquad (1.9.15)$$

Then,

$$y_n(+0) \neq 0, \qquad y_n^{I}(+0) \neq 0, \qquad \text{etc.}$$

Therefore, on the boundary of the given domain $(0, l)$ the following conditions derived from (1.9.14) must be satisfied:

$$y_0(+0) = -y_n(+0); \qquad y_0^{I}(+0) = -y_n^{I}(+0); \qquad (1.9.16)$$

$$y_0(l-0) = -y_n(l-0); \quad y_0^{I}(l-0) = -y_n^{I}(l-0).$$

The general solution of the homogeneous equation can be obtained as the solution of the following equation:

$$EIy_0^{IV}(x) + ky_0(x) = 0,$$

with the homogeneous boundary conditions:

$$y_0^{II}(-0) = \frac{M_0}{EI} \quad ; \quad y_0^{III}(-0) = \frac{P_0}{EI}$$

$$y_0^{II}(l+0) = \frac{M_l}{EI} \quad ; \quad y_0^{III}(l+0) = \frac{P_l}{EI} \ , \qquad (1.9.17)$$

and then in Eq. (1.9.16) the unknowns will be the boundary factors $P_0$, $P_l$, $M_0$, $M_l$ on which the solution $y_0(x)$ will linearly depend. Thus, (1.9.16) is the system of "boundary equations," since these equations are assigned at the boundary points of considering boundary-value problem.

It should be noted that the general solution for a homogeneous equation can be obtained by solving a nonhomogeneous equation with the homogeneous boundary conditions in (1.9.15) and the right side being other than zero outside of the interval $(0, l)$ only, in particular by force and moment, applied at the points $x = -0$, and $x = l + 0$. Thus, the general solution of the problem will consist of two solutions of nonhomogeneous equation with boundary conditions (1.9.15), as well as given and additional loads. Thus, the boundary equations contain solutions of unassigned problems (1.9.13, 1.9.14), and some other problem, (1.9.15) in the extended domain. Usually an extended domain and boundary conditions on its boundary are chosen in such a way that the corresponding solution could be obtained in the simplest way. In this case, instead of solving the homogeneous equation with boundary conditions (1.9.15), the Green's function can be used to obtain the solution of the given problem in a reliable and quite convenient form.

Therefore, the extension of the domain when considering a problem and obtaining solutions of nonhomogeneous equations in the extended domain and corresponding given load and load which is applied at the boundary points, allows to build boundary equations without determining the fundamental system. This circumstance is especially important when solving multidimensional problems.

A question about transition from the homogeneous differential equation with nonhomogeneous boundary conditions (1.9.17) to the nonhomogeneous differential equation with homogeneous boundary conditions (1.9.15) requires a special consideration. As it is known that concentrated impacts such as force, moment, jump in the angle of rotation, and jump in the displacement can be represented in a form of the generalized functions $\delta(x)$, $\delta^I(x)$, $\delta^{II}(x)/EI$, and $\delta^{III}(x)/EI$, respectively, where $\delta(x)$ – delta-function. Therefore, for such an equation as in (1.9.13) this transition does not cause difficulties. In the more general case, to pass from heterogeneous (or inhomogeneous) boundary conditions to the nonhomogeneous equation, it is possible to transfer heterogeneity from boundary conditions to the right part of the equation, for example, with the aid entered by A.G. Butkovsky [56] of the standardizing functions. In this paper [56] the given theorem proves boundary value problem for a differential equation of order $n$ in "nonstandard" form (in designation of [27])

$$Ly(x) = b_0(x)y^{(n)} + b_1(x)y^{(n-1)}(x) + \ldots + b_n(x)y(x) = 0,$$

$$\Gamma_i y = \sum_{k=1}^{n} \alpha_{ik} y^{(k-1)}(x_0) = g_i, \qquad (i = 1, 2, \ldots, r), \qquad (1.9.18)$$

$$\Gamma_i y = \sum_{k=1}^{n} \alpha_{jk} y^{(k-1)}(x_1) = g_j, \qquad (j = 2+1, \ 2+2, \ldots, n),$$

where $b_k(x)$ – are sufficiently smooth functions $[b_0(x) \neq 0$ everywhere on $(x_0, x_1)]$, equivalent to some other boundary value problem in "standard" form

$$Ly(x) = w(x), \qquad (1.9.19)$$

$$\Gamma_i y = 0, \quad (i = 1, 2, \dots, n),$$

where $\Gamma_i$ – linearly independent forms of boundary conditions, which cannot be equal to zero determinants:

$$\Delta_0 = \begin{vmatrix} \alpha_{1i_1} & \alpha_{1i_2} & \cdots & \alpha_{1i_r} \\ \alpha_{2i_1} & \alpha_{2i_2} & \cdots & \alpha_{2i_r} \\ \cdot & \cdot & & \cdot \\ \alpha_{ri_1} & \alpha_{ri_2} & \cdots & \alpha_{ri_r} \end{vmatrix}$$

$$\Delta_1 = \begin{vmatrix} \alpha_{r+1j_1} & \alpha_{r+1j_2} & \cdots & \alpha_{r+1j_{n-1}} \\ \alpha_{r+2j_1} & \alpha_{r+2j_2} & \cdots & \alpha_{r+2j_{n-1}} \\ \cdot & \cdot & & \cdot \\ \alpha_{nj_1} & \alpha_{nj_2} & \cdots & \alpha_{nj_{n-1}} \end{vmatrix}$$

The standardizing function $w(x)$ is calculate by the formula

$$w(x) = -\frac{1}{\Delta_1} \sum_{j_k \in J_{m+l=n-j_k}} \sum (-1)^m \Phi_{lm}(x_1)\Delta_{jk}$$

$$+ \frac{1}{\Delta_0} \sum_{i_k \in I_{m+l=n-i_k}} \sum (-1)^m \Phi_{lm}(x_0)\Delta_{ik}, \qquad (1.9.20)$$

where

$$\Phi_{lm} = b_l^{(m)}(t)\delta(x-t) - C_m^1 b_l^{(m-1)}(t)\delta^I(x-t) + \ldots + (-1)^k C_m^k b_l^{(m-k)}$$
$$(t)\delta^k(x-t) + \ldots + (-1)^m b_l(t)\delta^m(x-t),$$

The determinates $\Delta_{jk}$, and $\Delta_{ik}$ are obtained from the determinate $\Delta_1, \Delta_0$ by the replacement of rows $j_k$ and $i_k$ with the rows $(g_{r+1}, \ldots, g_n)$ and $(g_1, \ldots, g_r)$ respectively, and through the $I$ and $J$ designated sets $\{i_1, i_2, \ldots, i_r\}$ and $\{j_1, j_2, \ldots, j_{n-r}\}$, respectively. Standardizing a function is not unique. A simple way to build a standard form of a boundary-value problem can be obtained with the aid of the method of delta-transformation; A.I. Tseitlin, L.G. Petrosian [377, 378]. It should be noted, that the problem of reduction of a boundary value problem for a differential operator with inhomogeneous boundary conditions to an equivalent problem with homogeneous boundary conditions and additional right part of differential equation, arises not only during the building of boundary equations, but also when solving Cauchy problems, and setting up various inhomogeneous boundary-value problems. It should also be noted that until recently there was not enough rigorous research in this area which lead to misunderstanding and discussions in the scientific literature when the generalized functions started being widely used in solving boundary-value problems and problems of Cauchy [243].

First, let us examine a boundary-value problem. We will introduce the operator $L$, generated by differential operation (1.9.1) and inhomogeneous boundary conditions, which, for the sake of convenience, we will write with different indices:

$$A_i w(x)|_{x=a+0} = g_{i1}; \qquad (i = 1, 2, ..., n) \qquad (1.9.21)$$

$$A_j w(x)|_{x=b-0} = g_{j2}, \qquad (j = 1, 2, ..., n)$$

If the number of various forms in (1.9.21) $s < 2n$, then we will complement the system of boundary form $A_i$, $A_j$ just as in (1.9.2) with some linearly-independent forms to complete the system of boundary conditions, consisting of $2n$ forms.

We will consider the boundary-value problem for the operator $L$

$$lw(x) = 0 \qquad (1.9.22)$$

with inhomogeneous boundary conditions (1.9.21) to "standard" form as nonhomogeneous differential equation with homogeneous boundary conditions. For this purpose it is feasible to conduct delta-transformation in (1.9.22), by multiplying the left and right parts of Eq. (1.9.22) by delta-function and integrating over the certain domain $\Omega \ni [a, b]$, including $(a, b)$. We will assume that $w(x)$, and its $2n - 1$ derivatives, are continuous outside of $[a, b]$ and satisfy the homogeneous boundary conditions of the type in (1.9.21) on the ends of the interval $\Omega$. So, we have:

$$\int_\Omega \delta(x_* - x) lw(x) dx = 0 \qquad (1.9.23)$$

Further, using Lagrange-Green formula in (1.9.23), we will obtain

$$\int_\Omega w(x) l\delta(x_* - x) dx = \sum_{i=1}^{2n} \Delta A_i w(a) A_{2n-i+1} \delta(x_* - a)$$

$$+ \sum_{j=1}^{2n} \Delta A_j w(b) A_{2n-j+1} \delta(x_* - b), \qquad (1.9.24)$$

where
$$\Delta A_i w(x) = A_i w(x+0) - A_i w(x-0).$$

Under the integral in (1.9.24) the function $w(x)$ is continuous together with its derivatives up to $2n-1$ order inclusive in the entire domain $\Omega$ except of points $x = a$ and $x = b$. Therefore, using the property of delta-function:

$$\int_a^b \delta^{(n)}(x - x_*)w(x)dx = (-1)^n w^{(n)}(x_*),$$

we will obtain in each interval of the continuity:

$$lw(x_*) = \sum_{i=1}^{2n} \Delta A_i w(a) A_{2n-i+1}\delta(x_* - a) + \sum_{j=1}^{2n} \Delta A_j w(b) A_{2n-j+1}\delta(x_* - b)$$

$$(1.9.25)$$

The continuation of the function $w(x)$ out of the interval $(a, b)$ must be carried out with the observance of conditions of continuities at the boundary points $x = a$ and $x = b$. From $2n$ boundary forms at both boundary points to have the jumps can only on $n$ of forms $A$. It is evident, that these forms can only be those $A$ forms which are included in the boundary conditions (1.9.21). Therefore, continuing $w(x)$, we have to set the values of $\Delta A_i$ and $\Delta A_j$ to zero, not included in (1.9.21), and remaining jumps can be put in accordance with (1.9.21) equal $g_i$, and $g_j$. Thus, we will have, omitting for simplicity asterisk at $x$:

$$lw(x) = \sum_i g_i A_{2n-i+1}\delta(x - a) + \sum_j g_j A_{2n-j+1}\delta(x - b). \qquad (1.9.26)$$

Thus, the boundary value problem (1.9.22, 1.9.21) is equivalent to the boundary value problem with nonhomogeneous differential equation in (1.9.26) and the homogeneous boundary conditions at the ends of the interval $\Omega$. The most important is the case where $\Omega = [a, b]$. In this case

$$A_i w(a - 0) = 0$$

$$A_j w(b + 0) = 0 \qquad (1.9.27)$$

and the equivalent operator has a "standard" form (1.9.26, 1.9.27). As an example, consider the problem of bending of a cantilever beam by force $P_0$ and by moment $M_0$ (Figure 1.1a) [269].

The equation of equilibrium of a beam we will write in the form:

$$ly(x) = [EI(x)y^{II}(x)]^{II} = 0 \quad (a < x < l) \qquad (1.9.28)$$

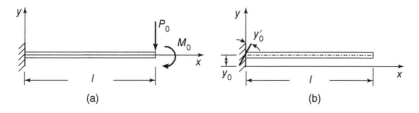

**Figure 1.1**   The scheme of a cantilever beam.

where $EI(x)$ – variable rigidity of the beam.
   The boundary conditions are:

$$y(+0) = 0; \qquad\qquad y^I(+0) = 0;$$

$$EI(x)y^{II}(x) = M_0; [EI(x)y^{II}(x)]^I = P_0; \quad (x = l - 0). \tag{1.9.29}$$

In this case $p_0(x) = EI(x)$, and $p_1(x) = p_2(x) = 0$. According to (1.9.6) when $\Omega = [0, l]$, the boundary-value problem (1.9.28, 1.9.29) is equivalent to the boundary-value problem for differential equation:

$$ly(x) = M_0[\delta^I(x - l)] + P_0\delta(x - l), \tag{1.9.30}$$

with the boundary conditions

$$y^{II}(x) = 0, \quad [EI(x)y^{II}(x)]^I = 0, \qquad\qquad (x = l + 0). \tag{1.9.31}$$

The boundary forms entering in (1.9.29) form the full system $2n = 4$ of boundary conditions. The inhomogeneous boundary-value problems in the form (1.9.30) are often used in applications. We note only that the correct consideration of the problem demands the correct statement of the boundary conditions in (1.9.29) and (1.9.31).
   In the case of inhomogeneous conditions at the left end (Figure 1.1b)

$$y(+0) = y_0, \qquad y^I(+0) = y_0^I, \tag{1.9.32}$$

and homogeneous conditions at the right end, we will have

$$ly(x) = y_0[EI(x)\delta^{II}(x - l)]^I + y_0^I EI(x)\delta^{II}(x - l) \tag{1.9.33}$$

or

$$ly(x) = y_0 EI(l)\delta^{III}(x - l) + [y_0 EI^I(l) + y_0^I EI(l)]\delta^{II}(x - l) \tag{1.9.34}$$

under boundary conditions

$$y(-0) = 0, \qquad y^I(-0) = 0. \tag{1.9.35}$$

The similar construction of the boundary equations can also be made for multidimensional problems. Let us further proceed to Cauchy's problems. In this case it is expedient to consider non-self-adjoint differential equation:

$$lu(t) = p_0(t)\frac{d^n u}{dx^n} + p_1(t)\frac{d^{n-1}u}{dx^{n-1}} + \ldots + p_n(t)u = 0, \qquad (1.9.36)$$

having applications, in particular, with the solution of dynamic problems, for mechanical, electrical, and other systems. The initial conditions can be written as:

$$A_i u(+0) = g_i, \qquad (i = 1, 2, \ldots, n) \qquad (1.9.37)$$

Let us again use delta-transformation procedure, having chosen as an extended interval the segment $[0, \infty)$. Since to the moment of $t = +0$, the system is in nondisturbed condition, we can assume that the function $u(t)$ and its derivatives up to $(n-1)$ order inclusive, equal to zero at $t = -0$. Therefore,

$$\int_0^\infty lu(t)\delta(t_* - t)dt = \int_0^\infty u(t)l^*\delta(t_* - t)dt + [u, \delta]_{+0}^\infty \qquad (1.9.38)$$

where $[u, \delta]_{+0}^\infty$ – Lagrange's bracket representing some bilinear form relative to $u(t)$ and $\delta(t_* - t)$; $l^*$ – adjoint differential operation

$$l^*\delta = (-1)^n(\overline{p}_0\delta^{(n)}) + (-1)^{n-1}(\overline{p}_1\delta^{(n-1)}) + \ldots + \overline{p}_n\delta.$$

In non-stationary problems for dissipative systems the function $u(t)$, and its derivatives equal to zero at infinity. Therefore, from (1.9.38) with the use of a property of delta-function we will obtain

$$lu(t_*) = -[u, \delta]_{t=+0}$$

with the initial conditions

$$A_i u(-0) = 0, \quad (i = 1, 2, \ldots, n).$$

As the simplest example, we will consider a problem of free oscillation of linear dissipative oscillator. The equation of oscillation can be written in the form of:

$$l(u) = m\frac{d^2 u}{dt^2} + c\frac{du}{dt} + ku$$

with the initial conditions

$$u(+0) = u_0, \quad \frac{du}{dt}(+0) = v_0.$$

For determining the equivalent nonhomogeneous equation we will build Lagrange's bracket. For this purpose we will use integration by parts in (1.9.38)

$$\int_0^\infty lu(t)\delta(t_* - t)dt = \int_0^\infty u(t)[m\delta^{II}(t_* - t) - c\delta^{I}(t_* - t) + k\delta(t_* - t)]$$

$$dt - [mu^I(+0)\delta(t_*) + mu(+0)\delta^I(t_*) + cu(+0)\delta(t_*)].$$

Considering, that

$$\delta^{(n)}(t_* - t) = (-1)^n \delta^{(n)}(t - t_*),$$

we will obtain

$$lu(t_*) = (mv_0 + cu_0)\delta(t_*) + mu_0\delta^I(t_*), \qquad (1.9.39)$$

it coincides with the result, given in [169]. The initial conditions for Eq. (1.9.39) will take the form:

$$u(-0) = 0; \qquad \frac{du}{dt}(-0) = 0.$$

Thus, the method of delta-transformation transformation makes it sufficiently simpl "standard" form of boundary-value problems [269]. For the Cauchy problems, such construction can be carried out even more simply, by introducing jumps directly into expressions for the unknown function and its derivatives, analogous to the technique presented in B. Ketch., P. Teodoresku [169]. However, this technique leads to the desired result only in case when $t < 0$, where the system is in rest. For the boundary-value problems, the method of delta-transformation allows for the simplest way to get the solution of the problem. The method of delta transformation provides the most efficient solution to boundary-value problems.

The given methods of transition from inhomogeneous boundary conditions to nonhomogeneous equation, i.e. the transfer of impacts from the boundary conditions into the load (right part of the equation) in a form of a delta-function and its derivatives, make it possible to build the solution of the corresponding problems in the form of Green's function (the reaction of system to the delta-function), and its derivatives by the parameter, which characterizes the point of the application of concentrated load. This question is central in the methods of boundary equations, since the use of Green's function in the extended domain provides simple and uniform boundary equations.

## 1.10　The Relationship of Green's Function with Homogeneous Solutions of the Method of Initial Parameters

Reduction of the differential equations to the standardized form allows establishing a close relationship between homogeneous solutions of the differential equation and Green's function. We will consider, as an example, two equivalent problems

$$lw(x) = 0, \qquad\qquad A_i w(x)|_{x=a+0} = g_i, \qquad (i = 1, 2n), \qquad (1.10.1)$$

$$lw(x) = \sum_{i=1}^{2n} g_i A_{2n-i+1} \delta(x - a), \quad A_i w(x)\bigg|_{x=a+0} = 0, \qquad (1.10.2)$$

the second of which is expressed in the standardized form. Let us consider the Cauchy's functions, i.e. solutions of homogeneous differential equation $l\varphi_k(x) = 0$, whose value at the point $x = a$ together with the values of differential operations $A_i \varphi_k(a)$ form a unit matrix. Thus the Cauchy's functions are solutions of the initial problem

$$l\varphi_k(x) = 0, A_i \varphi_k(a) = \delta_{ik}, \qquad (1.10.3)$$

where $\delta_{ik}$— Kronecker Delta. It is not difficult to see, that (1.10.3) is equivalent to (1.10.1) with $g_i = \delta_{ik}$. Therefore, the problem is equivalent to (1.10.3) and should look like (1.10.2) with $g_i = \delta_{ik}$. So, we have:

$$l\varphi_k(x) = A_{2n-k+1} \delta(x - a), \qquad A_i \varphi_k(x)|_{x=a+0} = 0. \qquad (1.10.4)$$

In many practical problems, operations of boundary conditions are such that with an accuracy to a constant coefficient the following conditions are satisfied

$$A_j A_i \varphi(x) = A_i A_j \varphi(x) = A_{i+j-1} \varphi(x), \qquad (1.10.5)$$

$$A_{2n+1} \varphi(x) = l\varphi(x).$$

For examples, for a homogeneous beam, operations of the boundary conditions are:

$$A_1 \varphi(x) = \varphi(x), \qquad A_2 \varphi(x) = \varphi^I(x),$$

$$A_3 \varphi(x) = \varphi^{II}(x), \qquad A_4 \varphi(x) = \varphi^{III}(x), \qquad A_5 \varphi(x) = \varphi^{IV}(x).$$

It is not difficult to see that the function $A_jA_i\varphi(x)$ is the solution of homogeneous equation $l\varphi(x) = 0$. This follows from the permutability of operations $A_j$ and $l = A_{2n}$. Actually, from $A_j l\varphi_i(x) = 0$ follows $l(A_j\varphi_i(x)) = 0$, from where $A_j\varphi_i(x)$ – homogeneous solution. In particular, we have

$$A_1A_i\varphi(x) = A_i\varphi(x), \quad A_2A_i\varphi(x) = A_{i+1}, \text{etc.}$$

We will define the generalized Green's function as a solution of the following problem

$$l\overline{G}_{2n-k+1}(x, a) = A_{2n-k+1}\delta(x - a), \quad A_i\overline{G}_{2n-k+1}(x, a)|_{x=a+0} = 0, \quad (1.10.6)$$

$$(i = \overline{1, 2n})$$

Comparing (1.10.4) and (1.10.6), we are convinced that

$$\overline{G}_{2n-k+1}(x, a) = \varphi_k(x) \tag{1.10.7}$$

Thus, the homogeneous solution of the Cauchy problem simultaneously determines the generalized Green's functions. Since the building of the homogeneous solutions does not usually present difficulties, the determination of the homogeneous Green's function with the aid of formula (1.10.7) is a very simple procedure.

Great interest is presented by generalized Green's function, satisfying the equation:

$$l G_{2n-k+1}(x, \xi) = A_{2n-k+1}\delta(x - \xi) \tag{1.10.8}$$

and some boundary conditions:

$$A_i G_{2n-k+1}(x)|_{x=a+0} = 0, \quad A_j G_{2n-k+1}(x)|_{x=b-0} = 0.$$

The generalized Green's functions have a physical meaning. They are the solutions of boundary-value problems which correspond to the load of contour (boundary) points by the force or kinematic parameters. We will show that in this case linear combination of the generalized Green's functions gives the homogeneous solutions of Cauchy problem, i.e. Cauchy's matrix can be built from the linear combinations of the generalized Green's functions. We will consider the left and right branches of Green's functions respectively for $x < \xi$, and $x > \xi$. Assume the Green's function as a linear combination of the functions of the Cauchy

$$G(x, \xi) = G_L(x, \xi) = \sum_{i=1}^{2n} a_i \varphi_i(x), \quad (x < \xi),$$

$$G(x, \xi) = G_R(x, \xi) = \sum_{i=1}^{2n} b_i \varphi_i(x), \quad (x > \xi)$$

According to the definition of the Green's function, it has at point $x = \xi$ continuous derivatives to $2n - 2$ orders and the unit jump of $2n - 1$ derivatives. The satisfaction of this condition leads to the following relations

$$b_i = a_i \quad (i = \overline{1, 2n - 2}), \qquad b_{2n} = a_{2n} + 1 \qquad (1.10.9)$$

Therefore,

$$G_L(x, \xi) = \sum_{i=1}^{2n} a_i \varphi_i(x),$$

$$G_R(x, \xi) = \sum_{i=1}^{2n-1} a_i \varphi_i(x) + (a_{2n} + 1)\varphi_{2n}(x)$$

From here

$$G_R(x, \xi) - G_L(x, \xi) = \varphi_{2n}(x) \qquad (1.10.10)$$

In the particular case of the operator $l = d^4/dx^4 + \lambda^4$ for the problem of bending of unlimited beam lying on the elastic Winkler foundation, the formula of the type (1.10.10) was obtained by B.G. Korenev [186, 188, 189].

Since $A_j l\varphi_k(x) = 0$ and $lA_j\varphi_k(x) = 0$, $A_j\varphi_k(x)$ – is homogeneous solution. At the same time it satisfies the condition

$$A_i A_j \varphi_k(x)|_{x=a} = A_{i+j-1}\varphi_k(x)|_{x=a} = \delta_{i+j-1,k} \qquad (1.10.11)$$

Let $A_j\varphi_k(x) = \varphi_m(x)$. Determine $m$ from (1.10.11)

$$A_i\varphi_m(x)|_{x=a} = \delta_{i,m} = \delta_{i+j-1,k}$$

If $m = i$ and $i + j - 1 = k$, then $m = k - j + 1$. Thus we have

$$A_j\varphi_k(x) = \varphi_{k-j+1} \qquad (1.10.12)$$

Thus, considering functions according to (1.10.5) and (1.10.12) we composed the Table 1.3 of the method of initial parameters with accuracy to the constant multipliers [278].

**Table 1.3** Homogeneous solutions of the initial parameters method.

| $\varphi_1$ | $\varphi_2$ | $\varphi_3$ | $\cdots$ | $\varphi_{2n}$ |
|---|---|---|---|---|
| $A_{2n}\varphi_{2n}$ | $A_{2n}\varphi_{2n+1}$ | $A_{2n}\varphi_{2n+2}$ | $\cdots$ | $A_{2n}\varphi_{4n-1}$ |
| $A_{2n-1}\varphi_{2n-1}$ | $A_{2n-1}\varphi_{2n}$ | $A_{2n-1}\varphi_{2n+1}$ | $\cdots$ | $A_{2n-1}\varphi_{4n-2}$ |
| $\cdot$ | $\cdot$ | $\cdot$ | $\cdot$ | $\cdot$ |
| $\cdot$ | $\cdot$ | $\cdot$ | $\cdot$ | $\cdot$ |
| $\cdot$ | $\cdot$ | $\cdot$ | $\cdot$ | $\cdot$ |
| $A_2\varphi_2$ | $A_2\varphi_3$ | $A_2\varphi_4$ | $\cdots$ | $A_{2n}\varphi_{2n+1}$ |
| $A_1\varphi_1$ | $A_1\varphi_2$ | $A_1\varphi_3$ | $\cdots$ | $A_1\varphi_{2n}$ |

Let us consider further the generalized Green's functions, which satisfy Eq. (1.10.8). We have

$$A_j G_L(x, \xi) = \sum_{i=1}^{2n} a_i(\xi) A_j \varphi_i(x),$$

$$A_j G_R(x, \xi) = \sum_{i=1}^{2n} a_i(\xi) A_j \varphi_i(x) + (a_{2n} + 1) A_j \varphi_{2n}(x)$$

As a result of (1.10.12)

$$A_j \varphi_{2n}(x) = \varphi_{2n-j+1}(x)$$

and hence

$$A_j[G_R(\xi, x) - G_L(\xi, x)] = \varphi_{2n-j+1}(x) \qquad (1.10.13)$$

Thus, the homogeneous solution of a method of initial parameters can be represented through the operations of the boundary conditions from Green's function.

## 1.11   The Spectral Method of Boundary Elements

As Section 1.6 shows, the application of generalized integral transformations (transforms) to solve boundary-value problems when eigenfunctions of operators close to the operator of the problem are used as kernels is reduced to boundary equations, we have one less dimensionality than the initial (original) equation of equilibrium or motion. In Sections 1.7 and 1.9 various generalizations of integral transformations associated with the use of classical transforms in non-canonical domains and with non-canonical boundary conditions, as well as transformations with close kernel and non-orthogonal kernels were considered. Below, the method of building of boundary integral and algebraic equations based on the generalized integral transformations will be presented in detail. Following G.J. Popov [291], we will examine orthogonal and bioorthogonal decompositions on eigenfunctions of close operators. Some extensions of this method, being the spectral analogue of boundary elements method (BEM) and method of delta transformation, are given. So we will call this method the Spectral Method of Boundary Elements (SMBE, L.G. Petrosian [270]). The essence of the SMBE for one-dimensional problems briefly presented in Sections 1.6 and 1.9 consists of building of boundary equations, not with the aid of Green's function, but on the basis of the spectral decomposition of the differential operators considered in the expanded area (domain). In particular for this purpose, the transformations of Fourier and Hankel, along with widely used structural mechanics decomposition on beam functions, may be used. The spectral method is especially convenient when constructing a Green's function is complicated or can be carried out only with the aid of an integrated or discrete decomposition. In general form the boundary equations of SMBE for the one-dimensional problem take the form of (1.6.22) and contain twice as many unknown jumps as the available boundary conditions for their determination. Depending on the choice of these jumps, as in the case of a delta-transformation method, we will obtain the spectral analogues of direct and indirect methods of boundary elements. In particular, if we assume that jumps of all differential forms, which contain derivatives up to the $n$th order, are equal to zero, then this approach will be equivalent to either an indirect method of the BEM or the method of potentials with lower multiplicity, corresponding to force impacts (for the equations of the second order simple layer). But if we assume that jumps of the differential forms, which contain the derivatives of order higher than $n-1$, are equal to zero, then we will obtain an analogue of the method of potential with the use of layers of the highest multiplicity (for the equations of the second order dual layer). With the retention of all jumps it is necessary to use an additional condition, for example, equating the displacements out of the interval $(a, b)$ to zero, ensuring for example, displacements out of the interval $(a, b)$, are equal to zero, which provide the sole solution of the problem. In this case SMBE is equivalent to the direct method of boundary elements or to the method of boundary integral equations. For the purpose of clarity, let us illustrate a scheme of a spectral method to solve a simplest model problem, in particular we will consider the problem described in Section 1.9, with bent free beam on elastic Winkler foundation. The boundary conditions can be written as:

$$y''(+0) = y'''(+0) = 0, \qquad y''(l-0) = y'''(l-0) = 0. \qquad (1.11.1)$$

In addition, we will assume the displacement of the beam, together with the first three derivatives as equal to zero at infinity. To solve the equation in (1.9.13) with the boundary conditions (1.11.1) we will use Fourier transformation. For this purpose we will prolong $y(x)$ and $q(x)$ to the entire axis, and will proceed in the left and right parts of the equation to the Fourier transforms. After multiplying (1.9.13) by $\frac{1}{\sqrt{2\pi}}e^{i\xi x}$ and integrating, we will obtain:

$$(EI\xi^4 + k)Y(\xi) == Q(\xi) - \frac{EI}{\sqrt{2\pi}}$$

$$[\Delta y'''(x) - i\xi\Delta y''(x) - \xi^2\Delta y'(x) + i\xi^3\Delta y(x)]_0^l,$$

where

$$Y(\xi) = \frac{1}{\sqrt{2\pi}}\int_{-\infty}^{\infty} y(x)e^{i\xi x}dx; \qquad Q(\xi) = \frac{1}{\sqrt{2\pi}}\int_{-\infty}^{\infty} q(x)e^{i\xi x}dx.$$

Similar to the method of delta-transformation and Green transformation, here we have twice as many unknown jumps as boundary conditions (eight unknown jumps with four boundary conditions (1.11.1)). Therefore there are variants of continuation of function $y(x)$ to the entire axis. We will consider the following two cases of choice of the jumps:

1. Let us assume that displacements and angles of rotations in cross-sections $x = 0$ and $x = l$ of an infinite beam in its expanded domain are continuous. Loading is achieved by force factors (force and moment). In this case $\Delta y(0) = \Delta y'(0) = \Delta y(l) = \Delta y'(l) = 0$ and

$$Y(\xi) = \frac{Q(\xi) - (Q_0 - i\xi M_0 - Q_l + i\xi M_l)/\sqrt{2\pi}}{EI\xi^4 + k}, \qquad (1.11.2)$$

where $M_0$, $M_l$, $Q_0$, and $Q_l$ – jumps of the moment and shear force in the central section and in the section $x = l$. Using the inversion formula we have:

$$y(x) = y_q(x) - \frac{(Q_0 - Q_l)}{\pi}\int_0^{\infty} \frac{\cos\xi x d\xi}{EI\xi^4 + k} - \frac{(M_0 - M_l)}{\pi}\int_0^{\infty} \frac{\xi\sin\xi x d\xi}{EI\xi^4 + k}, \qquad (1.11.3)$$

where

$$y_q(x) = \frac{1}{\sqrt{2\pi}} \int_{-\infty}^{+\infty} \frac{Q(\xi)e^{-i\xi x}d\xi}{EI\xi^4 + k}$$

is displacement of infinite beam due to load $q(x)$.

We will obtain boundary equations, satisfying the conditions (1.11.1).

$$y_q''(0) + \lim_{x \to +0} \left[ \frac{(Q_0 - Q_l)}{\pi} \int_0^\infty \frac{\xi^2 \cos \xi x d\xi}{EI\xi^4 + k} + \frac{(M_0 - M_l)}{\pi} \int_0^\infty \frac{\xi^3 \sin \xi x d\xi}{EI\xi^4 + k} \right]$$

$$= 0,$$

$$y_q''(l) + \lim_{x \to l-0} \left[ \frac{(Q_0 - Q_l)}{\pi} \int_0^\infty \frac{\xi^2 \cos \xi x d\xi}{EI\xi^4 + k} + \frac{(M_0 - M_l)}{\pi} \int_0^\infty \frac{\xi^3 \sin \xi x d\xi}{EI\xi^4 + k} \right]$$

$$= 0,$$

$$y_q'''(0) + \lim_{x \to +0} \left[ -\frac{(Q_0 - Q_l)}{\pi} \int_0^\infty \frac{\xi^3 \sin \xi x d\xi}{EI\xi^4 + k} + \frac{(M_0 - M_l)}{\pi} \int_0^\infty \frac{\xi^4 \cos \xi x d\xi}{EI\xi^4 + k} \right]$$

$$= 0$$

$$y_q'''(l) + \lim_{x \to l-0} \left[ -\frac{(Q_0 - Q_l)}{\pi} \int_0^\infty \frac{\xi^3 \sin \xi x d\xi}{EI\xi^4 + k} + \frac{(M_0 - M_l)}{\pi} \int_0^\infty \frac{\xi^4 \cos \xi x d\xi}{EI\xi^4 + k} \right]$$

$$= 0. \tag{1.11.4}$$

For calculation of the integrals in the boundary equations (1.11.4), let us note that with $q(x) = \delta(x)$, and $M_0 = Q_0 = 0$ solution (1.11.3) gives the expression of Green's function about the infinite beam problem.

$$G_0(x) = \frac{1}{\pi} \int_0^\infty \frac{\cos \xi x d\xi}{EI\xi^4 + k} \tag{1.11.5}$$

We will obtain the derivatives of Green's function $G_n(x) = \frac{d^n}{dx^n} G(x)$, by formal differentiation:

$$G_1(x) = -\frac{1}{\pi} \int_0^\infty \frac{\xi \sin \xi x d\xi}{EI\xi^4 + k},$$

$$G_2(x) = -\frac{1}{\pi} \int_0^\infty \frac{\xi^2 \cos \xi x d\xi}{EI\xi^4 + k},$$

$$G_3(x) = \frac{1}{\pi} \int_0^\infty \frac{\xi^3 \sin \xi x d\xi}{EI\xi^4 + k},$$

$$G_4(x) = \frac{1}{\pi} \int_0^\infty \frac{\xi^4 \cos \xi x d\xi}{EI\xi^4 + k}. \tag{1.11.6}$$

Integrals, representing functions $G_3(x)$ and $G_4(x)$, disperse and do not exist in a classical sense. Therefore, they have to be understood as the generalized functions, which consist in a general case of regular and irregular components. The last integral transforms as follows:

$$\frac{1}{\pi} \int_0^\infty \frac{\xi^4 \cos \xi x d\xi}{EI\xi^4 + k} = \frac{1}{\pi EI} \int_0^\infty \frac{(EI\xi^4 + k - k) \cos \xi x d\xi}{EI\xi^4 + k}$$

$$= \frac{1}{\pi EI} \int_0^\infty \cos \xi x d\xi - \frac{k}{\pi EI} \int_0^\infty \frac{\cos \xi x d\xi}{EI\xi^4 + k}$$

It is known from the theory of generalized functions [115], that

$$\int_0^\infty \cos \xi x d\xi = \pi \delta(x) \tag{1.11.7}$$

$$\int_0^\infty \frac{\sin \xi x}{\xi} d\xi = \frac{\pi}{2} signx \tag{1.11.8}$$

Multiplying and then dividing the integrand expression in (1.11.7) and (1.11.8) by $EI\xi^4 + k$, we will find

$$G_3(x) = -\frac{2\lambda^4}{k} signx e^{-\lambda|x|} \cos\lambda x,$$

$$G_4(x) = \frac{1}{EI}[\delta(x) - kG_0(x)] = \frac{1}{EI}\left[\delta(x) - \frac{\lambda}{2}e^{-\lambda|x|}(\cos\lambda x + \sin\lambda x)\right]. \tag{1.11.9}$$

Further satisfaction of the boundary conditions taking into account only regular components of $G_4(x)$, leads to boundary equations, which are similar to the equations of the method of delta-transformation considered in Section 1.7, and also to equations of the indirect method of boundary elements. It should be noted that the issues of improvement of convergence of integrals of the type (1.11.6), and similar integrals, which are encountered in the close problems for the beams, lying on linearly-deformed base, were considered by many authors, in particular B.G. Korenev [186], G.J. Popov [291, 292], V.I. Travush [362], and A.I. Tseitlin, L.G. Petrosian, D.R. and Atadjanov [378, 379].

2. Let us will assume that the force factors $\Delta y''(0) = \Delta y'''(0) = 0$ in sections $x = 0$ and $x = l$ are continuous, and the kinematic impacts $\Delta y(0) \neq 0$, $\Delta y'(0) \neq 0$ are applied. In this case

$$Y(\xi) = \frac{Q(\xi) + [i\xi(y_0 - y_l) - \xi^2(\varphi_0 - \varphi_l)]/\sqrt{2\pi}}{EI\xi^4 + k} \tag{1.11.10}$$

where $y_0$, $y_l$, $\varphi_0$, and $\varphi_l$ are jumps of displacement and angles of rotation in sections $x = 0$, $x = l$. By the inversion of (1.11.10), and taking into account (1.11.6), we will have:

$$y(x) = y_q(x) - \frac{(y_0 - y_l)}{\sqrt{2\pi}} G_3(x) + \frac{(\varphi_0 - \varphi_l)}{\sqrt{2\pi}} G_2(x) \tag{1.11.11}$$

The further submission (1.11.1) into (1.11.1) using the limiting transitions gives

$$y_q''(0) + \lim_{x \to +0} \left[ -\frac{y_0 - y_l}{\sqrt{2\pi}} G_3''(x) + \frac{\varphi_0 - \varphi_l}{\sqrt{2\pi}} G_2''(x) \right] = 0,$$

$$y_q''(l) + \lim_{x \to l-0} \left[ -\frac{y_0 - y_l}{\sqrt{2\pi}} G_3''(x) + \frac{\varphi_0 - \varphi_l}{\sqrt{2\pi}} G_2''(x) \right] = 0,$$

$$y_q'''(0) + \lim_{x \to +0} \left[ -\frac{y_0 - y_l}{\sqrt{2\pi}} G_3'''(x) + \frac{\varphi_0 - \varphi_l}{\sqrt{2\pi}} G_2'''(x) \right] = 0,$$

$$y_q'''(l) + \lim_{x \to l-0} \left[ -\frac{y_0 - y_l}{\sqrt{2\pi}} G_3'''(x) + \frac{\varphi_0 - \varphi_l}{\sqrt{2\pi}} G_2'''(x) \right] = 0, \tag{1.11.12}$$

Integrals in the expression (1.11.12) can be transformed in the same way as integral $G_4(x)$ in (1.11.6). In particular:

$$G_3''(x) = \frac{1}{\pi} \int_0^\infty \frac{\xi^5 \sin \xi x d\xi}{EI\xi^4 + k} = \frac{1}{\pi EI} \int_0^\infty \frac{\xi(EI\xi^4 + k - k) \sin \xi x d\xi}{EI\xi^4 + k}$$

$$= \frac{1}{\pi EI} \frac{d}{dx} \int_0^\infty \cos \xi x d\xi + \frac{k}{\pi EI} \frac{d}{dx} \int_0^\infty \frac{\cos \xi x d\xi}{EI\xi^4 + k} = -\frac{\delta'(x) - kG'(x)}{EI}.$$

Lowering singular components of delta-function and its derivatives we will obtain boundary equations of the type (1.11.8).

## 1.12   The Compensate Loads Method

The method of the compensating loads [186–188, 214] arose as a generalization of methods of structural mechanics associated with the consideration of the infinite primary system, as applied to the analysis of beams on elastic Winkler foundation. The method of the compensating loads proposed by B.G. Korenev provides precise satisfaction of boundary conditions on the entire surface of the construction with finite dimensions. So, we have Fredgolm equations of the first kind, and we also have Fredgolm equations of the second kind. However, it is impossible to obtain exact solutions even for canonical bodies because of mathematical difficulties. Nevertheless, the compensate loads method (CLM) is widely known as an approximate method. As a rule, researchers build the system of integral equations, and then they apply to it an approximate method of solution and reduce the problem to the system of linear algebraic equations. In the method of compensating loads the boundary impacts, which are the unknown functions of spatial coordinates (compensating loads), to be determined in the process of solving the problem, are assigned at the outline of domain in the form of forces and moments, including the moments of the highest orders corresponding to jumps of kinematic factors – displacements and angles of rotation. As an indirect method of boundary elements, the method of compensating loads demands expansion of the domain to some canonical, in particular unlimited, application on its boundary of some compensating loads. The solution of the problem is built as the sum of the basic and compensating solutions in the extended domain. Satisfaction of boundary conditions leads to the boundary equations. The method of the compensating loads was developed in two directions. The first direction is connected with the use of the compensating loads only on the border of the domain, occupied by construction. In this case, the extraction of special features on the border makes it possible to reduce the problem to regular and singular equations of second kind recorded for the boundary of the construction. The second direction [61, 393] is associated with a full realization of the opportunities presented by CLM, in particular with the arbitrary choice of the compensating loads and their locations relative to the boundary of the construction, and it is, in a certain sense, a generalization of the method. Then the boundary-value problem, generated by elliptical differential operator, can be reduced to the integral equations of the first kind with the continuous kernels relative to the compensating loads, located out of the considered construction. B.G. Korenev considered a wide range of questions relating to the method of compensating loads: solutions to one- and two-dimensional problems; the connection of the method of the compensating loads with the method of the initial parameters; the interpretation of the compensating loads of the high orders – which are the discrete analogues (analogs) of multiple layers; the calculation of the constructions by the method of the compensating loads taking into account plastic deformations; and the solution of dynamic problems – calculation of plates of arbitrary shape on an elastic foundation.

An important result of B.G. Korenev's proposal is the removal of compensating loads beyond the limits of the considered domain. In this case, this deteriorates the structure of boundary equations, i.e. on the way of practical application of several essential difficulties, associated with an incorrectness of the resolving equations are encountered, A.H. Tikhonov and V.J. Arsenin [356]. As a consequence, the need for the building of special regularizing algorithms for the numerical application of the method appears; however they possess properties of regularity that are very necessary for simplification of the calculations. Further works of mathematical justification and numerical realization of this modification of the method of the compensating loads belong to E.S. Ventsel [393].

Let us use the scheme of the method of the compensating loads for the solution of a three-dimensional problem of the theory of elasticity, for the purposes of its comparison to BEM and the method of potential. We will consider a more general linear boundary-value problem for the equation

$$Lu(p) = \varphi(p); \qquad\qquad p = (x_1, x_2) \in D \qquad\qquad (1.12.1)$$

after joining to it boundary conditions

$$\alpha u(p) + \beta lu(p) = f(p) \qquad\qquad (p \in \Gamma) \qquad\qquad (1.12.2)$$

where $\alpha$, $\beta$ – constants. With $\beta = 0$, $\alpha = 1$ from (1.12.2) we obtain the boundary conditions of the first basic problem, and with $\alpha = 0$, $\beta = 1$ – the second, where $L$ – elliptical operator on the order of $2n$. Any system of $2n$ linearly-independent solutions of the equation

$$Lu(p) = 0 \qquad\qquad (1.12.3)$$

is a fundamental system, and the fundamental solution of Eq. (1.12.1) is any solution of the equation:

$$LE(p,q) = \delta(p-q) \tag{1.12.4}$$

Here $l = (l_1, l_2, \dots, l_n,)$ – differential operator of boundary conditions; $\varphi = (\varphi_1, \varphi_2, \dots \varphi_n)$ – are known on the border $\Gamma$ vector functions.

We will introduce the expanded domain $\Omega$, which includes $D$ ($D \cup \Gamma \subseteq \Omega \subseteq R$) into space $R$. The minimum expansion of the domain $D$ is the accession to it of the border, and the maximum – the accession of the space $R$. In accordance with CLM, consider a wider domain $\Omega$ with boundary $\Gamma'$

$$d(\Gamma, \Gamma') = \inf|p-q| = d_\varepsilon > 0, \quad (p \in \Gamma, q \in \Gamma') \tag{1.12.5}$$

The approximate solution of the problems (1.12.1) and (1.12.2) is found in the form of

$$u_q(p) = \int_D E(p,q)\varphi(q)ds_q + \int_{\Gamma'} \sum_{i=0}^{n-1} \frac{\partial^i}{\partial n_q^i} E(p,q)X_{i+1}(q)dl_q, \tag{1.12.6}$$

where $n_q$ – normal to $\Gamma'$ at point $q$; and $X(q) = (X_1(q), X_2(q), \dots, X_n(q))$ – unknown vector function of compensating loads, determined at $\Gamma'$. The function $u_q(p)$ satisfies Eq. (1.12.1). Substituting it into boundary conditions (1.10.2), we obtain the resolving system of the integral equations of the first kind with the continuous kernels due to (1.12.5) relative to compensating loads $X(q)$

$$\sum_{i=1}^{n-1} \int_{\Gamma'} \left\{ \left[ \frac{\partial^i}{\partial n_q^i} E(p,q) \right] \right\} X_{i+1}(q)dl_q = f(p) - \int_D [lE(p,q)]\varphi(q)ds_q \tag{1.12.7}$$

or in the operator's form

$$E \cdot X = K \tag{1.12.8}$$

The numerical solution of resolving integral equations (1.12.7) is unstable and requires the application of the regularizing algorithms [356]. It should be noted that many publications in the recent decade considered the problems solved by the method of compensating loads with the presence of all specific stages and procedures for indirect BEM, including dividing (partition) of the boundary and satisfaction of boundary conditions in selected nodal points, considerably moved the scientific works on (BEM) ahead.

# Chapter 2
# Boundary Elements Methods (BEM) in the Multidimensional Problems

## 2.1 The Integral Equations of Boundary Elements Methods

As mentioned before, the boundary element method (BEM) as the apparatus of solution of the boundary-value problems using the *boundary integral equations* (BIEs) are divided into three main methods that are distinguished by the approach to the construction of the initial integral equations; direct, indirect, and semi-direct C.A. Brebbia [45–48], C.A. Brebbia, C. Walker [49], C.A. Brebbia, Zh. Tells, L. Wrobel [50], A.P. Calderon [57]. Indirect method finds its sources in the classical method of potentials Y.L. Bormot [42], Y.D. Burago, V.G. Mazya [53], M.I. Lazarev [209]. BIE methods in structural mechanics were well known through the works of authors such as A.J. Alexandrov [3], A.J. Alexandrov, B.Y. Lashchenikov, N.N. Shaposhnikov, V.A. Smirnov [4], S.M. Aleynikov [6], L.A. Alexeyeva [7], P.K. Banerjee, R. Buterfield [23, 27], P.K. Banerjee, S. Mukherjec [24], P.K. Banerjee [29], P.K. Banerjee, J.O. Watson [26], P.K. Banerjee, S.M. Mamoon [28], F.M. Besuner, D.U. Snow [31], D.E. Beskos [33, 34], A.V. Bitsadze [36], T.V. Burchuladze, T.G. Gegelya [54], T.V. Burchuladze, R.V. Rukhadze [55], D.M. Cole, D.D. Kosloff, J.B.A. Minstter [72], S.L. Crouch, A.M. Starfield [78], D.J. Forbes, A.R. Robinson [112], A. Frangi [109, 110], N.M. Gersevanov

*Static and Dynamic Analysis of Engineering Structures: Incorporating the Boundary Element Method*,
First Edition. Levon G. Petrosian and Vladimir A. Ambartsumian.
© 2020 John Wiley & Sons Ltd. Published 2020 by John Wiley & Sons Ltd.

[117], S.K. Godunov, V.S. Ryabenkiy [118], M.D. Greenberg [124, 125], E.I. Grigolyuk, V.M. Tolkachev [126], H.M. Gyugter [131], G.D. Hatzige-orgiou, D.E. Beskos [132–134], U. Heise [136–139], I. Herrera, F.J. Sabina [140], J.L. Hess, A.M.O. Smith [141], T. Horibe [143, 144], W.F. Hughes, W. Gayford [147], A.G. Ishkova [154], A.I. Kalandia [159], D.D. Kellar [165], S. Kobayaski [179], Y.D. Kopeikin [183], V.D. Kupradze [199], V.D. Kupradze, T.V. Burchuladze [200], V.D. Kupradze, T.G. Hegelia, M.O. Basheleyshvili, T.V. Burchuladze [201], S.M. Belotserovsky, I.K. Lifanov [32], J.C. Lachat [202], J.C. Lachat, J.O. Watson [203], M.I. Lazarev [206–208], A.M. Linkov [210], Ya.B. Lopatinsky [213], A.A. Love [213], A.I. Lurie [216], A.M. Lyapunov [218], W.J. Mansur, C.A. Brebbia [220], C. Massonnet [222], V.G. Mazya B.A. Plamenevsky [224], D.I. Sherman [326], D.J. Shippi [327], A.B. Zolotov [418], V.N. Sidorov, A.B. Zolotov [330], N.I. Muskelishvili [237, 238], S.G. Mikhlin [229–233], V.Z. Parton, P.I. Perlin [256–258], V.Z. Parton, B.A. Kudryavt-sev [259, 260], G.N. Polozhiy [289], C.P. Providakis, D.E. Beskos [294], V.G. Romanov [304], I.Z. Roytfarb, T.V. Kyung [307], V.S. Ryabenkiy [308–311], V.S. Sarkisian [316], R.T. Seeley [319], and Y.V. Veryuzhsky [395–397], Y.S. Uflyand [389], A.G. Ugodchikov, N.M. Khutoryansky [387], A.I. Vaydiner, V.V. Moskvitin [390], N.P. Vekua [391], V.S. Vladimirov [398, 399], D.V. Weinberg, A.L. Sinyavsky [407], S.S. Zargaryan [411, 412], A. Zigmund, A.P. Calderon [419]. The direct method can be traced back to V.D. Kupradze [199]. In 1963 M.A. Jaswon and A.R. Ponter [156] provided the general formulation of direct BIE method in engineering, and later M.A. Jaswon, M. Maiti, and M. Symm [157] conducted numerical be-harmonic analysis for some sample structures. The direct method, also called the *boundary integral equations method* (BIEM), is based on the integral identities of Green and Betty, and leads to the similar equations with the same set of kernels, as the indirect method. F.J. Rizzo [301–303] in 1967 presented the direct formulation of boundary value problems of clas-sical elasto-static, using Betty's and Somigliana's formulas. The semi-direct method operates with the equations relative to auxiliary functions (such as: type of stresses functions in a two-dimensional problem of the theory of elasticity) and is applied relatively rarely. In 1968 T.A. Cruse [79–81], T.A. Cruse, F.J. Rizzo [82–84], T.A. Cruse, J.L. Swedlov [85], T.A. Cruse, W. Van Buren [86] and P.K. Banergee, R Show [25][1] presented indirect and direct boundary element formulations respectively for electrodynamic

---

[1] Shaw R. *JASA*, **44**, 745 – 748 , 1968

problems. The main interest is represented by direct and indirect methods and, therefore, we will give their summary and analysis below. For the greater clarity, let us construct BIEs by these methods based on the example of spatial problem of the theory of elasticity. A first we will consider the main equations of the method of potential.

As we known, the vector of small elastic displacements of linear isotropic elastic body $u$ with the Cartesian components $u_1$, $u_2$, and $u_3$ at each point $p$ of three-dimensional (3D) space $R$, which has Cartesian coordinates $x_1$, $x_2$, and $x_3$, satisfies the equations of the equilibrium of Lamé, which in the absence of mass forces are written in the form of

$$\mathbf{V}^*u = \mu\nabla^2 u + (\lambda + \mu)grad \cdot divu = 0, \qquad (2.1.1)$$

where
$\nabla^2$ – three-dimensional Laplace operator in the Cartesian coordinates

$$\nabla^2 = \frac{\partial^2}{\partial x^2} + \frac{\partial^2}{\partial y^2} + \frac{\partial^2}{\partial z^2};$$

$\lambda$, $\mu$ – Lamé constants, related to the elastic modulus $E$ and Poisson's coefficient $\sigma$ ratio formulas

$$E = \frac{\mu(3\lambda + 2\mu)}{\lambda + \mu} \qquad \sigma = \frac{\lambda}{2(\lambda + \mu)} \qquad (2.1.2)$$

Let the same plane, whose position is fixed by normal $n$ with the directing cosines of $n_1$, $n_2$, and $n_3$ pass through point $p$. The vector of the stresses, operating in this plane, can be presented in the form

$$T_n^p u = 2\mu\frac{\partial u}{\partial n} + \lambda n\, divu + \mu(n \times rotu) \qquad (2.1.3)$$

Here and above use the following differential operators [215]:

• gradient of scalar function $f(r) = f(x, y, z)$

$$gradf(x,y,z) \equiv \nabla f(x,y,z) = \frac{\partial f}{\partial x}i - \frac{\partial f}{\partial y}j - \frac{\partial f}{\partial z}k;$$

- divergence and the rotor of the vector function $F(r)$

$$divF(x,y,z) \equiv \nabla \cdot F(x,y,z) = \frac{\partial F_x}{\partial x} + \frac{\partial F_y}{\partial y} + \frac{\partial F_z}{\partial z};$$

$$rotF(x,y,z) \equiv \nabla \times F(x,y,z)$$

$$= \left( \frac{\partial F_z}{\partial y} - \frac{\partial F_y}{\partial z} \right) i + \left( \frac{\partial F_x}{\partial z} - \frac{\partial F_z}{\partial x} \right) j + \left( \frac{\partial F_y}{\partial x} - \frac{\partial F_x}{\partial y} \right) k$$

$$= \begin{vmatrix} i & j & k \\ \frac{\partial}{\partial x} & \frac{\partial}{\partial y} & \frac{\partial}{\partial z} \\ F & F & F \end{vmatrix}$$

The gradient and rotor are vector functions, and divergence is the scalar function. The nabla-operator of Hamilton is defined in a rectangular system of coordinates with the unit vectors $i, j$, and $k$ by the symbolic vector

$$\nabla = \frac{\partial}{\partial x} i + \frac{\partial}{\partial y} j + \frac{\partial}{\partial z} k$$

Let us recall also, that the point between the operator nabla and vector function indicates scalar product, and the sign of multiplication indicates vector. 
   Let us assign some bounded domain $D^+$ with smooth boundary $\Gamma$ in the space $R$. For the unlimited domain we will use a designation $D^-$ which is in addition to $D^+ \cup \Gamma$ up to $R$. We will designate as $\Gamma^+$ a limited surface of $D^+$ when the points belonging to $D^+$ approaching the border $\Gamma$. Similarly, we will designate $\Gamma^-$ as the limited surface when the points in the unlimited domain $D^-$ are approaching the smooth boundary $\Gamma$. Let us consider in $D^+$ boundary-value problems:

$$\nabla^* u(p) = 0, \qquad (p \in D^+) \tag{2.1.4}$$

$$\text{(I)} \qquad u(p) = f(p), \qquad (p \in \Gamma^+) \tag{2.1.5}$$

$$\text{(II)} \qquad T_n^p u(p) = f(p), \qquad (p \in \Gamma^+) \tag{2.1.6}$$

The problem (2.1.4), (2.1.5) is the first main boundary-value problem in the theory of elasticity, when displacements on the border are assigned, and the problem (2.1.4), (2.1.6) is the second boundary-value problem when stresses on the border are assigned.

For solving formulated problems of theory of elasticity by the method of potential, we use Green's tensor of Eq. (2.1.4) in the space $R$, which is limited at the infinity solution of Eq. (2.1.4), corresponding to the displacement at the point $p(x_1, x_2, x_3)$ due to effect at the point $q(y_1, y_2, y_3)$, by concentrated force with the unit components. This solution is given by the matrix of Kelvin-Somigliana's $G(p, q)$. The matrix elements are of the form

$$G_{ij}(p, q) = \frac{1}{16\pi\mu(1-\sigma)}\left[\frac{(3-4\sigma)\delta_{ij}}{r} + \frac{(x_i - y_i)(x_j - y_j)}{r^3}\right], \qquad (2.1.7)$$

Where $r$ – distance between points $p$, and $q$.

$$r = \sqrt{(x_1 - y_1)^2 + (x_2 - y_2)^2 + (x_3 - y_3)^2},$$

$\delta_{ij}$ – delta Kronecker. Here one of the indexes corresponds to the coordinate of displacements, and the other to the coordinate of unit force. In the method of potential two more main kernels are used, which for simplicity will enter as the operators of stresses of $G(p, q)$, on the coordinates $p$ and $q$ respectively:

$$T_n^p G(p, q) = G_p(p, q);$$

$$T_n^q G(p, q) = G_q(p, q). \qquad (2.1.8)$$

Formulas in (2.1.8) are generalizations of the (three-dimensional) spatial case of formulas of differentiation of Green's function of one-dimensional differential operator on the given coordinate and on the coordinate of the unit force application. In the first case $G_p(p, q)$, we have stress at the point of $p$ of space $R$ due to concentrated force with unit components applied at the point $q$; in the second case $G_q(p, q)$ is the displacement at the point of $p$, due to unit moments applied at the point $q$ in the plane with normal $n$. The elements of the conjugate matrices $G_p(p, q)$ and $G_q(p, q)$ take the form below (V.Z. Parton, P.I. Perlin [258])

$$G_{pij}(p, q) = -\frac{1}{2r^3}\left\{\left[m\delta_{ij} + n\frac{(x_i - y_i)(x_j - y_j)}{r}\right]\sum_{k=1}^{3}(x_k - y_k)n_k(p)\right.$$

$$\left. - m[n_i(p)(x_j - y_j) - n_j(p)(x_i - y_i)]\right\}, \qquad (2.1.9)$$

with

$$G_{qij}(p,q) = G_{pji}(p,q). \tag{2.1.10}$$

Here

$$m = \frac{1}{2\pi}\frac{\mu}{\lambda + 2\mu}, \qquad n = \frac{3}{2\pi}\frac{\lambda + \mu}{\lambda + 2\mu}$$

Even though, in the method of potential, this direct introduction of kernels $G_p$, and $G_q$ is not carried out due to the need to justify the possibility of the transposition of operations $T_n$ and integration at $\Gamma$, we will assume the forgoing as substantiated, understanding however, that in certain instances, the integrals are taken as generalized functions [115]. As the basis of the solution of boundary-value problems by the method of potential we take a body contour on the border $\Gamma$, and loading it with some loads, distribution of which is done by satisfaction of boundary conditions. If the load of surface $\Gamma$ is carried out by forces, then solution of the problem should be sought in the form of the integral

$$\boldsymbol{u}(p) = \int_{\Gamma} G(p,q)\boldsymbol{\varphi}(q)d\Gamma_q = \boldsymbol{V}(p), \tag{2.1.11}$$

which, by analogy with the harmonic potential, is called the generalized elastic potential of a single layer. The density of this integral $\varphi(q)$ is to be determined from the boundary conditions. If the moments on the contour distributed under some condition $\varphi(q)$ are set, it is natural to seek a solution in the form of an integral

$$\boldsymbol{u}(p) = \int_{\Gamma} G_q(p,q)\boldsymbol{\varphi}(q)d\Gamma_q = \boldsymbol{W}(p), \tag{2.1.12}$$

which is called a generalized elastic potential of dual-layer. The properties of elastic potentials of $\boldsymbol{V}(p)$ and $\boldsymbol{W}(p)$ are studied in detail in scientific literature. Clearly, constructing of potentials satisfies the equations of Lamé in the entire space $R$, except for the surface $\Gamma$. The potential of a simple layer $\boldsymbol{V}(p)$ is a vector-function continuous in the entire space, the limit values of such vector-function on the surface $\Gamma$ [$\boldsymbol{V}^+(p)$, and $\boldsymbol{V}^-(p); (p \in \Gamma)$] are equal to each other, and therefore, equal to the direct value of $\boldsymbol{V}(p) = 0.5 [\boldsymbol{V}^+(p) + \boldsymbol{V}^-(p)]$ when $(p \in \Gamma)$. The stress tensor, which corresponds to the potential of a simple layer, $T_n^p \boldsymbol{V}(p)$, has different limits on the surface $\Gamma$, when point $p$ is approaching $\Gamma$ from inside and outside.

The direct value of the integral

$$T_n^p V(p) = \int_\Gamma G_p(p,q)\varphi(q)d\Gamma_q = \frac{[T_n^p V(p)]^+ + [T_n^p V(p)]^-}{2} \qquad (2.1.13)$$

should be considered in the sense of the regularized value, since matrix elements $G_p$ have special features of the second order. The difference between the limiting values of stresses (i.e. the jump of stresses on the surface $\Gamma$) is equal to the applied load

$$[T_n^p V(p)]^+ - [T_n^p V(p)]^- = \varphi(p), \qquad (p \in \Gamma) \qquad (2.1.14)$$

In the method of potential this fact is well substantiated. The potential of dual layer $W(p)$ also has breakages (gaps) on the surface $\Gamma$. The difference in the limiting values is equal

$$W^+(p) - W^-(p) = -\varphi(p) \qquad (2.1.15)$$

and the direct value

$$W(p) = \frac{W^+(p) + W^-(p)}{2} \qquad (2.1.16)$$

is singular integral. The stresses, corresponding to the potential of dual layer, are continuous on $\Gamma$

$$[T_n^p W(p)]^+ = [T_n^p W(p)]^- \qquad (p \in \Gamma) \qquad (2.1.17)$$

if they exist in the usual sense. Let us note, that from (2.1.13), (2.1.14) and (2.1.15), (2.1.16) follows

$$[T_n^p V(p)]^+ = \frac{\varphi(p)}{2} + T_n^p V(p), \qquad (p \in \Gamma)$$

$$[T_n^p V(p)]^- = -\frac{\varphi(p)}{2} + T_n^p V(p), \qquad (p \in \Gamma) \qquad (2.1.18)$$

and

$$W^+(p) = -\frac{\varphi(p)}{2} + W(p)$$

$$W^-(p) = \frac{\varphi(p)}{2} + W(p). \qquad (2.1.19)$$

The integral equations of the basic boundary-value problems of the theory of elasticity are obtained by means of determining displacements in the form of the potentials of single or dual layers (2.1.11, 2.1.12). Let us assume that the solution of Eq. (2.1.4) lies in the form of (2.1.11). Satisfying the boundary conditions (2.1.5, 2.1.6), we will obtain

$$\text{(I)} \quad u^+(p) = V^+(p) = f(p),$$

$$(p \in \Gamma) \tag{2.1.20}$$

$$\text{(II)} \quad T_n^p u^+(p) = [T_n^p V(p)]^+ = f(p)$$

Taking into account the continuity of the potential of the single layer (2.1.11), and the relation (2.1.18), we find the following integral equations

$$\text{(I)} \quad \int_\Gamma G_p(p, q) \varphi(q) d\Gamma_q = f(p), \tag{2.1.21}$$

$$\text{(II)} \quad \frac{\varphi(p)}{2} + \int_\Gamma G_p(p, q) \varphi(q) d\Gamma_q = f(p), \tag{2.1.22}$$

We will now search for the solution of the Eq. (2.1.4) in the form of the potential of dual-layer (2.1.12). After satisfaction of the boundary conditions, we will obtain

$$\text{(I)} \quad u^+(p) = W^+(p) = f(p),$$

$$(p \in \Gamma) \tag{2.1.23}$$

$$\text{(II)} \quad T_n^p u^+(p) = [T_n^p W(p)]^+ = f(p),$$

and taking into account (2.1.17, 2.1.19), we will find

$$\text{(I)} \quad \frac{\varphi(p)}{2} - \int_\Gamma G_q(p, q) \varphi(q) d\Gamma_q = -f(p), \tag{2.1.24}$$

$$\text{(II)} \quad \int_\Gamma G_{pq}(p, q) \varphi(q) d\Gamma_q = f(p), \tag{2.1.25}$$

where

$$G_{pq}(p, q) = T_n^p G_q(p, q).$$

The kernel $G_{pq}$ contains generalized functions and, therefore, the integral in (2.1.25) requires a special interpretation. This question is studied in detail in the works of R.T. Seeley [319], and M.I. Lazarev [206], where the expanded form of an integral of the type (2.1.25) is given. Here we will simply formally record the kernel $G_{pq}$, understanding that the integrals with this kernel are generalized functions.

Generally, in the scientific literature on the method of potential, equations like (2.1.25) are not considered due to the above-mentioned difficulties with substantiation of the transposition of the operator $T_n{}^P$, and operation of integration. The greatest interest in the given equations, the integral equations of the second kind (2.1.22) and (2.1.24), obtained by V. D. Kupradze [199] present the greatest interest. Similar equations can be obtained for the external problems, i.e. when boundary-value problem for the domain $D^-$ is solved.

Let us further consider two basic methods of boundary elements; direct and indirect. The indirect method is based on the load of the boundary plane of elastic body, i.e. surface $\Gamma$ with some fictitious forces, the density of which is determined by the point, within which the displacement is sought, when the point is approaching the internal boundary $\Gamma^+$. This method is equivalent to the method of potentials, if the solution of the problem is sought in the form of the elastic potential of simple layer. This method leads to Eqs. (2.1.21, 2.1.22) (in the absence of body forces), but it differs from the method of potential, perhaps, only in the terms of a more detailed mechanical interpretation of the solution. It is possible to expand the indirect method, by assigning forces (jumps of stresses), and jumps of displacements. The forgoing is equivalent to the application of elastic potentials of dual layer. In the event that unknown jumps of displacements $\varphi(p)$ are assigned on the border $\Gamma$, their density is determined from boundary conditions. From (2.1.15) it follows that the jump in the displacements gives an elastic potential of dual layer. Therefore, assuming $u(p) = W(p)$ and taking into account the above indicated properties of this potential, we will obtain the integral Eqs. (2.1.24) and (2.1.25), which can be the basis of algorithms of the indirect method of boundary elements.

Thus, analysis of the equations of the indirect method and its expansion due to the examination of the kinematic load of a contour, equivalent to an application of an elastic potential of dual layer, show the complete analogy between the indirect boundary elements method (IBEM) and method of the potentials.

The direct boundary element method (DBEM) and BIEM use other approaches than these two methods. The integral identification of Green or the formula of Betty is used for obtaining the corresponding integral equations. The formula of Betty can be written in the form:

$$\int_{D^+} [u\nabla^*v - v\nabla^*u]d\Omega = \int_{\Gamma^+} [uT_n^q v - vT_n^q u]d\Gamma_q \qquad (2.1.26)$$

Let us assume that $v = G(p, q)$. It is possible to show that Green's tensor satisfies the nonhomogeneous equation of Lamé

$$\nabla^* G(p, q) + \delta(p, q) = 0, \tag{2.1.27}$$

where delta-function possesses the following property

$$\int_\Omega g(q)\delta(p, q)d\Omega_q = g(p), \qquad (p \in \Omega). \tag{2.1.28}$$

Because of this fact, (2.1.4), and the basic property of scalar delta-function we have

$$\int_\Gamma [G(p, q)T_n^q u(q) - G_q(p, q)u(q)]d\Gamma_q = u(p), \qquad (p \in D^+), \tag{2.1.29}$$

$$\int_\Gamma [G(p, q)T_n^q u(q) - G_q(p, q)u(q)]d\Gamma_q = 0, \qquad (p \in D^-). \tag{2.1.30}$$

Upon $p$ approaching $\Gamma^+$, the formula (2.1.29) leads to the following integral equation (taking into account the properties of kernels of $G$, and $G_q$)

$$\frac{u^+(p)}{2} - \int_\Gamma G(p, q)T_n^q u(q)d\Gamma_q + \int_\Gamma G_q(p, q)u(q)d\Gamma_q = 0. \tag{2.1.31}$$

This equation is used for solving any correctly set problem of the theory of elasticity. In particular, for the first and second basic problems (tasks) we have

$$\text{(I)} \qquad \int_\Gamma G(p, q)T_n^q u(q)d\Gamma_q = \frac{f(p)}{2} + \int_\Gamma G_q(p, q)f(q)d\Gamma_q \tag{2.1.32}$$

$$\text{(II)} \qquad \frac{u^+(p)}{2} + \int_\Gamma G_q(p, q)u(q)d\Gamma_q = \int_\Gamma G(p, q)f(q)d\Gamma_q \tag{2.1.33}$$

In the DBEM the integral equation of the second kind (order) (2.1.33) is preferred, with the aid of which the solution of the second basic problem (task) of the theory of elasticity is obtained.

If we use the kernel $G_{qp}$, and set the corresponding integral as a generalized function, then instead of the equation of the first kind (order) (2.1.32), it is possible to obtain the integral equation of the second kind as a result of the transition (passage) in (2.1.31) to the stresses

$$\text{(II)} \qquad \frac{[T_n^p u]^+(p)}{2} - \int_\Gamma G_p(p, q)T_n^q u(q)d\Gamma_q = - \int_\Gamma G_{qp}(p, q)f(q)d\Gamma_q \tag{2.1.34}$$

Thus, the DBEM leads to the integral equations, which differ from the equations of the indirect method and method of potentials. At the same time a strong connection is outlined between the equations of these methods. Furthermore, as will be shown below, the equations have a direct analogy with the methods of structural mechanics: methods of forces, displacements, and initial parameters.

## 2.2   The Construction of Boundary Equations by the Delta-Transformation Technique

The method of delta-transformation described above allows us to construct all types of boundary equations obtained by the methods of boundary elements, potential, compensating loadings, the generalized solutions, extended domain, and others. We will apply the scheme of a method of delta transformation for the solution of a three-dimensional problem of the theory of elasticity in order to compare it to the BEM, and the method of potential. Let us examine a more general boundary-value problem for the Eq. (2.1.4), after setting the boundary conditions

$$\alpha u(p) + \beta T_n u(p) = f(p), \qquad (p \in \Gamma^+) \tag{2.2.1}$$

where $\alpha$, $\beta$ are constants. When $= 0$, $\alpha = 1$, from (2.2.1) we obtain the boundary conditions of the first fundamental problem, and when $\alpha = 0$, $\beta = 1$, the second fundamental problem of the theory of elasticity.

Let us introduce an extended domain $\Omega$, which includes $D^+(D^+ \cup \Gamma \subseteq \Omega \subseteq R)$ into space $R$. The minimum expansion of the domain $D^+$ is achieved by joining it to the boundary $\Gamma$, where the maximum space is $R$.

Let us continue somehow to insert $u(p)$ into the extended domain and carry out a delta-transformation, that is, we will multiply both sides of the Eq. (2.1.4) by $\delta(p, p^*)$ and integrate over the volume $\Omega$. We believe that in the $\Omega$, the components of the function $u$ are differentiated twice everywhere, except for on the boundary $\Gamma$. Then, using the formula of Green-Betty (2.1.26) in the domain of continuity $D^+$ and $\Omega \setminus D^+ \cup \Gamma$, we obtain

$$\int_{\Omega} u(p) \nabla^* \delta(p, p^*) d\Omega_p = \int_{\Gamma^+} - \int_{\Gamma^-} + \int_{\Gamma^*}$$

$$[u(q) T_n^q \delta(q, p^*) - \delta(q, p^*) T_n u(q)] d\Gamma_q. \tag{2.2.2}$$

Here $\Gamma^*$ – the external border of the domain $\Omega$.

Due to the properties of the delta-function, the last equation, examined in $\Omega - \Gamma$, where $u$ is differentiated twice, gives the following (for simplicity, we will the asterisks in $p^*$ going forward):

$$\nabla^* u(p) = \int_\Gamma [\Delta u(q) T_n^q \delta(q,p) - \Delta T_n u(q) \delta(q,p)] d\Gamma_q$$

$$+ \int_{\Gamma^*} [u(q) T_n^q \delta(q,p) - \delta(q,p) T_n u(q)] d\Gamma_q^*, \qquad (p \in \Omega), \quad (2.2.3)$$

where $\Delta$ — indicates the difference of the values of the corresponding vector-functions on $\Gamma^+$ and $\Gamma^-$, i.e. their jumps on $\Gamma$. Thus, instead of Eq. (2.1.4) in the domain $D^+$, we obtained Eq. (2.2.3) in the domain $\Omega$, such that the solution of both equations in $D^+$ coincides. In this case the right side of Eq. (2.2.3) contains additional members in the form of delta-functions and their derivatives, which are "compensating" forces and kinematic impacts (actions). For the solution of Eq. (2.2.3), we will use the tensor of Green $G(p, q)$ of the boundary-value problem in (2.1.4) in the extended domain $\Omega$. The vector-function $G(p, q)$ satisfies Eq. (2.1.27) and some boundary conditions on the border of $\Gamma^*$. Therefore, in the view of the permutability of operations $\nabla^*$ and $T_n$, tensor $G_q(p, q) = T_n{}^q G(p, q)$ satisfies the equation:

$$\nabla^* G_q(p,q) = -T_n{}^q \delta(p,q) \qquad (2.2.4)$$

Taking into account the parity of delta-function the solution of Eq. (2.2.2) can be represented in the form of

$$u(p) = \int_\Gamma [G(p,q) \Delta T_n^q u(q) - G_q(p,q) \Delta u(q)] d\Gamma_q$$

$$+ \int_{\Gamma^*} [G(p,q) T_n^q u(q) - G_q(p,q) u(q)] d\Gamma_q^*. \qquad (2.2.5)$$

The boundary conditions for Green's tensor on $\Gamma^*$ (and for $u$) are arbitrary; it is possible to select them is such a way that the building (construction) of Green's tensor would be achieved more simply. Since $u(p)$ and $G(p, q)$ on $\Gamma^*$ satisfy identical boundary conditions, the last integral in (2.2.5) becomes zero, and finally we obtain

$$u(p) = \int_\Gamma [G(p,q) \Delta T_n^q u(q) - G_q(p,q) \Delta u(q)] d\Gamma_q. \qquad (2.2.6)$$

Green's tensors of $G(p, q)$ and $G_q(p, q)$ are well studied for the entire space. The expression for Green's tensors, as well as the description of the properties of the corresponding integrals of these functions, are provided in the previous paragraph. It is also possible to apply other Green's tensors, for example; for a semi-continuum or finite domain (a cube, or a sphere) when their construction can be carried out with the help of eigenfunction decomposition.

Thus, we obtained the standard representation of the function $u(p)$, similar to the formula of Somigliana, but different from it in that the integrals on the boundary surface contain corresponding jumps and not displacements $u$ and stresses (tensions) of $T_n u$. Such formulation of boundary integrals allows us to provide a clear mechanical performance (2.2.6) within the framework of the classical methods of structural mechanics. In particular, assuming any jumps equal to zero, i.e. considering the displacements of stresses on the border of the domain as continuous, we arrive at the three-dimensional analog (analogue) of the method of forces or method of displacements, while the primary system is defined by the expanded domain and boundary conditions taken on the border. Depending on the retention of various jumps we can come to the indirect or direct methods of boundary elements, as well as to different equations of the method of potentials.

In order to review the comparison of the method of delta-transformation and the direct and indirect methods of boundary elements, let us pause at the solution of fundamental problems of the theory of elasticity with the use of an extended domain, which coincides with the entire space $R$.

First let us examine a case when $\Delta u = 0$, i.e. assuming that the displacements are continuous on $\Gamma$ and forces are applied giving rise to a jump of stresses. Then from (2.2.6) it follows:

$$u(p) = \int_{\Gamma} G(p,q)\Delta T_n^q u(q)\, d\Gamma_q, \qquad (p \in R\backslash\Gamma), \qquad (2.2.7)$$

Using the operator of boundary conditions (2.2.1), we obtain the integral equation

$$\alpha u^+(p) + \beta T_n^p u^+(p) = \int_{\Gamma} [\alpha G^+(p,q) + \beta G^+{}_p(p,q)]\Delta T_n^q u(q)\, d\Gamma_q, \quad (p\in\Gamma^+).$$

$$(2.2.8)$$

Similarly, we can write for $(p\in\Gamma^-)$

$$\alpha u^-(p) + \beta T_n^p u^-(p) = \int_{\Gamma} [\alpha G^-(p,q) + \beta G^-{}_p(p,q)]\Delta T_n^q u(q)\, d\Gamma_q, \quad (p\in\Gamma^+).$$

$$(2.2.9)$$

Add the left and right sides of Eqs. (2.2.8) and (2.2.9); after simple calculations, taking into account the continuity of displacements, we will obtain the integral equation

$$0,5\beta\Delta T_n^p u(p) + \int_\Gamma [\alpha G(p,q) + \beta G_p(p,q)]\Delta T_n^q u(q)\, d\Gamma_q = f(p). \quad (2.2.10)$$

For the first boundary-value problem ($\beta = 0$, $\alpha = 1$), and second boundary-value problem ($\alpha = 0$, $\beta = 1$, ) of the theory of elasticity, we come to the integral equations of the first and second kind, respectively:

$$\text{(I)} \quad \int_\Gamma G(p,q)\Delta T_n^q u(q)\, d\Gamma_q = f(p), \quad (2.2.11)$$

$$\text{(II)} \quad 0,5\Delta T_n^p u(p) + \int_\Gamma G_p(p,q)\Delta T_n^q u(q)\, d\Gamma_q = f(p), \quad (2.2.12)$$

Eqs. (2.2.11) and (2.2.12) coincide with the equations of the theory of potential (2.1.21), when a solution in a form of the potential of single layer is sought. The indirect method of boundary elements leads to the similar equations.

Let us move to the examination of the second case, when $\Delta T_n u = 0$, i.e. assuming that stresses are continuous on border $\Gamma$, and the kinematic impacts are applied giving jump in the displacements. In this case

$$u(p) = -\int_\Gamma G_q(p,q)\Delta u(q)\, d\Gamma_q, \quad (2.2.13)$$

Carrying out similar computations, we obtain the integral equation

$$0,5\alpha\Delta u(p) - \int_\Gamma [\alpha G_q(p,q) + \beta G_{qp}(p,q)]\Delta u(q)\, d\Gamma_q = f(p). \quad (p\in\Gamma) \quad (2.2.14)$$

The following integral equations are for the first and second boundary-value problems respectively

$$\text{(I)} \quad 0,5\alpha\Delta u(p) - \int_\Gamma G_q(p,q)\Delta u(q)d\Gamma_q = f(p), \quad (2.2.15)$$

$$\text{(II)} \quad -\int_\Gamma G_{qp}(p,q)\Delta u(q)d\Gamma_q = f(p), \quad (2.2.16)$$

which with $\varphi = -\Delta u$ coincide with the equations of the method of potential (2.1.24), (2.1.25).

In the case of external problems, the boundary conditions is assigned on $\Gamma^-$

$$\alpha u(p) + \beta T_n^p u(p) = f(p), \qquad (p \in \Gamma^-) \qquad (2.2.17)$$

Absolutely similar reasoning leads to the integral equations:

$$-0,5\beta\Delta T_n^p u(p) + \int_\Gamma [\alpha G(p,q) + \beta G_p(p,q)]\Delta T_n^q u(q)d\Gamma_q = f(p), \qquad (p \in \Gamma^-)$$

$$(2.2.18)$$

$$-0,5\alpha\Delta u(p) - \int_\Gamma [\alpha G_q(p,q) + \beta G_{qp}(p,q)]\Delta u(q)d\Gamma_q = f(p), \qquad (2.2.19)$$

which are reduced to the form below for the first and second problems of the theory of elasticity, respectively:

$$\text{(I)} \quad -0,5\Delta u(p) - \int_\Gamma G_q(p,q)\Delta u(q)d\Gamma_q = f(p), \qquad (2.2.20)$$

$$\text{(II)} \quad -0,5\Delta T_n^p u(p) + \int_\Gamma G_p(p,q)\,\Delta T_n^q u(q)d\Gamma_q = f(p), \qquad (2.2.21)$$

Thus, assuming jumps of stresses or jumps of displacements in the equations of the method of delta-transformation to be continuous, we come to the integral equations of the method of potential or indirect method of boundary elements. It is essential not only because of the clearness of mechanical performances, but also for the simplicity of obtaining equations without the need of sufficient and thorough studies of the properties of potentials. This circumstance, however doesn't matter for problems of the theory of elasticity, since the properties of elastic potentials are studied in sufficient detail. The specified advantage of the method of delta-transformation can appear quite significant during the solution of different boundary-value problems for other elliptic operators. The analysis of the formula (2.2.6) shows that in addition to the two cases ($\Delta u = 0$, $\Delta T_n u = 0$) two more versions are possible:

3.  $\Delta u \neq 0$, $\Delta T_n u \neq 0$ (this case corresponds to the prolongation of function $u(p)$ by zero into the extended domain $\Omega$);
4.  $\Delta u = 0$, $\Delta T_n u = 0$ (the absence of compensating impacts on the border $\Gamma$).

It is obvious that in the latter case compensating impacts (actions) must be applied outside of the boundary $\Gamma$ within the extended domain. Let us examine in more detail the case #3, when both jumps are not equal to zero, i.e. the distributed forces and kinematic impacts (actions) are applied on the border of the domain $D^+$. In this case, two unknown density jumps correspond to one boundary condition. Therefore, it is possible to enter an additional functional correlation between density of jumps of stresses and displacements in order to obtain the sole solution to the problem. As such correlation, let us assume the requirements of the self-equilibrium force and kinematic impacts (influences, actions) as set inside the domain $D^+$, as a result of which the outer boundary of domain $D^+$ remains undeformable

$$u(p) = T_n^p u(p) = 0, \qquad (p \in \Gamma^-), \qquad (2.2.22)$$

which corresponds to prolongation of $u(p)$ by zero into the extended domain. Then $\Delta u = u^+$, $\Delta T_n u = T_n u^+$ and formula (2.2.6) take the form of:

$$\frac{u^+(p)}{2} = \int_{\Gamma} [T_n^q u^+(q) G(p,q) - u^+(q) G_q(p,q)] d\Gamma_q, \qquad (p \in \Gamma^+), \quad (2.2.23)$$

which is characteristic of the direct method of boundary elements. Subsequent use of the boundary condition (2.2.1) gives

$$2 \int_{\Gamma} \left\{ \alpha T_n^q u^+(q) G(p,q) - f(q) G_q(p,q) + \beta T_n^q u(q) [G_q(p,q) + G_p(p,q)] \right.$$

$$\left. + \frac{\beta}{\alpha} [\beta T_n^q u^+(q) - f(q)] G_{qp}(p,q) \right\} d\Gamma_q = f(p), \qquad (p \in \Gamma^+) \quad (2.2.24)$$

$$2 \int_{\Gamma} \left[ \alpha f(q) G(p,q) - \frac{\alpha^2}{\beta} u^+(q) G(p,q) - \alpha u^+(q) G_q(p,q) + f(q) G_p(p,q) \right.$$

$$\left. - \alpha u^+(q) G_p(p,q) - \beta u^+(q) G_{qp}(p,q) \right] d\Gamma_q = f(p). \qquad (2.2.25)$$

From the equations in (2.2.24), (2.2.25) with the appropriate values of $\alpha$ and $\beta$ it is possible to obtain the equations of the direct method of boundary elements (2.1.32, 2.1.33).

Due to the fact that $u(p)$ is continued by zero into $D^-$ we can choose a certain quantity of points with the coordinates $p_i (i = 1, k)$ in the domain $D^-$, and from the condition $\boldsymbol{u}(p_i) = 0$, $T_n^p \boldsymbol{u}(p_i) = 0$, obtain a system of equations

$$\int_\Gamma [T_n^q \boldsymbol{u}^+(q) G(p_i, q) - \boldsymbol{u}^+(q) G_q(p_i, q)] \, d\Gamma_q = 0,$$

$$(i = \overline{1, k})$$

$$\int_\Gamma [T_n^q \boldsymbol{u}^+(q) G_p(p_i, q) - \boldsymbol{u}^+(q) G_{qp}(p_i, q)] \, d\Gamma_q = 0.$$

Expressing through (2.2.1)

$$\boldsymbol{u}^+(q) = -\frac{\beta}{\alpha} T_n^q \boldsymbol{u}^+(q) + \frac{1}{\alpha} f(q),$$

or

$$T_n^q \boldsymbol{u}^+(q) = -\frac{\alpha}{\beta} \boldsymbol{u}^+(q) + \frac{1}{\beta} f(q),$$

and substituting into the given equation, we obtain the system $k$ for the equations, sufficient for solving the boundary equations on $\Gamma$ by means of partition (dividing on, splitting into, breaking into) of boundary elements on $k$. Such equations are characteristic for the method of functional equations B.D. Kupradze [199].

Let us now move on to the consideration of the fourth option of the selection of the jumps of force and kinematic factors on the border of the domain, when all jumps are taken as equal to zero, i.e. there are no actions, either force or kinematic, that are assigned on $\Gamma$.

So in order for the solution to the problem not be trivial in this case, it is necessary to set the jumps on the outline within the limits of the extended domain $\Omega$. The advantage of this method is marked by the regularity of the obtained integral equation; the deficiency of the method is the impossibility of obtaining equations of the second kind. Let us designate the outer contour of the force and kinematic loading as $S$. For the best conditionality of the resolving equations, this outline contour is desirable to choose as close to $\Gamma$ as possible, and for simplification of the solution of integral equation – as simple and smooth as possible. Then by analogy (2.2.6) the solution of the three-dimensional problem of the theory of elasticity should be written down in the form of

$$\boldsymbol{u}(p) = \int_S [G(p, q) \Delta T_n^q \boldsymbol{u}(q) - G_q(p, q) \Delta \boldsymbol{u}(q)] \, dS_q, \quad (p \in \Omega \backslash S). \quad (2.2.26)$$

Similarly, further operations can be performed when $S = \Gamma$. Let us examine three basic cases.

1. The displacements are continuous on $S$ and $\Delta u = 0$. In this case, satisfying the boundary conditions (2.2.1), we will obtain:

$$\int_S [\alpha G(p,q) + \beta G_p(p,q)] \Delta T_n^q u(q) dS_q = f(p) \quad (p \in \Gamma).$$

In particular, the equations of the first and second fundamental problems of the theory of elasticity will take the form:

$$\text{(I)} \qquad \int_S G(p,q) \Delta T_n^q u(q) dS_q = f(p), \qquad (p \in \Gamma), \qquad (2.2.27)$$

$$\text{(II)} \qquad \int_S G_p(p,q) \Delta T_n^q u(q) dS_q = f(p), \qquad (p \in \Gamma), \qquad (2.2.28)$$

All these equations are regular but they do not lead to the equations of second kind.

2. The stresses are continuous on $S$ ($\Delta T_n^q u(q) = 0$), and we are searching for the jump of the displacements, determined by kinematic impact. In this case, the integral equations will take the form ($p \in \Gamma$)

$$\int_S [\alpha G_q(p,q) \Delta u(q) + \beta G_{pq}(p,q) \Delta u(q)] dS_q = -f(p) \qquad (2.2.29)$$

$$\text{(I)} \qquad \int_S G_q(p,q) \Delta u(q) dS_q = -f(p), \qquad (2.2.30)$$

$$\text{(II)} \qquad \int_S G_{pq}(p,q) \Delta u(q) dS_q = -f(p). \qquad (2.2.31)$$

Thus, the method of delta-transformation makes it possible to uniformly obtain all known BIEs, to classify them and to offer new equations, closing the entire system of building boundary equations of the fundamental types. The use of the method of delta-transformation in an extended domain for calculating of BIEs makes it possible to develop further generalizations of BEM, with the arbitrary choice of differential forms on the boundary. For the illustration, let us show this generalization based on the example of a three-dimensional self-adjoint (self-conjugate) boundary-value problem of elliptic type. In three-dimensional space $R$ we will consider the linear self-adjoint matrix differential operator $L$, generated by a differential operation of elliptic type in partial derivatives $2n$ – order in three coordinates $x, y,$ $z$ and also by some boundary conditions on the surface $\Gamma^+$. The surface $\Gamma^+$ is the internal boundary of the domain $D^+$. Let us examine the boundary-value problem:

$$Lu(p) = 0, \qquad (p \in D^+)$$

$$lu(p) = f(p). \qquad (p \in \Gamma^+) \tag{2.2.32}$$

The matrix operator $l$ of boundary conditions will set at the order not higher than $2n - 1$, and can be assigned (given, set) by various boundary forms in different intervals (sections, segments) of surface $\Gamma^+$. For the linear self-adjoint operator $L$ within the limits of the domain of the continuous vector functions $u, v$, and their derivatives of $2n - 1$ order we will use the generalized Green's formula (P.K. Banerjee, J.O. Watson [26]),

$$\int_D vLu d\Omega_p = \int_D uLv d\Omega_p + \int_{\Gamma^+} [MvNu - MuNv] d\Gamma, \tag{2.2.33}$$

where $M, N$ are some operators of boundary conditions, appearing during integration by parts of integral from $vLu$. It is known that the operators $M, N$ can be arbitrary but are connected with each other by conjugate, linear combinations of boundary forms. In particular, $M, N$ can explicitly contain the elements of the operator $l$, which corresponds to the boundary conditions in (2.2.1). We will extend the domain of $D^+$ to a certain domain $\Omega$, and will continue out of $D^+$ the vector-function $u(p)$; and if the right-side of (2.2.32) is present, will continue out that too. Let us search for solution of the problem $u(p)$ in the extended domain $\Omega$, which satisfies certain boundary conditions on the border of $\Gamma$ of domain $\Omega$.

$$l_* u(p) = 0, \qquad (p \in \Gamma_\Omega). \tag{2.2.34}$$

Let us perform in a delta-transformation over the left and right sides of Eq. (2.2.32) in $\Omega$. Using formula (2.2.33) in $D^+$ and $D^- = \Omega \backslash (D^+ \cup \Gamma)$, respectively, and by analogy (2.2.2) we will obtain:

$$\int_\Omega u(p)L\delta(p, p^*) d\Omega_p = \int_{\Gamma^+} - \int_{\Gamma^-} + \int_{\Gamma^*}$$

$$[Mu(q)N\delta(q, p^*) - M\delta(q, p^*)Nu(q)]$$

$$d\Gamma_q.$$

Considering that $u(q)$ is $2n-1$ times differentiable in $D^+$ and $D^-$ and taking into account the properties of the delta-function, we obtain

$$Lu(p^*) = \int_\Gamma [\Delta Mu(q)N\delta(q,p^*) - \Delta Nu(q)M\delta(q,p^*)]d\Gamma_q$$

$$+ \int_{\Gamma^*} [Mu(q)N\delta(q,p^*) - M\delta(q,p^*)Nu(q)]d\Gamma^*. \qquad (2.2.35)$$

The last formula can be considered as a mathematical analog of the method of the compensating loads, or the three-dimensional analog of "standardized" form of the multidimensional differential operator. Formula (2.2.35) contains the jumps of the arbitrary self-conjugate (self-adjoint) differential boundary forms and gives the interpretation of the generalized potentials. The boundary indignations, as we see, correspond to differential forms obtained from a delta-function, and the unknown densities of impacts (actions, influences, effects) and corresponding potentials are the jumps of differential forms obtained by the solution of the problem $u(p)$. The inversion of Eq. (2.2.35) can be realized in the domain $\Omega$ with the additional assumptions about the nature of jumps on $\Gamma$, providing uniqueness of the solution. In order to obtain the resolving integral equations for a given structure it is necessary, in (2.2.35), to preserve these or other jumps by selecting an auxiliary problem, for which the equation in the extended domain is solved. If the boundary conditions on the entire boundary $\Gamma$ are identical, then it is possible to accept $M = l$ or $N = l$. For the sake of certainty let $M = l$, then $N = \bar{l}$, where the $\bar{l}$ – conjugate operation. Let us assume in (2.2.35) $\Delta Mu = 0$, i.e. let us choose a continuation of $u$ in $D^-$, such that the boundary forms of $Mu$ are continuous. Searching for the solution of the equation in (2.2.35), which satisfies some boundary conditions on $\Gamma^*$ (not necessary coinciding with $Mu = 0$ or $Nu = 0$) we will obtain:

$$u(p^*) = -\int_\Gamma \Delta\bar{l}_q u(q)l_q G(q,p^*)d\Gamma_q. \qquad (2.2.36)$$

The integral on $\Gamma^*$ disappears because the boundary shape either from $u$ or from $G$ is equal to zero. Here $G(q, p^*)$ is tensor of Green of the given problem in the extended domain at the accepted continuation of $u$ and differential forms from $u$ through the boundary $\Gamma$. From (2.2.36) it follows (asterisk at $p$ is omitted)

$$l_p u^+(p) = -\int_\Gamma \Delta\bar{l}_q u(q)l_{qp}G^+(q,p)d\Gamma_q. \qquad (2.2.37)$$

Here $l_{qp}$ – sequential operations $l$ on variables $q$ and $p$. Similarly:

$$l_p u^-(p) = -\int_\Gamma \Delta \bar{l}_q u(q) l_{qp} G^-(q,p) d\Gamma_q. \tag{2.2.38}$$

Add the left and right sides of (2.2.37) and (2.2.38), then divide by 2, and take into account the determination of direct value

$$\frac{l_p u^+(p) + l_p u^-(p)}{2} + \int_\Gamma \Delta \bar{l}_q u(q) l_{qp} G(q,p) d\Gamma_q = 0.$$

Adding to both parts of the equation $l_p u^+ = -f$, we will obtain the integral equation of the first kind.

$$\int_\Gamma \Delta \bar{l}_q u(q) l_{qp} G(q,p) d\Gamma_q = -f(p) \tag{2.2.39}$$

Here, $\Delta lu = 0$. For obtaining the integral equation of the second kind, let us examine the same boundary-value problem but with the following boundary condition

$$\bar{l}_p u^+(q) = f(p), \qquad (p \in \Gamma^+) \tag{2.2.40}$$

From (2.2.36):

$$l_p u^+(p) = -\int_\Gamma \Delta \bar{l}_q u(q) \bar{l}_p l_q G^+(q,p) d\Gamma_q, \tag{2.2.41}$$

and

$$\bar{l}_p u^-(p) = -\int_\Gamma \Delta \bar{l}_q u(q) \bar{l}_p l_q G^-(q,p) d\Gamma_q, \tag{2.2.42}$$

adding the left and right parts of Eqs. (2.2.41) and (2.2.42), and then dividing by 2, we will obtain:

$$\frac{1}{2}\Delta \bar{l}_p u(p) - \int_\Gamma \Delta \bar{l}_q u(q) \bar{l}_p l_q G(q,p) d\Gamma_q = f(p). \tag{2.2.43}$$

Equation (2.2.43) is an integrated equation of the second kind, relative to the unknown density $\Delta l_q u(q)$. After determining $\Delta l_q u(q)$ by integration in (2.2.36), we obtain the solution of the problem. In each specific problem, properties of kernels $l_{qp} G(q,p)$ and $\bar{l}_p l_q G(q,p)$ can be studied. Thus, the overall diagram of the construction (build) of the equations of generalized BEM is obtained. For obtaining the integral equations of the second kind it is necessary to choose a "primary system," i.e. an extended domain, such that the differential operations conjugate to the operations of the boundary conditions of the given boundary-value problem are continuous on the border. The given generalization can be also extended to the direct BEM or the BIEM by means of continuation of vector-function $\boldsymbol{u}(p)$ by zero out of the domain $D^+$. In that case, we will obtain both: the generalization of (BIEM) of the spatial problem of the theory of elasticity, and the generalization of the functional equations of B.D. Kupradze (selection of points $p_i$), where integral relationships in the domain $D^-$ are satisfied. The generalizations of the method of the compensating loads with the jumps of differential forms placed outside of $\varGamma$ in the domain $D^-$ can be analogously built.

## 2.3   The Equivalence of Direct and Indirect BEM

Now we will briefly analyze three basic schemas of the construction of the BIEs by the method of delta-transformation, leading to different equations of the method of potential, direct, and indirect BEM. We will also prove the equivalence of the integral equations, obtained by direct and indirect BEM, or by the method of potential and BIEM, which is the same thing [272]. Let us introduce some formulas which will be necessary for further proof and analysis. We accept
$\boldsymbol{u} = \boldsymbol{G}(q,p),\ \boldsymbol{v} = \boldsymbol{G}(q,p_1)$ in Bettie's formula (2.1.26). Taking into account (2.1.27), we will obtain:

$$-\int_{D^+} \{G(p,q)\delta(q,p_1) - G(q,p)\delta(p,q)\}d\Omega$$

$$= \int_{\varGamma} \{G(p,q)G_q(q,p_1) - G(q,p_1)G_q(p,q)\}d\varGamma_q,$$

from where

$$\int_{\varGamma} G(p,q)G_q(q,p_1)d\varGamma_q = \int_{\varGamma} G_q(p,q)G(q,p_1)d\varGamma_q. \tag{2.3.1}$$

Now let us consider integral equations of the second kind, obtained by indirect and direct methods of boundary elements ($p \in \varGamma$). In the case of the second fundamental problem of theory of elasticity

$$\frac{\Delta T_n^p \boldsymbol{u}(p)}{2} + \int_{\varGamma} G_p(p,q)\Delta T_n^q \boldsymbol{u}(q)d\varGamma_q = [T_n^p \boldsymbol{u}]^+(p), \tag{2.3.2}$$

$$\frac{\boldsymbol{u}^+(p)}{2} + \int_\Gamma G_q(p,q)\boldsymbol{u}^+(q)d\Gamma_q = \int_\Gamma G(p,q)[T_n^q\boldsymbol{u}]^+(q)d\Gamma_q, \qquad (2.3.3)$$

Multiplying the left and right sides of Eq. (2.3.2) by $G(p_1,p)$ $(p_1 \in \Gamma)$, and integrating over the surface $\Gamma$, we will obtain:

$$\frac{1}{2}\int_\Gamma G(p_1,p)\Delta T_n^p \boldsymbol{u}(p)d\Gamma_p + \int_\Gamma \tilde{G}(p_1,q)\Delta T_n^q \boldsymbol{u}(q)d\Gamma_q$$

$$= \int_\Gamma G(p_1,p)T_n^p \boldsymbol{u}^+(p)d\Gamma_p, \qquad (2.3.4)$$

where

$$\tilde{G}(p_1,q) = \int_\Gamma G(p_1,p)G_p(p,q)d\Gamma_p$$

With an accuracy before changing variables, the kernel $\tilde{G}(p_1,q)$ is equal to left side of the equation in (2.3.1); therefore:

$$\tilde{G}(p_1,q) = \int_\Gamma G_p(p_1,p)G(p,q)d\Gamma_p. \qquad (2.3.5)$$

It is easy to see, that in light of the determination of the Green's function and absence of any kinematic impact ($\Delta \boldsymbol{u} = 0$)

$$\int_\Gamma G(p_1,p)\Delta T_n \boldsymbol{u}(p)d\Gamma_p = \boldsymbol{u}(p_1),$$

so taking into account (2.3.5), Eq. (2.3.4) takes the form,

$$\frac{\boldsymbol{u}(p_1)}{2} + \int_\Gamma G_p(p_1,p)\boldsymbol{u}(p)d\Gamma_p = \int_\Gamma G(p_1,p)T_n\boldsymbol{u}^+(p)d\Gamma_p, \qquad (2.3.6)$$

with the accuracy before changing variables, coinciding with (2.3.3).

For greater clarity of the connection among BIEs, obtained by various methods, let us combine all integral equations in Table 2.1 [278]. Designations in the table will follow P.K Banerjee, R. Butterfield [27], for the DBEM and IBEMs, and for the method of potentials with the use of a (single) layer potential method (SLPM) and (double) layer potential method (DLPM).

Analyzing Table 2.1, we first note that integral equations with the kernels $G_p(p, q)$ and $G_q(p, q)$ are singular due to the fact that the elements of these kernels have special features of the second order. These equations of the second kind are most convenient for calculations [272]. The integral equation with the kernel $G(p, q)$ is regular but its deficiency is the fact that it is of the first kind, and we run into a problem of sustainability (stability) of computational procedures when solving it. For a detailed study of solvability of computational procedures given in Table 2.1, see the monographs of V.Z. Parton and P.I. Perlin [257, 258].

The first internal and both external fundamental problems are solvable and have a single solution. The second internal problem has a solution, determined with the accuracy up to the vectors of rigid displacement, when and only when $f(p)$ is orthogonal to these vectors [208]. The integral equations considered in Table 2.1 [278] are widely used in BEM and the method of potential. Only two types of the equations, for which the table shows no analogy to other methods, have no practical use. This can be explained by the fact that they contain kernel $G_{pq}(p, q)$, which has singular component, and also by that fact that the indicated equations do not have advantages over other equations, in particular, the equations of B.D. Kupradze (the second and third row of Table 2.1).

The connection of the analyzed integral equations to the classical methods of structure mechanics is of interest to us. The two basic methods of structural mechanics are the force method and displacement method. During the application of these methods a given mechanical system is replaced with another system or a set of others; as a rule, rather simple

**Table 2.1** The basic types of boundary integral equations of the spatial problem of the theory of elasticity.

| Jumps on $\Gamma$ | Problems | Unknown function | Equation Kernel | Equations regularity | Type of equations | Analogy with BEM[1] |
|---|---|---|---|---|---|---|
| $\Delta u = 0$ | I | $\Delta T_n u(p)$ | $G(p, q)$ | Regular | First | SLPM, IBEM |
| $\Delta T_n u \neq 0$ | II | $\Delta T_n u(p)$ | $G_p(p, q)$ | Singular | Second | SLPM, IBEM |
| $\Delta u \neq 0$ | I | $\Delta u(p)$ | $G_q(p, q)$ | Singular | Second | DLPM, IBEM |
| $\Delta T_n u = 0$ | II | $\Delta u(p)$ | $G_{qp}(p, q)$ | Regular | First | —— |
| $\Delta u \neq 0$ | I | $[T_n u]^+(p)$ | $G_p(p, q)$ | Singular | Second | —— |
| $\Delta T_n u \neq 0$ | II | $u^+(p)$ | $G_q(p, q)$ | Singular | Second | DBEM |

[1]SLPM, *Single Layer Potential Method*; IBEM, *Indirect Boundary Element Method*; DLPM, *Double Layer Potential Method*; DBEM, *Direct Boundary Element Method*.

subsystems, calculations of which do not cause difficulties. This new system or a set of subsystems, obtained by a certain breakdown of a given system, is called a primary system. Although traditional interpretation of the force and displacement methods, accepted in the classical courses of structural mechanics, is somewhat different, it is possible, nevertheless, to find the sense of the force and displacement methods in finding such a force (by forces and the moments) or kinematic (by linear and angular displacements) impact, that converts the primary system or its part into a given system. It should be noted that in the case of the force method, the unknowns are the force factors and the canonical equations are the conditions of continuity of the displacements. In the case of the displacement method, the unknowns are kinematic factors, and canonical equations express the conditions of the equilibrium of corresponding nodes of the system. The mixed method of structural mechanics is based on the selection of different unknowns (force or kinematic) in different sections of the system. A similar pattern takes place when solving problems of the theory of elasticity using the given methods. This is clearly demonstrated in Table 2.1. So in the IBEM and the method of potential (with the use of potentials of a simple layer), the unknowns are stresses on the boundary domain ($\Delta T_n u$), and the condition of continuity of displacements is made possible by assuming $\Delta u = 0$. Hence, this method is similar to the force method. In this case the integral equation of the first kind is the analog of canonical equations and kernels $G(p, q)$ are analogous to the matrix of unit displacements. As far as the integral equations of the second kind are concerned, they are not common in the methods of structural mechanics, concerned with the breakdown of a system, and are much more common when extended systems are considered. Selection of unknown displacements corresponds to the method of potentials, when the potentials of double layer and the condition of continuity are used for stresses ($\Delta T_n u = 0$). In this case the kernel $G_{pq}$ is the analogues of the stiffness matrix or unit reactions, and we have an analogue of the method of displacements. In both cases the entire space is chosen as the primary system. The choice of various unknown impacts (influences) on various sections of the boundary corresponds to the mixed method of structural mechanics. The last two equations of Table 2.1 are standing to the side, one of which is the base of the direct BEM. This equation is built on the assumption that in the boundary sections the force and kinematic factors are accepted as unknowns.

## 2.4  The Spectral Method of Boundary Elements (SMBE) in Multidimensional Problems

The spectral method of boundary elements (SMBEs; [270]) can be built on the basis of the common scheme, which was used earlier in this chapter for the derivation of the formulas of the generalized method of boundary element. However, if for the derivation of that generalized method of boundary element we used a delta-transformation, then in SMBEs we will to use any integral transformation with the orthogonal kernel. As the kernel of transformation it is customary to choose eigenfunctions of a primary system, i.e. an extended region, taking into account the conditions of continuing solution through the boundary $\Gamma$. Let us examine the same boundary-value problem (2.2.32) in the three-dimensional space $R$, with assumptions of the differential operator $L$, accepted in Section 2.2. We will realize the same expansion of the domain to $\Omega$ and define the boundary conditions (2.2.34) on the border of the extended domain $\underline{\Gamma}^*$, i.e. search for a solution, which satisfies Eq. (2.2.32) in $\Omega$ and boundary conditions (2.2.34) on its border. At the same time we will choose any system of the eigenfunctions of operator $L$ in $\Omega$, which satisfying the equation $L\varphi = \lambda\varphi$ and conditions (2.2.34). This system can be discrete, if the spectrum of operator $L$ in $\Omega$ is discrete; and continuous with the continuous spectrum. In particular, if region $\Omega$ is limited, then the spectrum is discrete, and eigenfunctions form an infinite numerable set. The decomposition on eigenfunctions in this case has a form of a series. If the domain $\Omega$ is unlimited, then the spectrum of operator $L$ is continuous (in dynamic tasks with continuous mass), and decomposition on eigenfunctions has an integral mode (form, shape), characteristic for the classical integral transformation. The dual formulas, which determine eigenfunctions decompositions in a general case, can be written as follows:

$$F(\lambda) = \int_{\Omega} f(p)\varphi(\lambda, p)d\Omega_p;$$

$$f(p) = \int_{-\infty}^{\infty} F(\lambda)\varphi(\lambda, p)d\tau(\lambda), \tag{2.4.1}$$

where $\tau(\lambda)$ – the spectral function of distribution, which ensures the standardization of eigenfunctions. If $\Omega$ is unlimited domain and $\tau(\lambda)$ is continuous, then eigenfunctions generate integral transformation with the infinite limits (according to Ian Sneddon [343]), i.e. the integral decomposition. In the case of the bounded (constrained) domain $\Omega$ the spectral function of distribution $\tau(\lambda)$ is a step function with jumps in a numerable set of eigenvalues $\lambda_i$. In this case, the integral transformation with the finite limits (according to Ian Sneddon [343]) is equivalent to the formulas of decomposition in a series in terms of eigenfunctions

$$F(\lambda_i) = \int_\Omega f(p)\varphi(\lambda_i, p)d\Omega_p;$$

$$f(p) = \sum_i F(\lambda_i)\varphi(\lambda_i, p)\tau_i \tag{2.4.2}$$

where $\tau_i$ – normalizing multipliers, appearing as coefficients with the delta-functions, which arise out of the differentiation of the step function $\tau(\lambda)$. Generalizing the problem and omitting two basic cases, when domain $\Omega$ is not limited or final, we will use a single integral transformation, determined by the formulas in (2.4.2), bearing in mind that this integral transformation covers both cases and the bounded domain, taking $\tau(\lambda)$ as the function of the jumps. Let us carry out an integral transformation with the kernel $\varphi(\lambda, p)$ in an extended domain; we will multiply the left and right sides of Eq. (2.2.32) by $\varphi(\lambda, p)$ and integrate over the entire domain $\Omega$

$$\int_\Omega \varphi(\lambda, p) \cdot Lu(p)d\Omega_p = 0 \tag{2.4.3}$$

Let us use the further generalized Green's formula (2.2.33), we obtain:

$$\lambda \int_\Omega u(p)\varphi(\lambda, p)d\Omega_p = \int_{\Gamma+} - \int_{\Gamma-} + \int_{\Gamma*} [Mu(p)N\varphi(\lambda, p)$$

$$- M\varphi(\lambda, p)Nu(p)]d\Gamma_p. \tag{2.4.4}$$

Since $u$ and $\varphi$ satisfy the same boundary conditions *on $\Gamma^*$*, then the integral on $\Gamma^*$ disappears. As a result we arrive at the formula:

$$\lambda U(\lambda) = \int_\Gamma [\Delta Mu(p) \cdot N\varphi(\lambda, p) - \Delta Nu(p) \cdot M\varphi(\lambda, p)]d\Gamma_p, \tag{2.4.5}$$

where $U(\lambda)$ – is transformant

$$U(\lambda) = \int_\Omega u(p) \cdot \varphi(\lambda, p)d\Omega_p.$$

To determine (2.4.5) it is necessary, as in the method of delta-transformation, to choose the "primary system," in this case, the system of boundary actions. Let us assume, for example, as in Section 2.2, that $M = l$, $N = \bar{l}$, and $\Delta Mu(p) = \Delta lu(p) = 0$ with $p \in \Gamma$. Then we will have

$$U(\lambda) = -\lambda^{-1} \int_\Gamma \Delta \bar{l}u(p) \cdot l\varphi(\lambda, p) d\Gamma_p, \tag{2.4.6}$$

where, according to the inversion formula (2.4.1), we obtain:

$$u(p) = -\int_\Gamma \Delta \bar{l}u(q) \int_{-\infty}^\infty \lambda^{-1} l\varphi(\lambda, q) \cdot \varphi(\lambda, p) d\tau(\lambda) d\Gamma_q. \tag{2.4.7}$$

Here it is assumed that the operations of integration on $q$ and $\lambda$ are permutable. This circumstance is ensured in the necessary cases by the interpretation of the obtained integrals in the meaning of the generalized functions. Using the general (common) scheme

$$\frac{lu^+(p) + lu^-(p)}{2} = -\int_\Gamma \Delta \bar{l}u(q) \int_{-\infty}^\infty \lambda^{-1} l\varphi(\lambda, q) \cdot l\varphi(\lambda, p) d\tau(\lambda) d\Gamma_p,$$

where using boundary condition (2.2.32), we obtain the integral equation of the first kind, taking into account that $\Delta lu(p) = 0$ $(p \in \Gamma)$

$$\int_\Gamma \Delta \bar{l}u(q) \int_{-\infty}^\infty \lambda^{-1} l\varphi(\lambda, q) \cdot l\varphi(\lambda, p) d\tau(\lambda) d\Gamma_p = -f(p) \tag{2.4.8}$$

But if boundary conditions (2.2.40), are assigned on the border of the domain $D^+$, then by using similar operations, i.e. by extracting expressions for $\bar{l}u^+(p)$ and for $\bar{l}u^-(p)$ and summarizing them, we will obtain the integral equation of the second kind:

$$\frac{1}{2}\Delta \bar{l}u(p) - \int_\Gamma \Delta \bar{l}u(q) \int_{-\infty}^\infty \lambda^{-1} l\varphi(\lambda, q) \cdot \bar{l}\varphi(\lambda, p) d\tau(\lambda) d\Gamma_q = f(p). \tag{2.4.9}$$

Integral equations (2.2.8) and (2.2.9) are completely equivalent to the equations in (2.2.39) and (2.2.43), since they have identical kernels. Defining the function of Green as the solution of the equation

$$LG(q, p) = \delta(q, p) \tag{2.4.10}$$

and applying transformation with kernel $\varphi$ to (2.4.10), we obtain:

$$\int_\Omega \varphi(\lambda, p) LG(q, p) d\Omega_p = \varphi(\lambda, q).$$

Using Green's function (2.2.33), we will find

$$\int_\Omega G(q,p)\varphi(\lambda,p)d\Omega_p = \lambda^{-1}\varphi(\lambda,q). \qquad (2.4.11)$$

Let us note that the integrals on the boundaries $\Gamma$ and $\Gamma^*$ in (2.4.11) disappear, since differential forms from $G$ and $\varphi$ are continuous on $\Gamma$, and both functions satisfy the same boundary conditions on $\Gamma^*$. Inverting the formula (2.4.11) with the aid of (2.4.1), we will find:

$$G(q,p) = \int_{-\infty}^{\infty} \lambda^{-1}\varphi(\lambda,p) \cdot \varphi(\lambda,q)d\tau(\lambda). \qquad (2.4.12)$$

Consequently,

$$\int_{-\infty}^{\infty} \lambda^{-1}l\varphi(\lambda,p) \cdot l\varphi(\lambda,q)d\tau(\lambda) = l_{pq}G(q,p),$$

$$\int_{-\infty}^{\infty} \lambda^{-1}l\varphi(\lambda,p) \cdot l\varphi(\lambda,q)d\tau(\lambda) = \bar{l}_p l_q G(q,p),$$

providing the equivalence of Eqs. (2.4.8, 2.4.9) and (2.2.39, 2.2.43). Thus, in those cases when Green's function of the given problem is unknown, it is more convenient to apply the (spectral method of boundary elements) SMBE. Furthermore, the SMBE make it possible to obtain other integral equations, compared to the usual BEM. Thus, designating in (2.4.8)

$$\int_\Gamma \Delta\bar{l}u(q)l\varphi(\lambda,q)d\Gamma_q = R(\lambda),$$

we come to the integral equation

$$\int_{-\infty}^{\infty} R(\lambda)K(\lambda,p)d\tau(\lambda) = -f(p), \qquad (p \in \Gamma^+), \qquad (2.4.13)$$

with the kernel

$$K(\lambda,p) = \lambda^{-1}l\varphi(\lambda,p),$$

and unknown function $R(\lambda)$. After the determination of $R(\lambda)$ we will find the solution to the problem by the formula

$$u(p) = -\int_{-\infty}^{\infty} R(\lambda)\lambda^{-1}\varphi(\lambda,p)d\tau(\lambda),$$

Similar equations can be obtained, assuming $\Delta Nu(p) = \Delta \bar{l}u(p) = 0$, $(p \in \Gamma)$. From (2.4.5) we have:

$$U(\lambda) = \lambda^{-1} \int_\Gamma \Delta lu(p) \cdot \bar{l}\varphi(\lambda, p) d\Gamma_p,$$

and, according to the inversion formula (2.4.1)

$$u(p) = \int_\Gamma \Delta lu(q) \int_{-\infty}^{\infty} \lambda^{-1} \bar{l}\varphi(\lambda, q) \cdot \varphi(\lambda, p) d\tau(\lambda) d\Gamma_q. \qquad (2.4.14)$$

Defining operations $lu^+(p)$ and $lu^-(p)$ and taking into account the boundary condition in (2.2.32), we will obtain the integral equation of the second kind, as above

$$\Delta lu(p) + \int_\Gamma \Delta lu(q) \int_{-\infty}^{\infty} \lambda^{-1} \bar{l}\varphi(\lambda, q) \cdot l\varphi(\lambda, p) d\tau(\lambda) d\Gamma_q = f(p). \qquad (2.4.15)$$

If a boundary condition takes the form of (2.2.40), then by determining operations $\bar{l}u^+(p)$ and $\bar{l}u^-(p)$, we will obtain as a result of the simple transformations, an integral equation of first kind

$$\int_\Gamma \Delta lu(q) \int_{-\infty}^{\infty} \lambda^{-1} \bar{l}\varphi(\lambda, q) \cdot \bar{l}\varphi(\lambda, p) d\tau(\lambda) d\Gamma_q = f(p). \qquad (2.4.16)$$

Integral equation (2.4.16) can be also reduced to the form:

$$\int_{-\infty}^{\infty} \overline{R}(\lambda) \overline{K}(\lambda, p) d\tau(\lambda) = f(p), \qquad (p \in \Gamma^+), \qquad (2.4.17)$$

where

$$\overline{R}(\lambda) = \int_\Gamma \Delta lu(q) \cdot \bar{l}\varphi(\lambda, q) d\Gamma_q,$$

$$\overline{K}(\lambda, p) = \lambda^{-1} \bar{l}\varphi(\lambda, p).$$

Finally, assuming, as we did in a case of delta-transformation, that $\Delta Mu(p) = \Delta Nu(p) \neq 0$, $(p \in \Gamma)$, i.e. prolongation of the vector-function $u(p)$ with a zero in $D^-$, we can obtain one additional type of integral equations, analogues to the BIEM in the spectral BEM. It is obvious that in this case

$$\Delta Mu(p) = Mu^+(p), \quad \Delta Nu(p) = Nu^+(p).$$

Inverting the formula (2.4.5) and changing the order of integration, we obtain spectral representation of $u(p)$ through the boundary forms, similar to the standard BIEM:

$$u(p) = \int_\Gamma [Mu^+(q) \int_{-\infty}^{\infty} \lambda^{-1} N\varphi(\lambda, q) \cdot \varphi(\lambda, p) d\tau(\lambda)$$

$$- Nu^+(q) \int_{-\infty}^{\infty} \lambda^{-1} M\varphi(\lambda, q) \cdot \varphi(\lambda, p) d\tau(\lambda)] d\Gamma_q, \quad (p \in D^+) \quad (2.4.18)$$

If the boundary conditions $Mu^+(p) = lu^+(p) = f(p)$ are assigned on the border, then from (2.4.18), after taking operations $l$ on the left and right sides of the equation and letting $p$ approach the border $\Gamma^+$ from inside of the domain $D^+$, we obtain an integral equation of the first kind:

$$\int_\Gamma \bar{l}u^+(q) \int_{-\infty}^{\infty} \lambda^{-1} l\varphi(\lambda, q) \cdot l\varphi(\lambda, p) d\tau(\lambda) d\Gamma_q = F(p), \quad (2.4.19)$$

where

$$F(p) = \int_\Gamma f(q) \int_{-\infty}^{\infty} \lambda^{-1} \bar{l}\varphi(\lambda, q) \varphi(\lambda, p) d\tau(\lambda) d\Gamma_q - f(p).$$

An integral equation of the second kind can be obtained if we perform the operation of $N = \bar{l}$, and go to the limit values of $p$ on the border $\Gamma^+$ at the left and right sides of Eq. (2.4.18)

$$\int_\Gamma \bar{l}u^+(q) \int_{-\infty}^{\infty} \lambda^{-1} l\varphi(\lambda, q) \bar{l}\varphi(\lambda, p) d\tau(\lambda) d\Gamma_q + \bar{l}u^+(p) = F_1(p), \quad (2.4.20)$$

where

$$F_1(p) = \int_\Gamma f(p) \int_{-\infty}^{\infty} \lambda^{-1} l\varphi(\lambda, q) \bar{l}\varphi(\lambda, p) d\tau(\lambda) d\Gamma_q.$$

Similar equations can be obtained when boundary forms $Nu^+ = f(p)$, are assigned on the border. It is necessary to note that since formula (2.4.18) is obtained on the assumption that $u(p) = 0$, $p \in D^-$, the equation below follows directly from this proposal

$$\int_\Gamma Mu^+(q) \int_{-\infty}^{\infty} N\varphi(\lambda, q) \varphi(\lambda, p) d\tau(\lambda) d\Gamma_q$$

$$= \int_\Gamma Nu^+(q) \int_{-\infty}^{\infty} N\varphi(\lambda, q) \varphi(\lambda, p) d\tau(\lambda) d\Gamma_q, \quad (p \in D^-)$$

One of the forms $Mu^+$, $Nu^+$ is assumed to be known. The other form is written as an equation, which is a spectral analog of V.D. Kupradze's functional Eqs. [199]. One additional modification of the spectral BEM (SMBM) can be built under the assumption that

$$\Delta M u(p) = \Delta N u(p) = 0, \qquad (p \in \Gamma^*)$$

and the arrangement of the external perturbations, determining the jumps of differential forms outside of the domain $D^+$ and its boundary $\Gamma$, i.e. within the limits of domain $D^-$. In such case, as was indicated above, the integral equations will be regular; however, obtaining integral equations of the second kind here is impossible. Assume that external actions (compensating loads) are applied on the contour S. Substituting boundary $\Gamma$ with the contour S in the formulas given above we will obtain

$$u(p) = \int_S [\Delta M u(q) \int_{-\infty}^{\infty} N\varphi(\lambda, q) \cdot \varphi(\lambda, p) d\tau(\lambda)$$

$$- \Delta N u(q) \int_{-\infty}^{\infty} M\varphi(\lambda, q) \cdot \varphi(\lambda, p) d\tau(\lambda)] dS_q, \qquad (2.4.21)$$

Further, we can choose $n$ of various jumps on S and satisfy the boundary conditions of the problem. As a result, we obtain an integral equation of the first kind, typical for the method of compensating loads when these loads are applied outside of the contour. Thus, we obtain two types of integral equations of the SMBE. The first type of equation is characterized by the fact that these equations are the spectral analogs of BIEM, due to the fact that their kernels are the spectral decomposition of the corresponding Green's functions. These equations can be used in those cases when the explicit construction of Green's function runs into computational difficulties. The other type of the integral equation does not have analogs among the BIEs. The significant circumstance, in this case, is that the integration in the equation occurs not over the a boundary surface (of curved) $\Gamma$, but along the spectral axis $\lambda$. A characteristic example of the use of the SMBE is the application of Fourier transformation in Cartesian or Fourier-Hankel transformation in cylindrical coordinates.

## 2.5    The Problems Described by the Integro-Differential System of Equations

Along with the systems of differential equations in ordinary and partial derivatives, many problems of continuum mechanics are developed in the form of the integro-differential equations or systems. Such problems, in particular, include a very broad class of contact problems, which has wide applications in the theory of structure on the elastic foundation, soil mechanics, theory of design forms, dynamics and the strength of materials and machines, and so forth. Let us examine, as an example, a contact problem for a plane structure, whose stress-strained state in the domain $D^+$ is described by the linear differential operator $L$ in partial derivatives of order $2n$

$$Lu(x,y) = q(x,y) - p(x,y), \quad (x,y \in D^+), \tag{2.5.1}$$

by boundary conditions

$$l_j u(x,y) = 0, \quad (x,y \in \Gamma^+; \ j = \overline{1,n}) \tag{2.5.2}$$

and by conditions of contact with the linear-deformed medium.

$$\iint_D G(x,y;x_1,y_1)p(x_1,y_1)dx_1dy_1 = u(x,y), \quad (x,y \in D). \tag{2.5.3}$$

Here the operators $L$ and $l_j$ have the same properties as in the previous paragraphs; $u(x, y)$ – functions of displacement; $q(x, y)$, $p(x, y)$ – functions of external actions and contact stresses; $G(x, y; x_1, y_1)$ – Green's Functions of the linear-deformed medium; and $\Gamma^+$ – the internal boundary of domain $D$, occupied by the structure. Let the linear-deformed medium occupy region $R$, i.e. Green's function is built for $x$, $y$, $x_1$, $y_1 \in R$, and for the certain conditions of fixing on the boundaries $R$ (Figure 2.1). SMBE can be applied for solving the system of Eqs. (2.5.1–2.5.3), if in medium $R$, there is an integral

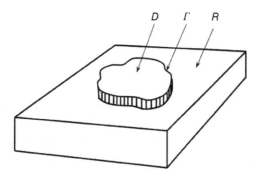

**Figure 2.1** The linearly deformed medium $R$, the area of contact $D$, and the boundary of the area occupied by the structure.

transformation, that

$$\iint_R \varphi(x,y;\xi,\eta)G(x,y;x_1,y_1)dxdy = \varphi(x_1,y_1;\xi,\eta)\overline{G}(\xi,\eta), \qquad (2.5.4)$$

and

$$\iint_R \varphi(x,y;\xi,\eta)Lu(x,y)dxdy = P(\xi,\eta)\overline{u}(\xi,\eta).$$

Here $\overline{G}(\xi,\eta), \overline{u}(\xi,\eta)$ – transforms of Green's function and displacements with the kernel $\varphi$, for example

$$\overline{G}(\xi,\eta) = \iint_R \varphi(x,y;\xi,\eta)G(x,y)dxdy$$

where $P(\xi,\eta)$ is some polynom.

It is obvious that for obtaining the transform $\overline{G}$, it is necessary to bring the Green's function to the dependence on only two parameters, $x$ and $y$, by the replacement of variables. Such transformations exist, in particular, for problems in the Cartesian and polar coordinate among others, when $R$ is the entire plane. For example, Green's function for a half-space when placing the points $x$, $y$, $x_1, y_1$ on its boundary (border) is a function of a difference between the coordinates $x - x_1$, $y - y_1$. Fourier transform satisfies the conditions (2.5.4)

$$\frac{1}{2\pi} \iint_{-\infty}^{\infty} e^{i(\xi x + \eta y)}G(x - x_1; y - y_1)dxdy = e^{i(\xi x_1 + \eta y_1)}\overline{G}(\xi,\eta),$$

where

$$\overline{G}(\xi,\eta) = \frac{1}{2\pi} \iint_{-\infty}^{\infty} e^{i(\xi u + \eta v)}G(u,v)dudv$$

Similarly, in the case of polar coordinates Hankel transform can be used. Apparently this kind of transformation can be determined also, for the cases when $R$ is the finite domain with some boundary conditions for the linear-deformed medium on the boundaries of this region (for example, rectangular domain with Navier boundary conditions, i.e. with only even or odd derivatives, given on the boundary). Let us choose $R$ as that extended region and continue into it $u(x, y)$ and $v(x, y)$ from $D$ in such a way that Eq. (2.5.1) would be satisfied in the extended domain:

$$Lu(x,y) = q(x,y). \qquad (2.5.5)$$

Similarly, continuing $u(x, y)$ in (2.5.3), and taking into account, that $p(x_1, y_1) = 0$ out of the domain of contact, we obtain:

$$\iint_D G(x, y; x_1, y_1) p(x_1, y_1) dx_1 dy_1 = u(x, y), \qquad (x, y \in R) \qquad (2.5.6)$$

With such continuation, the functions $u(x, y)$ and $q(x, y)$ $(x, y \in R \backslash D)$ are unknown, and the condition of the absence of stresses on the free surface of medium can be used for their determination. Let us assign some perturbations on the boundary $\Gamma$, such that differential forms from $u(x, y)$ on the boundary $\Gamma$ can have gaps. Performing the transformation with kernel $\varphi(x, y; \xi, \eta)$ in Eqs. (2.5.1) and (2.5.6), we obtain:

$$P(\xi, \eta) \bar{u}(\xi, \eta) = \bar{q}(\xi, \eta) - \bar{p}(\xi, \eta) + T(\xi, \eta),$$

$$\bar{G}(\xi, \eta) \bar{p}(\xi, \eta) = \bar{u}(\xi, \eta) \qquad (2.5.7)$$

where

$$T(\xi, \eta) = \int_\Gamma \sum_{j=1}^n [\Delta M_j u(x, y) N_j \varphi(x, y; \xi, \eta) - \Delta N_j u(x, y) M_j \varphi(x, y; \xi, \eta)] d\Gamma.$$

The scalar representation of the formula in (2.5.5) is used and designated here

$$\bar{p}(\xi, \eta) = \iint_R p(x, y) \varphi(x, y; \xi, \eta) dx dy = \iint_D p(x, y) \varphi(x, y; \xi, \eta) dx dy$$

Excluding from the system (2.5.7) the function $\bar{p}(\xi, \eta)$, we find:

$$\bar{u}(\xi, \eta) = \frac{\bar{q}(\xi, \eta) + T(\xi, \eta)}{P(\xi, \eta) + G^{-1}(\xi, \eta)} \qquad (2.5.8)$$

It is necessary to consider that the transform $\bar{q}(\xi, \eta)$ is an unknown function. So:

$$\bar{q}(\xi, \eta) = \iint_R q(x, y) \varphi(x, y; \xi, \eta) dx dy$$

$$= \iint_{D+} q(x, y) \varphi(x, y; \xi, \eta) dx dy + \iint_{D-} q(x, y) \varphi(x, y; \xi, \eta) dx dy$$

$$= \bar{q}_1(\xi, \eta) + \bar{q}_2(\xi, \eta), \qquad (2.5.9)$$

where $\bar{q}_1(\xi,\eta)$ is the known function, and $\bar{q}_2(\xi,\eta)$ is the unknown function. From (2.5.7) we can also obtain

$$\bar{p}(\xi,\eta) = \bar{u}(\xi,\eta)\bar{G}^{-1}(\xi,\eta) \tag{2.5.10}$$

With the aid of the inversion formula for $\varphi$ – transformation, let us invert formulas (2.5.8, 2.5.9)

$$u(x,y) = \iint_{-\infty}^{\infty} \frac{[\bar{q}_1(\xi,\eta) + \bar{q}_2(\xi,\eta) + T(\xi,\eta)]\varphi(x,y;\xi,\eta)}{P(\xi,\eta) + \bar{G}^{-1}(\xi,\eta)}$$

$$d\tau(\xi)d\sigma(\eta),$$

$$p(x,y) = \iint_{-\infty}^{\infty} \bar{G}^{-1}(\xi,\eta) \frac{[\bar{q}_1(\xi,\eta) + \bar{q}_2(\xi,\eta) + T(\xi,\eta)]\varphi(x,y;\xi,\eta)}{P(\xi,\eta) + \bar{G}^{-1}(\xi,\eta)}$$

$$d\tau(\xi)d\sigma(\eta), \tag{2.5.11}$$

Where $\tau(\xi)$, $\sigma(\eta)$ are the spectral functions of distribution of $\varphi$ – transformation. Selecting as we did above, $n$ – jumps of differential forms $\Delta Mu$, $\Delta Nu$ equal to zero, using the boundary conditions (2.5.2) and the condition of $(x,y) = 0$ $(x,y \in D^-)$, we obtain a system of resolving integral equations for determining $\bar{q}_2(\xi,\eta)$ and $T(\xi,\eta)$ from (2.5.11):

$$\iint_{-\infty}^{\infty} \frac{T(\xi,\eta)l\varphi(x,y;\xi,\eta)}{P(\xi,\eta) + \bar{G}^{-1}(\xi,\eta)} d\tau(\xi)d\sigma(\eta) = Q_1(x,y) - Q_2(x,y), \tag{2.5.12}$$

$$(x,y \in \Gamma),$$

$$\iint_{-\infty}^{\infty} \frac{T(\xi,\eta)\varphi(x,y;\xi,\eta)}{P(\xi,\eta)\bar{G}^{-1}(\xi,\eta) + 1} d\tau(\xi)d\sigma(\eta) = Q_1^*(x,y) - Q_2^*(x,y), \tag{2.5.13}$$

$$(x,y \in D^-),$$

where

$$Q_i(x,y) = -\iint_{-\infty}^{\infty} \frac{\bar{q}_i(\xi,\eta)l\varphi(x,y;\xi,\eta)}{P(\xi,\eta) + \bar{G}^{-1}(\xi,\eta)} d\tau(\xi)d\sigma(\eta), \quad (i = 1,2),$$

$$(x, y \in \Gamma)$$

$$Q_i^*(x,y) = -\iint_{-\infty}^{\infty} \frac{\bar{q}_i(\xi,\eta)\varphi(x,y;\xi,\eta)}{P(\xi,\eta)\bar{G}^{-1}(\xi,\eta) - 1} d\tau(\xi) d\sigma(\eta), \quad (i = 1, 2),$$

$$(x, y \in \Gamma).$$

The system of the equations in (2.5.12) and (2.5.13) has the same feature, so that the first integral equation can be solved by the BEM, while the second can be solved by the finite element methods (FEMs). The fact is that the first equation is an integral equation on the boundary of $\Gamma$. This follows from the transposition of the order of integration in (2.5.12) after recording $T(\xi, \eta)$ explicitly. Let us write down this equation, for simplicity reasons setting jumps $\Delta Nu$ equal to zero

$$\int_{\Gamma} \sum_{j=1}^{n} \iint_{-\infty}^{\infty} \frac{\Delta M_j u(x_*,y_*) N_j \varphi(x_*,y_*;\xi,\eta) l\varphi(x_*,y_*;\xi,\eta)}{P(\xi,\eta) + \bar{G}^{-1}(\xi,\eta)} d\tau(\xi) d\sigma(\eta)$$

$$= Q_1(x_*,y_*) - Q_2(x_*,y_*) \tag{2.5.14}$$

Here the asterisks in $x^*, y^*$ designate that the points with the coordinates $x^*, y^*$ are selected on $\Gamma$. The second equation must be satisfied in the entire domain $D^-$ and it requires the partition of this region and not just the boundary. As we see, for the contact problems and problems of design structure on elastic foundation described by coherent models, it is necessary to use the combined methods, which certainly have finite element and boundary element basis. However, there is a large class of contact problems for which Eq. (2.5.13) is satisfied identically and, therefore, BEM in its pure form can be used for their solution. This class of problems is characterized by the absence of noncontact zones and by placement of the structure, connected in one way or another throughout the entire region $R$. Such constructions are called not isolated [188, 189]. They are sufficiently widespread in engineering, and therefore their solutions present a great interest. Thus, the SMBE can be used for solving the integro-differential equations, which describe the stress-strained state of complex systems, and in particular, structures and media interacting with each other in some domain. The methods of potentials and BEM are not applicable in the overall scheme of solutions to such problems, and can be used only for calculating a separate structure or medium, and then by putting together separate solutions per contact zones, a complete solution of a problem can be obtained. In contrast to these methods, the SMBE gives a direct solution in its pure form for non-isolated structures and isolated separate structures, with additional integral equations for non-contact domain. Detailed examples of the application of SMBE to the design structures on elastic foundation are given in the subsequent chapters of this book. In particular, the application of SMBE to the solution of multidimensional problems is illustrated based on the example of the solution of a nonhomogeneous equation of Helmholtz in Chapter 7, Section 7. The problems of dynamic calculation of membranes, two-dimensional wave problems, problems of calculating the elastic foundation, described by models with two elastic characteristics, etc., are given in these types of equations.

# Chapter 3
# Oscillation of Bars and Arches

## 3.1 The Nonlinear Oscillations of Systems with One Degree of Freedom

Strictly speaking, the oscillations of mechanical systems are always nonlinear because they are described by nonlinear differential equations. In general, linear models are applicable only in a very restrictive domain, i.e. when the vibration amplitude is very small. Thus, to accurately identify and understand the dynamic behavior of a structural system under general loading conditions, it is essential for the nonlinearities present in the system also be modeled and studied [71]. In structural mechanics, nonlinearities can be broadly classified into the following two general categories: geometric nonlinearity, which exists in systems undergoing large deformations or deflections; and physical nonlinearity, which includes but is not limited to the kinetic energy of the system, which is the source of inertia nonlinearities. For example, damping is essentially a nonlinear phenomenon. Another nonlinear phenomenon is described by material nonlinearity (i.e. stresses and strains are nonlinear, restoring force and displacement are nonlinear etc). Nonlinearities can also appear in some boundary conditions such as spring supports.

However, in order to avoid difficulties, which appear during the solution of nonlinear differential equations, the linearization of these equations is produced, i.e. the initial nonlinear differential equation is substituted by a linear equation. Two types of linearization are distinguished; the first is

*Static and Dynamic Analysis of Engineering Structures: Incorporating the Boundary Element Method*,
First Edition. Levon G. Petrosian and Vladimir A. Ambartsumian.

a geometric linearization and it relates to the geometry of the body being deformed. It is assumed that dependences of deformations on the displacements are linear. The second type of linearization is physical linearization, since it concerns the physical properties of the body being deformed. It is assumed that dependences between the strains and stresses are linear.

We will examine oscillation of "physically nonlinear" mechanical systems, for which the displacements are small and dependences between the stresses and strains are described by the nonlinear functions. Nonlinear mechanical systems in comparison with the linear systems are characterized by new physical properties. If in the linear system natural frequency is well defined and independent of the initial conditions value, then in the nonlinear system natural frequency depends on an initial displacement, and varies in the dependence on a change of stiffness of the system.

A system is linear if for every element in the system, the response is proportional to the excitation. The dynamic properties of each element in the system can be represented by a set of linear differential equations with constant coefficients. For a linear system, using values of natural frequencies and modes of oscillations, we can determine the reactions of the system with the arbitrary forced oscillations In the nonlinear system that is impossible to do, since the principle of superposition of the separate (individual) modes of oscillations is not applicable. Nevertheless, the determination of frequencies and forms of the oscillations of nonlinear systems represents a significant interest because they characterize systems with single-frequency oscillations, and a similar motion can take place with forced oscillations. On the other hand the frequencies and modes of free nonlinear oscillations can be used for the "linearization" of nonlinear systems and thus for developing engineering methods of their calculation.

Let us examine free nonlinear oscillation of a system with one degree of freedom, for which the dependence of restoring force of the displacement takes the form of bilinear diagram. The equation of motion for this system takes the form

$$my'' + R(y) = 0, \tag{3.1.1}$$

$$R(y) = \begin{cases} ay & \text{with } y \le y_1 \\ a[y_1 + \gamma(y - y_1)] & \text{with } y \ge y_1, \end{cases}$$

where $m$ – value of the concentrated mass; $y, y''$ – respective displacement; the acceleration of mass, $R(y)$ – restoring force; $a$ – initial stiffness of system; $y_1$ – given value of displacement at which a change of the rigidity of the system takes place; and $\gamma$ – the ration of the changed (reduced) stiffness (rigidity) of the system to the initial condition. The solution of Eq. (3.1.1) with the zero initial conditions for both sections will take the following general form:

$$y(t) = \frac{c}{\omega^2\gamma} + \left(y(0) - \frac{c}{\omega^2\gamma}\right)\cos\omega\sqrt{\gamma}t + \frac{y'(0)}{\omega\sqrt{\gamma}}\sin\omega\sqrt{\gamma}t, \qquad (3.1.2)$$

$$c = \begin{cases} 0, & y \leq y_1, \\ -\omega^2(1-\gamma)y_1, & y \geq y_1, \end{cases}$$

Where $\omega = \sqrt{\frac{a}{m}}$ is the natural frequency of linear system; and $y(0)$, $y'(0)$ – the initial values of displacement and velocity. If we deflect the mass $m$ up to the distance $y_2$ ($y_2 > y_1$), and then allow it to oscillate without the initial velocity, then for displacing the system from (3.1.2) we will obtain

$$y(t) = -\frac{(1-\gamma)}{\gamma}y_1 + \left(y_2 + y_1\frac{(1-\gamma)}{\gamma}\right)\cos\omega\sqrt{\gamma}t. \qquad (3.1.3)$$

The time $t_{12}$, necessary to the return of mass $m$ from $y = y_2$ to $y = y_1$, is obtained from (3.1.3), if we substitute $y(t)$ with $y_1$:

$$t_{12} = \frac{1}{\omega\sqrt{\gamma}}\arccos\frac{1}{(\beta-1)\gamma+1}; \beta = \frac{y_2}{y_1}. \qquad (3.1.4)$$

The time $t_{01}$, in the course of which mass $m$ passes the distance from $y = y_1$ to $y = 0$, is obtained from the corresponding expression (3.1.2) with the substitution of $y(t) = 0$, $y(0) = y_1$, and $y'(0) = y_1'$:

$$t_{01} = \frac{1}{\omega}\arctan\frac{\sqrt{\gamma}}{[(\beta-1)\gamma+1]\sin\arccos\frac{1}{(\beta-1)\gamma+1}}. \qquad (3.1.5)$$

Since the dependence between the force and deflection is symmetrical relative to the center of coordinates, the value of the period of free nonlinear oscillations $T_{Nonlin}$ can be determined according to the formula

$$T_{Nonlinear} = 4(t_{01} + t_{02}) = \frac{2\pi}{\omega}\left\{\frac{2}{\pi\sqrt{\gamma}}\arccos\frac{1}{1+(\beta-1)\gamma}\right.$$

$$\left. + \frac{2}{\pi}\arctan\frac{\sqrt{\gamma}}{[1+(\beta-1)\gamma]\sin\arccos\frac{1}{1+(\beta-1)\gamma}}\right\}. \qquad (3.1.6)$$

In a similar manner, the period of the free nonlinear oscillation of the system can be determined, for which the loading occurs according to the bilinear law, and the unloading to the linear law (Figure 3.1).

In this case the formula is obtained:

$$T_{Nonlinear} = \frac{2\pi}{\omega} \left\{ \frac{1}{\pi} \arctan \frac{\sqrt{\gamma}}{[1 + (\beta - 1)\gamma] \sin \arccos \frac{1}{1+(\beta-1)\gamma}} \right. \tag{3.1.7}$$

$$\left. + \frac{1}{\pi\sqrt{\gamma}} \arccos \frac{1}{1 + (\beta - 1)\gamma} + \frac{1}{\pi} \right\}.$$

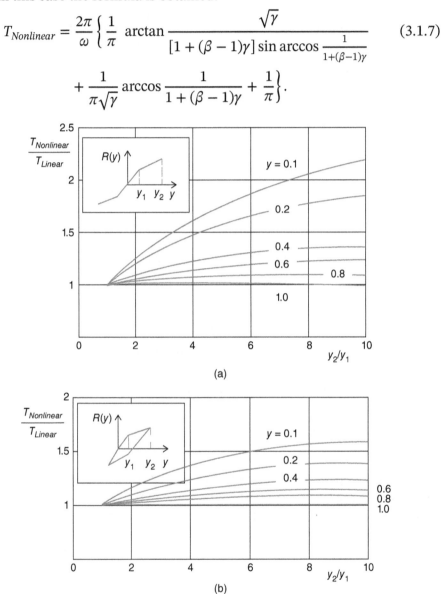

**Figure 3.1** Dependences of the relations of periods of non-linear and linear oscillation on relative displacements: (a) with the bilinear-elastic law of deformation; and (b) with the bilinear-hysteresis law of the deformation.

From the formulas in (3.1.6) and (3.1.7) with the substitution of $\beta = 1$ and $\gamma = 1$, the value of the period of linear oscillation is

$$T_{Linear} = \frac{2\pi}{\omega}$$

The expressions (3.1.6) and (3.1.7) show that the period of non-linear oscillations is a continuous function of the displacements, and for a non-linear system there is no longer a concept of natural frequency, since such frequencies already have an infinite set. The values of the relations of the periods of non-linear and linear oscillations $T_{Nonlinear}/T_{Linear}$, depending on relative displacement of $\beta = y_2/y_1$, are calculated with the aid of these expressions. The calculations were conducted for the values $\gamma = 1.0, 0.8, 0.6, 0.4, 0.2, 0.1$, and $\beta = 1, 2, 4, 6, 8, 10$. It follows from the graphs (Figure 3.1), that an increase in the initial displacement leads to an increase of the value of the periods of non-linear oscillations. At the same parameters $\gamma$ and $\beta = \frac{y_2}{y_1}$, this increase is greater with the elastic-nonlinear than with the hysteresis oscillations. The method, with the aid of which the value of the period of oscillations of bilinear system was obtained, is applicable for systems whose stiffness (rigidity) changes according to the piecewise-linear (piece-linear) law.

Let us examine another method of determining the period of non-linear oscillations with the aid of which precise values for some curvilinear diagrams of deformation [357, 358] were obtained. We will consider the algorithm of solving the problem for an arbitrary dependence $R(y)$ [358]. The equation of motion of the system presets in the form of

$$y'' + \omega^2 f(y) = 0, \tag{3.1.8}$$

where $\omega^2 = \frac{a}{m}$ represents the frequency of free oscillations of linear system that is at $f(y) = y$. The differential Eq. (3.1.8) can be represented in the form of:

$$\frac{1}{2} \frac{d(y')^2}{dy} + \omega^2 f(y) = 0 \tag{3.1.9}$$

Integrating Eq. (3.1.9) with the initial conditions $y_{(t=0)} = y_0$ and $y'_{(t=0)} = 0$, we obtain

$$\frac{1}{2} y'^2 = \omega^2 \int_y^{y_0} f(y)dy = 0 \tag{3.1.10}$$

Since $y' = \frac{dy}{dt}$ from (3.1.10) it follows:

$$dt = -\frac{dy}{\sqrt{2\omega^2 \int_y^{y_0} f(y)dy}} \qquad (3.1.11)$$

By repeated integration of (3.1.1), we obtain the time $t$ during which the system reaches the position $y = 0$ from the position $y_0$. The period of the system with the symmetrical diagram of restoring force will be equal to $T = 4t$. For the period $T_{Nonlinear}$ the following expression is obtained

$$T_{Nonlinear} = 4\int_0^{y_0} \frac{dy}{\sqrt{2\omega^2 \int_y^{y_0} f(y)dy}} \qquad (3.1.12)$$

With the explicit analytical expression for $T_{Nonlinear}$ is possible to obtain when $\omega^2 f(y)$ takes the form

$$\omega^2 f(y) = \omega^2(y - \varepsilon y^3) \qquad (3.1.13)$$

In this case, the period of oscillations of system $T_{Nonlinear}$ is expressed through elliptical integral of the first kind, $F\left(\sqrt{\frac{\theta}{2-\theta}}, \frac{\pi}{2}\right)$:

$$T_{Nonlinear} = \frac{4}{\omega}\sqrt{\frac{2}{2-\theta}} \cdot F\left(\sqrt{\frac{\theta}{2-\theta}}, \frac{\pi}{2}\right), \theta = \varepsilon y_0^2. \qquad (3.1.14)$$

Using the formula in (3.1.14) and applying the tables of elliptical integrals we determine the values of the period of nonlinear oscillations $T_{Nonlinear}$. The ratios $T_{Nonlinear}/T_{Linear}$ for values of $\theta = \varepsilon y_0^2$ are given in Table 3.1

In Figure 3.2 the dependences of the relation of the periods of nonlinear and linear oscillations, $T_{Nonlinear}/T_{Linear}$, on the displacement $y$ with different

Table 3.1    The Ratios $T_{Nonlinear}/T_{linear}$ for Values of $\theta = \varepsilon y_0^2$.

| $\theta$ | $T_{Nonlinear}/T_{Linear}$ | $\theta$ | $T_{Nonlinear}/T_{Linear}$ |
|---|---|---|---|
| 0 | 1 | 0.5 | 1.274 |
| 0.25 | 1.115 | 0.75 | 1.592 |
| 0.33 | 1.154 | 1 | $\infty$ |

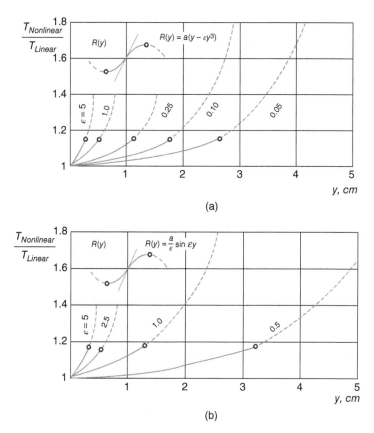

**Figure 3.2** Dependences of relations of periods of nonlinear and linear oscillations on displacements with the laws of the deformation: (a) on cubic parabola; and (b) on the sinusoid.

nonlinearity coefficient $\varepsilon$ are built. With the determination of the periods of real constructions we examine the ascent path of function $R(y)$. In Figure 3.2, this part of the dependence $T_{Nonlinear}/T_{Linear} \div y$ is shown by the solid line. For each curve there is a value $y$, with which $T_{Nonlinear}/T_{Linear} \to \infty$. This corresponds, as shown in Table 3.1, to the value $\theta = 1$. When considering only the ascending branches of the function, the maximum increase in the period of nonlinear oscillations occurs when $\theta = 1/3$ and comprises 1.15 times. A sinusoidal function is another characteristic of the restoring force, which allows the explicitly precise determination of the oscillatory period [9, 11]. In this case the function $f(y)$ is taken in the form

$$f(y) = \frac{1}{\varepsilon} \sin \varepsilon y \qquad (3.1.15)$$

The equation of motion in this case is the same as in the case of oscillation of the mathematical (simple) pendulum. The difference consists in the fact that in this case it is possible to use the obtained solution when $-\frac{\pi}{2\varepsilon} \leq y \leq \frac{\pi}{2\varepsilon}$. A similar assumption was used in the previous case. Substituting the expression in (3.1.15) in the formula (3.1.12), we obtain the value $T_{Nonlinear}$, expressed through the elliptical integral of the first kind:

$$T_{Nonlinear} = \frac{4}{\omega} \int_0^{\frac{\pi}{2}} \frac{d\theta}{\sqrt{1 - k^2 \sin^2 \theta}} \tag{3.1.16}$$

Where $k = \sin \frac{\varepsilon y_0}{2}$.

In Figure 3.2 the dependences for this function are built. If we examine only ascending branches of the function, then the maximum value of the period of nonlinear oscillations will comprise $T_{Nonlinear}/T_{Linear} = 1.18$, when $= \pi$, $T_{Nonlinear}/T_{Linear} \to \infty$.

The analytic expressions of the restoring force are characterized by special feature; the functions described by them have ascending and descending parts. In the descending part of the diagram the structural failure occurs, therefore, we introduced limitations at the maximum displacements and maximum values of the coefficient of non-linearity. But this somewhat limits the framework of the selection of the parameters. Therefore, the application of functions, increasing with the increase in the displacement, represents an interest, as examples of such functions are the following:

$$f(y) = \frac{1}{\varepsilon} \arctan \varepsilon y \tag{3.1.17}$$

$$f(y) = \frac{1}{\varepsilon} \tan \varepsilon y \tag{3.1.18}$$

In both cases $\varepsilon$ (1/cm) characterizes the degree of nonlinearity. With $\varepsilon = 0, f(y) = y$, i.e. a linear dependence occurs. Substituting these values into (3.1.12) and carrying out integration, we obtain the value of $T_{Linear}$. For the considered cases it was not possible to obtain the value of the period in an explicit form. The value of $T_{Nonlinear}$ is obtained in the following form:

for the function (3.1.17)

$$T_{Nonlinear} = 4 \int_0^{y_0} \frac{dy}{\sqrt{2\omega^2 \left[ y_0 \arctan \varepsilon y_0 - y \arctan \varepsilon y - \frac{1}{2\varepsilon} \ln \frac{1+\varepsilon^2 y_0^2}{1+\varepsilon^2 y^2} \right]}}; \tag{3.1.19}$$

for the function (3.1.18)

$$T_{Nonlinear} = 4 \int_0^{y_0} \frac{dy}{\sqrt{2\omega^2 \frac{1}{\varepsilon} \ln \frac{\cosh \varepsilon y_0}{\cosh \varepsilon y}}}. \tag{3.1.20}$$

The value of $T_{Nonlinear}$ can be determined by numerical calculation of the integral in each specific case. The application of a characteristic of the restoring force in the form

$$\omega^2 f(y) = \omega_r^2 y^{\frac{1}{r}}, \tag{3.1.21}$$

where $r$ — positive integer number, is also of interest. With $r = 1$ we have a linear dependence. This function is increasing monotonically. In this case the initial rigidity of the system is equal to infinity. This dependence does not correspond to the actual properties of deformation of structural materials but, because of the simplicity of expression, it frequently adapts in the engineering calculations, especially for describing the law of the deformation of metals. In the expressions (3.1.12), (3.1.15), (3.1.17), and (3.1.18) the value $\omega^2$ represents the value of the frequency of free linear oscillations which corresponds to the case $\varepsilon \to 0$. In expression (3.1.21) $\omega_r^2$ already does not have a dimensionality of frequency and represents a certain coefficient, dependent on the form of the accepted curve. The number $r$ characterizes the degree of nonlinearity. When $r$ is even, the function $f(y)$ with $y < 0$ should be selected in the form of

$$f(y) = \omega_r^2 |y|^{\frac{1}{r} - 1} y$$

We will next assume that $r$ is odd. Substituting the value $f(y)$ into (3.1.12), we obtain the value of the period of free nonlinear oscillations

$$T_{Nonlinear} = \frac{4}{\sqrt{2\omega_r^2 \frac{y_0^{\frac{\mu-1}{\mu+1}}}{\mu+1}}} \frac{\sqrt{\pi} \cdot \Gamma\left(\frac{1}{\mu+1}\right)}{\mu + 1 \cdot \Gamma\left(\frac{\mu+3}{2\mu+2}\right)}, \tag{3.1.22}$$

where $= \frac{1}{r}$, $\Gamma(\mu)$ – gamma-function. Tables of the values of gamma-functions are seen in [409].

With $\mu = 1$, $T_{Nonlinear} = 2\pi / \omega_1$;

with $\mu = 1/3$, $T_{Nonlinear} = \frac{4 \cdot 3.5944}{\omega_3 \sqrt{6}} y_0^{\frac{1}{3}}$ \tag{3.1.23}

From (3.1.23) it follows that the period of nonlinear oscillations depends on the amplitude and increases with an increase in the latter. To calculate the values of the period of nonlinear oscillations of a real system it is necessary to know the specific value of the coefficients of nonlinearity $\varepsilon$ in the formulas (3.1.16), (3.1.19), (3.1.20) or $\omega_r$, and $r$ in the formula (3.1.22). The values of these coefficients can be obtained with the use of results of structural testing. However, the dependence of the restoring force on the displacement can also be obtained using the results of material testing under uniaxial linear tension and compression. Let us assume that the dependence between the stresses and strains take the form of the exponential function:

$$\sigma = B|\varepsilon|^{\mu-1} \cdot \varepsilon, \mu = \frac{1}{r}, r = 1, 2, 3 \ldots \tag{3.1.24}$$

where $\sigma$ – stress; $\varepsilon$ – relative deformation; and $B$ – the coefficient, characterizing the deformation curve, with $r = 1$, $B = E$ ($E$ – *modulus of elasticity*). On the basis of (3.1.24), let us switch over to obtaining the dependence between the force and displacement. For this we will consider a one-story frame with the undeformed horizontal element of the frame (beam $EI = \infty$). Using the conventional assumption that the hypothesis of flat sections is also valid for nonlinear deformations, we will obtain a dependency between the displacement and statically loading (acting, impacting) force $P$:

$$y = \frac{P^r H^{r+2}}{(r+2)2^{r+1}\left(\sum_{i=1}^{s} BI_{ir}\right)^r} \tag{3.1.25}$$

where $H$ – is the height of the floor; $s$ – the number of floor columns; and $I_{ir}$ – generalized moment of inertia of the $i$ column [341, 342]. From (3.1.25) we have

$$P = a_r y^{\frac{1}{r}} \tag{3.1.26}$$

where

$$a_r = \frac{(r+2)^{\frac{1}{r}} \cdot 2^{\frac{r+1}{r}}}{H^{\frac{r+2}{r}}} \sum_{i=1}^{s} BI_{ir},$$

$a_r$ – rigidity of the system in the nonlinear deformation. When (at) $r = 1$

$$a_1 = 12 \sum_{i=1}^{s} \frac{EI_{i1}}{H^3}$$

The value of the generalized moment of inertia for different (various) types of sections is given in [347, 348]. For a rectangular cross-section with dimensions $b \times h$, value $I_r$ is determined by the expression

$$I_r = \frac{bh^{2+1/r}}{2^{2+1/r}\left(1 + \frac{1}{2r}\right)},$$

with $r = 1$,

$$I_1 = \frac{bh^3}{12}.$$

If we consider that the examined frame is a system with one degree of freedom and the mass is concentrated at the overlap level, then the equations of motion of this system can be obtained substituting $P = -my''(t)$ into (3.1.26):

$$my'' + a_r y^{\frac{1}{r}} = 0 \qquad\qquad (3.1.27)$$

Above we obtained the value of the period of nonlinear oscillations of a similar system. The period of oscillations of the system described by Eq. (3.1.27) can be obtained from (3.1.22), if instead of $\omega_r^2$ we substitute the value of $a_r/m$. Thus, knowing the values of $B$ and $r$ (3.1.24), from (3.1.26) we will be able to define $a_r$ and from (3.1.22) define the value of the period with different amplitudes $y_0$. In order to approximately estimate the values of $a_r$, we carried out an approximation of the dependency $\sigma - \varepsilon$ for steel of (CT-3) in the domain $0 \le \varepsilon \le 6.\ 10^{-3}$, obtained by uniaxial linear tension [290]. In the first approximation, when $r = 1/3$, we obtained $B = 2.1,\ 10^4 \kappa \Gamma c/cm^2$.

## 3.2   The Nonlinear Oscillations of Systems with Multiple-Degrees-of-Freedom

Let us examine free nonlinear oscillations of a system with multiple-degrees-of-freedom. The exact solutions of such problems are very rare in the literature of applied nature. In the works of R.M. Rosenberg [305, 306] the nonlinear oscillations of a multi-mass system with the aid of potential function and use of properties of geodetic lines in space are investigated. Multi-mass systems, which are design schemas of multistory frame buildings, will be investigated below. The periods and forms of oscillation are defined precisely by the direct integration of the equations of motion. The method used is actually the generalization of Duffing's method, which was adapted for systems with one degree of freedom. Let us examine free oscillations of the multi-story frame, whose mass is concentrated at the overlap levels. The dependence "stress-strain" takes the form (3.1.24). The equation of motion of this frame takes the form [8, 9]

$$m_i y_i'' + a_{ir}(y_i - y_{i-1})^{\frac{1}{r}} - a_{i+1,r}(y_{i+1} - y_i)^{\frac{1}{r}} = 0 \qquad (3.2.1)$$

where $m_i$ is the mass, concentrated at the level of $i$th floor; $y_i$, $y_i''$ – displacement and acceleration of the $i$th mass; and $a_{ir}$ – generalized rigidity of the $i$th of floor with the nonlinear deformation with the parameter $r$. The value of $a_{ir}$ is determined by the expression in (3.1.26). Let us determine the periods and corresponding forms of free oscillations of the nonlinear system. As in linear systems, the forms of oscillations of a nonlinear system are characterized by the following properties: all masses of the system pass through the position of equilibrium simultaneously and oscillate with one and the same period. To solve this problem let us represent (3.2.1) in the form of

$$\frac{d(y_i')^2}{dy_i} = -2\frac{a_{ir}}{m_i}(y_i - y_{i-1})^{\frac{1}{r}} + 2\frac{a_{i+1,r}}{m_i}(y_{i+1} - y_i)^{\frac{1}{r}} \qquad (3.2.2)$$

Integrating Eq. (2.2.2) with the initial conditions

$$y_i(0) = c_i \quad \text{and} \quad y_i'(0) = 0$$

we will obtain

$$y_i'^2 = -2\frac{a_{ir}}{m_i}\int_{c_i}^{y_i}(y_i - y_{i-1})^{\frac{1}{r}}dy_i + 2\frac{a_{i+1,r}}{m_i}\int_{c_i}^{y_i}(y_{i+1} - y_i)^{\frac{1}{r}}dy_i \qquad (3.2.3)$$

Designating $y_i/y_n = \gamma_i$ and taking into account that with single-frequency oscillations $\gamma_i$ is taken as constant, we obtain

$$y_i'^2 = -2\frac{a_{ir}}{m_i}\frac{r}{1+r}\left(1 - \frac{\gamma_{i-1}}{\gamma_i}\right)^{\frac{1}{r}}\left(y_i^{\frac{1}{r}+1} - c_i^{\frac{1}{r}+1}\right) \qquad (3.2.4)$$

$$+ 2\frac{a_{i+1,r}}{m_i}\frac{r}{1+r}\left(\frac{\gamma_{i+1}}{\gamma_i} - 1\right)^{\frac{1}{r}}\left(y_i^{\frac{1}{r}+1} - c_i^{\frac{1}{r}+1}\right)$$

Integrating the expression (3.2.4) from $t = 0$ to $T/4$, with corresponding displacement from $y_i = c_i$ to $y_i = 0$, we will obtain

$$T'_{j,Nonlinear} = \frac{1}{\sqrt{2\frac{r}{1+r}\left[\frac{a_{ir}}{m_i}\left(1 - \frac{\gamma_{i-1}}{\gamma_i}\right)^{\frac{1}{r}} - \frac{a_{i+1,r}}{m_i}\left(\frac{\gamma_{i+1}}{\gamma_i} - 1\right)^{\frac{1}{r}}\right]c^{\frac{1-r}{2r}}}}\int_0^1\frac{du}{1 - u^{\frac{1+r}{r}}}.$$

After designating

$$\int_0^1 \frac{du}{1 - u^{\frac{1+r}{r}}} = I, \quad a_{ir} = a_{1r}\alpha_i, \quad m_i = m_1\mu_i$$

and bearing in mind, that

$$c^{\frac{1-r}{2r}} = \gamma_i^{\frac{1-r}{2r}} c_n^{\frac{1-r}{2r}},$$

we will obtain

$$\frac{T_j\sqrt{2\frac{r}{1+r}}}{4Ic_n^{\frac{r-1}{2r}}} = \frac{1}{\sqrt{\frac{a_{1r}}{m_i}\left\{\frac{\alpha_i}{\mu_i}(\gamma_i - \gamma_{i-1})^{\frac{1}{r}} - \frac{\alpha_{i+1}}{\mu_i}(\gamma_{i+1} - \gamma_i)^{\frac{1}{r}}\right\}\frac{1}{\gamma_i}}}. \tag{3.2.5}$$

Designating

$$\left[\frac{T_j\sqrt{2\frac{r}{1+r}}}{4Ic_n^{\frac{r-1}{2r}}}\right]^r = \frac{1}{\lambda} \tag{3.2.6}$$

we will obtain

$$\lambda = \left[\frac{\alpha_i}{\mu_i}(\gamma_i - \gamma_{i-1})^{\frac{1}{r}} - \frac{\alpha_{i+1}}{\mu_i}(\gamma_{i+1} - \gamma_i)^{\frac{1}{r}}\right]\frac{1}{\gamma_i}. \tag{3.2.7}$$

We obtain the equation for determining the periods using the following condition:

$$\gamma_0 = y_0/y_n = 0, \tag{3.2.8}$$

From (3.2.7) and with $i = n$, we will obtain

$$\gamma_{n-1} = 1 - \lambda^r\left(\frac{\mu_n}{\alpha_n}\right)^r,$$

bearing in mind that $\gamma_n = 1$ and $\alpha_{n+1} = 0$. From the system (3.2.7) and using the recurrent method the following expression is obtained:

$$\gamma_{i-1} = \gamma_i - \left(\frac{\mu_i}{\alpha_i}\right)^r\left[\frac{\alpha_{i+1}}{\mu_i}(\gamma_{i+1} - \gamma_i)^{\frac{1}{r}} + \lambda\gamma_i\right]^r \tag{3.2.9}$$

Eq. (3.2.8) will take the open form:

$$\gamma_0 = \gamma_1 - \left(\frac{\mu_1}{\alpha_1}\right)^r\left[\frac{\alpha_2}{\mu_2}(\gamma_2 - \gamma_1)^{\frac{1}{r}} + \lambda\gamma_1\right]^r = 0 \tag{3.2.10}$$

where $\gamma_1$, $\gamma_2$ and $\gamma_i$ are defined using expression (3.2.9). With the aid of Eq. (3.2.10) the values of $\lambda_j$ are determined, through which the value of the period with using formula (3.2.6) is expressed:

$$T_j = 4I \sqrt{\frac{1 + r}{2r} \frac{m_1}{a_{1r}} \frac{1}{\lambda_j} c_n^{\frac{r-1}{2r}}} \tag{3.2.11}$$

After the determination of $\lambda_j$, the coefficients of form oscillations $\gamma_{ij}$ are determined by expression (3.2.9). The method of determining periods and forms of free oscillations presented is applicable to the linear oscillations, i.e. in the case $r = 1$. We determined the values $\lambda_j$ and $\gamma_{ij}, j = 1, 2, 3$ for multistory buildings with equal values of concentrated masses and stiffness of floors, i.e. with $\alpha_i = 1$, $\mu_i = 1$. The calculations were conducted with the value $r = 3$. The system with $n$ degrees of freedom with the linear oscillations has $n$ of values of periods, which are determined using the algebraic equation of $n$ degrees (3.2.10). In the case of linear oscillations, all roots of Eq. (3.2.10) are real numbers. With nonlinear oscillations, the degree of an equation for determining the periods is more than $n$ and depends on the parameter value of the nonlinear deformation $r$. A numerical study of Eq. (3.2.10) showed that there are $n$ valid solutions $\lambda_j, j = 1, 2, ... n$ and corresponding to them $n$ modes of oscillations. As with the linear oscillations, the first form does not have a nodal point, the second mode has one nodal point, and so on. The obtained values $\lambda_j$, and $\gamma_j, j = 1, 2, ... n$ for buildings with $n = 2 - 10$ floors are given in Tables 3.2 and 3.3. Let us calculate the values of the periods of nonlinear oscillations by referring to previously used parameters of

Table 3.2   The value of roots $\lambda_j$.

| Number of stories | $\lambda_1$ | Roots $\lambda_2$ | $\lambda_3$ |
|---|---|---|---|
| 2 | 0.50999 | 1.42000 | 1 |
| 3 | 0.33280 | 1.23591 | 1.47500 |
| 4 | 0.24009 | 0.90501 | 1.43870 |
| 5 | 0.18459 | 0.71180 | 1.25750 |
| 6 | 0.14809 | 0.61170 | 1.17560 |
| 7 | 0.12249 | 0.53120 | 0.92461 |
| 8 | 0.10399 | 0.43590 | 0.74900 |
| 9 | 0.08911 | 0.38899 | 0.64721 |
| 10 | 0.07800 | 0.31157 | 0.58998 |

**Table 3.3**   The value of the coefficient of the modes of oscillations $\gamma_{ij}^{(n)}$.

| Number of stories $n$ | $i$ | $\gamma_{i1}$ | $\gamma_{i2}$ | $\gamma_{i3}$ |
|---|---|---|---|---|
| 1 | 2 | 3 | 4 | 5 |
|   | 1 | 0.867 | −1.863 |   |
| 2 | 2 | 1 | 1 |   |
|   | 1 | 0.684 | −0.891 | 1.946 |
| 3 | 2 | 0.963 | −0.888 | 2.093 |
|   | 3 | 1 | 1 | 1 |
|   | 1 | 0.553 | −1.220 | 0.822 |
|   | 2 | 0.878 | −1.220 | 0.807 |
| 4 | 3 | 0.986 | 0.259 | 1.978 |
|   | 4 | 1 | 1 | 1 |
|   | 1 | 0.460 | −1.048 | 0.869 |
| 5 | 2 | 0.784 | −1.068 | 0.866 |
|   | 3 | 0.944 | −0.946 | −0.988 |
|   | 4 | 0.944 | 0.639 | −0.981 |
|   | 5 | 1 | 1 | 1 |
|   | 1 | 0.393 | −0.906 | 0.903 |
|   | 2 | 0.700 | −0.976 | 0.902 |
|   | 3 | 0.886 | −0.970 | −0.649 |
| 6 | 4 | 0.971 | −0.501 | −0.711 |
|   | 5 | 0.997 | 0.771 | −0.625 |
|   | 6 | 1 | 1 | 1 |
|   | 1 | 0.344 | −0.999 | 1.017 |
|   | 2 | 0.628 | −1.000 | 0.171 |
|   | 3 | 0.824 | −0.957 | −1.050 |
| 7 | 4 | 0.935 | −0.532 | −1.081 |
|   | 5 | 0.983 | 0.397 | −0.980 |
|   | 6 | 0.998 | 0.926 | 0.627 |
|   | 7 | 1 | 1 | 1 |
|   | 1 | 0.317 | −0.806 | 1.063 |
|   | 2 | 0.576 | −0.997 | 1.074 |
|   | 3 | 0.769 | −1.000 | 0.875 |
| 8 | 4 | 0.894 | −0.975 | −1.027 |
|   | 5 | 0.961 | −0.606 | −1.130 |
|   | 6 | 0.990 | 0.337 | −1.077 |
|   | 7 | 0.999 | 0.918 | 0.580 |
|   | 8 | 1 | 1 | 1 |

*(continued)*

Table 3.3 (*continued*)

| Number of stories $n$ | $i$ | $\gamma_{i1}$ | $\gamma_{i2}$ | $\gamma_{i3}$ |
|---|---|---|---|---|
| | 1 | 0.284 | −1.000 | 0.999 |
| | 2 | 0.524 | −0.999 | 0.833 |
| | 3 | 0.714 | −0.987 | −0.293 |
| | 4 | 0.848 | −0.848 | −0.947 |
| 9 | 5 | 0.931 | −0.402 | −0.977 |
| | 6 | 0.975 | 0.279 | −0.958 |
| | 7 | 0.994 | 0.790 | −0.387 |
| | 8 | 0.999 | 0.976 | 0.796 |
| | 9 | 1 | 1 | 1 |
| | 1 | 0.257 | −0.956 | 0.892 |
| | 2 | 0.480 | −0.999 | 0.951 |
| | 3 | 0.664 | −1.000 | 0.940 |
| | 4 | 0.802 | −0.988 | 0.394 |
| 10 | 5 | 0.896 | −0.852 | −0.825 |
| | 6 | 0.953 | −0.410 | −0.985 |
| | 7 | 0.983 | 0.273 | −0.624 |
| | 8 | 0.996 | 0.788 | 0.742 |
| | 9 | 0.999 | 0.976 | 0.742 |
| | 10 | 1 | 1 | 1 |

deformation. For $n = 5$, $H/h = 10$, $r = 3$ the following values for $T_j$ are obtained

$$T_j = T_{j,Linear}\left(\frac{c_5}{5H}\right)^{\frac{1}{3}} \cdot k_j \qquad (3.2.12)$$

$$k_1 = 5.710, \quad k_2 = 8.461, \quad k_3 = 9.911$$

As in the case of a system with one degree of freedom, the periods of nonlinear oscillations $T_j$ can be greater or smaller than the period of corresponding linear oscillations $T_{j, Linear}$. For example, in this system with $c_5/5H = 1/186$, $T_1 = T_{1, Linear}$, and with $c_n \lessgtr 1/186$, $T_1 \lessgtr T_{1, Linear}$. With the second and third modes, the periods of nonlinear oscillations become more than the periods of linear oscillations with $c_5/5H > 1/605$ and $c_5/5H > 1/973$, respectively.

The modes of the nonlinear vibrations of 10-story buildings are shown in Figure 3.3. The figure shows that the nature of the deformation of nonlinear and linear systems with single-frequency oscillations of the same tone is identical.

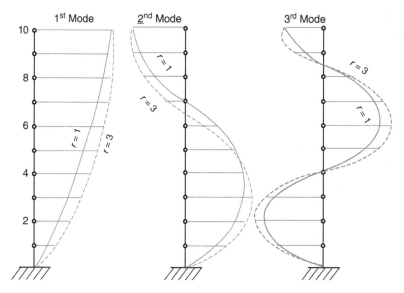

**Figure 3.3** Modes of free linear and nonlinear oscillations of the 10-story building.

The strain–stress diagrams of the deformation of the materials of structures with dynamic processes are characterized by different laws of loading and unloading. During loading, the restoring force is subordinate to the nonlinear law; while during unloading it is subordinated to the linear. In general, this leads to a significant complication when solving a problem of nonlinear vibrations of structures as compared to nonlinear tasks when encountered in other areas of technology; for example in theory of automatic regulation, radio engineering, etc. In the case of different laws of deformation during loading and unloading, the period of nonlinear oscillations can be determined according to the formula:

$$T_{j,Hysteresis} = \frac{1}{2}(T_{j,Linear} + T_{j,Nonlinear}) \qquad (3.2.13)$$

where $T_{j,\,Hysteresis}$ – period of $j$th form of the nonlinear hysteresis oscillations of system; $T_{j,\,Linear}$ – period of $j$th form of the linear oscillations; and $T_{j,\,Nonlinear}$ – period of $j$th modes of elastic-nonlinear oscillations.

In the above example, the forms of oscillations of a buildings during loading and unloading are different. During steady single-frequency hysteresis oscillations, the system will apparently oscillate in a mode, occupying the mid-position with respect to the modes of linear and elastic-nonlinear modes of oscillations.

In the case of the power law of deformation, we succeeded in obtaining the precise value of periods and forms of nonlinear oscillations for a multiple-degree-of-freedom system. In the case of other dependences, obtaining the exact solution of the problems proved to be impossible. The effective methods of study of nonlinear systems with the "small nonlinearity" are the methods of approximations; the asymptotic method of Krylov-Bogoliubov [37] in particular. Let us consider free nonlinear oscillations of multi-story frames with the undeformed horizontal elements of the frame. The equation of motion of this frame with the power law of deformation takes the form of (3.2.1). When $r = 1$, the linear oscillations take place. The equations of the linear oscillations for a frame are written in the form presented by E.E. Khachian particularly for analyzing an elasto-plastic system [171]

$$\sum_{k=1}^{s} m_k y_k'' + a_s(y_s - y_{s+1}) = 0, \quad s = 1, 2, .. \; n \qquad (3.2.14)$$

Here, the concentrated masses are numbered on top; for example, $m_1$ is concentrated mass of the upper most floor. Presence in each equation of one elastic reaction, corresponding to given floor, is the characteristic property of Eqs. (3.2.14). Let us examine a case when dependency between the restoring force and relative floor deformation takes the form of a cubic parabola. In this case the equations of elastic-nonlinear oscillations take the form [8, 9]

$$\sum_{k=1}^{s} m_k y_k'' + a_s[(y_s - y_{s+1}) - \varepsilon_s(y_s - y_{s+1})^3] = 0, \quad s = 1, 2, .. \; n \qquad (3.2.15)$$

where $\varepsilon_s$ is a small positive parameter, characterizing the nonlinearity of deformation $(n - s + 1)$th floor. In a multistory frame, the lower and upper levels are compressed to a different degree. This means that in Eqs. (3.2.15), the value of normal forces for the ground floors must be greater, as compared to the upper. However, we further neglect the influence of the normal forces and accept the following:

$$\varepsilon_1 = \varepsilon_2 = ... = \varepsilon_s = \varepsilon$$

On the basis of the physical representations, Eq. (3.2.15) can be used for such values $(y_s - y_{s+1})$, with which the restoring force grows in its absolute value. From the conditions

$$R_s'(y_s - y_{s+1}) = a_s[1 - 3\varepsilon(y_s - y_{s+1})^2] \geq 0,$$

we see that

$$\varepsilon \le \frac{1}{3(y_s - y_{s+1})^2_{max}} \tag{3.2.16}$$

For finding the frequency of the period of a nonlinear system, described by Eqs. (3.2.15), we will use the asymptotic method of Krylov-Bogoliubov [37]. We search for the general solution of Eq. (3.2.15), in the form of decomposition

$$y_s = C_s^{(1)} Y \cos(\omega_l^{(1)} t + \theta) + \varepsilon U_s^{(1)}(Y, \omega_\wedge^{(1)} t + \theta)$$
$$+ \varepsilon^2 U_s^{(2)}(Y, \omega_l^{(1)} t + \theta) + \cdots, \tag{3.2.17}$$

Where $Y$ – maximum displacement of the free end of the frame; $C_s^{(1)}$ – the particular solution of the linear system of Eqs. (3.2.14); $\omega_l^{(1)}$ – the angular frequency of the main form of linear oscillations; and $U_s^{(1)}$, $U_s^{(2)}$ – the periodical functions of angle $\psi = \omega_l^{(1)} t + \theta$ with period $2\pi$. The values of $Y$ and $\psi$, as the functions of time, are determined by differential equations:

$$\frac{dY}{dt} = \varepsilon A_1(Y) + \varepsilon^2 A_2(Y) + \varepsilon^3 \cdots,$$

$$\frac{d\psi}{dt} = \omega_l^{(1)} + \varepsilon B_1(Y) + \varepsilon^2 B_2(Y) + \varepsilon^3 \cdots$$

As the first approximation we assume:

$$y_s = C_s^{(1)} Y \cos \psi, \qquad s = 1, 2 \cdots n, \tag{3.2.18}$$

where

$$\frac{dY}{dt} = \varepsilon A_1(Y) \tag{3.2.19}$$

$$\frac{d\psi}{dt} = \omega(Y) = \omega_l^{(1)} + \varepsilon B_1(Y)$$

Here $\omega(Y)$ – the desired frequency of nonlinear oscillations. To find $B_1(Y)$, we will use one of the equations of harmonic balance [172]

$$\int_0^{2\pi} \sum_{s=1}^n C_s^{(1)} \left\{ \sum_{k=1}^s m_k y_k'' + a_s(y_s - y_{s+1}) - a_s \varepsilon(y_s - y_{s+1})^3 \right\} \cos \psi d\psi = 0 \tag{3.2.20}$$

where: $s = 1, 2, \cdots n$

Differentiating (3.2.18) and accounting for (3.2.19), with the accuracy to the values of the first-order of smallness, we find

$$y_s^{(1)} = C_s^{(1)}[-2\varepsilon\omega_l^{(1)} \sin\psi A_1(Y) - Y\omega_l^{(1)^2} \cos\psi$$
$$- 2\varepsilon Y\omega_l^{(1)} \cos\psi B_1(Y)] \tag{3.2.21}$$

After substituting (3.2.18) and (3.2.21) into Eq. (3.2.20), we will obtain:

$$\int_0^{2\pi} \sum_{r=1}^{n} C_r^{(1)} \left\{ \sum_{k=1}^{s} -m_k\, C_s^{(1)}[2\varepsilon\omega_l^{(1)} \sin\psi A_1(Y)\omega_l^{(1)^2} + Y\cos\psi \right.$$
$$+ 2\varepsilon Y\omega_l^{(1)} \cos\psi B_1(Y)] + a_s Y\cos\psi(C_s^{(1)} - C_{s+1}^{(1)}) \tag{3.2.22}$$
$$\left. - \varepsilon a_s Y^3 \cos^3\psi(C_s^{(1)} - C_{s+1}^{(1)})^3 \right\} \cos\psi\, d\psi = 0$$

After integration and appropriate conversion, we will obtain

$$- 2Y\omega_l^{(1)}\varepsilon B_1(Y)\pi \sum_{r,k=1}^{n,s} m_k C_r^{(1)} C_s^{(1)} + Y\pi \sum_{r,k=1}^{n,s} C_r^{(1)}[-m_k\omega_l^{(1)^2} C_s^{(1)}$$

$$+ a_s(C_s^{(1)} - C_{s+1}^{(1)})] = \frac{3}{4}\pi\varepsilon Y^3 \sum_{r,s=1}^{n} a_s C_r^{(1)}(C_s^{(1)} - C_{s+1}^{(1)})^3 \tag{3.2.23}$$

Assuming that all concentrated masses are equal and the rigidity (stiffness) of all floors are identical:

$$a_1 = a_2 = \cdots = a, m_1 = m_2 = \cdots = m$$

Keeping in mind that

$$\sum_{r,k=1}^{n,s} \{-m\omega_l^{(1)^2} C_s^{(1)} + a(C_s^{(1)} - C_{s+1}^{(1)})\} = 0, \quad s = 1, 2, \dots n,$$

and designating

$$\sum_{r,k=1}^{n,s} C_r^{(1)} C_s^{(1)} = M_1^{(n)}$$

and

$$\sum_{r,s=1}^{n} C_r^{(1)}(C_s^{(1)} - C_{s+1}^{(1)})^3 = \Delta_1^{(n)},$$

**Table 3.4**  Values of $M_1^{(n)}$, $\Delta_1^{(n)}$ and relations on nonlinear and linear frequencies for frame buildings up to 20 stories.

| $n$ | $M_1^{(n)}$ | $\Delta_1^{(n)}$ | $\omega(Y)/\omega_l^{(1)}$ |
|---|---|---|---|
| 1 | 1 | 1 | 1–0.375 |
| 2 | 4.235 | 0.869 | 1–0.109 |
| 3 | 11.328 | 0.316 | 1–0.0528 |
| 4 | 24.089 | 0.247 | 1–0.0317 |
| 5 | 43.542 | 0.197 | 1–0.0209 |
| 6 | 71.373 | 0.164 | 1–0.01479 |
| 7 | 109.336 | 0.141 | 1–0.01107 |
| 8 | 158.886 | 0.123 | 1–0.00912 |
| 9 | 221.977 | 0.111 | 1–0.00687 |
| 10 | 299.713 | 0.100 | 1–0.00562 |
| 15 | 961.594 | 0.068 | 1–0.00256 |
| 20 | 2224.004 | 0.051 | 1–0.00147 |

we will obtain

$$\varepsilon B_1 = -\frac{3}{8}\frac{\varepsilon Y^2}{\omega_l^{(1)}}\frac{a}{m}\frac{\Delta_1^{(n)}}{M_1^{(n)}} \tag{3.2.24}$$

The values of $C_s^{(1)}$ and $\lambda_1 = \omega_l^{(1)^2}\frac{m}{a}$ for $n = 2 - 20$ are presented in [299]. Using these data and calculating $\Delta_1^{(n)}$ and $M_1^{(n)}$, we will obtain $\varepsilon B_1$. Further substituting $\varepsilon B_1$ into the second equation of the system (3.2.19), we obtain the value of the frequency of nonlinear oscillations $\omega(Y)$ in the first approximation. Calculations were conducted for $n = 2$, 10, 15, and 20 stories. The obtained values $M_1^{(n)}$, $\Delta_1^{(n)}$, and $\omega(Y)$ are given in Table 3.4. The solution of Eqs. (3.2.15) in the second approximation has the form

$$y_s = C_s^{(1)}Y\cos\psi + \varepsilon U_s^{(1)}, \qquad s = 1, 2\cdots n \tag{3.2.25}$$

where $Y$ and $\psi$ are determined from the equations of the second approximation:

$$\frac{dY}{dt} = \varepsilon A_1(Y) + \varepsilon^2 A_2(Y), \frac{d\psi}{dt} = \omega_l^{(1)} + \varepsilon B_1(Y) + \varepsilon^2 B_2(Y) \tag{3.2.26}$$

The value $U_s^{(1)}$ in Eq. (3.2.25) is obtained as the amplitude of the forced oscillations, excited in the system (3.2.15) by the forces $a\varepsilon(y_s - y_{s+1})^3$, in which $y_s$ and $y_{s+1}$ are sinusoidal:

$$U_s^{(1)} = \sum_{j=1}^{n} C_s^{(j)} \frac{\sum_{p=1}^{n} f_p^{(0)} C_p^{(j)}}{mM_j \omega_j^2} + \sum_{j=2}^{n} C_s^{(j)} \frac{\sum_{p=1}^{n} (f_p^{(1)} \cos \psi + g_p^{(1)} \sin \psi) C_p^{(j)}}{mM_j(\omega_j^2 - \omega_l^{(1)^2})}$$

$$+ \sum_{i=2}^{\infty} \sum_{j=1}^{n} C_s^{(j)} \frac{\sum_{p=1}^{n} (f_p^{(i)} \cos i\psi + g_p^{(i)} \sin i\psi) C_p^{(j)}}{mM_j(\omega_j^2 - i^2\omega_l^{(1)^2})} \qquad (3.2.27)$$

where

$$\sum_{r,k=1}^{n,s} C_r^{(j)} C_s^{(j)} = M_j, \qquad (s,j = 1, 2 \cdots n)$$

$f_p^{(i)}$, $g_p^{(i)}$ – the coefficients of the decomposition function $a(y_p - y_{p+1})^3$ in Fourier series. When calculating $U_s^{(1)}$ in the sum of the third addend, three numbers were taken into account. Differentiating (3.2.25) and taking into account (3.2.26) with the accuracy to the values of the second-order of small-ness inclusively, and substituting $y_s$ and $y_s''$ in (3.2.20), we find $B_2(Y)$. After substituting the latter into the second equation of the system (3.2.26), we find value $\omega(Y)$ by conducting two approximations.

The values $B_2(Y)$ are calculated for $n = 1, 2, 3$ floors (stories). The values $\omega(Y)$ taking into account two approximations take the following form:

when $n = 1$ $\omega(Y) = \omega_l^{(1)}(1 - 0.375\varepsilon Y^2 - 0.058\varepsilon^2 Y^4)$;

when $n = 2$ $\omega(Y) = \omega_l^{(1)}(1 - 0.109\varepsilon Y^2 - 0.0039\varepsilon^2 Y^4)$;

when $n = 3$ $\omega(Y) = \omega_l^{(1)}(1 - 0.058\varepsilon Y^2 - 0.0008\varepsilon^2 Y^4)$.

For systems with degrees of freedom higher than three, the second approximation is not produced because of the bulkiness of calculations. Furthermore, as can be seen from Table 3.5, in the given interval of variation of coefficient of nonlinearity $\varepsilon$, the first approximation gives results with the accuracy sufficient for practical purposes. Quantitative change of the frequency of nonlinear oscillations $\omega(Y)$, depending on the number of floors and the coefficient of nonlinearity, is illustrated by the data of Table 3.5, where $(y_s - y_{s+1})_{max} = 1$ is set for the sake of simplicity.

From the conditions of (3.2.16), it turns out that $\varepsilon \leq 0.33$. To compare the influence of height of the building on the value of $\omega(Y)/\omega_l^{(1)}$ in the first approximation, we take that $Y = n(y_s - y_{s+1})_{max}$. As can be seen in Table 3.5, with the increasing height of the building, the ratio of $\omega(Y)/\omega_l^{(1)}$ decreases,

**Table 3.5**   Values of relations of nonlinear and linear frequencies depending on coefficient of nonlinearity and numbers of stories.

| | $\omega(Y)/\omega_l^{(1)}$ First approximation | | $\omega(Y)/\omega_l^{(1)}$ Second approximation | |
| --- | --- | --- | --- | --- |
| $n$ | $\varepsilon = 0.16$ | $\varepsilon = 0.33$ | $\varepsilon = 0.16$ | $\varepsilon = 0.33$ |
| 1 | 0.93 | 0.8763 | 0.9384 | 0.8700 |
| 2 | 0.9302 | 0.8562 | 0.9286 | 0.8501 |
| 3 | 0.9239 | 0.8432 | 0.9222 | 0.8362 |
| 4 | 0.9188 | 0.8327 | — | — |
| 5 | 0.9164 | 0.8276 | — | — |
| 6 | 0.9147 | 0.8242 | — | — |
| 7 | 0.9137 | 0.8222 | — | — |
| 8 | 0.9124 | 0.8195 | — | — |
| 9 | 0.9109 | 0.8164 | — | — |
| 10 | 0.9100 | 0.8446 | — | — |
| 15 | 0.9078 | 0.8100 | — | — |
| 20 | 0.9060 | 0.8060 | — | — |

and the difference between the frequencies of nonlinear and linear oscillations is more essential at the large values of $\varepsilon$. Data analysis of Table 3.5 showed that the values of $\omega(Y)/\omega_l^{(1)}$ were well approximated by the function

$$\omega(Y)/\omega_l^{(1)} = 1 - 0.375n^{-1.84}\varepsilon Y^2,$$

and for relation of the periods of oscillations $T(Y)/T_l^{(1)}$ with the accuracy of $\varepsilon$, we will have

$$T(Y)/T_l^{(1)} = 1 + 0.375n^{-1.84}\varepsilon Y^2 \tag{3.2.28}$$

After substituting $T_l^{(1)}$ in (3.2.8) with its value as given in [171], we will obtain a similar formula for determining the period of single-frequency elastic-nonlinear oscillations

$$T(Y) = 2\pi(0.367 + 0.633n)\sqrt{\frac{m}{a}}(1 + 0.375n^{-1.84}\varepsilon Y^2).$$

If $\varepsilon = 0$, then we obtain the value of the period of the main (fundamental) tone of free linear oscillations.

Now let us examine the case when restoring force under load is subordinate to the nonlinear law

$$R_s = a[(y_s - y_{s+1}) - \varepsilon(y_s - y_{s+1})^3],$$

and unloading occurs linearly. The angle of the slope of the unloading line with the axis $y_s - y_{s+1}$ is equal to $\alpha = \arctan a$. Using (3.2.13), we obtain the formula for determining the period of hysteresis oscillations:

$$T_{Hysteresis} = T_l^{(1)} \left[ 1 + \frac{1}{2} 0.375 n^{-1.84} \varepsilon Y^2 \right] \qquad (3.2.29)$$

When determining the period of hysteresis oscillations it was accepted that the rigidity of the system during the unloading was equal to the greatest initial rigidity of the system when loading. If we continuously reduce rigidity of the system during the unloading, then it is possible to ascertain that the period of hysteresis oscillations will approach the period of linear oscillations.

In calculating the periods of nonlinear oscillations of reinforced concrete structures according to formulas (3.2.28) and (3.2.29), it is possible to use the values of coefficient of nonlinearity $\varepsilon$, given in [236]. This work is dedicated to the experimental studies of the forced oscillations of the reinforced concrete bent elements. The patterns were tested with different values of perturbing force, until destruction. As the design scheme of the tested elements with adequate accuracy shows, it is possible to accept a system with one degree of freedom. According to the data of these experiments, the coefficient of nonlinearity $\varepsilon$ for the frame columns with the initial period $T = 0.15 - 0.40$ s was equal $0.0255 - 0.066$ cm$^{-2}$, and for the simple beam $T \leq 0.1$ s $- 0.10^{-2}$. The results of vibrational tests of two reinforced concrete monolithic and buildup frame buildings given in [410] make it possible to calculate coefficients of nonlinearity in this case for a building as a whole. The values of these coefficients were equal to $\varepsilon = 1$ and $\varepsilon = 1.8$, respectively. Comparatively large values of the coefficient of nonlinearity can be explained not only by nonlinear deformation of reinforced concrete, but also by the pliability of joints and the failure of the load bearing elements of the framework filling.

## 3.3    The Nonlinear Oscillations of Systems with Distributed Mass

Let us examine free nonlinear oscillations of systems with distributed mass. As an example of such systems we will examine the oscillations of bent beams with different supporting fastenings and shifted cantilever beam. To solve these problems we will use a method used in the work of

G. Kauder [163]. In a rectangular coordinate system, a prismatic bent bar with a symmetrical relative to $y$ and $z$ cross-section is examined. It is well established that the dependency between the normal stress $\sigma_x$ and relative strain $\theta_x$ takes the form

$$\sigma_x = E\theta_x(1 - \varepsilon E^2 \theta_x^2) \tag{3.3.1}$$

where $E$ – modulus of elasticity of material; and $\varepsilon$ – coefficient of nonlinearity of deformation.

To this oscillatory system we will apply the Hamilton's principle, according to which, in an interval of real motion of a system in the time interval $t_1 \div t_2$, the integral $\int_{t_1}^{t_2}(P - K)dt$ takes the extreme value

$$\delta \int_{t_1}^{t_2} (P - K)dt = 0 \tag{3.3.2}$$

where $P$ – potential energy; and $K$ – kinetic energy of the oscillatory system.

The potential energy of the bar is equal to

$$P = \frac{EI_0}{2} \int_0^l \left[ 1 - \frac{1}{2}\varepsilon E^2 \frac{I_2}{I_0}\left(\frac{\partial^2 \eta}{\partial x^2}\right)^2 \right]\left(\frac{\partial^2 \eta}{\partial x^2}\right)^2 dx, \tag{3.3.3}$$

and the kinetic energy is equal to

$$K = \frac{m}{2} \int_0^l \left(\frac{\partial \eta}{\partial t}\right)^2 dx. \tag{3.3.4}$$

In (3.3.3) and (3.3.4) the following designations are accepted: $m$ – the mass of the unit length of the bar; $I_0 = \iint_F y^2 dy dz$; $I_2 = \iint_F y^4 dy dz$–the area of cross-section; and $\eta = \eta(x, t)$ – deflection.

We will compose Hamilton's integral and move to dimensionless parameters according to the formulas

$$\xi = \pi\frac{x}{l}; \tau = \omega t,$$

where $\omega$ – the desired frequency of nonlinear oscillations. Hamilton's integral takes the form

$$F = \frac{mF_1}{2\pi\omega} \int_0^{2\pi} \int_0^{\pi} \left\{ a^2 \left[ 1 - \frac{\lambda}{6}\left(\frac{\partial^2 \eta}{\partial x^2}\right)^2 \right]\left(\frac{\partial^2 \eta}{\partial x^2}\right)^2 - \frac{\omega^2 \partial^2 \eta}{\partial t^2} \right\} d\xi d\tau,$$

where the following designations are accepted

$$a^2 = \frac{\pi^4 EI}{ml^4}; \qquad \lambda = \frac{3\pi^4 \varepsilon E^2 I_2}{l^4 I_0}.$$

It is assumed that the desired function $\eta(\xi, \tau)$ can be found in the following way

$$\eta(\xi, \tau) = X(\xi) \cdot Y(\tau)$$

Let us introduce the designations:

$$b_0 = \int_0^\pi X^2 d\xi; \quad v^2 = \frac{1}{b_0} \int_0^\pi X''^2 d\xi; \quad b_2 = \frac{1}{v^2 b_0} \int_0^\pi X''^4 d\xi. \qquad (3.3.5)$$

Integral $F$ takes the form

$$F = \frac{mF_1}{2\pi\omega} b_0 \int_0^{2\pi} \left[ v^2 a^2 Y^2 \left( 1 - \frac{\lambda}{6} b_2 Y^2 \right) - \omega^2 \left( \frac{\partial Y}{\partial \tau} \right)^2 \right] d\tau. \qquad (3.3.6)$$

Function $Y(\tau)$ must satisfy the Euler's equation, composed for the functional in (3.3.6), and having the form:

$$\frac{\partial^2 Y}{\partial \tau^2} + \frac{v^2 a^2}{\omega^2} \left( 1 - \frac{1}{3} \lambda b_2 Y^2 \right) Y = 0 \qquad (3.3.7)$$

The solution of Eq. (3.3.7) is given in [163]. Using it, we find the value of the period of nonlinear oscillations $T$

$$T = \frac{1}{va} \left[ 1 + \frac{3}{8} \frac{\lambda b_2}{3} Y_i^2 + \frac{57}{265} \left( \frac{\lambda b_2}{3} \right)^2 Y_i^4 + \frac{315}{2048} \left( \frac{\lambda b_2}{3} \right)^3 Y_i^6 + \cdots \right], \qquad (3.3.8)$$

where $Y_i$ – the fixed value of initial maximum displacement.

From the above information it is clear that in order to solve the problem it is necessary to set the function of $X(\xi)$. The best approximation for the "weakly nonlinear" systems can be obtained if we proceed from the corresponding fundamental functions, obtained at (with) $\varepsilon = 0$. Let us examine several types of beams with different supporting fastenings, i.e. with different boundary conditions.

### 3.3.1  Simply Supported Beams

The function, which determines the main form (first mode) of free linear oscillations has the form

$$X(\xi) = \sin \frac{k\xi}{\pi}, \quad \text{where } k = \pi$$

Constants in (3.3.5) will take the following values:

$$b_0 = \frac{\pi}{2}; \quad \nu^2 = 1; \quad b_2 = 0.75$$

The period of nonlinear oscillations according to (3.3.8) will be:

$$T = T_l[1 + 0.09375\lambda Y_i^2 + 0.013916\lambda^2 Y_i^4 + 0.002403\lambda^3 Y_i^6 + \cdots], \quad (3.3.9)$$

where $T_l = \frac{2\pi}{a}$ – period of free nonlinear oscillations. This result is obtained by G. Kauder [163].

### 3.3.2   Beams With Built-in Ends

As $X(\xi)$, we assign a corresponding fundamental function, determining the main mode of free linear oscillations [8, 348], with the value of maximum amplitude, equal to one:

$$X(\xi) = \frac{1}{1.61643}\left[\sin\frac{k\xi}{\pi} - 1.0178\cos\frac{k\xi}{\pi} - \sinh\frac{k\xi}{\pi} + 1.0178\cosh\frac{k\xi}{\pi}\right];$$

$$k = 4.73$$

After calculating integrals (3.3.5), we will obtain the following values:

$$b_0 = 1.24542; \nu^2 = 5.1384; \quad b_2 = 3.77213.$$

From (3.3.8), we will have

$$T = T_l[1 + 0.47152\lambda Y_i^2 + 0.35203\lambda^2 Y_i^4 + 0.30577\lambda^3 Y_i^6 + \cdots], \quad (3.3.10)$$

$$T_l = \frac{2\pi}{2.267a}$$

### 3.3.3   Beams With One End Hinged Support and Another End Built-in Support

For function $X(\xi)$, we have

$$X(\xi) = \frac{1}{1.06676}\left[\sin\frac{k\xi}{\pi} + 0.02785\sinh\frac{k\xi}{\pi}\right], \quad k = 3.927.$$

The maximum value of the function $X(\xi)$ is equal to one, and takes place at the distance $0.42l$ from the supported end.

$$b_0 = 1.37911; \quad v^2 = 2.4423; b_2 = 1.8116$$

$$T = T_l[1 + 0.22645\lambda Y_i^2 + 0.081193\lambda^2 Y_i^4 + 0.03387\lambda^3 Y_i^6 + \cdots], \quad (3.3.11)$$

$$T_l = \frac{2\pi}{1.563a}$$

### 3.3.4   The Cantilever Beam

$$X(\xi) = \frac{1}{2.7242}\left[\sin\frac{k\xi}{\pi} - 1.3622\cos\frac{k\xi}{\pi} - \sinh\frac{k\xi}{\pi} + 1.3622\cosh\frac{k\xi}{\pi}\right]$$

$$k = 1.875; \quad b_0 = 0.78536; \quad v^2 = 0.1269; \quad b_2 = 0.074514$$

$$T = T_l[1 + 0.009314\lambda Y_i^2 + 0.000137\lambda^2 Y_i^4 + 0.0000023\lambda^3 Y_i^6 + \cdots],$$
$$(3.3.12)$$

$$T_l = \frac{2\pi}{0.362a}$$

As an illustration, we conduct calculations of the periods of nonlinear oscillations of the reinforced concrete bent beams. Let us determine the periods of the nonlinear oscillations of beams, with the following geometric dimensions: $l = 300$ cm; $b = h = 40$ cm. On the basis of physical considerations, it is possible to use the dependency (3.3.1) with such values of $\theta$, for which $\sigma'(\theta) > 0$, i.e.

$$\sigma'(\theta) = E(1 - 3\varepsilon^2\theta^2) > 0$$

Hence, we have:

$$\varepsilon_{max} \leq \frac{1}{3E^2\theta^2_{max}}$$

Assuming, that the maximum relative deformation in the reinforced concrete bent beams equals to $\theta = 0.003$, we obtain the maximum value of coefficient $\lambda$:

$$\lambda_{max} \leq \frac{\pi^4}{l^4\theta^2_{max}}\frac{12}{80}h = 0.32$$

Based on this data Figure 3.4 shows the dependencies of periods $T/T_l$ and the initial displacement $Y_i$, the maximum value of which is equal to $Y_{i,max} = \frac{l}{200} = 1.5$ cm. As can be seen in Figure 3.4, the same initial displacement for

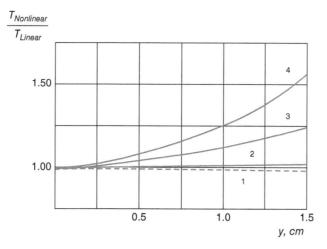

**Figure 3.4**   Dependencies of periods of nonlinear and linear oscillations of bent beams and displacement: (1) cantilever beam; (2) simply hinged supported beam; (3) beam with one end-hinged support and another with end built-in support; and (4) beam with built-in ends.

nonlinearity of deformation has the greatest effect on the period of oscilla-tions of the beam with the built-in ends.

If in the simple beam with $Y_i = 1.5$ cm, the period of nonlinear oscil-lations increases by 1.08 times, then in the beam with the built-in ends it increases by 1.6 times. Similarly, a dependency "period displacement" can be obtained for a shifted cantilever beam with distributed mass $m$. The relationship of sheering stress $\tau$ and angular deflection $\gamma$ is written in the form of:

$$\tau = \gamma G(1 - \varepsilon\gamma^2 G^2)$$

where $G$ – shear modulus; and $\varepsilon$ – coefficient of nonlinearity of deformation. We compose Hamilton's integral

$$F = \frac{1}{2}\int_0^{\frac{2\pi}{\omega}}\int_0^l \left\{ kF_1 G\left[1 - \frac{\varepsilon G^2}{2}\left(\frac{\partial\eta}{\partial x}\right)^2\right]\left(\frac{\partial\eta}{\partial x}\right)^2 - \left(\frac{\partial\eta}{\partial t}\right)^2\right\} dxdt, \quad (3.3.13)$$

where $k$ – coefficient for non-uniform distribution of shearing stresses on the height of the cross-section. We look for the unknown displacement in the form of

$$\eta(x, t) = X(x) \cdot T(t) \quad (3.3.14)$$

For the cantilever beam, the function $X(x)$ is set as one corresponding to the $i$th mode of the free linear oscillations, i.e.

$$X(x) = \sin \frac{2i-1}{2}\pi\frac{x}{l}; i = 1, 2 \dots$$

Let us introduce the following designations:

$$a_0 = \int_0^l X^2 dx; \quad a_1 = \int_0^l X'^2 dx; \quad a_2 = \int_0^l X'^4 dx; \qquad (3.3.15)$$

The calculations of coefficients for the cantilever beam gives:

$$a_0 = \frac{l}{2}; \quad a_1 = \frac{1}{2}\left(\frac{2i-1}{2}\pi\right)^2; \quad a_2 = \frac{3}{4l^3}\left(\frac{2i-1}{2}\pi\right)^4 \qquad (3.3.16)$$

Substituting (3.3.14) in (3.3.13), and then conducting integration for $X$, taking into account (3.3.16), and composing the Euler's equation for this functional, we obtain the following:

$$Y'' + G\frac{kF_1}{m}\left(\frac{2i-1}{2}\frac{\pi}{l}\right)^2 Y\left[1 - \frac{3}{4}\varepsilon G^2\left(\frac{2i-1}{2}\frac{\pi}{l}\right)^2 Y^2\right] = 0. \qquad (3.3.17)$$

Let us designate:

$$\chi_i^2 = \frac{kGF_1}{m}\left(\frac{2i-1}{2}\frac{\pi}{l}\right)^2; \quad \mu_i = \varepsilon G^2\left(\frac{2i-1}{2}\frac{\pi}{l}\right)^2\frac{kGF_1}{m}. \qquad (3.3.18)$$

The approximate solution of the nonlinear differential equation given above is known [163]. With the aid of that solution the following expression is obtained for the period of $i$th mode

$$T_i = \frac{2\pi}{\chi_i}[1 + 0.28125\mu_i Y_i^2 + 0.125244\mu_i Y_i^4 + 0.064888\mu_i^3 Y_i^6 +] \qquad (3.3.19)$$

As in [163], here, too, it is possible to show that the relation of the periods of nonlinear and linear oscillations with the identically stressed states does not depend on the number of the form of oscillations. If, in the first approximation we use linear theory, then the maximum shearing stress $\tau_{max}$ can be determined according to the formula:

$$\tau_{max} \approx kGX'(x)_{max} = \frac{(2i-1)\pi}{2}\frac{\pi}{2}xY_i \qquad (3.3.20)$$

Hence, we find

$$Y_i = \frac{\tau_{max}}{kG}\frac{2l}{(2i-1)\pi}$$

For the second term of the series in (3.3.19) we will obtain the expression:

$$0.28125\mu Y_i^2 = \varepsilon \frac{GF_1}{km}\tau_{max} \qquad (3.3.21)$$

The expression (3.3.21) shows that with the identical stressed state an increase in the period of nonlinear oscillations, as compared to the period of linear oscillations of the corresponding form, occurs to the identical degree.

If, in the examined systems where the distributed parameters hysteresis oscillations occur, then their periods are determined from formula (3.2.13)

## 3.4   The Oscillations of the Beam of the Variable Cross-sections

The solution to the problem of free oscillation of the wedge-shaped beam in the Bessel functions was obtained by Kirchhoff. Subsequently, the exact solutions for beams of variable sections, expressed through the special functions, were obtained in works [91, 187, and 204]. However, the set of problems that have exact solutions is very small. Therefore, when studying free oscillations of a bar of variable cross-sections, the methods of approximations are more frequently applied in particular the Ritz method and Galerkin method, as well as other methods of approximations. It is practically impossible to provide a review of all the research in which these methods are used. We will note only some basic analysis, which led to the mathematical foundation of these methods, for more complete bibliography see [187, 341, and 358]. The enumerated methods, being approximate, under specific conditions allow for obtaining solutions with predetermined accuracy. So, for example, with the aid of the Ritz's method it is possible to obtain the exact solution of a problem, if as the basic functions we select a system of linear independent functions, which satisfy boundary conditions and condition of completeness of energy [230]. In [194] the solution of the problem of free oscillations of a hinge-supported beam with linearly changing height of the cross-section is obtained by this method. Below, with the aid of Ritz's method, we examine the oscillations of the hinge-supported beam with the arbitrary changing height of the cross-section [13] (Figure 3.5).

The equation of the free oscillations of the beam with variable cross-section has the form

$$\frac{\partial^2}{\partial x^2}\left[EI(x)\frac{\partial^2 y}{\partial x^2}\right] + \rho F(x)\frac{\partial^2 y}{\partial t^2} = 0 \qquad (3.4.1)$$

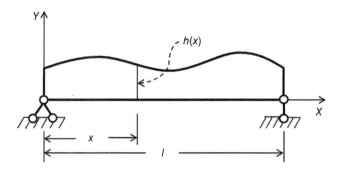

**Figure 3.5** Scheme of the beam.

Assuming harmonic oscillations with a frequency $\omega$,

$$Y(x,t) = Y(x)e^{-i\omega t}$$

and replacing the variable in the formula $\xi = \frac{x}{l}$, we will obtain:

$$\frac{d^2}{d\xi^2}\left[I(\xi)\frac{d^2Y}{d\xi^2}\right] - \lambda F(\xi)Y(\xi) = 0, \quad \lambda = \frac{\rho\omega^2}{E}l^4 \tag{3.4.2}$$

The boundary conditions for the hinge-supported beam are the following:

$$Y(0) = Y(l) = Y''(0) = Y''(l) = 0 \tag{3.4.3}$$

The functional for which the Euler equation is the equation of motion (3.4.2) has the form:

$$S = \frac{1}{2}\int_0^l \{I(\xi)[Y''(\xi)]^2 - \lambda F(\xi)[Y(\xi)]^2\}d\xi \tag{3.4.4}$$

The problem is solved for the case of the arbitrary law of variation in the cross section. It is accepted that cross-sectional area $F(\xi)$ and the moment of inertia of the cross-section $I(\xi)$ are given in the form of Fourier series:

$$F(\xi) = \frac{a_0}{2} + \sum_{j=1}^{\infty} a_j \cos j\pi\xi + b_j \sin j\pi\xi \tag{3.4.5}$$

$$I(\xi) = \frac{c_0}{2} + \sum_{j=1}^{\infty} c_j \cos j\pi\xi + d_j \sin j\pi\xi$$

The solution of Eq. (3.4.2) is sought in the form of the Fourier series, with members satisfying the boundary conditions of the problem:

$$Y(\xi) = \sum_{k=1}^{\infty} A_k \sin k\pi\xi \tag{3.4.6}$$

Substituting the expression for $Y(\xi)$ and $Y''(\xi)$, and also $F(\xi)$ and $I(\xi)$ in (3.4.4), we will obtain:

$$2S = \int_0^l \left\{ \sum_{i=1}^{\infty}\sum_{k=1}^{\infty} \left(\frac{c_0}{2}i^2k^2\pi^4 - \lambda\frac{a_0}{2}\right) A_iA_k \sin k\pi\xi \sin i\pi\xi \right.$$

$$+ \sum_{j=1}^{\infty}\sum_{i=1}^{\infty}\sum_{k=1}^{\infty}(c_ji^2k^2\pi^4 - \lambda a_j)A_iA_k \sin k\pi\xi \sin i\pi\xi \cos j\pi\xi \tag{3.4.7}$$

$$+ \left. \sum_{j=1}^{\infty}\sum_{i=1}^{\infty}\sum_{k=1}^{\infty}(d_ji^2k^2\pi^4 - \lambda b_j)A_iA_k \sin k\pi\xi \sin i\pi\xi \sin j\pi\xi \right\} d\xi.$$

Let us calculate the integrals, entering in (3.4.7)

$$I_0(k,i) = \int_0^l \sin k\pi\xi \sin i\pi\xi d\xi = \begin{cases} 0, i \neq k \\ 1/2, i = k \end{cases} \tag{3.4.8}$$

$$I_1(k,i,j) = \int_0^l \sin k\pi\xi \sin i\pi\xi \cos j\pi\xi \, d\xi = \begin{cases} 0, j \neq \pm(k \mp i) \\ 1/4, j = \pm(k - i) \\ -1/4, j = \pm(k + i) \end{cases}$$

$$I_2(k,i,j) = \int_0^l \sin k\pi\xi \sin i\pi\xi \sin j\pi\xi \, d\xi$$

$$= \begin{cases} \dfrac{j}{2\pi}\left[\dfrac{1-(-1)^{k-i+j}}{j^2-(k-i)^2} - \dfrac{1-(-1)^{k+i+j}}{j^2-(k+i)^2}\right] \\ \qquad k \mp i \mp j \neq 0 \\ 0, j = \pm(k \mp i) \end{cases}$$

Taking into account (3.4.8), the functional $2S$ is represented in the form of

$$2S = \frac{1}{4}\sum_{i=1}^{\infty}(c_0i^4\pi^4 - \lambda a_0)A_i^2 + \sum_{j=1}^{\infty}\sum_{i=1}^{\infty}\sum_{k=1}^{\infty} I_1(k,i,j)(c_ji^2k^2\pi^4 - \lambda a_j)A_iA_k$$

$$+ \sum_{j=1}^{\infty}\sum_{i=1}^{\infty}\sum_{k=1}^{\infty} I_2(k,i,j)(d_ji^2k^2\pi^4 - \lambda b_j)A_iA_k. \tag{3.4.9}$$

According to the Ritz method, the conditions of the minimum (minimal conditions) of functional (3.4.9) will be:

$$\frac{\partial S}{\partial A_i} = 0 \quad i = 1, 2, 3 \cdots \tag{3.4.10}$$

Eq. (3.4.10) in the expanded form will be:

$$\frac{1}{2}(c_0 i^4 \pi^4 - \lambda a_0)A_i + \frac{\partial}{\partial A_i}\sum_{i=1}^{\infty} f_{ji}^{(1)}A_i + \frac{\partial}{\partial A_i}\sum_{i=1}^{\infty} f_{ji}^{(2)}A_i = 0,$$

where

$$f_{ji}^{(1)} = \sum_{j=1}^{\infty}\sum_{k=1}^{\infty} I_1(j,i,k)(c_j i^2 k^2 \pi^4 - \lambda a_j)A_k, \tag{3.4.11}$$

$$f_{ji}^{(2)} = \sum_{j=1}^{\infty}\sum_{k=1}^{\infty} I_2(j,i,k)(d_j i^2 k^2 \pi^4 - \lambda b_j)A_k.$$

Bearing in mind that

$$\frac{\partial}{\partial A_i}\sum_{i=1}^{\infty} f_{ji}^{(1,2)}A_i = f_{ji}^{(1,2)} + \sum_{k=1}^{\infty} A_k\frac{\partial f_{ji}^{(1,2)}}{\partial A_i}, \quad I_{1,2}(j,i,k) = I_{1,2}(j,k,i),$$

we will obtain an infinite system of homogeneous equations, which has the form:

$$\sum_{k=1}^{\infty} A_k \mu_{ik} = 0, i = 1, 2, 3 \cdots \tag{3.4.12}$$

$$\mu_{ik} = 2\sum_{j=1}^{\infty} I_1(j,i,k)[c_j i^2 k^2 \pi^4 - \lambda a_j] + I_2(j,i,k)[d_j i^2 k^2 \pi^4 - \lambda b_j],$$

$$\mu_{ii} = 2\sum_{j=1}^{\infty}\{I_1(j,i,i)[c_j i^4 \pi^4 - \lambda a_j] + I_2(j,i,i)[d_j i^4 \pi^4 - \lambda b_j]\}$$

$$+ \frac{1}{2}(c_0 i^4 \pi^4 - \lambda a_0)$$

The Eigen (natural) frequencies are determined by the solution of the infinite determinant of the form:

$$|\mu_{ik}| = 0, i, k = 1, 2.3 \cdots \tag{3.4.13}$$

Since in the solving of the problem, the basis were taken as complete (total, full) energy functions, then according to [230], at $i, k \to \infty$, the dimensionless frequencies $\lambda$ have to approach their exact values. After determining Eigen frequencies the forms of oscillations can also be determined. If dimensionless frequencies $\lambda$ are determined from the truncated determinant of order $P$, the coefficients $A_i$ can be determined from the system of equations of $P - 1$ order, in which one of the coefficients is taken as the given one (for example $A_1 = 1$). The system of such equations will take the form:

$$\sum_{k=2}^{P} A_k \mu_{ik} = -\mu_{i1}, \ i = 2, 3, \cdots P. \qquad (3.4.14)$$

As an example, let us examine the free oscillations of a beam of a constant width $b$ and variable height $h(\xi)$:

$$h(\xi) = h_0 + h_1 \sin \pi \xi \qquad (3.4.15)$$

where $h_0, h_1$ – heights of the sections at $\xi = 0, 1/2$.
In this case, $F(\xi)$, and $I(\xi)$ are represented in the form of:

$$F(\xi) = F_0(1 + \gamma \sin \pi \xi) \qquad (3.4.16)$$

$$I(\xi) = I_0\{1 + 1.5\gamma^2 - 1.5\gamma^2 \cos 2\pi \xi + (3\gamma + 0.75\gamma^3) \sin \pi \xi - 025\gamma^3 \sin 3\pi \xi\}$$

where

$$F_0 = bh_0, \quad I_0 = \frac{bh_0^3}{12}, \quad \gamma = \frac{h_1}{h_0}$$

Comparing (3.4.16) and (3.4.5) we find that the coefficients of decomposition of the functions $F(\xi)$ and $I(\xi)$ in the Fourier series take the form:

$$\frac{a_0}{2} = F_0; \quad a_1 = a_2 = \cdots = 0; \quad b_1 = F_0\gamma; \quad b_2 = b_3 = \cdots = 0;$$

$$\frac{c_0}{2} = I_0(1 + 1.5\gamma^2); \quad c_1 = 0; \quad c_2 = -1.5\gamma^2 I_0; \quad c_3 = c_4 = \cdots = 0;$$

$$d_1 = I_0(3\gamma + 0.75\gamma^3); \quad d_2 = 0; \quad d_3 = -0.25\gamma^3 I_0; \quad d_4 = d_5 = \cdots = 0.$$

Eq. (3.4.13) was solved at $i, k = 1, 3, 5$. The speed of the convergence of the solution can be judged by the data in Table 3.6, where the values of dimensionless frequency $\sqrt{\lambda_i}$ were given by $\left(\sqrt{\lambda_i} = \omega_i \frac{l^2}{\pi^2} \sqrt{\frac{m}{EI_0}}\right)$.

**Table 3.6**  Values of $\sqrt{\lambda_i}$.

| | $i,k=1$ | $i,k=3$ | | | $i,k=5$ | | | | |
|---|---|---|---|---|---|---|---|---|---|
| $\gamma$ | $\sqrt{\lambda_1}$ | $\sqrt{\lambda_1}$ | $\sqrt{\lambda_2}$ | $\sqrt{\lambda_3}$ | $\sqrt{\lambda_1}$ | $\sqrt{\lambda_2}$ | $\sqrt{\lambda_3}$ | $\sqrt{\lambda_4}$ | $\sqrt{\lambda_5}$ |
| 0 | 1 | 1 | 4 | 9 | 1 | 4 | 9 | 16 | 25 |
| 0.1 | 1.085 | 1.084 | 4.274 | 9.599 | 1.084 | 4.268 | 9.585 | 17.053 | 26.638 |
| 0.25 | 1.215 | 1.209 | 4.691 | 10.528 | 1.209 | 4.660 | 10.453 | 18.693 | 29.192 |
| 0.75 | 1.666 | 1.623 | 6.122 | 13.798 | 1.618 | 5.910 | 13.301 | 24.494 | 38.251 |

As one would expect, the convergence of the process is correct, especially when determining the fundamental frequency and for small values of $\gamma$. The accuracy of the obtained results can be judged by the proximity of the results for the same frequencies, obtained during the solution of the determinants of gradually increasing order.

Figure 3.6 shows the dependencies of the first three natural frequencies of the beams, cross-sections of which change according to the law (3.4.15), depending on the parameter $\gamma$. As evident from Figure 3.6, the values of natural frequencies increase with an increase in $\gamma$.

In the examined interval of the variation $\gamma$ the obtained dependences are very close to the linear dependency.

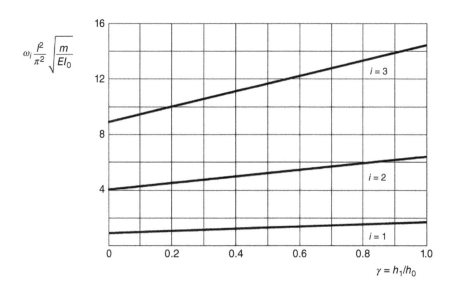

**Figure 3.6**  Dependencies of frequencies of free oscillation of a beam from a parameter characterizing changes of cross sections.

## 3.5   The Optimum Design of the Bar

The problems of the optimum design of bars are examined in the works [300, 350, 365, 385]. Let us examine a problem for an optimal design of a cantilever-shifted bar with distributed mass, having a concentrated mass at its free end. Lei determined the law of variation of cross-section of a bar, which has a minimum mass of its own, with a given value of natural (fundamental) frequency. A similar task for the longitudinal oscillations (vibrations) was examined by M. Turner [385]. The solution of the same problem, based on the energy considerations, is given by Zh. Taylor [350], the following is the solution of the problem, based on that work.

According to the Hamilton's principle, in a section of a real motion of a system in the time interval $t_1 \div t_2$, the integral $A = \int_{t_1}^{t_2}(K - P)dt$ takes extreme value. Here $K$, $P$ are kinetic and potential energy of the system, respectively, which, with the shift oscillations (vibrations), are determined by the expressions:

$$K = \frac{1}{2}\int_0^l m(x)\left(\frac{\partial y}{\partial t}\right)^2 dx + M\frac{\partial y(l)^2}{dt};  \qquad (3.5.1)$$

$$P = \frac{1}{2}\int_0^l \frac{kG}{\rho}m(x)\left(\frac{\partial y}{\partial x}\right)^2 dx,$$

where $m(x)$ – the intensity of the distributed mass; $M$ – the value of the mass, concentrated on the free end of the cantilever bar; $\rho$ – the mass of unit volume; and $k$ – coefficient, considering the nonuniform (uneven) distribution of shearing stresses. At free oscillations with frequency of $\omega$, displacement $y(x, t)$ takes the form:

$$y(x, t) = Y(x) \cdot \cos \omega t$$

The Hamilton's integral in this case takes the form $\overline{A} = \pi(\overline{K} - \overline{P})$, where

$$\overline{K} = \frac{\omega^2}{2}\left[\int_0^l m(x)Y^2(x)dx + MY^2(l)\right], \quad \overline{P} = \frac{1}{2}\int_0^l \frac{kG}{\rho}m(x)Y'^2(x)dx.$$

The problem is formulated as follows: to determine the minimum of the functional $(\overline{K} - \overline{P})$, depending on the variable $Y(x)$ and $m(x)$ with the fixed volume, i.e.;

$$\overline{M} = \int_0^l m(x)dx = const$$

The work [350] shows that such formulation of the problem statement is equivalent to the requirement of a minimum mass with the satisfaction of the equation of motion. However, in our case the course of solution proves to be quite simple. Let us compose a new functional

$$A^* = (\overline{K} - \overline{P}) + \mu \overline{P} = \frac{1}{2} \int_0^l \left[ \omega^2 m(x) Y^2(x) - \frac{kGm(x)}{2} Y'^2(x) + 2\mu m(x) \right] dx,$$

(3.5.2)

where $\mu$ – Lagrange's coefficient.

For this functional it is possible to compose two Euler equations relative to two variables $Y(x)$ and $m(x)$.

$$\omega m(x) Y(x) + \frac{kG}{\rho} [m(x) Y'(x)] = 0$$

(3.5.3)

$$\omega^2 Y^2 - \frac{kG}{\rho} Y'^2 + 2\mu = 0$$

with the boundary conditions:

$$Y(0) = 0$$

$$\frac{kGm(l)}{\rho} Y'(l) = \omega^2 M Y(l)$$

(3.5.4)

Differential Eq. (3.5.4) depends only on displacement and is solved explicitly. In order to determine constant of integration and undetermined coefficient of Lagrange, we use the boundary condition $Y(0) = 0$ and the condition of normalization (regulation) of displacement of $Y(l) = 1$.

The solution to (3.5.4) will be:

$$Y(x) = \frac{\sinh \beta x}{\sinh \beta l},$$

(3.5.5)

where $\beta^2 = \frac{\omega^2 \rho}{kG}$.

From (3.5.3) we have

$$\frac{m'(x)}{m(x)} = \frac{Y'' + \beta^2 Y}{Y'}; \quad \ln \frac{m(x)}{m(l)} = \int_x^l \frac{Y'' + \beta^2 Y}{Y'} dx$$

(3.5.6)

Using (3.5.5) and the second boundary condition from (3.5.4), we will obtain the law of variation of change of mass $m(x)$

$$m(x) = \beta M \frac{\sinh \beta l \cosh \beta l}{\cosh^2 \beta x}$$

(3.5.7)

The law of variation in the cross section $F(x)$ will take the form:

$$F(x) = \frac{m(x)}{\rho} \tag{3.5.8}$$

The mass of the entire bar (without concentrated mass) is equal to:

$$M_0 = \int_0^l m(x)dx = M \sinh^2 \beta l \tag{3.5.9}$$

Let us calculate the weight saving of the mass in an optimum bar as compared to a bar with a constant cross-section, with concentrated mass. Let us assume that the fundamental frequency of free oscillations of a bar is assigned and equal to: $\omega^2 = \left(\frac{0.86}{l}\right)^2 \frac{kG}{\rho}; \beta l = 0.86$. This corresponds to the case when the ratio of the mass of the bar to the concentrated mass equals to $M_0/M = 1$. If we synthesize the optimum bar with this frequency, then its mass will be equal $M_0 = M \sinh^2 \beta l = 0.935$. Therefore the mass of the optimum bar composes 0.935 from the mass of the bar with constant cross-section. When $\beta l$ is equal to 0.65 and 1.08, which corresponds to $M_0/M$ equaling to 0.5 and 2, then the savings of mass will be 0.97 and 0.845, respectively. It turns out that the savings in the mass of a construction is

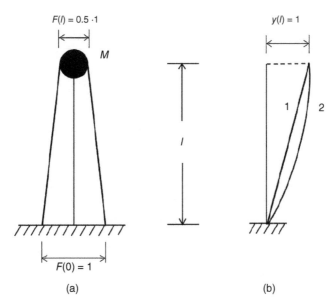

(a)                                                  (b)

**Figure 3.7**   (a) Form of optimally shifted bar with $\beta l = 0.86$; and (b) form of oscillations of a bar of an optimal configuration (1), and constant (2) cross section.

more with greater values of $\beta l$, which corresponds to a case when the mass of a bar itself is greater than the concentrated mass. This is understandable, since the value of the concentrated mass is fixed and the minimum of weight is obtained by the variation of the mass of bar.

Figure 3.7 shows the optimum configuration of a bar with $\beta l = 0.86$ and the modes of free oscillations of the fundamental tone of optimum and constant cross-sections of bars. The configuration of the bar depends on the value of $\beta l$, and the greater are the values of $\beta l$ the more pointed (peaked) it becomes.

## 3.6 The Oscillations of Flexural-Shifted (Bending-Shifted) Bars Under the Seismic Impacts

Extensive literature has been dedicated to the oscillation of bent beams, taking into account the shearing strain and rotary inertia (or the so-called beams of Timoshenko) [1, 146, and 388]. The majority of these works is devoted to the determination of frequencies and forms of free oscillations. The analytical solution of a problem with forced oscillations of Timoshenko's beam and simplified boundary conditions with simple external impact can be ascertained in [388]. In [170, 171] forced oscillations of a beam, taking into account only shearing strain, are examined. These works do not consider an influence of damping oscillations. The design diagrams of certain constructions are represented in a form of the cantilever bending-shifted bar with a constant or variable cross section. The seismic impact is represented in the form of a given law of the base motion [314]. Most frequently, the law of oscillation (vibration) of the basic (bearing) cross-section of cantilever bar is given in the form of function of acceleration from time accelerograms. Below we will examine a problem of determining displacements, bending moments, and shear forces of the bending-shifted cantilever bar taking, into account damping oscillations (vibrations) with the known accelerogram of the base motion. The equations of motion of the bending-shifted beam taking into account damping are represented in the form of:

$$kFG\left(\frac{\partial y}{\partial x} - \psi\right) + \alpha_2 kFG\left(\frac{\partial^2 y}{\partial x \partial t} - \frac{\partial \psi}{\partial t}\right) + EI\frac{\partial^2 \psi}{\partial x^2} + \alpha_1 EI\frac{\partial^3 \psi}{\partial x^2 \partial t} = 0 \quad (3.6.1)$$

$$\frac{q}{g}\frac{\partial^2}{\partial t^2}(y + y_0) - kFG\left(\frac{\partial^2 y}{\partial x^2} - \frac{\partial \psi}{\partial x}\right) - \alpha_2 kFG\left(\frac{\partial^3 y}{\partial x^2 \partial t} - \frac{\partial^2 \psi}{\partial x \partial t}\right) = 0$$

were $y$ – the displacement of beam; $\psi$ – the angle of tangent inclination to the elastic curve, which appears from the impact of the bending moment; $EI$ – the flexural (bending) rigidity of beam; $kFG$ – the shift stiffness (rigidity) of beam; $k$ – coefficient, which considers the nonuniform distribution (which takes into account the unequal distribution of tangential stresses) of shearing stresses ($k = 0.83$ for the rectangular cross section); $m = q/g$ – the mass of a unit length of the beam; $\alpha_1$, $\alpha_2$ – coefficients, accounting for damping of oscillations, and the corresponding bending, and shearing strains; and $y_0''(t)$ – the given accelerogram of the ground (base) motion.

Equations similar to (3.6.1) and corresponding to free oscillations, were obtained in [75]. We search for unknown displacements $y$ and $\psi$ in the form of

$$y = \sum_{j=1}^{\infty} X_j(x) q_j(t), \quad \psi = \sum_{j=1}^{\infty} \psi_j(x) q_j(t), \tag{3.6.2}$$

where $q_j(t)$ – the main coordinates of the given elastic system; and $X_j$, $\psi_j$ – the functions characterizing the $j$th form of free oscillations and determined by the solution to a problem with free oscillations:

$$X_j(x) = A_j k_j \cos k_j \frac{x}{l} - B_j k_j \sin k_j \frac{x}{l} + C_j \omega_j \cosh \omega_j \frac{x}{l} + D_j \omega_j \sinh \omega_j \frac{x}{l}$$

$$\psi_j(x) = -A_j \frac{k_j^2}{1 + k_j^2 v^2} \sin k_j \frac{x}{l} - B_j \frac{k_j^2}{1 + k_j^2 v^2} \cos k_j \frac{x}{l} \tag{3.6.3}$$

$$+ C_j \frac{\omega_j^2}{1 - v^2 \omega_j^2} \sinh \omega_j \frac{x}{l} + D_j \frac{\omega_j}{1 - v^2 \omega_j^2} \cosh \omega_j \frac{x}{l},$$

$$v^2 = \frac{EI}{kFGl^2}, \omega_j = k_j (1 + k_j^2 v^2)^{-\frac{1}{2}},$$

where $l$ – the length of beam; $\omega_j$ – the frequency of $j$th form of free oscillations, $A_j$, $B_j$, $C_j$; and $D_j$ – arbitrary constants. The equation for determining the natural frequencies of the cantilever beam takes the form [171]:

$$(\omega_j^4 + k_j^4) \cosh \omega_j \cos k_j - v^2 k_j^3 \omega_j^3 \sin k_j \sinh \omega_j + 2\omega_j^2 k_j^2 = 0 \tag{3.6.4}$$

The arbitrary constants, corresponding to the $j$th form, are bound by the relations:

$$A_j k_j = -C_j \omega_j$$

$$B_j k_j = -C_j \frac{k_j^3 \left( \omega_j^3 \cos k_j + k_j^2 \omega_j \cosh \omega_j \right)}{\omega_j^3 \left( k_j^3 \sin k_j + k_j^2 \omega_j \sinh \omega_j \right)} \tag{3.6.5}$$

$$D_j = -C_j \frac{k_j^3 \left( \omega_j^3 \cos k_j + \omega_j k_j^2 \cosh \omega_j \right)}{\omega_j^3 \left( \omega_j k_j^2 \sinh \omega_j + k_j^3 \sin k_j \right)}$$

To solve the problem of the forced oscillations we will first determine the main coordinates $q_j(t)$. For that we will use Lagrange's equation:

$$\frac{d}{dt} \left( \frac{\partial K}{\partial \dot{q}_j} \right) - \frac{\partial K}{\partial q_j} + \frac{\partial P}{\partial q_j} = 0, \tag{3.6.6}$$

where $K$, $P$ – are kinetic and potential energy of system respectively, having the form:

$$P = \int_0^l \frac{Q^2(x, t)}{2kFG} dx + \int_0^l \frac{M^2(x, t)}{2EI} dx; \tag{3.6.7}$$

$$K = \frac{1}{2} \int_0^l \frac{\partial}{\partial t} [y(x, t) + y_0(t)]^2 \frac{q}{g} dx$$

The value of kinetic energy is determined while accounting for the base motion. The bending moment $M(x, t)$ and shear force $Q(x, t)$, in (3.6.7), are bound to the unknown $y$ and $\psi$ by the expressions:

$$M = -EI \frac{\partial \psi}{\partial x} - \alpha_1 EI \frac{\partial^2 \psi}{\partial x \partial t}; \tag{3.6.8}$$

$$Q = kFG \left( \frac{\partial y}{\partial x} - \psi \right) + \alpha_2 kFG \left( \frac{\partial^2 y}{\partial x \partial t} - \frac{\partial \psi}{\partial t} \right).$$

Subsequently, the condition of orthogonality of the characteristic functions of $X_j$ and $\psi_j$ are used to determine $q_j$, which take the following form:

$$\int_0^l X_j(x)X_i(x)dx = \begin{cases} 0 \text{ when } i \neq j \\ \int_0^l X_j^2(x)dx; \text{ when } i = j \end{cases} \tag{3.6.9}$$

$$\int_0^l \{EI\psi_j'(x)\psi_i'(x) + kFG[X_j'(x) - \psi_j(x)][X_i'(x) - \psi_i(x)]\}dx =$$

$$\begin{cases} 0 \text{ when } i \neq j \\ \int_0^l \{EI\psi_j'^2(x) + kFG[X_j'(x) - \psi_j(x)]^2\}dx; \text{ when } i = j \end{cases}$$

Using the expressions in (3.6.8), (3.6.9), and (3.6.2) we calculate the values of potential and kinetic energies (3.6.7). Substituting them in the Lagrange Eq. (3.6.6), we will obtain the differential equation of the second order for determining $q_j$. Expressions for determining $y$ and $\psi$ are obtained in the form of:

$$y(x,t) = -\sum_{j=1}^{\infty} \eta_{x,j} \frac{1}{P_j} \int_0^t y_0''(\xi) e^{-\frac{1}{2}(\alpha_1 P_{j,bend}^2 + \alpha_2 P_{j,shift}^2)(t-\xi)} \cdot \sin P_j(t-\xi)d\xi,$$

$$\tag{3.6.10}$$

$$\psi(x,t) = -\sum_{j=1}^{\infty} \eta_{\psi,j} \frac{1}{P_j} \int_0^t y_0''(\xi) e^{-\frac{1}{2}(\alpha_1 P_{j,bend}^2 + \alpha_2 P_{j,shift}^2)(t-\xi)} \cdot \sin P_j(t-\xi)d\xi;$$

where the following designations are introduced:

$$\eta_{x,j} = X_j(x)\frac{\int_0^l X_j(x)dx}{\int_0^l X_j^2(x)dx}, \quad \eta_{\psi,j} = \psi_j(x)\frac{\int_0^l X_j(x)dx}{\int_0^l X_j^2(x)dx},$$

$$P_j^2 = \frac{\int_0^l \{EI\psi_j'^2(x) + kFG[X_j'(x) - \psi_j(x)]^2\}dx}{m\int_0^l X_j^2(x)dx}, \tag{3.6.11}$$

$$P_{j,bend}^2 = \frac{\int_0^l EI\psi_j'^2(x)dx}{m\int_0^l X_j^2(x)dx}, \quad P_{j,shift}^2 = \frac{\int_0^l kFG[X_j'(x) - \psi_j(x)]^2dx}{m\int_0^l X_j^2(x)dx}.$$

$P_j$ – the frequency of flexural-shift (bending-shift) oscillations; and $P_{j,\,bend}$, $P_{j,\,shift}$ – frequencies, corresponding to oscillations with predominantly bending or shearing strains. The reactions of the system, i.e. the values of the bending moments and of shear forces, are determined by the expressions:

$$M(x,t) = \frac{ql^2}{g} \sum_{j=1}^{\infty} \gamma_j(x) \frac{2\pi}{T_j} \int_0^l y_0''(\xi) e^{-\frac{1}{2}(\alpha_1 P_{j,bend}^2 + \alpha_2 P_{j,shift}^2)(t-\xi)} \cdot \sin \frac{2\pi}{T_j}(t-\xi)d\xi$$

$$\text{(3.6.12)}$$

$$Q(x,t) = \frac{ql}{g} \sum_{j=1}^{\infty} \gamma_j'(x) \frac{2\pi}{T_j} \int_0^l y_0''(\xi) e^{-\frac{1}{2}(\alpha_1 P_{j,bend}^2 + \alpha_2 P_{j,shift}^2)(t-\xi)} \cdot \sin \frac{2\pi}{T_j}(t-\xi)d\xi$$

where

$$\gamma_j(x) = \frac{l^2(1+v^2 k_j^2)}{k_j^4} \psi_j'\left(\frac{x}{l}\right) \frac{\int_0^l X_j\left(\frac{x}{l}\right)dx}{\int_0^l X_j^2\left(\frac{x}{l}\right)dx},$$

$$\gamma_j'(x) = \frac{l^3(1+v^2 k_j^3)}{k_j^4} \psi_j''\left(\frac{x}{l}\right) \frac{\int_0^l X_j\left(\frac{x}{l}\right)dx}{\int_0^l X_j^2\left(\frac{x}{l}\right)dx}.$$

With the aid of the obtained expressions, the stressed state of a construction with seismic impact and during the base motion on the accelerogram of a real earthquake were investigated and studied. The coefficients, characterizing damping $\alpha_1$ and $\alpha_2$, are set as inversely proportional to the corresponding frequency of oscillations. As is known, with single-frequency oscillations this leads to the hypothesis of Sorokin [345], but with non-stationary oscillations with a sufficient approximation, a damped process, not dependent on frequency, is described. Calculations are conducted for a construction which has the form of a rectangle with the following parameters:

$$G = 0.4E, \quad k = 0.8, \quad v^2 = 0.312\left(\frac{h}{l}\right)^2$$

Since at the present time there is no reliable data on the values of damping coefficients $\alpha_1$ and $\alpha_2$ with purely bending and purely shift oscillations, their values are accepted as equal. Reactions are determined taking into account three modes of oscillations. In Figure 3.8 the distributions of the maximum values $MT_1^* g/ql^2 2\pi v$ and $QT_1^* g/ql^2 2\pi v$ are built, where $T_1^*$ – the period of purely bending oscillations; and $v$ – speed of the base motion at

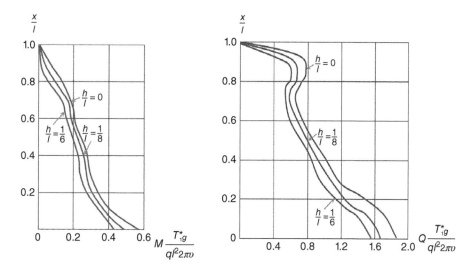

**Figure 3.8** Distribution of the maximum bending moments and shear forces with the seismic impact dependent on $h/l$ with $\alpha_1 = \alpha_2 = 0.1$.

seismic impact, depending on the ratio of height to length of the section $h/l$. As can be seen from Figure 3.8, the maximum bending moments and shear forces decrease upon consideration of the shifting strain, in this case the degree of the reduction depends on the ratio of $h/l$. The reduction of the maximum moment $M$ with the bending-shift oscillations in comparison to the case of flexural (bending) oscillations with $h/l = 1/8$ is equal to 7%, and with $\frac{h}{l} = 1/6$ equal to 12%. Similarly, the decrease of the shear force with the values of $h/l = 1/8$ and $1/6$ are equal to 6% and 10%, respectively.

Calculations were made on the accelerogram of an actual earthquake that occurred on September 3, 1949 in the city of Hollister, CA. The maximum acceleration of the ground composed 123 cm/s². The dependences of the maximum bending moments and shear forces $Mg/ql^2$ and $Qg/ql^2$ on the relation of $h/l$ are given in Figure 3.9. In this case the calculation of the shift strain (deformation) also leads to the decrease of the maximum bending moments and shear forces. The decrease of the bending moment when $h/l = 1/8$ and $1/6$ is 13% and 22%, respectively, and the corresponding shear force equals to 14% and 13%, respectively.

Thus, the conducted research showed that taking into account the shifting strain (deformation) when $h/l \leq 1/8$ can lead to substantial changes in the maximum values of the reactions of a bent construction.

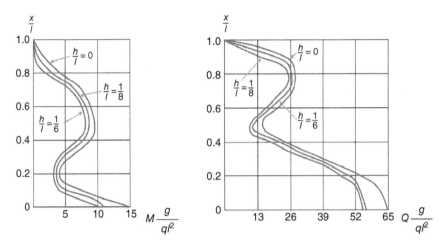

**Figure 3.9**   Distribution of the maximum bending moments and shear forces on the accelerogram of actual earthquake dependent on $h/l$ with $\alpha_1 = \alpha_2 = 0.1$.

## 3.7   Oscillations of Circular Rings and Arches

Let us examine flat bending (flexural) oscillations of circular rings following the work [298]. The diagram of the forces which act on the element of rings is shown in Figure 3.10.

Differential equations of the angular, radial, and tangential motion of elements take the form (Figure 3.10)

$$\frac{\partial M}{\partial \theta} + RQ = \gamma \frac{IR}{g}\frac{\partial^2 \Phi}{\partial t^2}$$

$$\frac{\partial Q}{\partial \theta} + N = \gamma \frac{FR}{g}\frac{\partial^2 U}{\partial t^2} \tag{3.7.1}$$

$$\frac{\partial N}{\partial \theta} - Q = \gamma \frac{FR}{g}\frac{\partial^2 W}{\partial t^2}$$

$M, Q$, and $N$ – bending moment, shear force, and the normal forces, acting in the section with the polar coordinate $\theta$; $F, I$ – area and the moment of inertia of transverse cross-section of the ring; $\gamma$ – the volume weight of material of the ring; $g$ – the acceleration of gravity; $R$ – a radius of the ring; $t$ – time; $U$, $W$ – radial and tangential displacements; and $\Phi$ – angle of rotation of the element of the ring with negligent of shifting strain. Dependency between the bending moment and displacements is represented in the form of:

$$M = \frac{EI}{R^2}\left(\frac{\partial^2 U}{\partial \theta^2} + \frac{\partial W}{\partial \theta}\right) \tag{3.7.2}$$

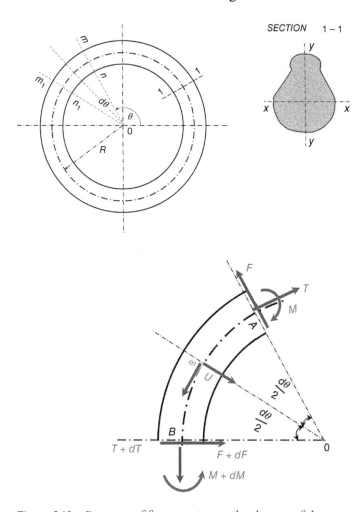

**Figure 3.10**    Diagram of forces acting on the element of the ring.

$E$ – the modulus of elasticity. The angle of rotation of the sections $\psi$ is represented in the form of

$$\Psi = \Phi + \beta \tag{3.7.3}$$

$\beta$ – the angular deformation (deflection), caused by the shift. From the other side, $\psi$ is expressed as displacements $U$ and $W$ according to the expression:

$$\psi = \frac{1}{R}\frac{\partial U}{\partial \theta} + \frac{W}{R} \tag{3.7.4}$$

We will obtain from (3.7.3) and (3.7.4) the following:

$$\beta = \frac{1}{R}\left(\frac{\partial U}{\partial \theta} + W - R\Phi\right) \tag{3.7.5}$$

Dependency of the shear force on $\beta$ is given in the form of:

$$Q = \beta k F G \tag{3.7.6}$$

$G$ – shift module; and $k$ – coefficient, accounting for uneven distribution of shearing stresses on the cross section. In the first approximation the value of $k$ is recommended to be set the same as in the straight beam [355]. Substituting (3.7.5) in (3.7.6) we will obtain:

$$Q = \frac{kFG}{R}\left(\frac{\partial U}{\partial \theta} + W - R\Phi\right) \tag{3.7.7}$$

Accepting the condition of the inextensibility (non-extensibility) of the ring as:

$$\frac{\partial W}{\partial \theta} = U \tag{3.7.8}$$

From (3.7.1), and taking into account (3.7.2), (3.7.6), and (3.7.8), it is possible to obtain the equation of motion of the ring relative to the tangential displacement of ring $W$:

$$\frac{\partial^6 W}{\partial \theta^6} + 2\frac{\partial^4 W}{\partial \theta^4} + \frac{\partial^2 W}{\partial \theta^2} = \left(\frac{\gamma R^2}{Eg} + \frac{\gamma R^2}{gkG}\right)\frac{\partial^6 W}{\partial \theta^4 \partial t^2}$$

$$-\frac{\gamma^2 R^4}{g^2 kEG}\frac{\partial^4 W}{\partial \theta^2 \partial t^2} + \left(2\frac{\gamma R^2}{Eg} - \frac{\gamma R^2}{gkG} - \frac{F\gamma R^4}{EIg}\right)\frac{\partial^4 W}{\partial \theta^2 \partial t^2} \tag{3.7.9}$$

$$+\frac{\gamma^2 R^4}{g^2 kEG}\frac{\partial^4 W}{\partial t^4} + \left(\frac{\gamma R^2}{Eg} + \frac{F\gamma R^4}{EIg}\right)\frac{\partial^2 W}{\partial t^2}$$

With purely bending (flexural) oscillations Eq. (3.7.9) takes the form: [358]:

$$\frac{\partial^6 W}{\partial \theta^6} + 2\frac{\partial^4 W}{\partial \theta^4} + \frac{\partial^2 W}{\partial \theta^2} = -\frac{F\gamma R^4}{EIg}\left(\frac{\partial^4 W}{\partial t^2 \partial \theta^2} - \frac{\partial^2 W}{\partial t^2}\right) \tag{3.7.10}$$

From (3.7.9) with $R \to \infty$ it is possible to obtain the equations of the flexural oscillations (vibrations) of a straight beam taking into account the shift and inertia of rotation. Let us examine the oscillations of a free ring, by assuming, that:

$$W(\theta) = A_1 \sin(n\theta + A_2)\sin e^{i\omega t} \tag{3.7.11}$$

We will obtain from (3.7.9):

$$\Omega^2(-n^2\alpha_2^2\alpha_1 - \alpha_2^2\alpha_1) + \Omega(n^6\alpha_2 + n^4\alpha_2\alpha_1 - 2\alpha_2 n^2$$

$$+ n^2\alpha_2\alpha_1 + n^2 + \alpha_2 + 1) + (-n^6 + 2n^4 - n^2) = 0, \qquad (3.7.12)$$

where

$$\Omega = \frac{\gamma F R^4 \omega^2}{EIg}; \alpha_1 = \frac{E}{kG}; \alpha_2 = \frac{I}{FR^2}.$$

In the case of a closed ring, the value $n$ is an integer because the circumference fits an integer number of waves. The case $n = 1$ corresponds to the oscillations of a ring as an absolutely rigid body. The quadratic equation is obtained relative to $\Omega$. Each value of $n$ in (3.7.12) corresponds to two frequencies; the smaller of them corresponds to the bending (flexural) shape oscillations, and the greater to the shift mode of oscillations. Let us examine the ring, resting on several hinged supports, which are evenly located on a circle. Assuming that

$$W(\theta, t) = W(\theta)e^{i\omega t} \qquad (3.7.13)$$

we will obtain from (3.7.9):

$$\frac{d^6 W}{d\theta^6} + \beta_1 \frac{d^4 W}{d\theta^4} + \beta_2 \frac{d^2 W}{d\theta^2} + \beta_3 W = 0, \qquad (3.7.14)$$

where

$$\beta_1 = 2 + \frac{\gamma R^2 \omega^2}{Eg} + \frac{\gamma R^2 \omega^2}{gkG}$$

$$\beta_2 = 1 + \frac{\gamma^2 R^4 \omega^4}{g^2 kGE} + 2\frac{\gamma R^2 \omega^2}{Eg} - \frac{\gamma R^2 \omega^2}{gkG} - \frac{\gamma F R^4 \omega^2}{EIg}$$

$$\beta_3 = \frac{\gamma R^2 \omega^2}{Eg} + \frac{\gamma F R^4 \omega^2}{EIg} - \frac{\gamma^2 R^4 \omega^4}{g^2 kGE}$$

The solution (3.7.14) can be found in the form of:

$$W(\theta) = \sum_{i=1}^{6} C_i e^{S_i \theta} \qquad (3.7.15)$$

where $C_i$ – constants; and $S_i$ – roots of the following algebraic equation:

$$S^6 + \beta_1 S^4 + \beta_2 S^2 + \beta_3 = 0 \qquad (3.7.16)$$

The displacement in each span $N$ of the ring we accept as:

$$W_k(\theta) = \sum_{i=1}^{6} C_{ik} e^{S_i \theta} \quad (k = 1, 2, \ldots N) \tag{3.7.17}$$

where $k$ – the number of spans (flight). From (3.7.8) we find that the radial displacement $U_k$ of $k$ – span, will take the form:

$$U_k(\theta) = \sum_{i=1}^{6} S_i C_{ik} e^{S_i \theta} \quad (k = 1, 2 \ldots N) \tag{3.7.18}$$

The boundary conditions for the $k$ – support are the following:

$$W_k(\theta) = 0 \text{ when } \theta = 0, \frac{2\pi}{N}$$

$$U_k(\theta) = 0 \text{ when } \theta = 0, \frac{2\pi}{N}$$

$$\frac{d^2 W_k}{d\theta^2}\left(\theta = \frac{2\pi}{N}\right) = \frac{d^2 W_{k+1}}{d\theta^2}(\theta = 0) \tag{3.7.19}$$

$$\frac{d^3 W_k}{d\theta^3}\left(\theta = \frac{2\pi}{N}\right) = \frac{d^3 W}{d\theta^3}(\theta = 0)$$

The first two conditions are requirements to the fact that the tangential and radial displacements of the $k$ support are equal to zero. The remaining equations are requirements to the continuity of the angle of the slope (inclination) and bending moment upon transfer through $k$ – support. For $N$ span ring we will obtain $6N$ boundary conditions. Substituting expressions (3.7.17) and (3.7.18) in (3.7.19) we will obtain $6N$ homogeneous equations relative to the unknown $C_{ik}$. Setting the determinant of matrix of coefficients equal to zero, we obtain the equation for determining the frequencies. The calculated frequencies of free oscillations of rings with different numbers of supports are given in Table 3.7. At the same time it is accepted:

$$\frac{I}{FR^2} = 0.0025; \quad \frac{E}{kG} = 3.0.$$

The frequencies are determined using equations (3.7.9) and (3.7.10). As seen in Table 3.7, the frequencies obtained taking into account shifting deformations and rotational inertia are less than when considering only bending (flexural) deformation. The solutions above give the opportunity

**Table 3.7**   The frequencies of free oscillations of circular rings.

| Frequencies | Bending (flexural) oscillations | | | | | Oscillations taking into account shift and rotational inertia | | | | |
| --- | --- | --- | --- | --- | --- | --- | --- | --- | --- | --- |
| | Free ring Supports | Ring 1 | Ring 2 | Ring 3 | Ring 4 | Free ring 0 | Ring 1 | Ring 2 | Ring 3 | Ring 4 |
| $\omega_1 \sqrt{\dfrac{\gamma F R^4}{EIg}}$ | 2.680 | 3.180 | 3.295 | — | — | 2.631 | 2.948 | 3.00 | — | — |
| $\omega_2 \sqrt{\dfrac{\gamma F R^4}{EIg}}$ | 7.600 | 8.090 | 8.38 | 8.575 | — | 7.292 | 7.74 | 7.99 | 8.15 | — |
| $\omega_3 \sqrt{\dfrac{\gamma F R^4}{EIg}}$ | 14.550 | 16.300 | 16.99 | — | 18.19 | 13.57 | 14.57 | 15.24 | — | 16.4 |

to investigate the oscillations of circular arches. We investigate the oscillations of arches taking into account only bending strain (deformation). Substituting (3.7.13) in (3.7.10) we will obtain:

$$\frac{d^6 W}{d\theta^6} + 2\frac{d^4 W}{d\theta^4} + a_1\frac{d^2 W}{d\theta^2} + a_2 W = 0, \qquad (3.7.20)$$

where

$$a_1 = 1 - \frac{\gamma F R^4 \omega^2}{EIg}; a_2 = \frac{\gamma F R^4 \omega^2}{EIg}$$

We search for a solution in the form of (3.7.15), in which $S_i$ – roots of the following algebraic equation:

$$S^6 + 2S^4 + a_1 S^2 + a_2 = 0 \qquad (3.7.21)$$

The boundary conditions for the circular arch with the fixed ends are the following:

$$U(0) = U(\alpha) = W(0) = W(\alpha) = 0; \qquad (3.7.22)$$

$$\frac{dU}{d\theta}(\theta = 0) = \frac{dU}{d\theta}(\theta = \alpha) = 0,$$

where $\alpha$ – the angle of the thrust of arch. For the circular arch with hinge-supported ends we have the following boundary conditions:

$$U(0) = U(\alpha) = W(0) = W(\alpha) = 0; \qquad (3.7.23)$$

$$\frac{d^3 W}{d\theta^3}(0) = \frac{d^3 W}{d\theta^3}(\alpha) = 0.$$

Table 3.8   Frequencies of free oscillations of circular arches.

| Angle of the thrust of arch (degrees) | Boundary conditions of arch | $\omega_i\sqrt{\dfrac{\gamma F R^4}{E I g}}$ |
|---|---|---|
| 80 | Hinge support | 17.610 |
| 180 | | 2.105 |
| 240 | | 0.725 |
| 360 | | 0 |
| 100 | Fixed ends | 17.85 |
| 270 | | 1.245 |
| 320 | | 0.813 |
| 360 | | 0.547 |

Substituting (3.7.15) in (3.7.22) and in (3.7.23) we will obtain the systems of homogeneous equations relative to $C_i$, and then by equating the corresponding determinants to zero, we will obtain frequency equations. The values of the dimensionless free (natural) frequencies of circular arches $\omega_i\sqrt{\dfrac{\gamma F R^4}{E I g}}$ are given in Table 3.8.

## 3.8   The Free Oscillations of System "Flexible Arch-Rigid Beam"

The problem of the dynamics of combined systems, which are subject to the impact of moving loads, is an actual problem of structural mechanics. Free oscillations of combined systems were investigated by many scientists: one of the first works was that of P.R. Nelasov. Free and forced oscillations of combined systems are investigated also by A.Ph. Smirnov [341], S.I. Konashenko [182], V.A. Smirnov [342], G.Y. Kunnos [198], and others. We made an attempt to determine the free (natural) oscillations of combined systems, taking into account deformations, not only of the lowest frequencies but also of a range of frequencies for a combined system, i.e. a flexible arch and rigid beam, containing the spectrum of frequencies [263, 280]. In these systems the interest lies in discerning in which cases the lowest frequency will be arch – reverse-symmetrical; and in which cases the lowest frequency will be beam – symmetrical. The obtained theoretical results were tested on models and structures (Figure 3.11).

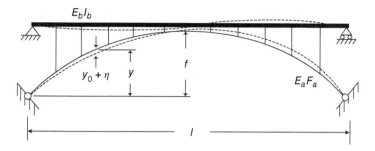

**Figure 3.11**   The scheme of a combined system of the rigid-beam type and flexible-arch with the ride on the top.

Let us examine the free oscillations of this combined system [280]. The integro-differential equation of the oscillations takes the form:

$$\frac{\partial^4 \eta}{dx^4} + \frac{H}{E_b I_b} \cdot \frac{\partial^2 \eta}{dx^2} - \frac{\Delta H}{E_b I_b} \cdot \frac{\partial^2 y}{dx^2} + \frac{m}{E_b I_b} \cdot \frac{\partial^2 \eta}{dt^2} = 0 \qquad (3.8.1)$$

This equation is obtained with the following prerequisites (preconditions): the axis of the arch is outlined on the square parabola; the posts are not under tensile force (inextensible, nonextensible) and they are not compressed; rigidity of the beam is constant on the entire span; and the mass of the bridge span is distributed along the length of the rigid beam. The dynamic part of the thrust, which is obtained with the examination of the strained state of the element of arch, takes the following form:

$$\Delta H = \frac{\frac{\partial^2 y}{dx^2}}{\frac{l}{E_a I_a}(1 + 8n^2)} \int_0^l \eta \, dx, \qquad (3.8.2)$$

Where $n = f/l$, and the arch axis is outlined on the square parabola

$$y = \frac{4f}{l^2} x(l - x)$$

For solving integral-differential Eq. (3.8.1) we assume:

$$\eta = \eta_x A \sin(pt + v), \frac{\partial^2 y}{dx^2} = -\frac{q}{H}. \qquad (3.8.3)$$

From (3.8.2) and (3.8.3) $\Delta H(t)$ is expressed in the form of:

$$\Delta H(t) = -\left( \frac{q E_a F_a}{H l (1 + 8n^2)} A \int_0^l \eta_x dx \right) \sin(pt + v) = \Delta H A \sin(pt + v)$$

$$(3.8.4)$$

Now let us examine the equation for $\eta_x$. From (3.8.1), (3.8.2), and (3.8.4) we obtain:

$$E_b I_b \frac{\partial^4 \eta_x}{dx^4} + H \frac{\partial^2 \eta_x}{dx^2} - \frac{q^2 E_a F_a}{H^2 l(1 + 8n^2)} \int_0^l \eta_x dx - mp^2 \eta_x = 0. \qquad (3.8.5)$$

The form of the standing wave is approximated by Fourier series:

$$\eta_x = \sum_{j=1}^{\infty} C_j \frac{\sqrt{2l}}{j\pi} \sin j\frac{\pi x}{l} \qquad (3.8.6)$$

Then multiplying (3.8.5) by $\sin i\frac{\pi x}{l}$ and integrating from 0 to $l$, we obtain

$$E_b I_b C_i \frac{\sqrt{2l}}{j\pi}\left(\frac{i\pi}{l}\right)^4 \int_0^l \sin^2\frac{i\pi x}{l} dx - HC_i \left(\frac{i\pi}{l}\right)^2 \frac{\sqrt{2l}}{i\pi} \int_0^l \sin^2\frac{i\pi x}{l} dx$$

$$- \frac{q^2 E_a F_a}{H^2 l(1 + 8n^2)} \sum_{j=1}^{\infty} C_j \frac{\sqrt{2l}}{j\pi} \left(\int_0^l \sin\frac{j\pi x}{l} dx\right)$$

$$\times \int_0^l \sin\frac{i\pi x}{l} dx - mp^2 C_i \frac{\sqrt{2l}}{i\pi} \int_0^l \sin^2\frac{i\pi x}{l} dx = 0 \qquad (3.8.7)$$

Bearing in mind the values of the following integrals

$$\int_0^l \sin^2\frac{i\pi x}{l} dx = \frac{1}{2}l, \quad \text{and} \quad \int_0^l \sin\frac{i\pi x}{l} dx = -\frac{l[(-1)^i - 1]}{i\pi}$$

Eq. (3.8.7) takes the form:

$$E_b I_b C_i \left(\frac{i\pi}{l}\right)^4 \frac{l}{2} - HC_i\left(\frac{i\pi}{l}\right)^2 \frac{l}{2} - \frac{q^2 E_a F_a}{H^2 l(1 + 8n^2)}$$

$$\times \sum_{j=1}^{\infty} C_j \frac{l^2[(-1)^i - 1]}{j\pi} \frac{[(-1)^i - 1]}{i\pi} - mp^2 C_i \frac{l}{2} = 0$$

or after simple conversions (transformations),

$$E_b I_b C_i \left(\frac{i\pi}{l}\right)^4 - HC_i\left(\frac{i\pi}{l}\right)^2 - \frac{2q^2 E_a F_a}{H^2(1 + 8n^2)} \frac{[(-1)^i - 1]}{i\pi}$$

$$\times \sum_{j=1}^{\infty} C_j \frac{[(-1)^i - 1]}{j\pi} - mp^2 C_i = 0. \qquad (3.8.8)$$

For solving Eq. (3.8.8) we will introduce the following designations:

$$A = E_b I_b \left(\frac{\pi}{l}\right)^4;$$

(3.8.9)

$$B = H\left(\frac{\pi}{l}\right)^2; \quad C = \frac{2q^2 E_a F_a}{H^2(1 + 8n^2)};$$

(3.8.10)

$$a_i = \frac{1 - (-1)^i}{i\pi}.$$

(3.8.11)

By increasing the degree of strengthening of the beam by the arch, parameter $C$ is increased, and a simple beam corresponds to the value of $C = 0$. Then (3.8.8) is rewritten as (where $i$ – are odd numbers)

$$Ai^4 C_i - C_i Bi^2 - Ca_i \sum_{j=1}^{\infty} C_j a_j - mp^2 C_i = 0.$$

Let us designate:

$$x = Ca_i \sum_{j=1}^{\infty} C_j a_j,$$

then

$$C_i Ai^4 - C_i Bi^2 - Ca_i x - mp^2 C_i = 0$$

$$C_i = \frac{Ca_i x}{Ai^4 - Bi^2 - mp^2}.$$

Then we have (all $i$ – are odd numbers, otherwise $a_i = 0$)

$$x = \sum_{j=1}^{\infty} C_{2j-1} a_{2j-1} = C_x \sum_{j=1}^{\infty} \frac{a^2_{2j-1}}{A(2j-1)^4 - B(2j-1)^2 - mp^2}$$

or $x = 0$, then $C_{2j-1} A(2j-1)^4 - C_{2j-1} B(2j-1)^2 = mp^2 C_{2j-1}$, i.e. $C_{2j-1}$ differs from zero for one two values of $j$. If for one, then $x \neq 0$, therefore it is necessary to determine whether the expression $A(2j-1)^4 - B(2j-1)^2$ can have an identical positive value with (in) two different natural $j$. It depends on $A$ and $B$, but it is impossible, since according to the Vieta's formula it can be written:

$$\frac{mp^2}{A} = -(2j_1 - 1)^2 \cdot (2j_2 - 1)^2 < 0,$$

where $j_1$ and $j_2$ are sought values of $j$, and means $x \neq 0$, and

$$1 = C \sum_{j=1}^{\infty} \frac{a^2_{2j-1}}{A(2j-1)^4 - B(2j-1)^2 - mp^2} \tag{3.8.12}$$

Let us calculate the sum

$$\sum_{j=1}^{\infty} \frac{a^2_{2j-1}}{A(2j-1)^4 - B(2j-1)^2 - mp^2} \tag{3.8.13}$$

According to (3.8.11) this sum takes the form:

$$\frac{4}{\pi^2} \sum_{j=1}^{\infty} \frac{1}{(2j-1)^2 [A(2j-1)^4 - B(2j-1)^2 - mp^2]} \tag{3.8.14}$$

We will designate $y = (2j-1)^2$, then we will obtain

$$\frac{1}{y^2(Ay^2 - By - mp^2)} = \frac{\gamma_1}{y - y_1} + \frac{\gamma_2}{y - y_2} + \frac{\gamma_3}{y},$$

where $\gamma_1$, $\gamma_2$, and $\gamma_3$ are some coefficients, and $y_1$ and $y_2$ are the roots of the equation

$$Ay^2 - By - mp^2 = 0. \tag{3.8.15}$$

After reduction to the common denominator, let us write

$$1 = \gamma_1(y - y_2)y + \gamma_2(y - y_1)y + \gamma_3(y - y_1)(y - y_2).$$

Taking into account that $(y - y_1)(y - y_2) = y^2 - \frac{B}{A}y - \frac{mp^2}{A}$, and equating coefficients with identical degrees of $y$ on the left and right sides, we will obtain:

$$y^2 \gamma_1 + \gamma_2 + \gamma_3 = 0;$$

$$y - \gamma_1 y_1 - \gamma_2 y_2 - \gamma_3 \frac{B}{A} = 0; \quad \begin{cases} \gamma_1 + \gamma_2 = -\dfrac{A}{mp^2}; \\ y_1\gamma_1 - y_2\gamma_2 = \dfrac{B}{mp^2} \end{cases}$$

$$1 - \gamma_3 \frac{mp^2}{A} = 1; \gamma_2(y_1 - y_2) = \frac{Ay_1 - B}{mp^2}$$

$$\gamma_3 = -\frac{A}{mp^2}; \gamma_2 = \frac{Ay_1 - B}{(y_1 - y_2)mp^2}.$$

Finally:

$$\gamma_3 = -\frac{A}{mp^2}, \gamma_2 = \frac{Ay_1 - B}{(y_1 - y_2)mp^2}, \gamma_1 = \frac{Ay_2 - B}{(y_2 - y_1)mp^2} \qquad (3.8.16)$$

Now the sum (3.8.14) takes the form:

$$\frac{4}{\pi^2}\left(\gamma_1 \sum_{j=1}^{\infty} \frac{1}{(2j-1)^2 - y_1} + \gamma_2 \sum_{j=1}^{\infty} \frac{1}{(2j-1)^2 - y_2} + \gamma_3 \sum_{j=1}^{\infty} \frac{1}{(2j-1)^2}\right)$$

$$(3.8.17)$$

For arbitrary $z$ let us calculate the sum

$$\sum_{j=1}^{\infty} \frac{1}{(2j-1)^2 - z^2} = \Phi(z) \qquad (3.8.18)$$

For the following cases:

1. $\Phi(z)$ has simple poles at the points $\pm(2j-1), j = 1, \dots$, the function $\tan\frac{\pi}{2}z$ at the same points has poles;

2. $\Phi(z) = \frac{1}{2z}\sum_{j=1}^{\infty}\left(\frac{1}{z-(2j-1)} + \frac{1}{z+(2j-1)}\right)$.

From here it is visible that the function $\Phi(z)2z$ is periodic with period 2, and at the points of the form of $2j - 1, j = 0, \pm 1, \dots$ has deductions, equal to one. Function $\tan\frac{\pi}{2}z$ at point $z = 2j - 1$ has a deduction $\lim_{z \to 2j-1}\left(\tan\frac{\pi}{2}z\right)\cdot(z - 2j - 1)$, equal to $\lim_{z \to 2j-1} -\frac{1}{\frac{\pi}{2}\frac{1}{\sin^2\frac{\pi}{2}z}} = -\frac{2}{\pi}$, moreover the function $\tan\frac{\pi}{2}z$ is periodic with the period 2.

From the above three statements it follows that the function $2z\Phi(z) + \frac{\pi}{2}\tan\frac{\pi}{2}z$ does not have poles (they are reduced). This function is periodic with period 2 and the strip $0 < R_cz < 2$ is limited, therefore it is limited in the entire complex plane. According to the theorem of Liuvilya whereby a bounded function, analytical in the entire complex plane is a constant, we will obtain:

$$2z\Phi(z) + \frac{\pi}{2}\tan\frac{\pi}{2}z = const$$

and since at the point $z = 0$ the left side of the equation becomes zero, the constant $= 0$, i.e. $\Phi(z) = \frac{\pi}{4z}\tan\frac{\pi}{2}z$, now (3.8.17) takes the form:

$$\frac{1}{\pi}\left(\gamma_1\frac{\tan\frac{\pi}{2}\sqrt{y_1}}{\sqrt{y_1}} + \gamma_2\frac{\tan\frac{\pi}{2}\sqrt{y_2}}{\sqrt{y_2}} + \gamma_3\frac{\pi}{2}\right)$$

and Eq. (3.8.12) is rewritten in the following form:

$$-\frac{\pi}{C} = \gamma_1 \frac{\tan \frac{\pi}{2}\sqrt{y_1}}{\sqrt{y_1}} + \gamma_2 \frac{\tan \frac{\pi}{2}\sqrt{y_2}}{\sqrt{y_2}} + \gamma_3 \frac{\pi}{2}. \tag{3.8.19}$$

This transcendental equation is relative to $p_i$. We convert Eq. (3.8.19) for more convenient use.

$$-\frac{\pi}{C}\left(\cos\frac{\pi}{2}\sqrt{y_1}\right) = \gamma_1 \frac{\sin\frac{\pi}{2}\sqrt{y_1}}{\sqrt{y_1}} + \cos\frac{\pi}{2}\sqrt{y_1}\left[\gamma_2\frac{\tan\frac{\pi}{2}\sqrt{|y_2|}}{\sqrt{|y_2|}} + \gamma_3\frac{\pi}{2}\right] \tag{3.8.20}$$

Substituting into (3.8.20) the explicit expressions for $y_1$ and $y_2$, the roots of Eq. (3.8.15), the coefficients $A$, $B$, and $C$ from (3.8.9) and (3.8.10), and also $\gamma_1$, $\gamma_2$, and $\gamma_3$ from (3.8.16) will obtain the exact values for $p_i$.

As an example, let us determine the free vertical oscillations of a steel combined bridge of the system: rigid beam – flexible arch with the rigid beam riding on top. The bridge being investigated overlaps the canyon river Arpa in the city of Jermuk, Armenia. The basic design parameters of the bridge are given below:

$$l = 120 \text{ m}; f = 18 \text{ m}; E_a = E_b = 210\,GPA; H = 4025.3 \text{ kN}$$

$$F_a = 0.0527 \text{ m}^2; F_b = 0.13373 \text{ m}^2; I_b = 0.04356 \text{ m}^4$$

The calculation of Eq. (3.8.20) was produced using an algorithmic language "FORTRAN-IV." Table 3.9 gives the values of the first seven frequencies for considering this example, calculated according to the proposed

Table 3.9    Frequency of vertical oscillations.

| Number of frequency | Frequency of vertical oscillations $p_i$ (s$^{-1}$) | |
|---|---|---|
| | Proposed formula (3.8.20) | Formula from S.I. Konashenko [182] |
| 1 | 3.850 | 4.102 |
| 2 | 7.504 | — |
| 3 | 10.504 | 9.968 |
| 4 | 25.409 | 25.930 |
| 5 | 49.592 | 50.351 |
| 6 | 82.834 | 83.190 |
| 7 | 123.899 | 124.233 |

formula in (3.8.20) and according to the formula proposed by S.I. Konashenko [182]. The table shows that including the calculations of deformations somewhat reduces the natural frequencies of the vertical oscillations [280].

The modes of free oscillations were represented by expression (3.8.6), in the form of trigonometric series

$$\eta = \sum_{j=1}^{\infty} C_j \frac{\sqrt{2l}}{j\pi} \sin \frac{j\pi x}{l} = \sum_{j=1}^{\infty} C_j \varphi_j(x) \tag{3.8.21}$$

Here $j$ – integers. Of course, the more inclusive the sum (3.8.21), the greater the accuracy of $\eta$ amplitides. Coefficients $C_j$ depend on parameters $A$, $B$, and $C$, and circular frequency $p_i$.

We will consider the modes of free oscillations, limited by two members of the sum of (3.8.21). Then for the first mode of oscillations we have:

$$\eta_1 = C_1 \varphi_1(x) + C_3 \varphi_3(x) = C_1 \left[ \varphi_1(x) + \frac{C_3}{C_1} \varphi_3(x) \right] \tag{3.8.22}$$

For different values of frequencies $p_i$ we will have different values of $C_{1i}/C_{3i}$. Obviously, without the initial conditions for the motion it is not possible to find $C_1$ and $C_3$; only their ratios $C_1/C_3$ or $C_3/C_1$, the so-called characteristic of the main modes of oscillations, can be determined. For the given combined system, the characteristic of the first mode of oscillations was determined using Eq. (3.8.8), which has the following value:

$$\left( \frac{C_3}{C_1} \right) = -3.910$$

Consequently, Eq. (3.8.22) takes the following form:

$$\eta_1 = C_1 \frac{\sqrt{2l}}{\pi} \left[ \sin \frac{\pi x}{l} - 1.303 \sin \frac{3\pi x}{l} \right] \tag{3.8.23}$$

The crossing of the curve of oscillations with the axis of abscissa is determined from the condition of equality zero of $\eta_1$, which leads to the equation:

$$x = \frac{l}{2\pi} \sqrt{3 \left( 1 + \frac{C_1}{C_3} \right)}$$

For the given combined system, $x = 0.237l$. The first symmetrical mode of natural vertical oscillations is shown graphically in Figure 3.12.

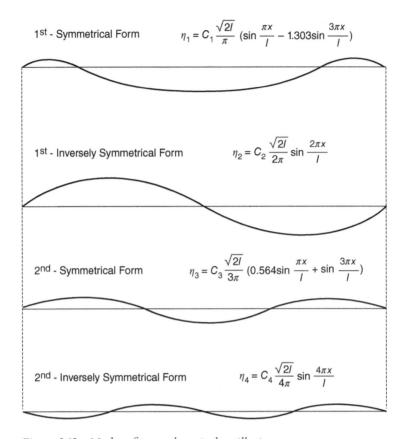

1st - Symmetrical Form $\qquad \eta_1 = C_1 \dfrac{\sqrt{2l}}{\pi} \left(\sin \dfrac{\pi x}{l} - 1.303\sin \dfrac{3\pi x}{l}\right)$

1st - Inversely Symmetrical Form $\qquad \eta_2 = C_2 \dfrac{\sqrt{2l}}{2\pi} \sin \dfrac{2\pi x}{l}$

2nd - Symmetrical Form $\qquad \eta_3 = C_3 \dfrac{\sqrt{2l}}{3\pi} \left(0.564\sin \dfrac{\pi x}{l} + \sin \dfrac{3\pi x}{l}\right)$

2nd - Inversely Symmetrical Form $\qquad \eta_4 = C_4 \dfrac{\sqrt{2l}}{4\pi} \sin \dfrac{4\pi x}{l}$

**Figure 3.12**   Modes of natural vertical oscillations.

The calculations showed that with the decrease in values of $B/A$ and $/A$, the modes of oscillations of the first and second symmetrical tones approached beamed modes. With an increasing the rigidity of the arch, i.e. with the presence of rigid arch, the modes of natural oscillations will be identical with the arched (arch) modes.

Let us examine the free oscillations of a combined system, when the arch is outlined on the circular arc and is very gently sloping, i.e. the ratio of the arrow of lifting arch to the flight (the ratio of the boom lift to the arch span) is $f/l = 0.15$. The basic preconditions for differential equation of the oscillations for this system are the same as for the static design: posts are not compressed, rigidity (stiffness) of the beam $E_b I_b$ is constant throughout the entire length; the mass of the bridge span is evenly distributed along the length of the beam; the mass of the arch (parts of a bridge trusses: posts, joints, upper and lower chords, diagonals, etc.) and the mass of posts (pillars) and top braces (pillars and upper joints) are related to the beam stiffness; and horizontal inertial forces are negligible.

Since the arch is gently sloping, the difference in the forces in the posts (pillars) is small and can therefore be disregarded. Thus, we obtain a system with the infinitely large number of degrees of freedom. Its position is determined by an elastic curve, which with the dynamic loads is the function of two variables – coordinate $x$ and time $t$. We start with the integro-differrential equation of oscillations (3.8.1) for the system the rigid beam – flexible arch with the ride on top. We will write down the equation of the axis of the arch $\sqrt{R^2 - (x - l/2)^2} + (R - f)$. There is no doubt that with the oscillations of arched systems it is necessary to take into account the dynamicity of thrust. Here we will show how to determine the dynamic part of the thrust – the dynamic "additive." The dynamic part of the thrust is obtained with examination of the strained state of an element of the arch

$$d\xi = \frac{ds}{dx}\Delta ds - \frac{dy}{dx}\Delta d\eta. \tag{3.8.24}$$

Since

$$\Delta ds = -\frac{\Delta H}{E_a F_a \cos\varphi} ds, \tag{3.8.25}$$

we will obtain

$$d\xi_a = -\frac{\Delta H}{E_a F_a \cos^3\varphi}\frac{dx}{} - \frac{dy}{dx}d\eta,$$

$\int_0^l d\xi_a$ is the divergence (represents a discrepancy) of the thrust sections and is equal to zero.

$$\xi_a = \int_0^l d\xi_a = 0 \tag{3.8.26}$$

Substituting (3.8.25) into (3.8.26) we will obtain:

$$\frac{\Delta H}{E_a F_a}\int_0^l \frac{dx}{\cos^3\varphi} + \int_0^l \frac{dy}{dx}d\eta = 0. \tag{3.8.27}$$

The value of the first integral:

$$\frac{\Delta H}{E_a F_a}\int_0^l \frac{dx}{\cos^3\varphi} = \frac{2R\Delta H}{E_a F_a}\cdot\frac{l}{\sqrt{4R^2 - l^2}}$$

The value of the second integral:

$$\int_0^l \frac{dy}{dx}d\eta = \frac{dy}{dx}\eta\Big|_0^l - \int_0^l \frac{d^2y}{dx^2}\eta dx = -\int_0^l \frac{d^2y}{dx^2}\eta dx$$

Substituting the values of these integrals into (3.8.27), we will obtain:

$$\frac{2R\Delta H}{E_a F_a} \cdot \frac{l}{\sqrt{4R^2 - l^2}} - \int_0^l \frac{d^2 y}{dx^2} \eta \, dx = 0$$

The dynamic part of the thrust:

$$\Delta H = \frac{1}{\dfrac{2R}{E_a F_a} \cdot \dfrac{l}{\sqrt{4R^2 - l^2}}} \int_0^l \frac{d^2 y}{dx^2} \eta \, dx \tag{3.8.28}$$

For solving Eq. (3.8.1), similarly to the beginning of Section 3.8, we assume:

$$\eta(x, t) = \eta_x A \sin(pt + v); \tag{3.8.29}$$

$$\Delta H(t) = \Delta H A \sin(pt + v),$$

where $p$ – the natural frequency of oscillations; and $v$ – the initial phase of oscillations. We approximate the form of standing wave of oscillations by Fourier series,

$$\eta_x = \sum_1^\infty b_n \sin \frac{n\pi x}{l}, \tag{3.8.30}$$

where $b_n$ – coefficient, which can be expressed as follows according to the recommendations of S.A. Ilyasevich [151, 152]:

$$b_n = c_n \frac{\sqrt{2l}}{n\pi}$$

where $c_n$ – the coefficient, dependent on load. Substituting (3.8.29) into (3.8.1) we will obtain:

$$E_b I_b \frac{d^4 \eta_x}{dx^4} + H \frac{d^2 \eta_x}{dx^2} - \Delta H \frac{d^2 y}{dx^2} - mp^2 \eta_x = 0 \tag{3.8.31}$$

Next, substituting $m$ (3.8.28) and (3.8.30) into (3.8.31), we will obtain:

$$E_b I_b \sum_{n=1}^\infty b_n \frac{n^4 \pi^4}{l^4} \sin \frac{n\pi x}{l} - H \sum_{n=1}^\infty b_n \frac{n^2 \pi^2}{l^2} \sin \frac{n\pi x}{l}$$

$$+ \frac{y_0 \int_0^l y'' \left( \sum_{n=1}^\infty b_n \sin \frac{n\pi x}{l} \right) dx}{\dfrac{l2R}{E_a F_a \sqrt{4R^2 - l^2}}} - mp^2 \sum_{n=1}^\infty b_n \sin \frac{n\pi x}{l} = 0 \tag{3.8.32}$$

We introduce the following designations:

$$B = \frac{2R}{E_a F_a \sqrt{4R^2 - l^2}}$$

$$\int_0^l y'' \left( \sum_{n=1}^\infty b_n \sin \frac{n\pi x}{l} \right) dx = \sum_{n=1}^\infty b_n \int_0^l y'' \sin \frac{n\pi x}{l} = \sum_{n=1}^\infty b_n a_n,$$

Where

$$a_n = \int_0^l y'' \sin \frac{n\pi x}{l} dx.$$

Then Eq. (3.8.32) takes the following form:

$$E_b I_b \sum_{n=1}^\infty b_n \frac{n^4 \pi^4}{l^4} \sin \frac{n\pi x}{l} - H \sum_{n=1}^\infty b_n \frac{n^2 \pi^2}{l^2} \sin \frac{n\pi x}{l}$$

$$+ \frac{y'' \sum_{n=1}^\infty b_n a_n}{lB} - mp^2 \sum_{n=1}^\infty b_n \sin \frac{n\pi x}{l} = 0. \tag{3.8.33}$$

Multiplying both parts of Eq. (3.8.33) by $\sin \frac{n\pi x}{l}$, integrating from 0 to $l$, and taking into account, that:

$$\int_0^l \sin \frac{n\pi x}{l} \sin \frac{i\pi x}{l} dx = \begin{cases} 0 \text{ when } i \neq n; \\ \frac{l}{2} \text{ when } i = n, \end{cases}$$

we will obtain:

$$E_b I_b b_i \frac{i^4 \pi^4}{l^4} \cdot \frac{l}{2} - H b_i \frac{i^2 \pi^2}{l^2} \cdot \frac{l}{2} + \frac{a_i \sum_{n=1}^\infty a_n b_n}{lB} - mp^2 b_i \frac{l}{2} = 0 \tag{3.8.34}$$

Instead of $\sum_{n=1}^\infty a_n b_n$, let us limit ourselves by the first four members of series, and taking into account that the even coefficients of $a_{2n} = 0$, we come to the system of homogeneous linear equations relative to $b_1$, $b_2$, $b_3$, and $b_4$. From the first and third equations we will obtain:

$$P_{1,3} = \frac{\pi^2}{l^2} \sqrt{\frac{E_b I_b g}{q} \left[ 41 - 5\overline{H}_1 + (a_1{}^2 - a_3{}^2)S_1 \pm 40 \sqrt{\frac{\left[ 1 + 0.1\overline{H}_1 + \frac{(a_3{}^2 - a_1{}^2)}{40} \right]^2}{+ \frac{a_1{}^2 a_3{}^2 S_1{}^2}{400}}} \right]}$$

$$\tag{3.8.35}$$

where

$$\overline{H}_1 = \frac{l^2}{\pi^2 E_b I_b} H; \quad S_1 = \frac{E_a F_a}{E_b I_b} \cdot \frac{l^2 \sqrt{4R^2 - l^2}}{2R};$$

$$a_1 = \frac{R^2 \int_0^l \sin\frac{\pi x}{l} dx}{[R^2 - (x - l/2)^2]^{3/2}}; \quad a_3 = R^2 \int_0^l \frac{\sin\frac{3\pi x}{l} dx}{[R^2 - (x - l/2)^2]^{3/2}}. \quad (3.8.36)$$

From the second and fourth equations we will determine:

$$p_2 = \frac{4\pi^2}{l^2} \sqrt{\frac{E_b I_b g}{q}(1 - \overline{H}_2)}; \quad (3.8.37)$$

$$p_4 = \frac{16\pi^2}{l^2} \sqrt{\frac{E_b I_b g}{q}(1 - \overline{H}_4)},$$

where

$$\overline{H}_2 = \frac{l^2}{4\pi^2 E_b I_b} H, \quad \overline{H}_4 = \frac{l^2}{16\pi^2 E_b I_b} H.$$

In order to compare the results of the calculation, carried out according to formulas (3.8.20), (3.8.35), and (3.8.37), let us determine the natural (free) frequencies for a combined system with the geometric and mass parameters, which coincide with the construction of the Jermuk bridge. The list of necessary design data is given at the beginning of this paragraph. Using the necessary data we can determine from the formula in (3.8.36) the values for $a_1$, and $a_3$:

$$a_1 = 109^2 \cdot 0.0000650 = 0.772265, \quad a_1^2 = 0{,}569$$

$$a_3 = 109^2 \cdot 0.0000300 = 0.356430, \quad a_3^2 = 0{,}127$$

and values for $S_1$ and $\overline{H}_1$:

$$S_1 = 108.746, \quad \overline{H}_1 = 0.4665.$$

Further, using (3.8.35) and (3.8.37), we can determine frequencies of free (natural) oscillations of the given combined system

$$p_1 = 4.5215\ s^{-1}; p_2 = 9.016\ s^{-1}; p_3 = 16.08\ s^{-1}; p_2 = 29.255\ s^{-1}.$$

With the calculation of $p_2$ and $p_4$ the value $\overline{H}_2 = 0.118$, and the value $\overline{H}_4 = 0.00024$.

## 3.9   The Results of Dynamic Testing Model of Combined System Rigid-beam and Flexible Arch

The study of the oscillations of constructions under the influence of moving loads has a great practical value. Currently, loads and speeds of traffic continuously grow, and therefore increase the dynamic impacts on constructions. An increase in traffic volume, the weight of traveling load, and the simultaneous decrease in the mass of bridges is the contemporary tendency in bridge construction. Because of this the need for using precise dynamic design schemas, diagrams, and models increases.

Traffic excitation causes non-stationary oscillations of bridge spans. In general the oscillations of bridge spans represent a transient process. i.e. the combination of forced and accompanying free oscillations. The degree of manifestation of the latter depends in essence on the inelastic resistance of the material of a construction. A large number of studies are devoted to the oscillations of bridge spans. Significant contributions to the solution of general problems of oscillations of constructions and bridge construction under the influence of mobile loadings were made by A.N. Krylov [193], S.P. Timoshenko [358], C.E.M. Becker, C.E. Inglis [153], S.T. Odman, A. Schallenkamp, H. Steuding, G. Stokes, R. Willis, W. Witting, V.V. Bolotin [39, 40], A.A. Alexandrov, N.G. Bondar [41], I.N. Goldenblat [119], I.A. Kolesnik [181], S. I. Konashenko [182], B.G. Korenev, A.B. Morgaevsky [235], G.B. Muravsky, I.G. Panovko, A.A. Petropavlovsky, V.A. Smirnoff [342], A.P. Philipov [284], I.I. Kazey [164], and others. The number of experimental studies, when compared to the number of theoretical works on the dynamics of steel and reinforced concrete beam and arch bridges with movable mobile loads, is considerably smaller. The first works devoted to this question were experimental (M. Becker) or experimentally theoretical (R. Willis and G. Stokes). After earlier theoretical and experimental studies analyzed the dynamic behavior of bridges due to moving loads, as performed by C.E. Inglis [153] in 1934, various newly developed and revised moving loads and bridge models were applied to the moving load analysis. D.S. Carder (1937) conducted a field investigation and observed vibrations of the San Francisco–Oakland Bay Bridge and the Golden Gate Bridge. These studies were conducted to determine the probability of damage caused by resonance during seismic excitation. An early summary of dynamic testing of highway bridges in the USA performed between 1948 and 1965 was presented by R.F. Varney and C.F. Galambos (1966).

T. Iwasaki, J Penzien, and R. Clough (1972) summarized tests performed in Japan between 1958 and 1969 to determine the dynamic properties of bridge structures. Another summary of field and laboratory model tests on bridge systems was presented by H.S. Ganga Rao. In 1984, Wang carried out research on the impact of freight trains on concrete railway bridges. C.R. Farrar, W.E. Baker, T.M. Bell, et al. (1994) report ambient vibration tests on the I-40 Bridge over the Rio Grande prior to a damage detection study. Traffic excitation on the bridge of interest was first used as the ambient vibration source. After traffic had been removed from the bridge, traffic on an adjacent bridge was used as the excitation source. O.S. Salawu and C. Williams (1995) provided a review of full-scale dynamic testing of bridges where methods of excitation were examined and reasons for performing dynamic tests were summarized. The analysis of experimental studies performed in the USS, were conducted by N.G. Bondar, I.I. Kazey, Y.G. Kozmin, and B.F. Lesokhin [41]. The extensive use of testing in the evaluation of bridges has resulted in the American Society of Civil Engineers (ASCE) Committee on Bridge Safety publishing a guide for field-testing of bridges in 1980. This guide includes an extensive reference list of papers summarizing previous bridge tests. Also discussed are static and dynamic load application methods, instrumentation, data acquisition, and methods for measuring material characteristics. The presented standards addressed both static and dynamic testing of bridges. Topics that are summarized include categories of load tests, test preparation, test procedures, evaluation of load steps, and evaluation of load tests and reporting of results. G.P. Tilley discusses different methods of measuring damping in bridges. Field test methods discussed include excitation methods such as driving a test vehicle over the bridge, step-relaxation, single-pulse loading (eccentric mass shakers, people jumping in unison, and pulling on ropes), and ambient excitation. C.R. Farrar, T.A. Duffey, and P. J. Cornwell [101], along with S.W. Doebling, presented a report in 1999 which summarized methods of excitation that have been used in past dynamic testing of bridge structures. The attributes and difficulties associated with the various excitation methods have been discussed in a very general manner. Although there does not appear to be consensus that one particular method is better than another, for large bridges the ambient excitation methods are the only practical method of exciting the structure. The report contains a more thorough list of references pertaining to the excitation methods and dynamic testing of bridges, in general. Despite the large number of publications on vibrations (oscillations) of bar structures (systems), bridge structures under the influence of moving loads, this issue still requires further development. In our opinion, the most expedient methods of analysis of the dynamic effect of mobile inertial loads are the methods of C.E. Inglis [153], V.V. Bolotin [39, 40], A.P. Philipov [284], S.S. Kokhmanyuk, O.A. Goroshko, and the method of A. Schallenkamp, I.I. Kazey [164]. It is necessary to improve the existing methods of dynamic analysis and to establish the limits of their applicability for solving specific problems put forward by the practice of using numerical methods. The study of the dynamic processes, revealed by traveling loads, can be conducted in two directions:

1. a study of the behavior of a bridge structure under a given load, and
2. a behavioral inquiry into an integrated system of a bridge and vehicle, i.e. the study of an interaction of a bridge span and a traveling load, and development of parameters needed to model the dynamics of a bridge-vehicle interaction. Despite the large number of publications on the oscillations of bar structures and combined bridge structures such as rigid beam–flexible arch under the influence of moving loads, this problem still requires further development.

Research on the dynamic analysis of a vehicle-bridge coupled vibration system is an important issue in civil engineering. In general, three methods are used to address this problem: the theoretical derivation; dynamic loading test; and numerical simulation. In recent years, with the development of computers, numerical simulation studies are widely used.

Analytical studies of the dynamic interaction of a span-vehicle coupled model for complicated bridges, such as combined systems "rigid beam–flexible arch" with various types of running vehicles, present enormous difficulties and in most cases do not provide sufficiently reliable results for practice. That is why the experimental theoretical method of study is the most appropriate method of research. The basis of this method is the assumption of a design schema (mathematical model), whose parameters are determined by comparing analytical results to experimental data.

A unique long-span (large-span) steel bridge in the form of a combined system "rigid beam (stiffening girder)–flexible arch" with the travel on top and the overlapping canyon of the river Arpa in the city Jermuk, Armenia, was put in operation in 1973 (Figure 3.13).

The complex configuration of the arch bridge span and the adjacent overpasses, with significant static indeterminacy, leads to a considerable difficulty in obtaining a precise dynamic calculation of this structure. Under such circumstances, using models necessary to calculate dynamic characteristics and parameters, considering the joint work of a bridge

**Figure 3.13**   Overall view of Arpa Bridge in Jermuk, Armenia.

span structure and trestle approaches, is preferable and expedient. From structural, design, and methodological considerations the model of Jermuk bridge was made geometrically similar to the original, on scale 1:20 (Figure 3.14).

It is known that the observance of geometric similarity provides the similarity of the original and the model in the qualitative deformations, forms of natural oscillations, and distribution of the natural oscillatory periods in structures made of a homogeneous material; A.G. Nazarov [244], V.A. Zakaryan [410]. In this case, homogeneity of the material of the original structure is disrupted only by paving the roadway of the bridge, possessing a lower modulus of elasticity. Therefore, the plate of the roadway together with the pavement in the model is realized in the form of a plate of rectangular cross-section, width, and thickness, selected from the conditions equivalent to the original structure with respect to weight and lateral stiffness. This condition together with the geometric similarity of the entire bridge structure provides the required mechanical similarity of the original and model.

The geometric characteristics of the model are given in Table 3.10. The axis of the arch is outlined on the square parabola: $y = \frac{4f}{l^2}x(l - x)$. Static tests were conducted for the purpose of development of the character of vertical and lateral deformations of a combined system of "rigid beam–flexible arch," and also to determine the influence of separate elements of a structure on the hardness in vertical and transverse directions.

(a)

(b)

**Figure 3.14**    Overall view of model of Arpa Bridge in Jermuk, Armenia.

A load in the vertical direction is applied to the central cross section of the arch and, with the aid of a side guy-rope on a pulley, a transverse horizontal force was created. The vertical and horizontal displacements of the points were determined by gauges. In order to exclude the possibility of random errors and to obtain reliable data, each test was conducted not less

**Table 3.10**   Geometric characteristics of the bridge model.

| 1  Design span of arches | $l = 6$ m |
|---|---|
| 2  The distance between axes of arches | $B = 0.42$ m |
| 3  The arrow of the rise (lifting) of the arch | $f = 0.93$ m |
| 4  The slope of the arch | $f/l = 0.15$ |
| 5  The length of the panel of above – arch structure | $K = 0.4$ m |
| 6  The cross-sectional area of the beam | $F_b = 5.14 \cdot 10^{-4}$ m$^2$ |
| 7  The cross-sectional area of the arch | $F_a = 1 \cdot 10^{-4}$ m$^2$ |
| 8  The moment of inertia of beam | $I_b = 42.73 \cdot 10^{-8}$ m$^4$ |
| 9  The moment of inertia of arch | $I_a = 1.15 \cdot 10^{-8}$ m$^4$ |
| 10  The modulus of elasticity of beam and arch | $E = 210$ GPa |
| 11  The ratio of the moment of inertia beam to the arch  $I_b/I_a = 37.16$ | |

than 10 times, and for the final results of the dynamic characteristics of the bridge model, the averaged values obtained by careful processing of a series of oscillograms were accepted [264]. The graphs of the displacements of the characteristic points of the system are given in Figure 3.15.

To evaluate the influence of rigidity of separate elements of a bridge structure, a model testing was conducted without orthotropic deck, without an arch, without the struts (cross-bars), and without a beam structure. Tests on the model showed that the orthotropic deck substantially influences the vertical and especially the lateral stiffness of the bridge span structure.

The stiffness of the model decreased approximately four times in the transverse direction, and at removal of the struts (cross-bars) by 15%. The lateral stiffness of separate flexible arches is approximately 25 times less than the stiffness of bridge span structures. With the dynamic tests, free horizontal lateral oscillations of the model were excited by pulling out and a sudden release of the central section of the arches. The oscillations were also caused by the symmetrical dropping of loads in one direction for the excitement of symmetrical oscillations, and in the opposite directions for the excitement of antisymmetric oscillations. During the model tests the same scheme of measurements of dynamic characteristics was used, as with the full-scale field bridge testing L.G. Petrosian [264] (See Figure 3.20 a, and b). The task of the full-scale field bridge tests was the verification of the dynamic characteristics of an operable bridge of the type of "rigid beam–flexible arch" with the travel on the top, and also a comparison with the results,

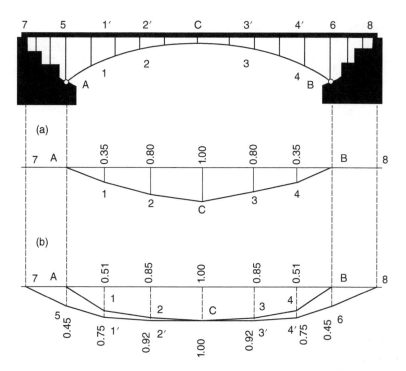

**Figure 3.15** Graphs of static (a) vertical; and (b) horizontal transverse displacements of the points of a model due to forces applied at the crown section of the arch in vertical and transverse the directions (ordinates are represented in arbitrary units).

obtained with the theoretical analysis of the given bridge structure. The obtained oscillograms showed that the characteristic points of the model executed synchronous oscillations and the amplitude ratios are constant for various points. The mode shapes of oscillations of the fundamental tone are represented in Figure 3.16, and Table 3.11 gives the values of their periods.

It is evident from Table 3.11 that the periods of free oscillations of the arch are somewhat lower than for the bridge span. It is explained by the

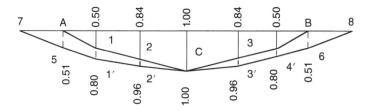

**Figure 3.16** Mode of natural lateral (transverse) oscillations of model.

Table 3.11  Periods of free oscillations.

| # | Studied structure | Periods of free oscillations | |
|---|---|---|---|
| | | Vertical, (s) | Lateral, (s) |
| 1 | Bridge span | 0.0434 | 0.07 |
| 2 | Bridge span without orthotropic plate | 0.039 | 0.065 |
| 3 | Free arch | 0.0375 | 0.0563 |

dominating influence on the periods and the increase in the inert mass due to the part of the bridge structure, located above the arch. Experiments confirm that the first frequency of spatial horizontal oscillations of free flexible arches is approximately 1.5 times lower than the frequency of the vertical oscillations.

In order to find the influence of trestle parts on the dynamic characteristics of the main bridge span, testing was carried out in four stages (Table 3.12). Tabulated data show that the frequency of free vertical oscillations (vibrations) diminishes with the decrease of the hardness of trestle approaches, and its value with the complete removal of the overpasses (see Table 3.12, position 4) with the adequate accuracy coincide with the theoretical calculations [280]. In the absence of trestle approaches to the main bridge span structure, the angular frequency of free vertical oscillation of the model is equal to $P = 100.48s^{-1}$, but the frequency, calculated for the model according to formula (3.8.20) gives the close result of $P = 102.5s^{-1}$.

In the case when the flexible arch with the rigid beam has trestle approaches, the elastic fixed-ends seemingly appear in the sections of the contiguity of the overpasses to the main bridge span. In order to use formula (3.8.20) the influence of trestle parts on the dynamic characteristics of the entire bridge structure is not considered to increase the frequency coefficient $\beta$, which makes it possible to approximately determine the frequency of free vertical oscillations for such systems. Let us replace the multi-span overpass with adjustable single-span, creating the elastic fixed-end of the same intensity as the multi-span. Table 3.13 depicts the computed values for the given adjustable spans for two and three equal spans of trestle approaches. For the various ratios of $l_0/l$, where $l$ is the main span, the fundamental frequencies of oscillations are determined.

The analysis of the experiments showed that for the ratio $l_0/l = 1/6 \div 1/8$, which is more frequently encountered in practice, and with $EI_b/EI_a \geq 30$, the frequency of fundamental tone depends little on these relations. On the basis of this, from the results of the experiments on the models, a curve

**Table 3.12** Angular frequency and period of free oscillations.

| # | Schemes of bridge model type rigid beam flexible arch | Vertical vibration | | Horizontal vibration | |
|---|---|---|---|---|---|
| | | $P(s^{-1})$ | $T(s)$ | $P(s^{-1})$ | $T(s)$ |
| 1 | | 144.69 | 0.0434 | 89.71 | 0.0700 |
| 2 | | 132.60 | 0.0474 | 87.50 | 0.0720 |
| 3 | | 116.80 | 0.0538 | 85.50 | 0.0734 |
| 4 | | 100.48 | 0.0625 | 92.50 | 0.0680 |

**Table 3.13**   Adjustable overpass single-span.

| Trestle bays $l_{trestle}$, (m) | Adjustable overpass single-span $l_0$, (m) | |
|---|---|---|
| | Two-bay overpass | Three-bay overpass |
| 4 | 3.504 | 3.468 |
| 6 | 5.256 | 5.202 |
| 8 | 7.008 | 6.936 |
| 10 | 8.760 | 8.670 |
| 12 | 10.512 | 10.404 |
| 14 | 12.264 | 12.138 |
| 16 | 14.016 | 13.872 |
| 18 | 15.768 | 15.606 |

for determining the coefficient of an increase the frequency $\beta$ (Figure 3.17) is built. As an example let us determine the fundamental frequency of oscillations of a bridge, which has trestle approaches from both ends of the main span. The angular frequency of the free vertical oscillations, calculated according to formula (3.8.20), is equal to $P = 7.50s^{-1}$. For three overpasses with equal bays $l_{trestle} = 8m$, let us determine the value of the given adjustable span $l_0 = 6.936m$, according to Table 3.13 and we will determine the coefficient $\beta = 1.44$ according to the graph in Figure 3.17. Then for the entire bridge structure the angular frequency of free vertical oscillations will be $P = 10.8s^{-1}$, with the corresponding period of the oscillations $T = 0.58s$. The period of free vertical oscillations with the full-scale field tests was $T = 0.57s$.

The precise calculations with the application of computer programs show that the approximate determination of dynamic characteristics from the proposed method gives a certain overstating of the natural frequencies up to 10%. In Figures 3.18 and 3.19, separate fragments of the tested model of the bridge are shown.

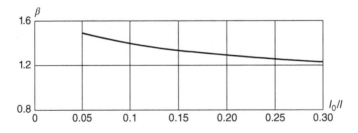

**Figure 3.17**   The dependence of coefficient $\beta$ on $l_0/l$.

**Figure 3.18**   Bridge model under dynamic testing.

**Figure 3.19**   Part of a bridge model.

Summarizing the results of dynamic testing of the model of a combined system "rigid beam–flexible arch," we come to the following conclusions (L.G. Petrosian [263]):

1. The dependence of the frequencies of the natural vertical oscillations of a combined system of the type "rigid beam–flexible arch" with the travel on top on elastic, geometric, and mass parameters is established taking into account the deformation of the entire system. The generated formula (3.8.20) makes it possible to determine the frequency spectrum of the free vertical oscillations for combined systems. Analysis of Eq. (3.8.20) shows that the enforcement of the beam structure by an arch

(a)

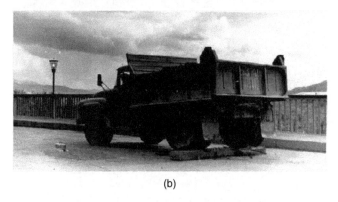

(b)

**Figure 3.20** (a) The motion of the test load along the bridge with various speeds. (b) The moment of impact in the characteristic sections of the main arched span of bridge.

substantially influences the first frequencies of the symmetrical mode shape of oscillations and free vibrations. The frequencies, which correspond to the antisymmetric mode shape of oscillations do not depend on the degree of the enforcement (strengthening) beam by an arch and the rigid beam works as a simple beam.

(a) Taking into account the deformations of a combined system results in lower frequencies of free vertical oscillations when deformation is not accounted for. Moreover, the greatest deviation is obtained with the first frequency to the corresponding anti-symmetric mode of oscillations. For large spans of 100m and more, the divergence is 7–12%.

(b) We developed a method that makes it possible to determine the fundamental tone of frequencies of vertical oscillations of combined systems with at least 90% accuracy. The method uses the coefficient of "increased" frequency and takes into account joint dynamic work of the main bridge span with adjacent trestle approaches.

(c) Of the two factors influencing the change periods of oscillations: (i) an increase in rigidity due to the part located above the arch structure, which decreases the oscillations period; (ii) and an increase in the inert mass (mass of inertia) due to the part located above the arch structure which increases the period of oscillations, the latter proved to be the dominant factor.

## 3.10   The Oscillations of the Combined System Taking into Account its Extent at a Given Harmonic Motion Base

According to the current methodology of calculations of buildings and constructions to the seismic excitation, all points of base oscillate (vibrate) synchronously with the same shift and acceleration. However, for buildings and structures, the methodology that the length in the direction of the propagation of excitation is commensurate with the length of seismic waves, is based on assumption that can lead to significant errors with the determination of the reactions of these systems. The long-span arched structures of bridges present complex design systems, which consist both of rectilinear "frame" elements and curvilinear "arched" elements. The seismic calculation of such constructions is associated with complex and labor-intensive computational operations. Therefore, in certain cases it is expedient to replace the curvilinear elements with rectilinear elements, and to thereby simplify the design scheme of a structure.

Guarantees of the reliability of residential, public, and industrial objects and facilities under seismic action are an important national, economic, and public safety imperative. When designing both the objects of mass construction and special construction social, technical, safety, and economic factors must be considered. This in particular relates to the questions of the design of construction undertaken in seismic regions. Our increasing ability to quantify real inertial forces, which appear in extreme natural events such as earthquakes, allows for a more economic and more reliable design. This relates especially to bridges and tunnels, in view of the fact that their damages due to earthquakes cause tragic and material losses, leading to severe consequences, including extended traffic disruption. The latter aggravates the disaster for the victims in the affected area. Restoration requires complex and labor-intensive work and becomes one of the first recovery priorities. Therefore, in many countries the design standards for bridges and other road constructions provide heightened requirements for seismic resistance. The task of the seismic resistance for road engineering structures should be considered as an independent problem, which has a number of special features, namely, the presence of long-span massive elements, possessing significant inertia features, and capable of accomplishing horizontal and spatial oscillations. It is also necessary to take into account the differences in the dynamic features of bridge spans and supports, considerable extension in the plan, and the essential role of the lateral seismic pressure of ground on the abutment of a bridge. Thus, all above enumerated features of road engineering structures require the development of specific methods of design and calculation, and they happen to be the component parts of the general theory of seismic resistance (seismic stability). The calculation of the seismic resistance of bridge structures as in all engineering constructions, is produced according to the spectral method adapted by the current seismic building code. For a unique and especially important construction calculation can be produced according to the real acceleragrams (ground acceleration, peak ground acceleration), seismograms, or ground velocity (peak ground velocity), which correspond to the seismicity of the region. The accuracy of such calculations, in many respects, depends on the right selection of a dynamic design scheme, which can be represented in the form of the system with continual or discrete parameters. Schemas in the form of the system with a finite number of degrees of freedom are more convenient with the use of contemporary methods of computer technology, and therefore, are used more often in practical calculations. Only the maximum seismic forces for separate forms of oscillations calculate according to the spectral curve $\beta$. The estimated maximum value of seismic forces is identified as middle-quadratic from the stresses, which appear with various forms of oscillations. In this case all stresses except the maximum are introduced with the coefficient 0.7. Usually in the calculation the first three or five modes of oscillations are considered. When applying the methods of calculation using the accelerograms of strong earthquakes, the displacements of the system for a discrete scheme are determined from the following formula [162, 171]:

$$y_k(t) = -\sum_{i=1}^{n} \frac{\eta_{ik} T_i}{2\pi} \int_0^t \ddot{Y}_0(\tau) e^{-\frac{\gamma\pi}{T_i}(t-\tau)} \sin \frac{2\pi}{T_i}(t-\tau) d\tau, \qquad (3.10.1)$$

where $k = 1, 2, \ldots, n$.

For a system with distributed continuous parameters, the displacements are determined from the following formula:

$$y(x, t) = -\sum_{i=1}^{n} \frac{T_i}{2\pi} \eta_{i(x)} \int_0^t \ddot{Y}(\tau) e^{-\frac{\gamma\pi}{T_i}(t-\tau)} \sin \frac{2\pi}{T_i}(t - \tau)d\tau. \qquad (3.10.2)$$

In the given formulas $T_i$ – period of the $i$th mode of the free oscillations of system; $\gamma = \frac{\delta}{\pi}$ – coefficient of the inelastic resistance of the system; $\delta$ – the logarithmic decrement of oscillations; and $\ddot{Y}(t)$ – the given accelerogram of an earthquake.

In calculations of the bridges exposed to seismic excitations according to the real accelerograms, the value of the coefficient of inelastic resistance starts within the limit (0.025–0.08) [162]. In this case, the lowest value of the coefficient of inelastic resistance corresponds to steel bridge structures, and the maximum value corresponds to reinforced concrete bridge constructions. With the study of the seismic oscillations of girder (beam, frame) bridges separate transverse and longitudinal oscillations are considered. In the preliminary calculations there we are able to calculate oscillations only on the first main (basic) mode. In order to determine seismic forces both in the longitudinal and in transverse directions it is recommended to consider a temporary load rigidly fastened with the construction and not to consider the possible phenomenon of slippage. The mode coefficient of $\eta$ on the level of bridge spans of girder (beam, frame) bridges varies within the limits 1.1–1.3 [162]. The calculations of seismic oscillations of bridges in the longitudinal direction is quite difficult because it is necessary to take into account the presence of friction in the mobile supporting parts of the bridge spans, asynchronous oscillations, ground soil under the individual bridge supports, and so forth. The practical design methodology must be based on the use of experimental values of natural frequencies and thus, indirectly on account of the work of nodal (joint) links and friction of bearing supports with the oscillations. The horizontal and spatial oscillations of arch bridges with the ride on the top as complex oscillatory systems are of difficult theoretical analysis. The dynamic interaction of bridge spans with supports is one of the difficult factors to be taken into account [119, 375]. The practical design recommendations, which are concerned with the calculations of such bridges to the seismic resistance, have an approximate character and will subsequently be refined in the future in proportion to the accumulation by sufficient theoretical and experimental information. With the seismic design of long-span bridges in the longitudinal direction it is necessary to take into account the asynchronous oscillations of the outermost supports. However, the existing design standards do not provide practical recommendations to incorporate this factor, since they are based on the assumption that all points of the base oscillate synchronously with one and the same displacement and the same acceleration. Meanwhile, accounting for asynchronous oscillations of supports, or in other words taking into account the extent of the structure, is especially necessary in cases when the length of a seismic wave in the direction of propagation excitement is commensurate with the length of construction itself. And therefore, this assumption leads to significant errors with the determination of the reactions of these systems. The influence of the extent of construction of the values of seismic loads was investigated for the first time by G. Hausner [135]. This question was considered by I.L. Korchinsky [184], and V.A. Grossman [128]. Assuming that the law of the motion of the base is represented in the form of the sum of the damped sinusoids, it was found that the effect of the extent is greater with rigid systems than with flexible. The influence of the extent with the harmonic oscillations of a base with different design diagrams of construction was investigated in the works of V.K. Egupova and T.A. Komandrina [98], Sh.G. Napetvaridze [241, 242], A.P. Sinitsin [334, 335], and E.E. Khachian et al. [175]. The statistic aspects of this problem are examined by M.Ph. Barshtein [30]. The translational-rotational oscillations vibrations, of constructions extended in the plan are investigated in the work of N.A. Amiraslanov and M.Ph. Barshtein [19]. The cases of match and mismatch between the centers of mass and stiffness of structures are examined. The displacements and accelerations of the base are represented as stationary random processes. For one-storied frame below they obtained concrete results about the influence of the accounting of structure extent on the magnitude of the seismic forces. So for example, for a building with sides of 72m × 144m with the wave of propagation velocity of $v = 600$ m/s, the results take into account a structure extent leads to decrease of inertial force in one direction to 25.5%, and in the other to 8%.

The reactions of the extent structures with the use of instrumental records of earthquakes are investigated in the work of G.L. Nikiporets [247]. The design model of the extended structure is selected in the form of an absolutely rigid beam on elastic supports. Taking simplified records of the equations of motion, the author built the spectra of translational and rotational (torsional) acceleration of constructions with different parameters, which characterize the dimensions of construction in the plan, velocity of propagation of seismic waves, and nature of the distribution of the stiffness of construction. The spectra of reactions are built for six accelerograms of strong earthquakes. The obtained spectra of reactions show the significant influence of the accounting of the extent of construction, especially for comparatively rigid constructions. The research work has important practical value in view of the use of records of real accelerogram. In this limited work it was only possible to consider the simplified form of initial equations of motion, as well as the fact that they were not able to examine the oscillations of structures with multi-degrees-of-freedom. Another study by the same author (G.L. Nikiporets), proposes the approximation formulas for determining the seismic load, which acts on the extent structures. It is noted that seismic load can be determined utilizing spectra of reactions, built for non-extent structures in the plan. The extent is considered with the aid of two coefficients, depending on the characteristics of construction and ground. In the research work of I.L. Korchinsky [184] the seismic oscillations on the extent structures of a one-story single-span frame with the absolutely rigid cross-bar (collar beam) are examined. The design scheme is given as a system with one-degree-of-freedom. Also investigated was the oscillations of systems in cases when base oscillates according to the law of damped sinusoid. The basic conclusion of the study is reduced to the fact that accounting for extent leads to the decrease of seismic forces in the constructions, the natural period of oscillations is more than 0.15s. For rigid structures, especially for girder bridges, the account of the length of structure can lead to an increase in the seismic load. It is noted in [184] that the given analysis and conclusions based on this do not purport to the quantitative assessment of the phenomenon. Nevertheless, the main conclusion of the work finds that the extent system should not be designed to be rigid. In another work [185] it is proposed to take into account the extent of construction by the calculation of the average value of acceleration throughout the entire length of the building. It accounts for when the direction of the acceleration of the oscillations of the ground is perpendicular to the direction of the motion of seismic wave. In this case, in addition to progressive inertial forces, rotational forces will also act on the structure. Simple formulas for enumerating these loads are proposed. The research work of A.B. Puhovsky and V.A. Gordin [296] examines the seismic resistance of the arch. The design schema is considered as a system with a finite number of degrees of freedom. The frequencies and modes of the free oscillations of an arch are determined by the solution of the corresponding equations on the computer. The internal forces of the structure from the seismic loads are determined by the current norms and standard; "Recommended Provisions for Seismic Regulations for New Buildings and Structures" [299], taking into account three and six modes of oscillations. It is shown that the internal forces in the structure can be determined with adequate accuracy taking into account three modes of oscillations. A method of determining design forces taking into account the highest modes of oscillations is proposed. The seismic oscillations of the extent girder, arched, and suspension bridges are studied in considerably smaller degree. The vertical oscillations of the suspension bridge with the horizontal harmonic oscillations of the base have been examined by K. Kubo [196]. In this work it is shown that under the action of the running harmonic wave on the supports of long-span suspension bridges the essential influence renders commensurably the span of structure with the wave length. The nonstationary random oscillations of suspension bridges with the action on the supports of the running seismic wave are examined in the work of A.A. Petrov and S.V. Basilevsky [282]. Horizontal seismic acceleration that propagates with a constant velocity at an arbitrary angle to the longitudinal axis of construction is considered. The impact is given in the form of the amplitude of a non-stationary random process. The equations of forced oscillations of the beam of a rigid bridge are given. The decomposition in terms of the natural oscillation modes is used for the solution of the problem. The expressions are obtained for the horizontal seismic force and torque moment, which are close to the expressions given in the design standards. These formulas contain the coefficients, accounting for commensurability of the span of bridge suspension with the length of seismic wave and small damping of the oscillations of construction and non-stationary seismic excitation. The authors give the approximate expressions for determining the coefficient that takes into account the extent of the construction. In the works [265, 267] the seismic loads acting on the arch bridge across the river Arpa, (Armenia) with different schemas for the motion of supports are determined. The motion of supports is considered as the statistical independent random processes with one and the same energy spectrum. The seismic loads are determined on the basis of the current design standards. The distribution of the seismic forces with different modes of oscillations is given.

The authors carried out a survey of research work on the accounting of extent of a structure with the determination of seismic loads which showed that, despite the numerous works conducted by scientists, this problem is still far from a form that is necessary for practical design and construction. In particular, this is relevant to the work taking into account the extent of seismic oscillations on girder bridges, but is even more relevant to the arched long-span bridges. Meanwhile, promising direction of research studies should include the refinement of the buildings of three-dimensional, spatial equations of motion of construction, taking into account the final velocity of propagation of seismic waves, and the actual distribution of mass and stiffness of the system. The substantiated analysis of this phenomenon assumes knowledge of the necessary seismological information, for instance, the velocity of propagation of seismic waves on the surface of ground, multicomponent seismograms and accelerograms of earthquakes, and also of specific dynamic characteristics of the construction itself; periods and modes of free oscillations and the value of decrement of oscillations. The aforesaid relates to long-span bridge constructions, for which the account of extent has fundamental importance and can lead to a substantial change in the condition (state) of stress. Based on this, we have set up the precise equations of motion for the extent of multi-span and multi-story frames and long-span arched bridges, representing systems with multiple degrees-of-freedom. The dynamic characteristics of long-span arched steel bridge are determined experimentally for the purpose of comparison with the analytical calculations, and for checking the authenticity of the design schemas given. Thus, in light of the above, we conducted studies in the following directions:

- seismic oscillations of long-span arched bridges taking into account their extent;
- seismic oscillations of multi-span frame bridges taking into account their extent; and
- seismic oscillations of single-span, one-story frames taking into account their extent.

When designing objects of large-scale construction and special construction, we must consider sociological, technical, environmental, and economic factors. This in particular relates to the questions of design of construction in seismic zones. The determination of the real inertial forces, which appear with earthquakes, will contribute to more economical, safer, and more reliable design. Long-span combined bridge systems are unique construction, and therefore, seismic loads for them should be determined not only on the acting normative standards and provisions, but also on the real seismograms or accelerograms of recorded strong earthquakes. Furthermore, in the construction of considerable range, whose length with the impacts of the length of seismic wave is commensurate, the displacements of the individual supports of construction in the general case will be generally shifted on a phase. These structures can be subject to very different motions along their length due to the spatial variability of the input seismic motion. Therefore, during calculations of bridge constructions we should consider the asynchronicity of the oscillations (vibrations) of individual supports.

The influence of the extent of construction of the values of seismic loads was investigated initially by G. Hausner [135]. Observations conducted by Hausner during earthquakes have shown that seismic ground motions can vary significantly over a distance of the same order of magnitude as the dimensions of these extended structures. To the influence of extent with the harmonic oscillations of ground with different design diagrams of construction were dedicated the works of: I.L. Korchinsky, A.A. Grill, A.P. Sinitsina, J.E. Bogdanoff, S.D. Werner, M. Shinozuka, M. Hoshiya, W.J. Hall, N.M. Newmark, M.J. Priestley, R. Richards, D.G. Elms, A.D. Kiureghian, R.S. Harichandran, E.E. Khachian, A. Kareem, L.C. Lee, L.H. Wong, O. Ramadan, E.H. Vanmarcke, J.-N. Yang, J. Wang, J.E. Luco, H.L. Wong, C.E. Taylor, A. Zerva, and others. According to S.D. Werner et al. [408] research analyses, the differences in the input ground motions applied to bridge foundations can have significant effects on the bridge response. Also, the nature of structural response to these traveling waves is strongly dependent on the direction of incidence as well as on the excitation frequency of the seismic waves. A study of these author's works suggests that the effect of the spatial variation of the seismic ground motions on the response of extended structures, such as beam (girder) and arched bridges, with the use of real records of the motion of ground and dynamic characteristics of the structure is a promising trend. Let us propose a method for calculating seismic resistance of combined bridge systems by using real records of the ground motion of strong earthquakes. In Figure 3.21 the dynamic design scheme and the scheme of seismic excitation are shown. The dynamic design scheme (model) of the long-span combined bridge system of the type rigid beam–flexible arch is taken as a system with a finite number of degrees of freedom.

In this case each concentrated mass possesses three degrees of freedom, and the position of $i$th mass in the space is characterized by the displacements of the $X_i$, $Y_i$ in the plane of the bridge, and displacement $Z_i$ in the direction perpendicular to the plane of the bridge.

It is assumed, that the seismograms of the oscillations of support A in three mutually perpendicular directions $X_A(t)$, $Y_A(t)$, and $Z_A(t)$ are known and so are the value of the velocity of propagation of waves of excitation on the surface of the ground $V$. The seismograms of the oscillations of support $B$ will be:

$$X_B(t) = X_A\left(t - \frac{L_\alpha}{V}\right); Y_B(t) = Y_A\left(t - \frac{L_\alpha}{V}\right); Z_B(t) = Z_A\left(t - \frac{L_\alpha}{V}\right).$$

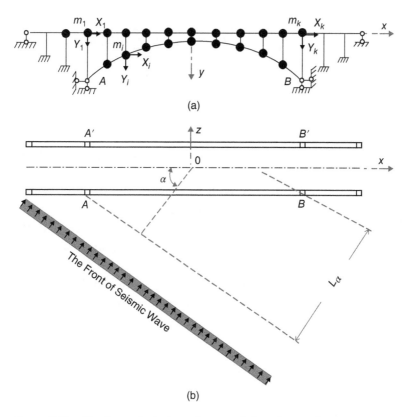

**Figure 3.21**    The dynamic design scheme and the scheme of seismic excitation.

It is assumed that the front of seismic excitation is the plane propagating with velocity $V$, under angle $\alpha$ to the longitudinal axis of the bridge.

The equations of motions of this oscillatory system with asynchronous motions of supports in the general three-dimensional case of oscillations (vibrations) can be represented in the form of:

$$X_j(t) = -\sum_{i=1}^{k} m_i X_i'' \delta_{ji}^{(xx)} - \sum_{i=1}^{k} m_i Y_i'' \delta_{ji}^{(xy)} - \sum_{i=1}^{k} m_i Z_i'' \delta_{ji}^{(xz)} + \delta_{jA}^{(xx)} X_A$$

$$+ \delta_{jB}^{(xx)} X_B + \delta_{jA}^{(xy)} Y_A + \delta_{jB}^{(xy)} Y_B + \delta_{jA}^{(xz)} Z_A + \delta_{jB}^{(xz)} Z_B$$

$$Y_j(t) = -\sum_{i=1}^{k} m_i X_i'' \delta_{ji}^{(yx)} - \sum_{i=1}^{k} m_i Y_i'' \delta_{ji}^{(yy)} - \sum_{i=1}^{k} m_i Z_i'' \delta_{ji}^{(yz)} + \delta_{jA}^{(yx)} X_A$$

$$+ \delta_{jB}^{(yx)} X_B + \delta_{jA}^{(yy)} Y_A + \delta_{jB}^{(yy)} Y_B + \delta_{jA}^{(yz)} Z_A + \delta_{jB}^{(yz)} Z_B \qquad (3.10.3)$$

$$Z_j(t) = -\sum_{i=1}^{k} m_i X_i'' \delta_{ji}^{(zx)} - \sum_{i=1}^{k} m_i Y_i'' \delta_{ji}^{(zy)} - \sum_{i=1}^{k} m_i Z_i'' \delta_{ji}^{(zz)} + \delta_{jA}^{(zx)} X_A$$

$$+ \delta_{jB}^{(zx)} X_B + \delta_{jA}^{(zy)} Y_A + \delta_{jB}^{(zy)} Y_B + \delta_{jA}^{(zz)} Z_A + \delta_{jB}^{(zz)} Z_B$$

where

- $m_i$ – mass concentrated in the $i$th section; $k$ – a quantity (number) of concentrated masses;
- $X_j, Y_j, Z_j, X_j'', Y_j'', Z_j''$ – respectively displacements and the acceleration of $j$th mass in three mutually perpendicular directions;
- $\delta_{ji}^{(xx)}, \delta_{ji}^{(xy)}, \delta_{ji}^{(xz)}$ – displacements of $j$th mass in the direction of the $X$ unit forces applied (acting) at the point $i$, in the directions $X$, $Y$, and $Z$ respectively;
- $\delta_{jA}^{(xx)}, \delta_{jA}^{(xy)}, \delta_{jA}^{(xz)}$ – displacements of $j$th mass in the directions of the $X$ due to unit displacement of the support $A$ in the directions $X$, $Y$, and $Z$ respectively;
- $\delta_{jB}^{(xx)}, \delta_{jB}^{(xy)}, \delta_{jB}^{(xz)}$ – displacements of $j$th mass in the directions of the $X$ due to unit displacement of the support $B$ in the directions $X$, $Y$, and $Z$ respectively.

The determination of seismic loads is reduced to the integration of the system of differential Eq. (3.10.1), and defining (to determination) of the maximum values:

$m_i X_i''(t)$,  $m_i Y_i''(t)$, and $m_i Z_i''(t)$

With the use of the known determined law of the ground motion, the seismic loads and the displacements of system can be determined with the aid of the analytical solution of the system of the linear differential equations. In the case of using the real accelerograms, assigned in tabular form, use the numerical solutions of equations by the methods of Runge-Kutta or Adams. Let us examine a concrete example of the determination of the seismic loads taking into account the extent of construction, operating in a large-span (long-span, wide-span) metallic bridge, which overlaps the canyon of the river in Armenia. We will determine the reactions of construction taking into account its extent with assigned horizontal harmonic motion of the base. The influence of asynchronic oscillation of the outer (extreme) supports will be most essential, when the front of a seismic wave is perpendicular to the longitudinal axis of bridge $x$. The dynamic design scheme of construction is taken in the form of a system with three concentrated masses as those applied in the characteristic sections of the combined system of the type of "rigid beam–flexible arch." Since each mass has two degrees of freedom, the entire system will have six degrees of freedom. The equation of motion of this oscillatory system with asynchronous oscillations of supports can be represented in the following form:

$$X_j(t) = -\sum_{i=1}^{3} m_i X_i'' \delta_{ji}^{(xx)} - \sum_{i=1}^{3} m_i Y_i'' \delta_{ji}^{(xy)} + \delta_{jA}^{(xx)} X_A$$

$$+ \delta_{jB}^{(xx)} X_B + \frac{\alpha}{\omega_1}(X_j - X_A)' = 0; \qquad (3.10.4)$$

$$Y_j(t) = -\sum_{i=1}^{3} m_i X_i'' \delta_{ji}^{(yx)} - \sum_{i=1}^{3} m_i Y_i'' \delta_{ji}^{(yy)} + \delta_{jA}^{(yx)} X_A$$

$$+ \delta_{jB}^{(yx)} X_B + \frac{\alpha}{\omega_1}(Y_j - Y_A)' = 0;$$

$$(j = 1, 2, \dots 3; Y_A = 0)$$

We have accepted upon consideration of internal friction, that with the asynchronous oscillations the damping is the same as with the synchronous. This assumption can be considered acceptable, having the low value of damping coefficient. In the equations of motion (3.10.1), the damping is taken into account on the equivalent hypothesis of Voight [175] by addition to the left side of the equations, the following members (terms):

$$\frac{\alpha}{\omega_1}(\Delta X_j)'; \frac{\alpha}{\omega_1}(\Delta Y_j)'; \frac{\alpha}{\omega_1}(\Delta Z_j)'.$$

Here $\Delta X_j$, $\Delta Y_j$, $\Delta Z_j$ represent the corresponding deformations of the jth cross-section in the directions x, y, and z; $\alpha$ – the damping coefficient; and $\omega_1$ – fundamental frequency of free oscillations. The deformation of jth cross-section $\Delta X_j$ can be presented in the following form:

$$\Delta X_j = X_j - \delta_{jA}^{(xx)} X_A - \delta_{jB}^{(xx)} X_B - \delta_{jA}^{(xy)} - \delta_{jB}^{(xy)} - \delta_{jA}^{(xz)} - \delta_{jB}^{(xz)}. \qquad (3.10.5)$$

The expression for $\Delta Y_j$ and $\Delta Z_j$ have a similar appearance. If the oscillations of supports occur synchronously ($X_A = X_B$), and only in the plane of xoy, then from (3.10.3) follows: $\Delta X_j = X_j - X_A$.

Eqs. (3.10.2) are built only with the assigned horizontal motion of the base $-X_A(t)$, $X_B(t)$. Determining the seismic reactions of construction requires the preliminary determination of the constant coefficients, entering the system of Eqs. (3.10.2); unit displacements $\delta_{ji}$, the values of the concentrated masses $m_i$, and fundamental tone of frequency $\omega_1$. It is also necessary to have the characteristic of the internal friction of the material of the bridge, the coefficient of damping $\alpha$ and the characteristics of the influence of seismic action (effect) seismogram of the ground motion, and velocity of propagation of waves of excitation of surface of the ground. The displacements in the multiple statically indeterminate system are determined by the application of Moore's formula taking into account the influence of the bending moment and normal force. We will determine each of these parameters separately. The bars system is a rigid beam – "stiffening girder" – being supported by the flexible arch. The calculation of this system is connected with the essential difficulties. In the cases examined by us, rigidity of posts and arches are small in comparison with the rigidity of a beam, therefore the problem can be considerably simplified by, following the recommendation of A.Ph. Smirnov [341], accepting the design scheme in the form shown in Figure 3.22. The beam of stiffness of the given system works as a seven-span continuous beam (one support is usually considered as an immovable hinge while the other supports are hinges on rollers), resting on supports of the trestle (overpass). For simplification in the calculations we accept the three-span scheme in which the first and last spans

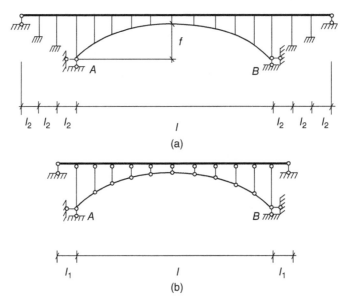

**Figure 3.22** Structural and design schemes of the bridge: (a) structural schema; and (b) design scheme.

of the overpass adjacent to the main arch span create the built-in support of the same stiffness as seven-spans. Here the rotation on the support of a three-span continuous beam with the spans $l_2$ due to a unit moment must be equal to the rotation angle on the support of a single-span beam with the span $l_1$ with the identical rigidity. Equating the angles of the rotations of the single-span and three-span of continuous beams, we obtain the value of the reduced length $l_1$, $l_1 = 0.87l_2 = 0.7m$. Represented in this form, the system becomes three times statically indeterminate. We will obtain a simple solution of the problem, if as the redundant unknowns we select the thrust of one of the supports and as the group of unknowns, concentrated moments at the sections $A - A'$ and $B - B'$ (Figure 3.22).

The unit displacements, entering the canonical equations for determination of "redundant" unknowns, are determined by the application of Moore's formula, taking into account the influence of the bending moment and normal force. For determining the displacement $\delta_{ik}$ of a statically indeterminate system, first the redundant unknowns due to force $P = 1$, applied in the direction $k$ are determined. Then the values $M_i$ and $N_i$ due to force $P = 1$, applied in the direction $i$ are determined. Thus $M_i$ and $N_i$ can be calculated for any geometrically invariable (unchangeable, immovable), statically indeterminate system, and in particular for this primary system. The following values of the geometrical dimensions and characteristics of a construction were taken, then calculating:

$$l = 120m;\ f = 18m; l_2 = 8m; I_{Arch} = 0.0012745m^4; I_{Beam} = 0.04356m^4$$

$$F_{Arch} = 0.05276m^2; F_{Beam} = 0.13373m^2.$$

The results of the static analysis and calculations of a bridge, including the values of unit displacements, are given in the research works of L.G. Petrosian [265, 280]. The equations of motion (3.10.1) can be solved both analytically and through using numerical integration with a computer. The decomposition in the natural modes of oscillation is most frequently used during the analytical solution. The knowledge of the fundamental natural frequency is also necessary for numerical integration. Thus, it is necessary to determine natural frequencies and modes of oscillations. The equations of the free oscillations of a system are obtained, if we consider that in (3.10.2) $X_A = X_B = 0;\ \alpha = 0$, and that $X_j = \overline{X}_j \sin \omega t; Y_j = \overline{Y}_j \sin \omega t$. The natural frequencies $\omega$ are determined from the frequency equation:

$$\left| A - \frac{1}{\omega^2} E \right| = 0, \tag{3.10.6}$$

where $A = \|m_i \delta_{ij}\|$ – square matrix of unit displacements; and $E$ – unit matrix.

The coefficients of the modes of oscillations $\overline{X}_j, \overline{Y}_j$ are determined by the solution of the system of the equations:

$$-\overline{X}_j + \sum_{i=1}^{k} m_i \omega^2 X_j \delta_{ji}^{(xx)} + \sum_{i=1}^{k} m_i \omega^2 Y_j \delta_{ji}^{(xy)} = 0; \qquad (3.10.7)$$

$$-\overline{Y}_j + \sum_{i=1}^{k} m_i \omega^2 X_j \delta_{ji}^{(yx)} + \sum_{i=1}^{k} m_i \omega^2 Y_j \delta_{ji}^{(yy)} = 0,$$

$$(j = 1, 2, \dots k)$$

in which one of the amplitudes of oscillations is taken as the given one, for example equal to unit (one), and the system is solved with $(2n-1)$ equations. For determining $\omega_i$, $X_i$, $Y_i$ it is necessary to have values of the concentrated masses $m_i$. These values, calculated according to the working drawings of the bridge and taken with the corresponding reduction coefficients, proved to be equal to:

$$m_1 = m_3 = m_4 = m_6 = 15.1 \frac{kNs^2}{m}; m_2 = m_5 = 35.3\frac{kNs^2}{m}.$$

In this case, most significant is the mass of the rigid beam, which is taken with the reduction coefficient, equal to 0.5; A.P. Philipov [284]. The values of the first six periods of oscillations for the given bridge, obtained by the solution of frequency Eq. (3.10.6), are:

$$T_1 = 0.650s; T_2 = 0.608s; T_3 = 0.261s;$$

$$T_4 = 0.242s; T_5 = 0.104s; T_6 = 0.092s.$$

The values of the coefficients of the modes of oscillations of the bridge are given in Table 3.14.

The results of dynamic analysis, including the geometrical construction of the first six modes of free oscillations which correspond to the symmetrical and inversely symmetrical modes of vertical oscillations, are given in the research works of L.G. Petrosian [263, 265]. The determination of the reactions of the extensive bridge under seismic influences was realized by the numerical solution of differential equations of motion (3.10.7). For solving the system of Eq. (3.10.7) with the arbitrary law of the ground motion,

**Table 3.14** Values of the coefficients of free oscillation modes of the bridge.

| Modes coefficients | Modes # | | | | | |
|---|---|---|---|---|---|---|
| | 1 | 2 | 3 | 4 | 5 | 6 |
| $X_1$ | 0.892 | 0.082 | −0.074 | 0.568 | −0.038 | 0.408 |
| $X_2$ | 1.000 | 0 | 0 | 1 | 0 | 1 |
| $X_3$ | 0.901 | −0.083 | 0.075 | 0.567 | 0.038 | 0.409 |
| $Y_1$ | 0.129 | −0.824 | −1.430 | 0.081 | −1.261 | 0.062 |
| $Y_2$ | 0 | −1.000 | 1 | 0 | 1 | 0 |
| $Y_3$ | −0.130 | −0.824 | 1.430 | −0.082 | 1.259 | −0.062 |

in accordance with the seismogram of the $X_A(t)$, $X_B(t)$ (the program of calculation in the algorithmic language FORTRAN–IV was developed prior to
using this program), the system of the equations is reduce to its normal form.
For this purpose it is necessary to first solve Eq. (3.10.7) relative to the second derivatives. For simplicity in this case, designating $Y_j(t) = X_{j+3}(t)$, $j = 1$,
2, 3, we convert Eq. (3.10.7) to the form:

$$X'_{k+6} = X_{k+6}; k = 1, 2, \ldots, 6;$$

$$X'_{k+6} = -\frac{1}{m_k} \left\{ \sum_{j=1}^{6} C_{kj} \left[ X_j + \frac{\alpha}{\omega_1}(X_j - X_A)' \right] - d_{kA}X_A - d_{kB}X_B \right\} \quad (3.10.8)$$

$$k = 1, 2, \ldots, 6.$$

Where the following designations are accepted:

$$C_{kj} = (-1)^{j+k}\frac{M_{jk}}{\Delta};$$

$\Delta = \|\delta_{kj}\|_1^6$ – matrix of unit displacements; and $M_{jk}$ – minor of element $\delta_{kj}$
of determinant $|\Delta|$;

$$d_{kA} = \frac{1}{\Delta} \sum_{i=1}^{6} \delta_{iA}(-1)^{i+k} \cdot M_{ik};$$

$$d_{kB} = \frac{1}{\Delta} \sum_{i=1}^{6} \delta_{iB}(-1)^{i+k} \cdot M_{ik};$$

$$k = 1, 2, \ldots, 6.$$

The program provides for the determination of horizontal and vertical accelerations of the concentrated masses, i.e. the values of $X'_{k+6}$; $k = 1, 2,$ ..., 6; and also the relative displacements of masses – $(X_k - \delta_{kA}X_k - \delta_{kB}X_{kB})$; $k = 1, 2, ..., 6$.

Eqs. (3.10.8) were solved with harmonic oscillations of the ground with parameters corresponding to the real seismic influence. The predominant periods of oscillations of ground during strong earthquakes most frequently vary with the limits: 0.1s ÷ 0.4s [175]. The laws of the motion of bridge supports were taken in the form of:

$$X_A = A \sin \omega_0 t; \quad X_B = A \sin \omega_0 \left(t - \frac{l}{v_0}\right)$$

Option I:

$$\omega_0 = 31.41\text{s}^{-1}; \quad T_0 = \frac{2\pi}{\omega_0} = 0.2\text{s}; \quad A = 0.031\text{m};$$

Option II:

$$\omega_0 = 15.70\text{s}^{-1}; \quad T_0 = \frac{2\pi}{\omega_0} = 0.4\text{s}; \quad A = 0.031\text{m}.$$

Three types of ground were examined with the values of the velocities of propagation of the surface waves, equaling to: $152.5\frac{\text{m}}{\text{s}}$; $305\frac{\text{m}}{\text{s}}$; $610$ m/s [337] respectively. The taken amplitude value of the acceleration of the seismic influence approximately corresponds to the maximum amplitude values of the seismograms of actual strong earthquakes. The values of velocities 305 m/s and 610 m/s, correspond to the velocities of surface waves in sandy loams, loams, and clay soil. The ground soil conditions in the canyon of river Arpa are approximately similar. For comparison, the softest soil and filled ground for which $v = 152.5$ m/s were also considered. The decrement of the oscillations of steel bridges, obtained by us according to the results of field full-scale tests, was set equal to $\delta = 0.023$. The coefficient $\alpha$, in the equations of motion, is equal to $= \delta/\pi$. The results of dynamic analysis including geometrical construction of the dependences of accelerations of concentrated masses on time, are given in the research work of L.G. Petrosian [265]. The dependences of the accelerations of concentrated masses on the time are obtained ($T = 0.2$ s, $T = 0.4$ s), and the values of maximum accelerations with the first option of the external seismic excitation ($T = 0.2$ s) are given in Tables 3.15 and 3.16.

The given results show that the calculation of extent leads to the decrease of the horizontal inertial forces. With the first option ($T = 0.2$ s)

Table 3.15 Values of the maximum horizontal and vertical accelerations, acting on the long-span steel bridge.

| (m/s²) Accelerations | 0 | 0.1967 | $l/v$ (s) 0.3934 | 0.7868 |
|---|---|---|---|---|
| $X_1''$ | 0.46 | 0.42 | 0.35 | 0.31 |
| $X_2''$ | 12.00 | 11.05 | 9.94 | 8.97 |
| $X_3''$ | 0.47 | 0.41 | 0.35 | 0.32 |
| $Y_1''$ | 0.17 | 0.20 | 0.23 | 0.25 |
| $Y_2''$ | 1.19 | 1.33 | 1.42 | 1.59 |
| $Y_3''$ | 0.18 | 0.21 | 0.22 | 0.28 |

Table 3.16 Values of accelerations for the second option of the external seismic excitation ($T = 0.4$s).

| (m/s²) Accelerations | 0 | 0.1967 | $l/v$ (s) 0.3934 | 0.7868 |
|---|---|---|---|---|
| $X_1''$ | 1.02 | 0.93 | 0.71 | 0.62 |
| $X_2''$ | 18.10 | 16.07 | 11.98 | 8.04 |
| $X_3''$ | 1.04 | 0.91 | 0.72 | 0.61 |
| $Y_1''$ | 0.42 | 0.53 | 0.65 | 0.84 |
| $Y_2''$ | 0.98 | 1.15 | 1.67 | 2.21 |
| $Y_3''$ | 0.41 | 0.53 | 0.62 | 0.82 |

the maximum horizontal accelerations, calculated taking into account the extent of the bridge construction, significantly decrease. In ground with the values of the velocities of $v = 152.5$ m/s, $v = 305$ m/s, and $v = 610$ m/s, accelerations of the $X_2''$ decrease by $\frac{12.00}{11.05} = 1.09$; $\frac{12.00}{9.94} = 1.21$, and $\frac{12.00}{8.97} = 1.34$ times, respectively. In the second option ($T = 0.4$ s) the influence of extent on the degree of the decrease of inertial forces is somewhat more. That is possible to explain by the fact that in the second case the fundamental frequency is close to the frequency of the oscillation of the base. From the given data, another characteristic: regularity – increases in the vertical accelerations of the bridge with an increase at the asynchronous oscillations of the supports. However, the absolute value of the amplitude of the vertical acceleration, resulting from the horizontal oscillations, is significantly small compared to the amplitudes of the horizontal accelerations.

Thus, the analysis showed that the inclusion of the asynchronous oscillations of supports, when considering bridge spans and ground conditions, result in a reduction of maximum inertial loads, which increases the resistance of the structure to seismic excitation. Summarizing the results of the oscillations of a combined bridge structure rigid beam–flexible arch, taking into account its extent under seismic excitation, we come to the following conclusions:

1. The design procedure of seismic calculations of long-span bridge structures taking into account their extent with the application of real seismograms of strong earthquakes is proposed.
2. General equations of motion for the system with finite number of degrees of freedom at spatial deformation of constructions are built.
3. The analysis of the seismic reaction of bridges showed that taking into account the extent, in ground of medium hardness, can lead to the decrease of inertia loads on the average of $1.2 \div 1.5$ times.
4. Vertical inertia loads, which appear with the asynchronous horizontal oscillations of supports, are small and can be disregarded.

## 3.11   The Determination of the Reactions of Multiple Spans Frame Bridges, Extended Buildings, and Structures Taking into Account the Initial Phase of Passing (Propagation) of the Seismic Wave

According to the current methodology of calculating the impact of seismic excitation in buildings and other construction, all points of base oscillate synchronously with the same shift and acceleration. However, for buildings and structures with a length in the direction of propagation of excitation commensurate to the length of seismic waves, this methodology can lead to significant errors when determining the reactions of these systems. Long-span arched constructions of bridges present complex design systems, which consist both of the rectilinear frame elements and of curvilinear arched elements. Seismic calculation of such constructions is associated with complex and labor-intensive computational operations. Therefore, in certain cases it is expedient to replace curvilinear elements with rectilinear elements, and thus, simplify the design diagram (scheme) of a structure [266].

Let us examine the oscillations of a multi-story, multi-span frame structure of considerable extent in the plan of Figure 3.23. In this particular case,

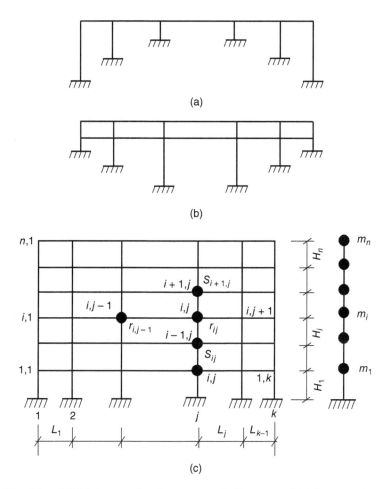

**Figure 3.23** (a) and (b) Schemes of multi-span trestle frames; and (c) dynamic design scheme of the multi-story, multi-span frame structure.

it is possible to obtain the equations of motion of trestle frames from the derivation of the equation of motion. The methodology of derivation of the equations of motion can be extended to frame-arched structures, in which rectilinear elements are substituted for curvilinear elements. The derived equations of motion can be used in the analysis of a stressed state of an over-arch pass-through span structure of a frame-arch system of a bridge. In particular, it is possible to consider an over-arch pass-through span structure of one- or two-tier frame as a special case of a multistoried, multi-span frame Figure 3.23.

The dynamic design model is accepted as a system with finite number of degrees of freedom. The equations of seismic oscillations of an elastic frame system taking into account extent can be obtained on the basis of the equations of the displacement method with the static horizontal shifting of supports, and taking into account the inertial forces in accordance with the Dalember's principle.

The equations obtained in this way for a multi-story frame, where each support has a certain displacement, can be represented in the following common form:

1. $2S_{ij}\varphi_{i-1} + 2r_{i,j-1}\varphi_{i,j-1} + 4(S_{ij} + r_{i,j-1} + r_{ij} + S_{i+1,j})\varphi_{ij} + 2r_{ij}\varphi_{i,j+1}$
$$+ 2S_{i+1,j}\varphi_{i+1,j} - 6\frac{S_{ij}}{H_i}(y_i - y_{i-1}) - 6\frac{S_{i+1,j}}{H_{i+1}}(y_{i+1} - y_i) = 0;$$

$$i = 2, 3, \ldots, n; j = 1, 2, \ldots, k; r_{i0} = r_{ik} = S_{n+1,j} = y_{n+1} = 0;$$

2. $2r_{1,j-1} + 4(S_{1j} + r_{1,j-1} + r_{1j} + S_{2j})\varphi_{1j} + 2r_{1j}\varphi_{1,j+1} + 2S_{2j}\varphi_{2j}$
$$- 6\frac{S_{1j}}{H_1}(y_1 - y_{0j}) - 6\frac{S_{2j}}{H_2}(y_2 - y_1) = 0;$$

$$j = 1, 2, \ldots, k; r_{1,0} = r_{1,k} = 0; \qquad\qquad (3.11.1)$$

3. $6\sum_{j=1}^{k}\frac{S_{ij}}{H_i}(\varphi_{ij} + \varphi_{i-1,j}) - 12\sum_{j=1}^{k}\frac{S_{ij}}{H_i^2}(y_i - y_{i-1}) + \sum_{\xi=i}^{n}(-m_\xi y_\xi'') = 0;$

$$i = 2, 3, \ldots, n;$$

4. $6\sum_{j=1}^{k}\frac{S_{1j}}{H_i}\varphi_{1j} - 12\sum_{j=1}^{k}\frac{S_{1j}}{H_1^2}(y_1 - y_{0j}) + \sum_{\xi=1}^{n}(-m_\xi y_\xi'') = 0;$

where

$S_{ij} = \dfrac{EI_{ij}^{(s)}}{H_i}$ – linear stiffness of $j$th column $i$th floor;

$r_{ij} = \dfrac{EI_{ij}^{(r)}}{L_i}$ – linear stiffness of $j$th beam $i$th floor;

$\varphi_{ij}$ – angle of rotation of the node with the coordinates $i$ and $j$;

$y_i$ – displacement of the $i$th floor;

$E$ – modulus of elasticity of material;

$I_{ij}^{(s)}, I_{ij}^{(r)}$ – the moments of inertia, of $j$th column or beam $i$th floor respectively;

$H_i$ – height of $i$th floor;

$L_j$ – length of $j$th span;

$m_i$ – mass concentrated at the level of the $i$th floor;

$y_j''$ – acceleration of the mass of the $i$th floor;

$n$ – numbers of floors;

$k$ – numbers of floor columns;

$y_{0j}$ – displacement of $j$th support.

   The total number of Eq. (3.11.1) is equal to $kn + n$. Since the system has $n$ degrees of freedom, then the number of unknowns, which determining the deformed state of the system, must be equal to $n$. If we solve the part of the system (3.11.1), representing the conditions of equality of moments in the nodes with respect to $\varphi_{ij}(i = 1, 2, \ldots, n; j = 1, 2, \ldots, k)$, and substitute the dependences obtained $\varphi_{ij} = f_{ij}y(y_i)$, $(i = 1, 2, \ldots, n; j = 1, 2, \ldots, k)$ into the rest of the equations, then we obtain $n$ equations of motion. The equations of motion of a multi-span one-story (single-tier) frame are a special case of the general Eq. (3.11.1), and it is possible to present them in the following form:

$$4(S_1 + r_1)\varphi_1 + 2r_1\varphi_2 - \frac{6S_1}{H}(y - y_{01}) = 0;$$

$$2r_1\varphi_1 + 4(r_1 + r_2 + S_2)\varphi_2 + -\frac{6S_2}{H}(y - y_{02}) = 0; \qquad (3.11.2)$$

$$2r_{j-1}\varphi_{j-1} + 4(r_{j-1} + S_j)\varphi_j + -\frac{6S_j}{H}(y - y_{0j}) = 0;$$

$$-my'' + \sum_{j=1}^{k}\frac{6S_j}{H}\varphi_j - \sum_{j=1}^{k}\frac{12S_j}{H^2}(y - y_{0j}) = 0,$$

with

$$t \geq t_{k-1} = \sum_{q=1}^{k-1}\frac{L_q}{v_0}.$$

   The equations of motion of the frame (3.11.1) are built for a case when each support has a certain displacement $y_{0j}$. However, a more detailed analysis of the oscillations of the system should also consider the initial moment of passage of seismic waves [175]. With the propagation of excitation, for example from left to right, the first support will move first, followed by the second support, and after a certain time the remainder of the supports will be involved in the movement. Here it is assumed that the seismic excitation propagates with a constant velocity $v_0$, dependent on the soil/ground conditions. Hence, it follows that the displacements of individual supports of a structure have the following dependence:

$$y_{0j}(t) = y_{01}\left(t - \sum_{q=1}^{j}\frac{L_q}{v_0}\right). \qquad (3.11.3)$$

During motion of only the first support, the behavior of the system is described by the system of Eq. (3.11.1), in which the remaining displacements of supports are taken as equal to zero, i.e. $y_{0j} = 0; j > 1$. The solution of this system is correct with $0 < t < L_1/v_0$. With $L_1/v_0 < t < (L_1 + L_2)/v_0$ when the second support moves, the behavior of the system is described by the system of Eq. (3.11.1), in which $y_{0j} = 0$; for $j > 2$. Finally, when all supports are involved in the motion, i.e. with

$$t > \sum_{q=1}^{k-1} \frac{L_q}{v_0},$$

the equations of motions will take the form (3.11.1), where all displacements of supports $y_{0j}$ are already different from zero. Thus, at the initial moment of the passage of the seismic wave from the first support to the last support, the motion of the system will be described by $k$ different systems of equations. With $0 < t < L_1/v_0$ the system of equations of motion must be solved with the initial conditions set to zero. With $L_1/v_0 < t < (L_1 + L_2)/v_0$ the initial parameters are those values of deformations and velocities which came out during the solution of the first system of equations at point $t = L_1/v_0$. Similarly, the remaining systems of equations must be solved. The solution of a system of equations of motion with a larger number of supports increases the volume of computational operations. However, as will be shown below, with the determination of the reactions of multi-span frames satisfactory results can be obtained if with the initial passage of a seismic wave we take into account only outer supports, i.e. we accept that the motion of the system is described only by two different systems of equations with

$$0 < t < \sum_{q=1}^{k-1} \frac{L_q}{v_0}, \quad \text{and} \quad t > \sum_{q=1}^{k-1} \frac{L_q}{v_0}.$$

In the case when the stiffness of a superstructure is several times greater than the stiffness of supports (columns, piers), it is possible to assume that the superstructure is not deformed. The equations of motion for this system will take a comparatively simple form. Assuming in (3.11.1) $\varphi_{ij} = 0$, we obtain:

$$\sum_{\xi=i}^{n} m_\xi y_\xi'' + 12 \sum_{j=1}^{k} \frac{S_{ij}}{H_i^2}(y_i - y_{i-1}) = 0$$

$$(i = 2, 3, \dots, n) \tag{3.11.4}$$

$$\sum_{\xi=1}^{n} m_\xi y_\xi'' + 12 \sum_{j=1}^{k} \frac{S_{1j}}{H_1^2}(y_1 - y_{0j}) = 0$$

By simple algebraic conversions we have:

$$m_i y_i'' - a_{i+1}(y_{i+1} - y_i) + a_i(y_i - y_{i+1}) = 0$$

$$(i = 2, 3, \ldots, n) \tag{3.11.5}$$

$$m_1 y_1'' - a_2(y_2 - y_1) + \sum_{j=1}^{k} a_{1j}(y_1 - y_{0j}) = 0$$

Where

$$a_i = \sum_{j=1}^{k} \frac{S_{ij}}{H_i^2}, \qquad (i = 1, 2, \ldots, n).$$

It is sometimes more convenient in Eq. (3.11.5) to substitute the variables. If we take into account the dumping per the equivalent hypothesis of Voigt, and designate $\bar{y}_i = y_i - y_{01}$, then it is possible to obtain:

$$m_i(\bar{y}_i + y_{01})'' - a_{i+1}(\bar{y}_{i+1} - \bar{y}_i) - \frac{a_{i+1}}{\omega} \alpha_{i+1}(\bar{y}_{i+1} - \bar{y}_i)'$$

$$+ a_i(\bar{y}_i - \bar{y}_{i-1}) + \frac{a_i}{\omega} \alpha_i(\bar{y}_i - \bar{y}_{i-1})' = 0; \tag{3.11.6}$$

$$m_1(\bar{y}_1 + y_{01})'' - a_2(\bar{y}_2 - \bar{y}_1) - \frac{a_2}{\omega} \alpha_2(\bar{y}_2 - \bar{y}_1)'$$

$$+ \sum_{j=1}^{k} a_{1j}(\bar{y}_1 + y_{01} - y_{0j}) + a_1 \frac{\alpha_1}{\omega} \bar{y}_1' = 0,$$

where
$\alpha_k$ – coefficient of damping of $k$th floor (story, tier); and $\omega$ – the fundamental frequency (the frequency of the main tone) of free oscillations.

In Eq. (3.11.6) it is assumed that the energy absorption of the first floor at asynchronous oscillations supports (columns) is the same as with the synchronous oscillations. This assumption leads to an unessential change in damping of the entire system. However, a significant simplification of the problem with the numerical solution of the equations of motion, when the seismogram or accelerogram of earthquake is given, is obtained. Thus the equation of motion solves both the absolute displacements $y_i$ and relative displacements $\bar{y}_i$. In the first case we use an accelerogram, and in the second case a seismogram and accelerogram of an earthquake. From (3.11.6) with $y_{01} = y_{0j} = y_0$ the equations of motion with the synchronous oscillations of supports are obtained [171]

In the solution of Eq. (3.11.6) we search for in the form of:

$$y_i = \sum_{r=1}^{\infty} C_{ir} q_r(t),$$

where $q_r(t)$ – generalized coordinate; $C_{ir}$ – coefficient, which characterizes the amplitude of free oscillations, of $i$th mass with $r$th mode of oscillations. Substituting the value of $y_i$ into (3.11.6), and using the orthogonality of the modes of oscillations $C_{ir}$, we obtain a differential equation relative to generalized coordinate $q_r(t)$. Solving this differential equation, we first find $q_r(t)$, and then $y_i(t)$. The seismic force $S_i$ is found using the expression:

$$S_i = m_i(\bar{y}_i + y_{01})''$$

For simplicity, the countdown of time starts for each interval equal to zero. Finally, we will obtain that in the interval:

$$\sum_{q=1}^{j} \frac{L_q}{v_0} < t_j < \sum_{q=1}^{j+1} \frac{L_q}{v_0},$$

where $j$ – is the number of oscillating supports; and the seismic force of $i$th floor/tiers $S_i$ is determined by the expression:

$$S_i = - m_i \sum_{r=1}^{n} e^{-\frac{\alpha_r}{2}\omega_r t} \left( \omega_r C_{ir} \frac{|C_{ir}|^*}{|C_{ir}|} \sin \omega_r t + \omega_r^2 \frac{C_{ir}|C_{ir}|^{**}}{|C_{ir}|} \cos \omega_r t \right)$$

$$+ m_i \sum_{r=1}^{n} \eta_{ir} \omega_r \int_0^t y_0''(\xi) e^{-\frac{\alpha_r}{2}\omega_r(t-\xi)} \cdot \sin \omega_r(t-\xi)d\xi \qquad (3.11.7)$$

$$+ m_i \sum_{r=1}^{n} \bar{\eta}_{ir} \omega_r \sum_{q=1}^{j} a_{iq} \int_0^t [y_{0q}(\xi) - y_{01}(\xi)] \cdot e^{-\frac{\alpha_r}{2}\omega_r(t-\xi)} \cdot \sin \omega_r(t-\xi)d\xi,$$

where $C_{ir}$ – normalized values of displacements of $i$th floor/tiers with the free oscillations of $r$th mode; $\eta_{ir}$ – coefficient of modes of oscillations taken by building codes

$$\eta_{ir} = \frac{C_{ir} \sum_{i=1}^{n} m_i C_{ir}}{\sum_{i=1}^{n} m_i C_{ir}^2} ; \bar{\eta}_{ir} = C_{ir} \frac{C_{ir}}{\sum_{i=1}^{n} m_i C_{ir}^2} ;$$

$|C_{ir}|$ – determinant of the corresponding coefficients ($i, r = 1, 2, ..., n$); and $|C_{ir}|^*$, $|C_{ir}|^{**}$ – determinants in which the $r$th column is replaced with the numbers $y_1(0), ... y_n(0); y_1'(0), ... y_n'(0)$ respectively. With the synchronous oscillations of supports from (3.11.7), we obtained the well-known formula for determining seismic forces [171]. As was noted above, the equations of motion of the extended construction can be set up by two methods. In one case the seismogram enters into the equations of motion, while in the other, both the seismogram and accelerogram of the earthquake enter into the equations of motion. During strong earthquakes the most frequently obtained records of oscillations of ground instrumentally are in the form of accelerograms. In order to use them with a numerical solution of Eq. (3.11.7), it is necessary to integrate them twice to obtain the corresponding seismogram, which is not always possible to successfully obtain because of the large number of errors introduced by the recording mechanism. Therefore the question arises: is it possible to set up the equations of motion in such way to use only an accelerogram to obtain the solution? Below, the opportunity of the accomplishment of this task is analyzed based on the simplest example of a system with one degree of freedom. The equations of motion, obtained from (3.11.2) and (3.11.4) in this case will be:

$$y'' + \omega^2 y = f(t);$$

$$f(t) = \begin{cases} \dfrac{1}{2}\omega^2 y_0(t) & 0 < t < t_1 = \dfrac{L}{v_0}; \\ \dfrac{1}{2}\omega^2 y_0(t) + \dfrac{\omega^2}{2} y_0(t - t_1)\sigma_0(t - t_1) & t > t_1, \end{cases} \qquad (3.11.8)$$

where $\sigma_0(t - t_1)$ – unit retarded (late) function [215]. The solution of Eq. (3.11.8) with $t > t_1$ takes the form:

$$y(t) = \frac{\omega}{2} \int_0^l y_0(\xi) \sin \omega(t - \xi) d\xi + \frac{\omega}{2} \int_{t_1}^l y_0(\xi - t_1) \sin \omega(t - \xi) d\xi. \quad (3.11.9)$$

In order to use the accelerogram of real earthquakes, we will differentiate Eq. (3.11.8) twice, and obtain:

$$y^{IV} + \omega^2 y'' = f''(t);$$

$$f''(t) = \begin{cases} \dfrac{1}{2}\omega^2 y_0''(t); \\ \dfrac{1}{2}\omega^2 y_0''(t) + \dfrac{\omega^2}{2} y_0''(t - t_1)\sigma_0(t - t_1). \end{cases} \qquad (3.11.10)$$

The obtained differential equation of the forth degree, is generally not equivalent to the initial Eq. (3.11.8). For its solution it is necessary to have four initial conditions. Two "extra" initial conditions should be selected so that the solutions of Eqs. (3.11.8) and (3.11.10) match. Let us solve Eq.

(3.11.10) using the Laplace-Carson transformation. The equation relative to images with $t > t_1$ takes the form:

$$Y(p) = \frac{\omega^2 y_0''(p)}{2p^2(p^2 + \omega^2)} + \frac{\omega^2 y_0''}{2p^2(p^2 + \omega^2)} e^{-pt_1} + \frac{p^2 y''(0) + py'''(0)}{p^2(p^2 + \omega^2)}. \quad (3.11.11)$$

Passing to the initial functions, we will obtain:

$$y(t) = \frac{1}{2} \int_0^t y_0''(\xi)(t - \xi)d\xi + \frac{1}{2} \int_{t_1}^t y_0''(\xi - t_1)(t - \xi)d\xi$$

$$- \frac{1}{2\omega} \int_0^t y_0''(\xi) \sin \omega(t - \xi)d\xi - \frac{1}{2\omega} \int_{t_1}^t y_0''(\xi - t_1) \sin \omega(t - \xi)d\xi$$

$$(3.11.12)$$

$$+ \frac{y''(0)}{\omega^2}(1 - \cos \omega t) + \frac{y'''(0)}{\omega^2}\left(t - \frac{\sin \omega t}{\omega}\right).$$

As can be seen from (3.11.10), the obtained solution of Eq. (3.11.10) depends thus far on the yet unknown values of the second and third derivative, and on the displacement at initial point, i.e. of $y_0''(t)$ and $y_0'''(t)$. Two other initial conditions with the derivation of formula (3.11.12) were assumed equal to zero, i.e. $y(0) = y'(0) = 0$. Integrating by parts twice the integrals in (3.11.12), and comparing the obtained expression $y(t)$ with (3.11.9), we can see that both solutions are the same, if the condition is satisfied:

$$\frac{y_0(0)}{2}\{[1 - \cos \omega t] + [1 - \cos \omega(t - t_1)]\} = \frac{y''(0)}{\omega^2}(1 - \cos \omega t); \quad (3.11.13)$$

$$\frac{y_0'(0)}{2}\left\{\left[1 - \frac{\sin \omega t}{\omega}\right] + \left[(t - t_1) - \frac{\sin \omega(t - t_1)}{\omega}\right]\right\} = \frac{y'''(0)}{\omega^2}\left(t - \frac{\sin \omega t}{\omega}\right)$$

Substituting the obtained from (3.11.13) value of $y''(0)$ and $y'''(0)$ into (3.11.12) we finally obtain the expression $y(t)$ with $t > t_1$:

$$y(t) = \frac{1}{2} \int_0^t y_0''(\xi)(t - \xi)d\xi - \frac{1}{2\omega} \int_0^t y_0''(\xi) \sin \omega(t - \xi)d\xi$$

$$+ \frac{1}{2} \int_{t_1}^t y_0''(\xi - t_1)(t - \xi)d\xi - \frac{1}{2\omega} \int_{t_1}^t y_0''(\xi - t_1) \sin \omega(t - \xi)d\xi$$

$$(3.11.14)$$

$$+ \frac{y_0(0)}{2}\{[1 - \cos \omega t] + [1 - \cos \omega(t - t_1)]\}$$

$$+ \frac{y_0'(0)}{2}\left\{\left[t - \frac{\sin \omega t}{\omega}\right] + \left[(t - t_1) - \frac{\sin \omega(t - t_1)}{\omega}\right]\right\}$$

With $0 < t < t_1$, $y(t)$ it is determined by the expression:

$$y(t) = \frac{1}{2} \int_0^t y_0''(\xi)(t - \xi)d\xi - \frac{1}{2\omega} \int_0^t y_0''(\xi) \sin \omega(t - \xi)d\xi \qquad (3.11.15)$$

$$+ \frac{y_0(0)}{2}(1 - \cos \omega t) + \frac{y_0'(0)}{2}\left(t - \frac{\sin \omega t}{\omega}\right).$$

The seismic force will be equal to $my''(t)$. As can be seen from expressions (3.11.14) and (3.11.15), for determining the reactions of a system with a given accelerogram it is necessary to know only the initial values, corresponding seismograms and peak ground velocity (PGV), which can be determined by the approximation of the initial section of an accelerogram by an analytic function and by numerical integration in that section only.

When determining seismic forces acting on structures of significant length of bridges, trestles, single-span industrial buildings and structures, overpasses, and so forth, it is necessary to take into account the influence of extent on the values of the reactions of these systems. Thorough analysis of the influence of extent must be executed by integration of the precise equations of the motion of these systems during the motion of the base according to the law close to the actual seismic excitation. In this paragraph the derivation of the equations of motion of one-story one-deck frame structures, taking into account their extent with the arbitrary ground motion, is given. For the derivation of these equations the method of deformation (displacement method) is used, which in this case leads to significant simplifications. With the general calculations of construction during seismic excitation as the design dynamic model, a cantilever bar with the concentrated masses is taken. Let us examine horizontal oscillations of a single-span frame, where a system with one degree of freedom is taken as the dynamic design scheme. It is assumed that the law of motion of base and the values of the velocity of propagation of seismic excitation over the surface of ground is given. Let us build the equations of motion of frame in the case, when the seismogram of ground $y_0(t)$ and the velocity of the motion of wave $v$ are given. It is more convenient to obtain the equation of motion of a frame using the appropriate static equations of equilibrium with the motion of supports with the addition of inertial terms. With the displacement of the left support on $y_{01}$, the equation of motion will take the following form:

$$4(S_1 + r_1)\varphi_1 + 2r_1\varphi_2 - \frac{6S_1}{H}(y - y_{01}) = 0;$$

$$2r_1\varphi_1 + 4(r_1 + S_2)\varphi_2 - \frac{6S_2}{H}y = 0; \qquad (3.11.16)$$

$$-my'' + \frac{6S_1}{H}\varphi_1 + \frac{6S_2}{H}\varphi_2 - \frac{12S_1}{H^2}(y - y_{01}) - \frac{12S_2}{H^2}y = 0,$$

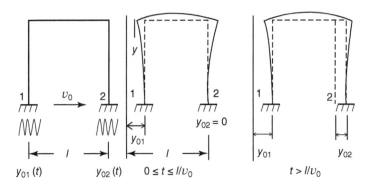

**Figure 3.24**  Scheme of motion on extended one story, single-span frame.

where

$S_1, S_2$ – linear stiffness of column/support of frame;

$r_1$ – linear stiffness of beam;

$\varphi_1, \varphi_2$ – angles of rotation of the frame nodes;

$m$ – mass concentrated at the level of the 1th floor;

$y, y''$ – displacement and acceleration of the concentrated mass

$y_{01}$ – assigned displacement of right column (support);

$H$ – height of the frame;

$L$ – length of span frame. In Figure 3.24 scheme of motion on extended one story, single-span frame is shown.

Eq. (3.11.16) is valid with $0 < t < t_1 = l/v$, where $l$ is the distance between the supports. Eq. (3.11.16) is more convenient to modify and to reduce to one differential equation relative to $y$. For this of the first two equations of the system (3.11.16) we determine $\varphi_1$ and $\varphi_2$, and substituting into the third equation we find:

$$my'' + y(a_1 + a_2) - y_{01}a_2 = 0, \tag{3.11.17}$$

where

$$a_1 = \frac{(12S_1S_2^2 + 12S_2^2r_1 + 66S_1S_2r_1 + 36S_2r_1^2)}{H^2(4S_1r_1 + 4S_1S_2 + 3r_1^2 + 4r_1S_2)}; \tag{3.11.18}$$

$$a_2 = \frac{(12S_1^2r_1 + 12S_1^2S_2 + 66S_1S_2r_1 + 36S_1r_1^2)}{H^2(4S_1r_1 + 4S_1S_2 + 3r_1^2 + 4r_1S_2)}.$$

During the motion of both supports on $y_{01}$ and $y_{02}$ respectively, the equations of motion will take the form:

$$4(S_1 + r_1)\varphi_1 + 2r_1\varphi_2 - \frac{6S_1}{H}(y - y_{01}) = 0;$$

$$2r_1\varphi_1 + 4(r_1 + S_2)\varphi_2 - \frac{6S_2}{H}(y - y_{02}) = 0; \qquad (3.11.19)$$

$$-my'' + \frac{6S_1}{H}\varphi_1 + \frac{6S_2}{H}\varphi_2 + \frac{12S_1}{H^2}(y - y_{01}) - \frac{12S_2}{H^2}(y - y_{02}) = 0,$$

when at $t > t_1$. Excluding $\varphi_1$, and $\varphi_2$ from (3.11.18) we will obtain:

$$my'' + (a_1 + a_2)y - a_1y_{01} - a_2y_{02} = 0. \qquad (3.11.20)$$

From (3.11.18) it is possible to obtain the values $a_1$ and $a_2$ with an absolutely rigid beam Figure 3.25. After dividing numerator and denominator of expression (3.11.18) by $r^2$, and then assuming that $r^2 = \infty$ we will obtain:

$$a_1 = \frac{36S_2}{3H^2} = \frac{12S_2}{H^2}; a_2 = \frac{36S_1}{3H^2} = \frac{12S_1}{H^2}, \qquad (3.11.21)$$

where $a_1$ and $a_2$ – stiffness of the first and second support/column respectively. Eqs. (3.11.17) and (3.11.19) describe the motion of the frame of a significant extent in the plan with the given seismogram of ground $y_0(t)$. If the motion of the left support occurs according to the law $y_{01}(t)$, then the motion of the right support can be taken as equal:

$$y_{02}(t) = y_{01}\left(t - \frac{l}{v}\right). \qquad (3.11.22)$$

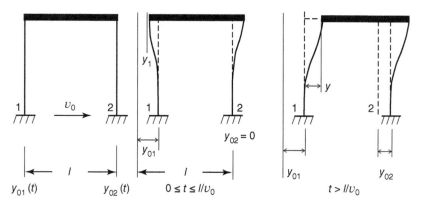

**Figure 3.25**   Scheme of motion on extended one story, single-span frame with absolute rigid beam.

Here $v$ represents the velocity of propagation (motion) of seismic wave over the surface of the ground. If there is a real measured value of velocity, then it is possible to use these data, but in general the value $v$ can be accepted as equal to the velocity of propagation of transverse waves – the waves of shift (shifting).

In some cases it is more convenient to replace variables in Eq. (3.11.20). Designating $\bar{y} = y - y_{01}$, and taking into account the damping in Voigt's hypothesis (into somewhat modified type) [175], we will obtain:

$$m(\bar{y}_1 + y_{01})'' + a_1\bar{y} + a_2(\bar{y} + y_{01} - y_{02}) + a_1\frac{\alpha_1}{\omega}\bar{y}' = 0, \qquad (3.11.23)$$

where $\alpha_1$ – damping coefficient; and $\omega$ – the fundamental frequency of free oscillations.

In Eq. (3.11.23) it is accepted that the energy absorption with the asynchronous oscillations of columns (supports) is the same as with the synchronous oscillations. This assumption leads to an insignificant change in damping of the entire system, at the same time the problem is simplified with the numerical solution of equation of motion, where assigned seismogram or accelerogram of earthquakes are given. Thus, the equations of motion can be solved relative to both the absolute displacements $y_j$, and displacements $\bar{y}_j$. In the first case the seismogram is used, and in the second case both the seismogram and the accelerogram of earthquake are used. The equation of motion with the synchronous oscillations of supports is obtained from (3.11.23) when $y_{01} = y_{0j} - y_0$.

For obtaining the spectrum of the seismic reactions, taking into account the extent of a structure, let us examine a single-span, one-story frame – a system with one degree of freedom. The equation of motion, which is calculated for the relative displacement $\bar{y}$ for a frame having columns of equal stiffness, will be:

$$\bar{y}'' + \left(\frac{2\pi}{T}\right)^2\bar{y} + \frac{2\pi}{T}\alpha\bar{y}' = y_0''(t) - \frac{1}{2}\left(\frac{2\pi}{T}\right)^2\left[y_0\left(t - \frac{L}{v_0}\right) - y_0(t)\right]. \qquad (3.11.24)$$

this equation is valid with $t > L/v_0$ and is found at zero initial conditions. Eq. (3.11.24) was integrated with the assigned law of the motion of the base:

$$y_0''(t) = 100\sin 10\pi t. \qquad (3.11.25)$$

The period and amplitude of the oscillation (vibration) of the base are respectively equal:

$$T = 0.2\,\text{s}; y_{max}'' = 1\,\text{m/s}^2.$$

The parameters of oscillations of the base in the first approximation simulate the real accelerograms of strong earthquakes [172, 265]. The damping coefficient is accepted as equal to $\alpha = 0.08$, which corresponds to a reinforced concrete structure. The constructed spectrum of accelerations shows that accounting for the extent leads to a substantial change in the nature of the resonance curve; Figure 3.26. For the periods $T < 0.2$ s, accounting for the extent increases the dynamic coefficient up to two times and more. For the periods $0.2$ s $< T < 0.4$ s the dynamic coefficient decreases $1.5 - 1.7$ times, and with $T > 0.4$ s, the influence of the extent is not substantial. For some values $L/v_0$ the dynamic coefficient can be equal to one (1); Figure 3.27. Hence it follows that with the harmonic oscillations of the ground for the given soil conditions, it is possible to choose a length of structure so that the phase shifting of supports would be equal:

$$\frac{L}{v_0} = \frac{T_0}{2} + kT_0, (k = 0, 1, 2, \ldots) \tag{3.11.26}$$

where $T_0$ – the predominant period of oscillation of the ground. In this case the construction will not undergo a dynamic impact, since the dynamic coefficient is approaching 1.

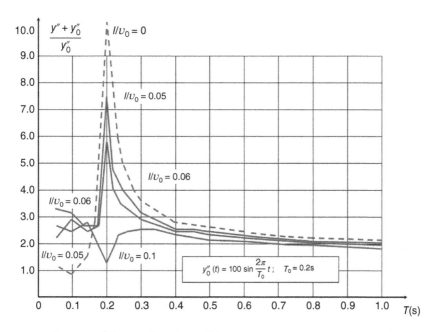

**Figure 3.26** Spectra of the accelerations of the extent structures with harmonic motion of base.

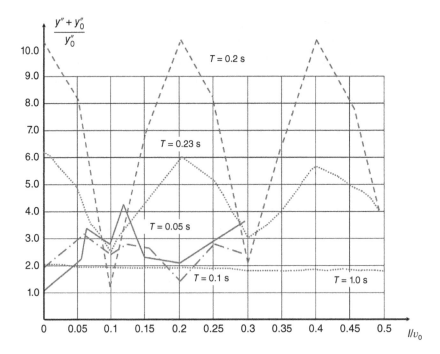

**Figure 3.27** Dependence of dynamic coefficient on the extent structures and ground conditions.

Using the formula (3.11.26) as a first approximation, it is possible to obtain the optimum value of the spans of construction for different types of ground/soil with the predominant periods of base oscillation. Table 3.17 provides optimal values of the extent spans of trestle frames or one-story, singe-span frames.

From Table 3.17 it follows that in ground, with the velocity of propagation of the seismic waves of $v_0 = 200$ m/s with the predominant period of $T_0 = 0.2$ s, the optimum structure is one whose length equals to 20, 60, and 100 m. The influence of the extent on the values of the reactions of multi-span, one-story, one-deck frames was investigated. Also examined were one-, two-, three-, four-, and five-span frames, with the same length and the same ground/soil conditions. It shows that the number of intermediate supports/columns has little influence on the magnitude of the seismic forces, and that the essential factor is the overall length of the structure. The results of the calculations are given in Table 3.18.

In some cases the problem of the calculation of the extent of a construction is possible to solve by averaging the acceleration of the ground and its

Table 3.17   Optimum values of the span length.

| Period of ground oscillation $T_0$ (s) | | Velocity of propagation of seismic waves $v_0$ | | | |
|---|---|---|---|---|---|
| | | 100 (m/s) | 200 (m/s) | 500 (m/s) | 1000 (m/s) |
| $T_0 = 0.1$ | $k = 0$ | 5 | 10 | 25 | 50 |
| | $k = 1$ | 15 | 30 | 75 | 150 |
| | $k = 2$ | 25 | 50 | 125 | 250 |
| $T_0 = 0.2$ | $k = 0$ | 10 | 20 | 50 | 100 |
| | $k = 1$ | 30 | 60 | 150 | 300 |
| | $k = 2$ | 50 | 100 | 250 | 500 |
| $T_0 = 0.3$ | $k = 0$ | 15 | 30 | 75 | 150 |
| | $k = 1$ | 45 | 90 | 225 | 450 |
| | $k = 2$ | 75 | 150 | 375 | 750 |
| $T_0 = 0.4$ | $k = 0$ | 20 | 40 | 100 | 200 |
| | $k = 1$ | 60 | 120 | 300 | 600 |
| | $k = 2$ | 100 | 200 | 500 | 1000 |

Table 3.18   Magnitude of seismic forces and number of intermediate supports.

| | Values $(y_0'' + \bar{y}'')_{max}$ for single-tier, one-story frame with the numbers of supports (columns) | | | | |
|---|---|---|---|---|---|
| Period (s) | 1 | 2 | 3 | 4 | 5 |
| 0.2 | 2.05 | 2.49 | 2.05 | 1.99 | 2.27 |
| 0.4 | 2.32 | 2.33 | 2.32 | 2.31 | 2.31 |
| 0.6 | 2.26 | 2.26 | 2.26 | 2.26 | 2.26 |
| 1.0 | 2.14 | 2.14 | 2.14 | 2.14 | 2.14 |

extension to the entire base of construction, i.e. as the initial acceleration of the ground to the following expression is accepted [185, 283]:

$$y_0''^{(average)}(t) = \frac{v_0}{L} \int_0^{\frac{L}{v_0}} y_0''(t - \xi)d\xi. \tag{3.11.27}$$

Using the accelerograms of some earthquakes we determined the values $y_0''^{(average)}(t)$ by numerical integration using expression (3.11.27). According to the obtained results in Figure 3.28, the dependencies of the relations of the averaged maximum and actual maximum values of the accelerations value $y_{0,max}''^{(average)}/y_{0,max}''$ from the length of construction $L$ with different values of the velocities of propagation of seismic wave $v$ are built. Structures with length up to 100 m are considered, and grounds/soils for which $v_0 = 100 \div 1000$ m/s. Let us recall that the minimum speed corresponds to the softest soil (soft, broken shale) and bulk soils, and the maximum speed to the sedimentary rock and medium hard rock grounds. These dependencies are built for four accelerograms of actual earthquakes with maximum accelerations of 0.93 m/s², 1.23 m/s², 2.42 m/s², and 0.66 m/s². The characteristics of these accelerogram are given in [171, 172, 265]. The given results show that the influence of the extent is substantial for such construction and ground, for which $v_0/L < 10$.

Since the frequency characteristics of accelerograms change insignificantly with the averaging, the reactions of structures undergo a similar phenomenon.

Summarizing the results of the oscillations of multiple-span frame bridges, extended buildings, and structures, and taking into account their extent under seismic excitation, we come to the following conclusions:

1. We derived the equations of motion of extent multi-tiered, multi-storied structures with asynchronous oscillations of supports. From the general equations as a special case the equations of oscillations of trestle frames are obtained.
2. The possible applications of the records of acceleration (accelerograms) of real earthquakes are analyzed for investigating the reactions of the extent system.

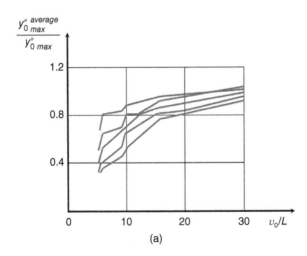

(a)

**Figure 3.28** (a) Averaged values of the accelerations of the real accelerogram of five strong earthquakes. (b) Dependencies of relations of the averaged and maximum accelerations on the length of structure.

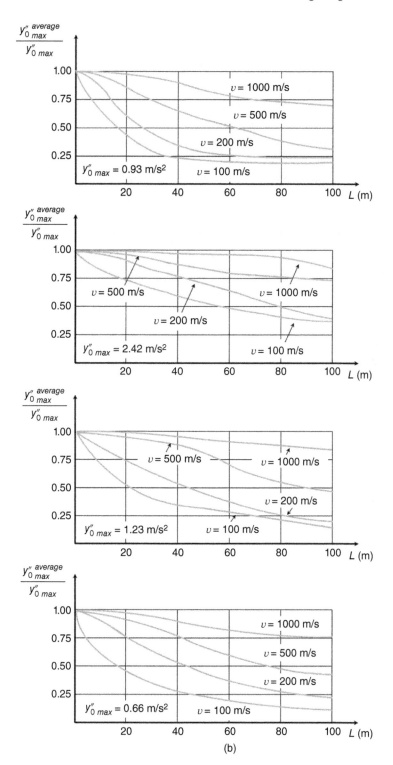

Figure 3.28 (Continued)

3. Analytical values of the reactions of multi-mass systems, taking into account the final velocity of propagation of seismic waves, are obtained. The spectra of accelerations of systems in harmonic influence, having parameters corresponding to real seismic excitation, are constructed.

4. For the very stiff systems $T < 0.2$ s, the accounting of extent can increase the value of dynamic coefficient. For the periods $0.2$ s $< T < 0.4$ s the dynamic coefficient decreases $1.5 - 1.7$ times, and for $T > 0.4$ s, the influence of the extent is not substantial.

5. In the first approximation, we obtained the optimal ratio of length of the construction to the value of velocity of seismic wave propagation on the surface of the ground (Table 3.17).

6. We investigated the possibility of taking into account the influence of the extent of a structure on its reaction by using the averaged acceleration of the ground. The given results show that the influence of the extent is substantial for such construction and ground, when $v_0/L < 10$.

# Chapter 4

# Oscillation of Plates and Shells

## 4.1 The Design of the Cantilever Plate of Minimal Mass Working on the Shift with the Assigned Fundamental Frequency

The questions of the optimum design of the oscillating plates are analyzed in the monograph by J.L. Arman [22]. This work offers the necessary conditions of the optimality of systems with distributed parameters, in the more general (common) case these were obtained in the work of K.A. Lurie [217]. Constructions subjected to complex dynamic effects with a known resonance range are widely applied in practice; one of the most important stages of their design is the resonance state. On one hand the resonance phenomena with ordinary materials sharply increase the dynamic load. From the point of view of strength, measures should be taken to remove them; for example, by widening the zones of operation without resonance which is especially necessary for building structures. On the other hand, however, many elements and structures operate in a resonance state as the main operation mode; that is why it is desirable to draw the natural frequencies nearer to those of the dynamic effects. Bearing in mind the main and increasingly important requirement for the economy of the contemporary structures, it follows that both problems presented above are dynamic optimization problems. They can be solved by a distribution of

*Static and Dynamic Analysis of Engineering Structures: Incorporating the Boundary Element Method*,
First Edition. Levon G. Petrosian and Vladimir A. Ambartsumian.
© 2020 John Wiley & Sons Ltd. Published 2020 by John Wiley & Sons Ltd.

structure parameters called design variables, where its natural frequencies take a state, determined in relation to the dynamic effects frequencies.

J.L. Arman [22] obtained the solution of a problem with the optimum design of a free-supported and shifting plate of the minimum mass, with the assigned fundamental natural frequency of free oscillations. Setting the conditions that the structures operate in a zone without resonance, the optimization problems are most often solved for natural frequency spectrum control [22], or for forced oscillations according to the stressed and deformed state criteria [E.J. Hang, J.S. Arora].[1] Below, using the method applied in the work [22], we consider a similar problem with optimal height of the cross section of a cantilever plate of the minimum mass, working on shift, and making harmonic oscillations with the natural fundamental frequency. Let us note that the design schemas of constructions in the form of cantilever beams, which work only on the shift, apply with the description of the oscillations of stone and precast panel buildings under a seismic influence [172, 173]. In certain cases, it is necessary to examine design schemas in the form of cantilever plates working on the shift. Therefore, we assume that the results of the research regarding the optimum forms of the cantilever-shifted plates can be used with the design of seismically resistant constructions. Let us examine a cantilever, rectangular plate, which works on the shift [22]. It is assumed that there is a constant fraction of unconstructive mass $\delta_2$ and thickness $h(x, y)$ which is expressed by the formula

$$h(x,y) = \delta_1 h^*(x,y) + \delta_2; \qquad \delta_1 + \delta_2 = 1 \qquad (4.1.1)$$

where $h^*(x, y)$ – the varied thickness of the plate.

The differential equation of motion of the steady single-frequency oscillations of the plate takes the form [22]:

$$\frac{\partial}{\partial x}\left(h^*\frac{\partial w}{\partial x}\right) + \frac{\partial}{\partial y}\left(h^*\frac{\partial w}{\partial y}\right) + \frac{\rho}{G}\omega^2(\delta_1 h^* + \delta_2) = 0 \qquad (4.1.2)$$

where $w$ – the displacement of plate; $\omega$ – natural frequency of the main mode of oscillation; $\rho$ – the mass of the unit volume; and $G$ – the shear modulus. The boundary conditions for the cantilever plate are the following:

$$w = 0 \text{ with } x = 0, \ h^*\frac{\partial w}{\partial x} = 0 \text{ with } x = a, \ h^*\frac{\partial w}{\partial y} = 0 \text{ with } y = 0, \ b$$

$$(4.1.3)$$

---

[1] Hang E.J., Arora J.S. *Applied optimal design. Mechanical and Structural systems.* Wiley, 1979.

The varied thickness $h^*$ is limited by a certain value $h_0^*$, i.e. the following condition takes place:

$$h^* - h_0^* \geq 0 \qquad (4.1.4)$$

The task is to determine the optimum thickness $h$, which minimizes the mass of the plate, expressed by the integral

$$I = \int_0^b \int_0^a h(x,y)dxdy \qquad (4.1.5)$$

with the fulfillment of the limitation (4.1.2) with the boundary conditions (4.1.3), and the inequality (4.1.4). The value of fundamental natural oscillations (vibrations) frequency was accepted as the one given. We will transform it (4.1.2) into a system of differential first order equations. Such a transformation is possible in various ways. Of course, in the general case, the preference should be given to the standard form of writing of equations taking into account parametric variables, proposed in the work [217]. However, in this case we used the equations proposed in the work [22], which give the opportunity to determine the values of the unknowns in a simpler way. The basic unknowns are accepted in the form of:

$$z_1 = w; \qquad z_2 = h^*\frac{\partial w}{\partial x}; \qquad z_3 = h^*\frac{\partial w}{\partial y}$$

Eq. (4.1.2) is written in the form of:

$$\frac{\partial z_1}{\partial x} = \frac{z_2}{h^*}; \qquad \frac{\partial z_1}{\partial y} = \frac{z_3}{h^*}; \qquad \frac{\partial z_2}{\partial x} = u_1; \qquad \frac{\partial z_2}{\partial y} = u_2 \qquad (4.1.6)$$

$$\frac{\partial z_3}{\partial x} = u_3; \qquad \frac{\partial z_3}{\partial y} = -u_1 - \frac{\rho}{G}\omega^2(\delta_1 h^* + \delta_2)z_1$$

The boundary conditions will be:

$$x = 0, \quad z_1 = 0, \quad x = a, \quad z_2 = 0, \quad y = 0, \quad y = b, \quad z_3 = 0 \qquad (4.1.7)$$

The expression of Hamiltonian $H$ is comprised:

$$H = h^* + \lambda_1\frac{z_2}{h^*} + \lambda_2 u_1 + \lambda_3 u_3 + \mu_1\frac{z_3}{h^*} + \mu_2 u_2$$

$$+ \mu_3\left[-u_1 - \frac{\rho}{G}\omega^2(\delta_1 h^* + \delta_2)\right] + \xi(h^* - h_0^*) \qquad (4.1.8)$$

where with (at) $\xi \leq 0$, $h^* = h_0^*$, with (at) $\xi = 0$, $h^* \geq h_0^*$. The necessary conditions of extremity are expressed in the form:

$$\frac{\partial \lambda_1}{\partial x} + \frac{\partial \mu_1}{\partial y} = \frac{\rho}{G}\omega^2(\delta_1 h^* + \delta_2)\mu_3, \quad \frac{\partial \lambda_2}{\partial x} + \frac{\partial \mu_2}{\partial y} = -\frac{\lambda_1}{h^*}$$

$$\frac{\partial \lambda_3}{\partial x} + \frac{\partial \mu_3}{\partial y} = -\frac{\mu_1}{h^*}, \quad \lambda_2 - \mu_3 = 0, \quad \mu_2 = 0, \quad \lambda_3 = 0 \qquad (4.1.9)$$

$$1 - \frac{\lambda_1 z_2}{h^{*2}} - \mu_1 \frac{z_3}{h^{*2}} - \frac{\rho}{G}\omega^2 \delta_1 z_1 \mu_3 + \xi = 0.$$

For the cantilever plate, the boundary conditions for the system (4.1.9) are the natural boundary conditions (4.1.7) and transversal conditions:

$$\text{with } x = a, \quad y = 0, \quad b\left[\lambda_1 \beta'(s) - \mu_1 \alpha'(s)\right] = 0$$

$$\text{with } x = 0, \quad y = 0, \quad b[\lambda_2 \beta'(s) - \mu_2 \alpha'(s)] = 0 \qquad (4.1.10)$$

$$\text{with } x = 0, \quad a[\lambda_3 \beta'(s) - \mu_3 \alpha'(s)] = 0$$

where the equation of the boundary of plate is assigned in the parametric form:

$$x = \alpha(s), \quad y = \beta(s). \qquad (4.1.11)$$

From Eqs. (4.1.9) and (4.1.10) the functions: $\lambda_i, \mu_i, i = 1, 2, 3$ are derived. From (4.1.9) we have $\mu_2 = \lambda_3 = 0$. From the first equation of (4.1.10) we obtain: with $x = a$, $\alpha' = 0$, $\beta' \neq 0$, $\lambda_1(a, y) = 0$, with $y = 0$, $b$, $\beta' = 0$, $\alpha' \neq 0$ and therefore $\mu_1(x, 0) = \mu_1(x, b)$. From the second Eq. (4.1.10) we obtain $\lambda_2 \beta'(s) = 0$, with $x = 0$, $\beta'(s) \neq 0$, $(0, y) = 0$. Since $\mu_3 = \lambda_2$, then $\mu_3(0, y) = 0$. Comparisons of (4.1.6) and (4.1.9) show, that

$$\lambda_1 = \frac{z_2}{\alpha}, \quad \mu_1 = \frac{z_3}{\alpha}, \quad \lambda_2 = \mu_3 = -\frac{z_1}{\alpha}, \qquad (4.1.12)$$

where $\alpha$ – an arbitrary constant. It is possible to ascertain that the above specified values $\lambda$ and $\mu$ satisfy the system (4.1.12). So, for example, $\lambda_1(a, y) = z_2(a, y) = 0$, and so on. Taking into account (4.1.12) and the designation of the basic unknowns $z_i$, $i = 1, 2, 3$, the last equation of the system (4.1.9) is reduced to the form:

$$\left(\frac{\partial w}{\partial x}\right)^2 + \left(\frac{\partial w}{\partial y}\right)^2 = c^2 + k^2 w^2 + \xi \qquad (4.1.13)$$

where

$$\alpha = c^2, \quad k^2 = \frac{\rho}{G}\omega^2\delta_1.$$

Equation (4.1.13) should be solved with satisfaction of the boundary conditions. As shown in [22], Eq. (4.1.13) is a sufficient condition of optimality. Before moving to solving the presented problem, let us examine the oscillations of a plate of a constant thickness, and solutions of similar problems without the observance of the limitation (4.1.14). With the oscillation of the rectangular plate of a constant thickness Eq. (4.1.2) takes the form

$$\frac{\partial^2 w}{\partial x^2} + \frac{\partial^2 w}{\partial y^2} + \alpha^2 w = 0 \tag{4.1.14}$$

where $\alpha^2 = \omega^2 \rho / G$.

The solution of (4.1.14) is represented as follows:

$$w = A_1 \sin \beta x \sin \gamma y + A_2 \sin \beta x \cos \gamma y$$

$$+ A_3 \cos \beta x \sin \gamma y + A_4 \cos \beta x \cos \gamma y \tag{4.1.15}$$

where $A_1, A_2, A_3$, and $A_4$ – constants; and $\beta, \gamma$ – constants, connected with $\alpha$ by the relationship

$$\beta^2 + \gamma^2 = \alpha^2. \tag{4.1.16}$$

Satisfying the boundary conditions (4.1.3), we obtain

$$\cos \beta a = 0, \quad \beta a = \frac{\pi}{2}(2m - 1), \quad m = 1, 2 \ldots, \tag{4.1.17}$$

$$\sin \gamma b = 0, \quad \gamma b = n\pi, \quad n = 1, 2 \ldots,$$

Taking into account (4.1.17) we find the values of the natural frequencies

$$\omega_{\min} = \sqrt{\left[\left(\frac{\pi(2m-1)}{2a}\right)^2 + \left(\frac{n\pi}{b}\right)^2\right]\frac{G}{\rho}}. \tag{4.1.18}$$

The modes of oscillations of the cantilever plate are determined by expression

$$w(x,y) = Amn \sin \frac{\pi(2m-1)}{2a}x \cos \frac{n\pi}{b}y. \tag{4.1.19}$$

Figure 4.1 shows the character of deformation and distribution of the transverse shear forces $Q_x = h\frac{\partial w}{\partial x}$, $Q_y = h\frac{\partial w}{\partial y}$ of a plate with the oscillation in the main mode (it is accepted, that $A_{11} = 1$, $h = 1$).

Now, let us examine the optimum problem without taking into account the limitation (4.1.14). Equation (4.1.13) should be solved with $\xi = 0$.

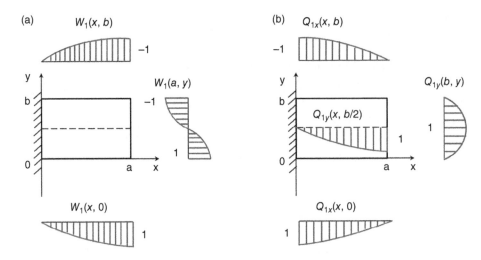

**Figure 4.1**   Scheme of cantilever plate: (a) deflection diagram; (b) shear forces diagram.

As in [22], introducing an unknown function $\theta$

$$\frac{\partial w}{\partial x} = \sqrt{c^2 + k^2 w^2}\, \cos\theta, \quad \frac{\partial w}{\partial y} = \sqrt{c^2 + k^2 w^2}\, \sin\theta, \tag{4.1.20}$$

we see that the characteristics are straight lines, along which constant value is preserved. Near the boundary of $x = 0$ the characteristics are perpendicular to the line of $x = 0$. Meanwhile $\theta$ determines the angle which the characteristic line composes with the axis of $ox$ (Figure 4.1). On other sides, the boundary conditions are satisfied by an appropriate selection of the function $h$ or corresponding derivatives $\partial w/\partial x$ or $\partial w/\partial y$. In the latter case, the characteristics, as it follows from (4.1.21), are straight lines, parallel to the corresponding side. The solution (4.1.13) is presented in the form of [160]

$$w = \frac{c}{k}\sinh k\frac{Ax + By}{\pm\sqrt{A^2 + B^2}}, \tag{4.1.21}$$

where $A$, $B$ – are constants. Figure 4.2 shows the partition of the plate into domains, where the displacement and thickness of the plate have different analytical expressions. In the case $a > b/2$ the deflection $w$ in the domains, shown in Figure 4.2, is represented in the form:

$$w_1 = \frac{c}{k}\sinh kx; \quad w_3 = -\frac{c}{k}\sinh k\left(y - \frac{b}{2}\right); \tag{4.1.22}$$

$$w_2 = w_5 = \frac{c}{k}\sinh k\left(y - \frac{b}{2}\right); \quad w_4 = w_6 = \frac{c}{k}\sinh k\left(y - \frac{b}{2}\right).$$

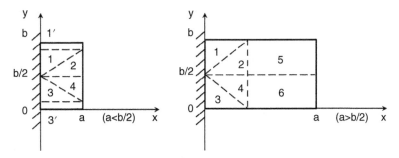

**Figure 4.2**   Partition of the area of plate on the subdomain, in which the optimum height has different analytical expressions.

The function $w$ is asymmetric relative to axis $y = b/2$. From (4.1.2) we find the unknown distribution $h^*(x, y)$ in domains 1, 2, 5.

$$h_1^*(x,y) = -\frac{\delta_2}{2\delta_1} + \frac{f_1(y)}{\cosh^2 kx}; \quad h_2^*(x,y) = -\frac{\delta_2}{2\delta_1} + \frac{f_2(x)}{\cosh^2 k(y - b/2)} \tag{4.1.23}$$

$$h_5^*(x,y) = -\frac{\delta_2}{2\delta_1} + \frac{f_5(x)}{\cosh^2 k(y - b/2)}$$

The function $h^*(x, y)$ is symmetric relative to the axis $y = b/2$. The unknown functions $f_1(y)$ and $f_2(y)$ are determined from the condition of continuity of transverse forces on the border of areas 1, 2. The function $f_5(y)$ is determined with the satisfaction of boundary condition on the line $y = b$. The resultant expressions for the total thickness of the plate will be:

$$h_1(x,y) = \frac{\delta_2}{2}\left[1 + \frac{\cosh^2 k(y - b/2)}{\cosh^2 kx}\right];$$

$$h_2(x,y) = \frac{\delta_2}{2}\left[1 + \frac{\cosh^2 kx}{\cosh^2 k(y - b/2)}\right]; \tag{4.1.24}$$

$$h_5(x,y) = \frac{\delta_2}{2}\left[1 + \frac{\cosh^2 kb/2}{\cosh^2 k(y - b/2)}\right].$$

The total mass $M$ of the optimal plate is determined by the expression:

$$M = \alpha\rho \int_{b/2}^{b} \int_{0}^{y-b/2} h_1(x,y)dxdy + \int_{0}^{b/2} \int_{b/2}^{x+b/2} h_2(x,y)dxdy$$

$$+ \int_{b/2}^{a} \int_{b/2}^{b} h_5(x,y)dxdy \tag{4.1.25}$$

The value of the optimal relative mass $m$, which is the ratio of the total mass to the mass of the plate of constant thickness and identical frequency, is calculated

$$m = \delta_2 \left[ \frac{0.125}{\lambda} + \frac{\cosh \pi\sqrt{\delta_1\left(1+\frac{0.25}{\lambda^2}\right)} - 1}{\pi^2\lambda\delta_1\left(1+\frac{0.25}{\lambda^2}\right)} + 0.25\left(1 - \frac{1}{2\lambda}\right) \right.$$

$$\left. + 0.5\left(1 - \frac{1}{2\lambda}\right) \frac{\cosh\frac{\pi}{2}\sqrt{\delta_1\left(1+\frac{0.25}{\lambda^2}\right)}\sinh\frac{\pi}{2}\sqrt{\delta_1\left(1+\frac{0.25}{\lambda^2}\right)}}{\pi\sqrt{\delta_1\left(1+\frac{0.25}{\lambda^2}\right)}} \right]. \tag{4.1.26}$$

Similar expressions are derived when $a < b/2$. The displacements in the domain $1'$ are defined by the same expression, as in domain 1. The resultant expressions for the total thickness of the plates $h_1(x,y)$ and $h_2(x,y)$ are determined by formula (4.1.24), and $h_1(x,y)$ by the expression:

$$h_1(x,y) = \frac{\delta}{2}\left[1 + \frac{\cosh^2 ka}{\cosh^2 kx}\right] \tag{4.1.27}$$

In this case the relative mass is determined by the formula:

$$m = \delta_2 \left[ \frac{1}{2} + \frac{\cosh 2\pi\sqrt{\delta_1\left(\lambda^2+\frac{1}{4}\right)} - 1}{2\pi^2\delta_1\left(\lambda+\frac{1}{4\lambda}\right)} \right.$$

$$\left. + \left(\frac{1}{2} - \lambda\right) \frac{\cosh\pi\sqrt{\delta_1\left(\lambda^2+\frac{1}{4}\right)}\sinh\pi\sqrt{\delta_1\left(\lambda^2+\frac{1}{4}\right)}}{\pi\sqrt{\delta_1(\lambda^2+0.25)}} \right] \tag{4.1.28}$$

From the aforementioned results, using the limit transition it is possible to obtain the corresponding optimal solution for the shifted beam. The beam is characterized by the unchanged stress-strained state in the direction of the $y$ axis. Therefore, by using the conditions $\lambda = \frac{a}{b} \to 0$, we obtain from (4.1.29):

$$m = \frac{\delta_2}{2}\left[1 + \frac{\sinh\pi\sqrt{\frac{\delta_1}{4}}\cosh\pi\sqrt{\frac{\delta_1}{4}}}{\pi\sqrt{\frac{\delta_1}{4}}}\right] \qquad (4.1.29)$$

Using (4.1.24) and (4.1.1), we will find the optimum height for the beam:

$$h(x,y) = \frac{\delta_2}{2}\left[1 + \frac{\cosh^2\pi\sqrt{\frac{\delta_1}{4}}}{\cosh^2\pi\sqrt{\frac{\delta_1}{4}\frac{x}{a}}}\right] \qquad (4.1.30)$$

Let us note that a mathematically equivalent problem of the optimum design of a longitudinally oscillating bar was examined in [385]. In this work no structural mass was applied on the free end.

There is an extensive literature [10, 15, 127, 250, 300, 365] discussing the problem of the optimal design of beams, beam systems, and other structures [318, 406]. In Table 4.1 the calculated values of the relative mass depending on the characteristic of the value of structural mass $\delta_1$ with different ratios of the sides $\lambda = a/b$ are given.

The table above shows that the mass of a plate with an optimum change in the height is actually less than the mass of the same plate with constant height and the same fundamental frequency. Additionally, from these data it follows that with the fixed ratio of sides, with an increase in the fraction (proportion, parties) of the mass of the plate structure, the value of $m$ decreases. The table cites the data, which correspond to the shifted beam ($\lambda = 0$). Figure 4.3 shows the character of an optimum change in the thickness of the plate ($\lambda = 1$, $\delta_2 = 0.6$) and the height of structural

**Table 4.1**   The value of relative mass depending of the characteristic of the value of structural mass $\delta_1$ with different ratios of the sides $\lambda = a/b$.

| $\lambda$ | 0 | 0.2 | 0.4 | 0.4 | 0.8 | 1 |
|---|---|---|---|---|---|---|
| 0 | 1 | 0.945 | 0.839 | 0.657 | 0.394 | 0 |
| 0.25 | 1 | 0.937 | 0.829 | 0.835 | 0.391 | 0 |
| 1 | 1 | 0.835 | 0.749 | 0.597 | 0.356 | 0 |
| 4 | 1 | 0.556 | 0.490 | 0.402 | 0.226 | 0 |

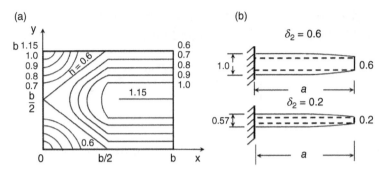

**Figure 4.3** Topography of the optimum construction with disregard for limitations: (a) topography of the surface of a square plate; (b) a change of cross- section of the shifted beam.

section. Now let us examine a problem of identifying the optimal thickness of a plate taking into account the limitation (4.1.4). In this part of the domain, in which the condition $h^* > h_0^*$, $\xi = 0$ is observed and deflections $w$ are determined by solution (4.1.13). For this domain the formulas (4.1.22), (4.1.23), and (4.1.24) are valid. There, where the minimum allowed value $h^* = h_0^*$, is reached, Eq. (4.1.2) is reduced to (4.1.14) in which

$$\alpha^2 = w^2 \frac{\rho}{G}\left(\delta_1 + \frac{\delta_2}{h_0^*}\right) \tag{4.1.31}$$

Let us find the form of the curve $\eta = \psi(\xi)$, dividing the two domains. The function $w$ and its derivatives must be continuous on this line. These conditions, recorded for this branch of this function, located in domain 1 (Figure 4.2), have the form

$$\sinh k\xi = A_1 \sin \beta\xi \sin \beta(\eta - b/2) + A_2[\sin \beta\xi \cos \beta(\eta - b/2)$$
$$+ \cos \beta\xi \sin \beta(\eta - b/2)]$$

$$0 = A_1 \sin \beta\xi \cos \beta(\eta - b/2) + A_2[- \sin \beta\xi \sin \beta(\eta - b/2)$$
$$+ \cos \beta\xi \cos \beta(\eta - b/2)] \tag{4.1.32}$$

$$\frac{k}{\beta} \cosh k\xi = A_1 \cos \beta\xi \sin \beta(\eta - b/2) - A_2[\cos \beta\xi \cos \beta(\eta - b/2)$$
$$- \sin \beta\xi \sin \beta(\eta - b/2)]$$

where $\beta = \frac{\sqrt{2}}{2}\alpha$.

The first equation expresses the condition of the equality of displacements on the dividing line. The expression for the displacements in the domain with a constant thickness is obtained from (4.1.16), taking into account the symmetry relative to the line $y = x + b/2$ and the condition $w = 0$ with $= 0$, $y = b/2$. From the system (4.1.33) we find the equation for determining the dependence $= \psi(\xi)$.

$$\sin 2\beta(\eta - b/2) = \sin 2\beta\xi - \frac{2k}{\beta}\frac{\sin^2\beta\xi}{\tanh k\xi} \tag{4.1.33}$$

Similarly, the equation for determining the dividing line of sub-domains with constant and variable thickness in domain 2 is found (Figure 4.2)

$$\sin 2\beta\xi = \sin 2\beta(\eta - b/2) - \frac{2k}{\beta}\frac{\sin^2\beta(\eta - b/2)}{\tanh k(\eta - b/2)} \tag{4.1.34}$$

From (4.1.34) and (4.1.35) follows that two branches of the function $\eta = \psi(\xi)$, determined from these equations, are symmetrical relative to the line $y = x + b/2$. In the domain 5, the dividing line is a straight line (Figures 4.2 and 4.3a). The functions $\eta = \psi(\xi)$, $\xi = \phi(\eta)$ are identical to the lines of a constant thickness, shown in Figure 4.3a. The law of variation of the thickness of a plate in the domains, where $h^* > h_0^*$, is determined with the aid of expressions (4.1.23). The functions $f_1, f_2$ , and $f_5$ are determined from the conditions of $h^* = h_0^*$, which occur on the dividing lines $\eta = \psi(\xi)$, $\xi = \phi(\eta)$. It is obtained that

$$h_1 = \frac{\delta_2}{2} + \left(\delta_1 h_0^* + \frac{\delta_2}{2}\right)\frac{\cosh^2 k\phi(y)}{\cosh^2 kx};$$

$$h_2 = \frac{\delta_2}{2} + \left(\delta_1 h_0^* + \frac{\delta_2}{2}\right)\frac{\cosh^2 k\psi(x)}{\cosh^2 k(y - b/2)}; \tag{4.1.35}$$

$$h_5 = \frac{\delta_2}{2} + \left(\delta_1 h_0^* + \frac{\delta_2}{2}\right)\frac{\cosh^2 k(\eta_1 - b/2)}{\cosh^2 k(y - b/2)}.$$

With formula (4.1.25) we determine the value of $M$, and then $m$. The optimal thickness of the shifted plate taking into account limitations (4.1.4) is determined, with $\lambda = 1$, $\delta_1 = 0.4$, $\delta_2 = 0.6$, $h^* = 0.25$. The minimally permissible thickness is given as: $h = 0.25 \cdot 0.4 + 0.6 = 0.7$. Figure 4.4 shows the topography of the optimum plate.

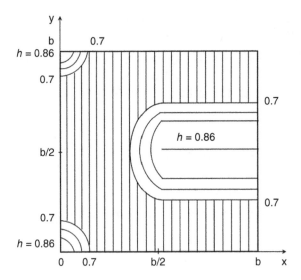

**Figure 4.4**   Topography of the optimum plate with the limitation: $h \geq 0.7$.

The reduced mass for this version is equal to $m = 0.75$. For comparison, let us note that solving this problem without the limitation, the value $m = 0.749$ (see Table 4.1).

## 4.2   The Experimental and Theoretical Research of Oscillation of a Cantilever Plate with Rectangular Openings

These works are dedicated to the studies of oscillations (vibrations) of thin rectangular plates with openings,[2] [293]. To determine the frequencies of the free oscillations of thin plates with openings, the variational methods the methods of Rayleigh – Ritz, and the Galerkin methods are most frequently applied. In the work [293] the Galerkin method determines fundamental (i.e. lowest) frequencies of hinged-supported plates with circular openings. In this case a model of a reduced rectangular plate with variable bending rigidity and distributed mass is used. To describe the variable characteristics, the method uses two-dimensional generalized functions, which describe the contour outlines of the openings of an arbitrary shape.

---

[2] Oscillations and stability of multi-connected thin-walled systems. Moscow, Mir, 1984, 312.

Below, following the methodology of [293], the fundamental frequencies of cantilever plates with rectangular openings bearing concentrated masses, are determined. Experimental research to determine the fundamental frequency of oscillations using the resonance method is conducted. The results of the experiment are compared with the theoretical data.

The equation of the free oscillations (vibrations) of a thin plate of variable rigidity (stiffness) takes the form [293]:

$$L(w) = D\left(\frac{\partial^4 w}{\partial x^4} + 2\frac{\partial^4 w}{\partial x^2 \partial y^2} + \frac{\partial^4 w}{\partial y^4}\right) + \frac{\partial^2 D}{\partial x^2}\left(\frac{\partial^2 w}{\partial x^2} + \mu\frac{\partial^2 w}{\partial y^2}\right)$$

$$+ 2\frac{\partial D}{\partial x}\left(\frac{\partial^3 w}{\partial x^3} + \frac{\partial^3 w}{\partial x \partial y^2}\right) + 2\frac{\partial D}{\partial y}\left(\frac{\partial^3 w}{\partial x^2 \partial y} + \frac{\partial^3 w}{\partial y^3}\right) \tag{4.2.1}$$

$$+ \frac{\partial^2 D}{\partial y^2}\left(\frac{\partial^2 w}{\partial y^2} + \mu\frac{\partial^2 w}{\partial x^2}\right) + 2(1-\mu)\frac{\partial^2 D}{\partial x \partial y} \cdot \frac{\partial^2 w}{\partial x \partial y} - \gamma\frac{h\omega^2}{g}w = 0,$$

where: $w$ – the transverse (lateral) displacement of plate; $\mu$ – Poisson's coefficient; $h$ – thickness of plate; $g$ – acceleration of gravity; $\omega$ – circular frequency of free oscillations (vibrations); and $D, \gamma$ – the parameters, characterizing the change in the bending rigidity and density of the plate.

$$D = D_0\lambda_1(x,y), \qquad \gamma = \gamma_0\lambda_2(x,y), \quad D_0 = \frac{E_0}{12}\frac{h^3}{(1-\mu^2)}, \tag{4.2.2}$$

$E_0, \gamma_0$ – modulus of elasticity and plate density.

The functions $\lambda_1(x, y)$, $\lambda_2(x, y)$ describing the changes to stiffness and density respectively in the case of rectangular openings (Figure 4.5), are presented in the form of:

$$\lambda_1(x,y) = \sum_{i,j=1,3} \{1 - \Gamma_0(x - x_i, y - y_j) + \Gamma_0(x - x_{i+1}, y - y_j)$$

$$+ \Gamma_0(x - x_i, y - y_{j+1}) - \Gamma_0(x - x_{i+1}, y - y_{j+1})\}, \tag{4.2.3}$$

$$\lambda_2(x,y) = \lambda_1(x,y) + \sum_{k=1,2}^{S} \frac{g}{h\gamma_0}M_k\Gamma_1(x - x_k, y - y_k)$$

where

$$\Gamma_0(x - x_i, y - y_j) = \Gamma_0(x - x_i)\,\Gamma_0(y - y_j)$$

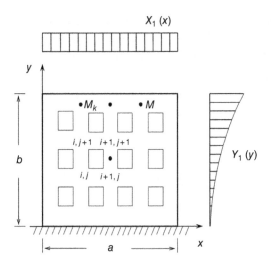

**Figure 4.5**    Cantilever plate with rectangular openings.

$M_k$ – the value of mass, concentrated at the point with coordinates $x_k$, $y_k$; and $\Gamma_0(x-x_i)$, $\Gamma_0(y-y_j)$ – are unit functions of Heaviside, depending on $x$ and $y$, respectively.

$$\Gamma_0(x - x_i) = \begin{cases} 1; & x \geq x_i \\ 0; & x < x_i \end{cases}, \qquad \Gamma_0(y - y_i) = \begin{cases} 1; & y \geq y_i \\ 0; & y < y_i \end{cases}$$

$$\Gamma_1(x - x_k, y - y_k) = \Gamma_1(x - x_k)\,\Gamma_1(y - y_k)$$

$\Gamma_1(x - x_k)$ – delta function.    $\Gamma_1(x - x_k) = \dfrac{\partial}{\partial x}\Gamma_0(x - x_k)$.
Galerkin's equation takes the form:

$$\int_0^a \int_0^b L(w)w\,dxdy = 0 \tag{4.2.4}$$

In the general case of the unknown displacement $w$ is presented in the form of:

$$w = \sum_{n=1}^{\infty} \sum_{m=0}^{\infty} f_{mn}X_m(x)Y_n(y). \tag{4.2.5}$$

As for the functions $X_m(x)$, $Y_n(y)$ we assume the eigenfunctions of beam. The $X_m(x)$ is the eigenfunction of the beam with free ends; and $Y_n(y)$ – the eigenfunction of the beam with one fixed end and another free end. These functions approximately satisfy the boundary conditions of cantilever plates [348]. These functions take the form:

$$X_m(x) = -A_m(\sinh \alpha_m x + \sin \alpha_m x) + B_m(\cosh \alpha_m x + \cos \alpha_m x)$$

$$A_m = \cosh \alpha_m a - \cos \alpha_m a); \quad B_m = -(\sinh \alpha_m a - \sin \alpha_m a); \quad m = 0, 1, 2 \ldots$$
$$\text{(4.2.6)}$$

$$Y_n(y) = C_n(\cosh \beta_n y - \cos \beta_n y) - D_n(\sinh \beta_n y - \sin \beta_n y); \quad n = 1, 2 \ldots$$

$$C_n = \sinh \beta_1 b + \sin \beta_1 b); \quad D_n = \cosh \beta_1 b + \cos \beta_1 b.$$

The functions corresponding to the first mode of oscillation of the plate:

$$X_0(x) = 2B_0$$

$$Y_1(y) = C_1(\cosh \beta_1 y - \cos \beta_1 y) - D_1(\sinh \beta_1 y - \sin \beta_1 y); \quad \text{(4.2.7)}$$

$$C_1 = 4.138, \quad \beta = \frac{1.875}{b}, \quad D_1 = 3.038.$$

For determining the fundamental frequency we assume in (4.2.5) that $= 0$, $n = 1$, $X_0 = 1$. The function $w_1(x, y)$ with $y = 0$, $b$ satisfies the exact boundary conditions, and $x = 0$ satisfies the "beam" boundary conditions. After substituting (4.2.1) and (4.2.7) into (4.2.4), let us calculate the obtained integrals:

$$\int_0^a \int_0^b D \frac{\partial^4 w_1}{\partial y^4} w_1 \, dxdy$$

$$= \sum_{i,j=1,3} D_0 \int_0^a \int_0^b \beta_1^4 Y_1^2(y)\lambda_1(x,y)dxdy$$

$$= D_0 \sum_{i,j=1,3} a \int_0^b Y_1^2(y)dy + (x_i - x_{i+1}) \int_{y_j}^b Y_1^2(y)dy - (x_i - x_{i+1}) \int_{y_{j+1}}^b Y_1^2(y)dy$$

$$= D_0 \beta_1^4 \left\{ a \int_0^b Y_1^2(y)dy - \sum_{i,j=1,3} (x_{i+1} - x_i) \int_{y_j}^{y_{j+1}} Y_1^2(y)dy \right\}. \quad \text{(4.2.8)}$$

$$2 \int_0^a \int_0^b \frac{\partial D}{\partial y} \frac{\partial^3 w_1}{\partial y^3} w_1 \, dxdy = 2D_0 \sum_{i,j=1,3} (x_{i+1} - x_i)\{Y_1'''(y_{j+1})Y_1(y_{j+1})$$

$$- Y_1'''(y_j)Y_1(y_j)\},$$

$$\int_0^a \int_0^b \frac{\partial^2 w_1}{\partial y^2} \frac{\partial^2 w_1}{\partial y^2} w_1 \, dxdy = D_0 \sum_{i,j=1,3} (x_{i+1} - x_i)\{[Y_1''(y)Y_1(y)]_{y=y_{j+1}}$$

$$- [Y_1''(y)Y_1(y)]_{y=y_j}\}.$$

$$\gamma_0 \frac{h\omega^2}{g} \int_0^a \int_0^b w_1^2 \lambda_2(x,y) dxdy$$

$$= \gamma_0 \frac{h\omega^2}{g} \left\{ a \int_0^b Y_1^2 dy - \sum_{i,j=1,3} (x_{i+1} - x_i) \int_{y_j}^{y_{j+1}} Y_1^2 dy + \sum_{k=1}^{S} \frac{g}{h\gamma_0} M_k Y_1^2(x_k y_k) \right\}.$$

The equality was used with the calculation of the integrals of the generalized functions [115]:

$$\int_0^b f(y)\Gamma_1^{(n)}(y - y_j)dy = (-1)^n f(y_j) \qquad (4.2.9)$$

In addition, when calculating the integrals of the eigenfunctions of a cantilever beam the following expressions taking place [358]:

$$\int_0^b Y_1^2(y)dy = \frac{b}{4} Y_1^2(b), \qquad (4.2.10)$$

$$\int_{y_j}^b Y_1^2(y)dy = \frac{y}{4} \left[ Y_1^2(y) - \frac{2}{\beta_1^4} Y_1'(y)Y_1'''(y) + Y_1''^2(y) \right]_{y_j}^b.$$

For the fundamental frequency of plate the following expression is obtained:

$$\omega = \frac{\Delta_1}{\Delta}$$

$$\Delta_1 = D_0\beta_1^4 \left\{ \frac{ab}{4} Y_1^2(b) - \sum_{i,j=1,3} (x_{i+1} - x_i) \left[ \frac{y}{4} (Y_1^2(y) - \frac{2}{\beta_1^4} Y_1'(y)Y_1'''(y) \right. \right.$$

$$
\left. \begin{array}{c} + \dfrac{1}{\beta_1^4} Y_1''^2(y) \Big|_{y_j}^{y_{j+1}} \end{array} \right\} + 2D_0 \sum_{i,j=1,3} (x_{i+1} - x_i)\{Y_1'''(y_{j+1})Y_1(y_{j+1})
$$

$$
- Y_1'''(y_j)Y_1(y_j)\} - D_0 \sum_{i,j=1,3} (x_{i+1} - x_i)\{[Y_1''(y)Y_1(y)]_{y=y_{j+1}}'
$$

$$
- [Y_1''(y)Y_1(y)]_{y=y_j}'\}. \tag{4.2.11}
$$

$$
\Delta = \gamma_0 \dfrac{h}{g} \left\{ \dfrac{ab}{4} Y_1^2(b) - \sum_{i,j=1,3} (x_{i+1} - x_i)\left[ \dfrac{y}{4} (Y_1^2(y)) - \dfrac{2}{\beta_1^4} Y_1'(y)Y_1'''(y) \right. \right.
$$

$$
\left. \left. + Y_1''^2(y)\dfrac{M}{M} \Big|_{y_j}^{y_{j+1}} \right\} + \sum_{k=1}^{S} M_k Y_1^2(x_k,y_k). \right.
$$

For the purpose of verifying the accuracy of the obtained formula (4.2.11), experimental studies were carried out using the resonance method, whereby the fundamental frequencies of cantilever plates with rectangular openings were determined.

Studies were conducted on the vibration stand of the excited harmonic oscillations in the range of frequencies from 0.1 to 40Hz (Figure 4.6).

The block diagram of connection of the measuring equipment is given in Figure 4.7. Four square plates were subjected to the test: continuous and with 1, 4, and 16 square openings (Figure 4.8). The sizes of the plates are: $a = b = 200$ mm, $h = 1$ mm. The material of the plates are steel

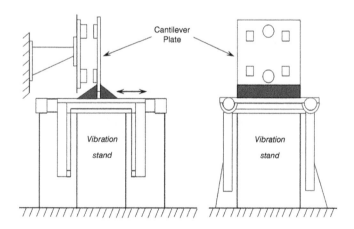

**Figure 4.6**   Diagram of the vibration stand.

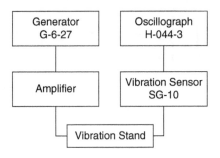

**Figure 4.7**   Circuit of test measuring equipment.

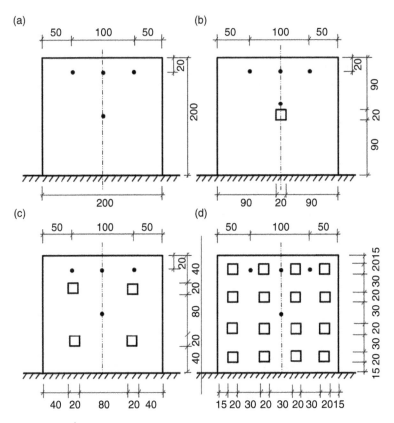

**Figure 4.8**   Cantilever plates: (a) continuous (solid) section; (b) with 1 opening; (c) with 4 openings; (d) with 16 openings.

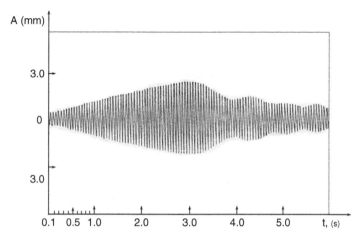

**Figure 4.9**   Record of the shift of oscillation of the plate with the continuous (solid) section.

with: $E = 2.1 \times 10^6 \, k\Gamma/\text{cm}^2$, and $\gamma_0 = 7.8 \times 10 \, k\Gamma/\text{cm}^3$. The values of the concentrated masses are identical and equal: $M = \dfrac{Q}{g} = 980^{-1} \times 10^{-2} k\Gamma f^2/\text{cm}$.

The resonance frequency of the plates were determined by the measurements of the displacements, recorded at the points marked in Figure 4.8. The records of oscillations (vibrations), obtained at the upper midpoint of a continuous plate, are shown in Figure 4.9.

The fundamental resonance frequency was defined as the frequency at which all points of the plate oscillate synchronously and simultaneously arrive at equilibrium. At the same time the mode of oscillation does not depend on time [248, 249]. The measurements show that the mode corresponding to the fundamental frequency is close to (4.2.7), which is accepted in the theoretical determination of frequency. Table 4.2 gives the results of the theoretical and experimental determination of the fundamental frequency of free oscillations (vibrations) of cantilever plates.

**Table 4.2**   Value of the fundamental frequencies of the free oscillations of cantilever plates.

| Method of determination | The frequency of the oscillations of plate $\omega$ (Hz) | | | |
| --- | --- | --- | --- | --- |
| | Continuous | One opening | Four openings | Sixteen openings |
| Theoretical (using a formula) | 18.79 | 18.77 | 18.64 | 17.37 |
| Experimental | 18 | 17.5 | 17 | 15 |

The difference in theoretical and experimental determinations of the frequencies for the continuous plate composes 4.4%, and for the plates with 1, 4, and 16 openings 7.26%, 9.65%, and 15.8% respectively. The greatest divergence occurs for the plate with 16 openings. The difference in theoretical and experimental determination of frequencies can be even smaller if we determine the frequency by the Galerkin method, taking into account the highest approximations.

## 4.3   The Oscillations (Vibrations) of Spherical Shells

An axisymmetric membrane – bending (flexural) oscillations of thin elastic spherical shells, are considered [239]. The motion of a shell is described by the system of two differential equations relative to normal displacement $w(\varphi, t)$, and functions of stress $F(\varphi, t)$.

$$\left[ R^2 \nabla^2 + (1 - \nu) - \frac{\rho(1 - \nu^2)}{E} \frac{\partial^2}{\partial t^2} \right] F + \left[ (1 - \nu)\frac{D}{R} \left( \nabla^2 + \frac{2}{R^2} \right) - \frac{Eh}{R} \right] w = 0$$

$$\left[ R^2 \left( \nabla^2 + \frac{2}{R^2} \right) + \frac{\rho(1 + \nu)}{E} R^2 \left( \frac{h}{12} \nabla^2 - 2 \right) \frac{\partial^2}{\partial t^2} \right] F+ \qquad (4.3.1)$$

$$+ \left[ DR \left( \nabla^2 + \frac{2}{R^2} \right) \left( \nabla^2 + \frac{2}{R^2} \right) + \rho hR \frac{\partial^2}{\partial t^2} \right] w = 0$$

where

$$\nabla^2 F = \frac{1}{R^2} \left[ \frac{\partial^2 F}{\partial \varphi^2} + \coth \varphi \frac{\partial F}{\partial \varphi} \right]; \qquad D = \frac{Eh^3}{12}(1 - \nu^2)$$

$\varphi$ – the spherical coordinate; $t$ – time; $R$ – the radius of curvature of a spherical median surface; $E$ – the module of elasticity; $\nu$ – Poisson's coefficient; $h$ – thickness of a shell; and $\rho$ – density.

Let us examine harmonic oscillations of a shell, proceeding with the frequency $\omega$. Assuming, that

$$w = Re[\overline{W}e^{i\omega t}]; \qquad F = Re[\overline{F}e^{i\omega t}], \qquad (4.3.2)$$

and designating

$$\Omega^2 = \rho\omega^2 \frac{R^2}{E},$$

it is possible to exclude the function of stresses from Eq. (4.3.1), and to obtain one differential equation of the sixth order relative to $\overline{W}$.

$$\nabla^6 \overline{W} + \frac{1}{R^2}[4 + (1 - v^2)\Omega^2]\nabla^4 \overline{W} + \frac{12(1 - v^2)}{h^2 R^2}(1 - \Omega^2)\nabla^2 \overline{W} \qquad (4.3.3)$$

$$+ \frac{12(1 - v^2)}{h^2 R^4}\{2 + \Omega^2[1 + 3v - (1 - \Omega^2)\Omega^2]\}\overline{W} = 0$$

The general solution of Eq. (4.3.3) is represented in the form of:

$$\overline{W} = \sum_{\alpha=1}^{3} W_\alpha, \qquad (4.3.4)$$

where

$$W_\alpha = A_\alpha P_{n_\alpha}(\cos \varphi) + B_\alpha Q_{n_\alpha}(\cos \varphi), \qquad (4.3.5)$$

$$n_\alpha = \left[\frac{1}{4} + \lambda_\alpha\right]^{1/2} - \frac{1}{2} \qquad (4.3.6)$$

$P_n(\cos \varphi)$, $Q_n(\cos \varphi)$ – Legendre's polynomials of the first and second kind, respectively. The function of stresses can be obtained from the differential equations of motion (4.3.1). The function of stresses $\overline{F}$ is represented in the form of:

$$\overline{F} = \frac{Eh}{R}\sum_{\alpha=1}^{3}\left[\frac{\Omega^2 - \frac{\left(\frac{h}{R}\right)^2}{12(1-v^2)}(2 - \lambda_\alpha)^2}{\frac{1+v}{12}\Omega^2\left[\left(\frac{h}{R}\right)^2\lambda_\alpha + 24\right] + 2 - \lambda_\alpha}\right]W_\alpha \qquad (4.3.7)$$

The parameters $\lambda_\alpha$ satisfy the equation:

$$\lambda^3 - [4 + (1 - v^2)\Omega^2]\lambda^2 + 12(1 - v^2)\left(\frac{R}{h}\right)^2(1 - v^2)\lambda \qquad (4.3.8)$$

$$-12(1 - v^2)\left(\frac{R}{h}\right)^2\{2 + \Omega^2[1 + 3v - (1 - v^2)\Omega^2]\} = 0$$

The values $\lambda_\alpha$ are represented in the form of:

$$\lambda_1 = A + B + \frac{1}{3}[4 + (1 - v^2)\Omega^2] \qquad (4.3.9)$$

$$\lambda_{2,3} = -\frac{1}{2}(A - B) + \frac{1}{3}[4 + (1 - v^2)\Omega^2] \pm i\frac{\sqrt{3}}{2}(A + B),$$

where

$$A = \left[-\frac{b}{2} + \Delta^{1/2}\right]^{1/3}; \quad B = \left[-\frac{b}{2} - \Delta^{1/2}\right]^{1/3}; \quad \Delta = \left(\frac{b}{2}\right)^2 + \left(\frac{a}{3}\right)^3;$$

$$a = \frac{1 - v^2}{3}\left[36\left(\frac{R}{h}\right)^2 - 36\left(\frac{R}{h}\right)^2 \Omega^2 - (1 - v^2)\Omega^4\right];$$

$$b = -\frac{2(2 - v^2)}{27}\left[(1 - v^2)\Omega^6 - 108(1 - v^2)\left(\frac{R}{h}\right)^2 \Omega^4\right.$$

$$\left. + 54\,(6 + 9v + v^2)\left(\frac{R}{h}\right)^2 \Omega^2 + 108\left(\frac{R}{h}\right)^2\right].$$

With the aid of the obtained solutions the oscillations (vibrations) of both closed and open spherical shells can be investigated. If the domain includes the top of the shell ($\varphi = 0$), then from the requirement of the continuity of the solution $B_\alpha = 0$ in (4.3.10) should be assumed. In the case of a hemisphere with free edges, the boundary conditions are the following

$$N_\varphi = M_\varphi = Q_\varphi = 0 \qquad (4.3.10)$$

here $N_\varphi$, $M_\varphi$, and $Q_\varphi$ – are designated as the intensity of forces, moments, and shear forces, respectively. Using the dependencies $N_\varphi$, $M_\varphi$, $Q_\varphi$ on $\overline{W}$ and $\overline{F}$ [239, 404], from the boundary conditions we obtain a system of three homogeneous equations relative to the coefficients $A_\alpha$. Equating the determinant, composed of the coefficients $A_\alpha$, to zero, we will obtain the equation of frequencies in the form of:

$$|C_{i\alpha}| = 0, \quad i, \alpha = 1, 2, 3 \qquad (4.3.11)$$

where

$$C_{1\alpha} = \left[12(1 + v) - \frac{h^2}{R^2}C_\alpha \lambda_\alpha\right] P_{n\alpha}(0),$$

$$C_{2\alpha} = (1 + v - \lambda_\alpha) P_{n\alpha}(0), \qquad (4.3.12)$$

$$C_{3\alpha} = \left[\lambda_\alpha - 2 - \frac{h^3}{12R^2}C_\alpha(\lambda_\alpha + 1 - v)\right] P_{n,\alpha+1}(0),$$

$$C_\alpha = \frac{\lambda_\alpha + 12(1 + v)\left(\frac{R}{h}\right)^2}{\lambda_\alpha - 1 + v - (1 - v^2)\Omega^2}.$$

To calculate the values of Legendre's functions it is possible to use the representation as a series or a definite integral [409]. The derivatives of Legendre's functions are determined with the aid of the known recursion relations. The first natural frequencies $\Omega$ of a hemispherical shell, obtained by the solving Eq. (4.3.11) with different $h/R$ are the following:

$\Omega = 0.875$, $0.919$, $0.975$, $1.100$, with respect to $h/R = 0.01$, $0.02$, $0.03$, $0.05$. The value $\Omega = 0.875\,(h/R = 0.01)$ is very close to the frequency obtained by A.A. Love [213], who examined the axisymmetrical membrane oscillations of the shell.

## 4.4   The Application of the Spectral Method of Boundary Element (SMBE) to the Oscillation of the Plates on Elastic Foundation

The method of integral transformation is especially effective with the solution of the multidimensional and dynamic problems; therefore, it is reasonable to expect that spectral method of boundary elements (SMBE) also can be successfully applied to dynamic calculation of structures [270].

As an example of the application of the SMBE to dynamic problems, let us consider the oscillations of a plate of an arbitrary form, which lies on the elastic Winklerian base, and which is under the action of a load $q(x, y, t)$. The equation of oscillations we will write in the form of:

$$DV^2V^2w(x,y,t) + m\ddot{w}(x,y,t) + kw(x,y,t) = q(x,y,t) \qquad (4.4.1)$$

where the designations are conventional. We will accept the boundary conditions on a contour of a plate as corresponding to those at the fixed end (build in) the edge

$$w(x,y,t) = 0,$$

$$(x,y \in \Gamma^+), \qquad (4.4.2)$$

$$\frac{\partial w}{\partial n}(x,y,t) = \beta_n(x,y,t) = 0,$$

where $n$ – normal to the contour of the plate. This choice of boundary conditions does not harm the generality of the consideration, since other boundary conditions can absolutely similarly be taken into account. We assume that the initial conditions are zero. We will also consider that the contour of the plate $\Gamma$ is the boundary of the domain $D^+$, which is formed by a smooth curve with no angular points. The presence of angular points does not introduce fundamental difficulties into the solution of the problem, but only increases the number of the unknowns in algebraic equations due to the angular forces [27], therefore, for simplicity, we will not take them into account. Furthermore, the rounding of plate angles is possible, eliminating the angular forces.

Let us designate through $n_x$ and $n_y$ the directing cosines of $n$ normal with the axes $x$, and $y$; $S_x$, $S_y$ – the directing cosines of tangential vector $S$ (Figures 4.10 and 4.11).

The expressions for normal $M_n$ moment, and for tangential $M_s$ moment will take the form [49]:

$$M_n = n_x M_1 + n_y M_2, \qquad (4.4.3)$$

$$M_S = -n_y M_1 + n_x M_2$$

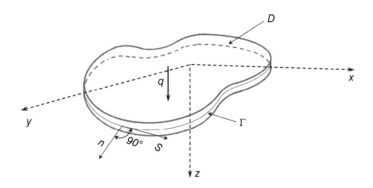

**Figure 4.10**    Scheme of a plate.

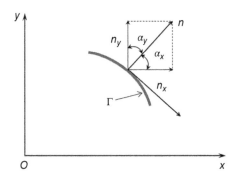

**Figure 4.11**    Element of the boundary of a plate.

where

$$M_1 = n_x M_x + n_y M_{xy},$$

$$M_2 = n_x M_{xy} + n_y M_y,$$

$M_x$, $M_y$ – bending moments; $M_{xy}$ – torque moment.
Let us also introduce the designations for the angle of rotations

$$\beta_x = -\frac{dw}{dx}; \ \beta_y = -\frac{dw}{dy}; \ \beta_n = -\frac{dw}{dn}; \ \beta_S = -\frac{dw}{dS}, \qquad (4.4.4)$$

and the given shear force

$$H = n_x Q_x + n_y Q_y + \frac{dM_S}{dS}. \qquad (4.4.5)$$

Here $Q_x$, $Q_y$ – shear forces. The moments and forces are determined using the standard formulas of the technical theory of plates

$$M_x = -D\left(\frac{\partial^2 w}{\partial x^2} + v\frac{\partial^2 w}{\partial y^2}\right), \ M_x = -D\left(\frac{\partial^2 w}{\partial y^2} + v\frac{\partial^2 w}{\partial x^2}\right),$$

$$M_{xy} = -D(1 - v)\frac{\partial^2 w}{\partial x \partial y}$$

$$Q_x = -D\frac{\partial}{\partial x}\nabla^2 w, \ \ Q_y = -D\frac{\partial}{\partial y}\nabla^2 w,$$

where $v$ – Poisson's coefficient (ratio). At the indicated designations the generalized formula of Green-Betty is fair [49, 50]

$$D\int_\Omega \nabla^2\nabla^2 w w^* d\Omega = D\int_\Omega \nabla^2\nabla^2 w^* w d\Omega + \int_\Gamma (M_n \beta_n^* + M_S \beta_S^*$$

$$+ \ Hw^* - M_n^* \beta_n - M_S^* \beta_S - H^* w)d\Gamma \qquad (4.4.6)$$

Let us apply a dual Fourier transformation of the spatial coordinates $x$, $y$ to the left and right sides of Eq. (4.4.1)

$$\frac{D}{2\pi}\iint_{-\infty}^{\infty} e^{i(\xi x + \eta y)}\nabla^2\nabla^2 w(x, y, t)dxdy + m\ddot{W}(\xi, \eta, t) + kW(\xi, \eta, t) = Q(\xi, \eta, t),$$

where $W(\xi, \eta, t)$ and $Q(\xi, \eta, t)$ – Fourier transforms of functions $W$ and $q$. Since on the plate border $\Gamma$, the function $w(x, y, t)$ and its derivatives can have gaps, apply the formula of Green-Betty (4.3.6) in the domain $D^+$ and $D^-$ separately. We will obtain as a result of the addition of both results

$$[D(\xi^2 + \eta^2)^2 + k]W(\xi, \eta, t) + m\ddot{W}(\xi, \eta, t)$$

$$= Q(\xi, \eta, t) + \int_\Gamma (M_n^*\Delta\beta_n + H^*\Delta w - \Delta M_n\beta_n^* - \Delta Hw^*)d\Gamma. \qquad (4.4.7)$$

Here we must accept

$$w^* = \frac{1}{2\pi}e^{i(\xi x + \eta y)}$$

and respectively:

$$\beta_n^* = -\beta_x^* n_x = -\frac{in_x\xi}{2\pi}e^{i(\xi x + \eta y)};$$

$$\beta_S^* = -\beta_x^* S_x = -\frac{iS_x\xi}{2\pi}e^{i(\xi x + \eta y)};$$

$$M_n^* = D[n_x^2(\xi^2 + v\eta^2) + 2n_x n_y(1 - v)\xi\eta + n_y^2(\eta^2 + v\xi^2)]e^{i(\xi x + \eta y)};$$

$$H^* = D[i\xi n_x(\xi^2 + \eta^2) + i\eta n_y(\xi^2 + \eta^2) + S_x(\xi^2 + v\eta^2)].$$

Here, it is taken into account also, that

$$\frac{dM_S}{dS} = M_x S_x$$

The solution of Eq. (4.4.7) with zero initial conditions takes the form:

$$W(\xi, \eta, t) = \frac{1}{mp(\xi, \eta)} \int_0^t F(\xi, \eta, \tau) \sin p(\xi, \eta)(t - \tau)d\tau, \qquad (4.4.8)$$

where

$$p(\xi, \eta) = \sqrt{[D(\xi^2 + \eta^2) + k]m^{-1}},$$

$$F(\xi, \eta, t) = Q(\xi, \eta, t) - \int_\Gamma (M_n^*\Delta\beta_n + H^*\Delta w - \Delta M_n\beta^* - \Delta Hw^*)d\Gamma.$$

Using the formula of conversion for the dual Fourier transform, we obtain

$$w(x,y,t) = \frac{1}{2\pi m} \iint_{-\infty}^{\infty} p^{-1} \int_0^t e^{-i(\xi x + \eta y)} F(\xi, \eta, \tau) \sin p(\xi, \eta)(t - \tau) d\tau d\xi d\eta$$

Let us satisfy the boundary conditions of the problem, having preliminary selected any jumps on $\Gamma$. For example, we will consider that the displacements and angle of rotations are continuous on $\Gamma$. Then $F(\xi, \eta, t)$ will take the form:

$$F(\xi, \eta, t) = Q(\xi, \eta, t) + F_1(\xi, \eta, t), \tag{4.4.9}$$

where

$$F_1(\xi, \eta, t) = \int_\Gamma (\Delta M_n \beta_n^* + \Delta H w^*) d\Gamma.$$

For determination of two unknown functions $\Delta M_n(\xi, \eta, t)$ and $\Delta H(\xi, \eta, t)$ we have two equations, obtained as a result of the satisfaction of the boundary conditions

$$\frac{1}{2\pi m} \iint_{-\infty}^{\infty} \int_0^t p^{-1} e^{-i(\xi x + \eta y)} F_1(\xi, \eta, \tau) \sin p(t - \tau) d\tau d\xi d\eta = -w_q(x, y, t),$$
$$\tag{4.4.10}$$

$$-\frac{n_x i}{2\pi m} \iint_{-\infty}^{\infty} \int_0^t \xi e^{-i(\xi x + \eta y)} p^{-1} F_1(\xi, \eta, \tau) \sin p(t - \tau) d\tau d\xi d\eta =$$

$$= -n_x \frac{\partial}{\partial x} w_q(x, y, t), \qquad (x, y \in \Gamma).$$

Here

$$w_q(x,y,t) = \frac{1}{2\pi m} \iint_{-\infty}^{\infty} \int_0^t e^{-i(\xi x + \eta y)} Q(\xi, \eta, \tau) \sin p(t - \tau) d\tau d\xi d\eta.$$

Equation (4.4.10) must be satisfied with each $t$. By dividing the interval $(0, t)$ into sections, in the limits of which we can disregard the changes in the integrand function, or use any interpolation, it is possible to obtain the step-by-step procedure for solving the problem by the method of sequential solutions of the quasi-stationary system of boundary Eq. (4.4.10). The solution of this system may be preferable and less time consuming than the use of transform of Laplace or Fourier over time, for example. Despite the apparent cumbersomeness of Eq. (4.4.10), which, after the substitution of the expression $F(\xi, \eta, t)$ must be converted by the replacement of the order of integration, there are opportunities of significant simplifications, based on the calculations of some integrals on $\xi$ or on $\eta$ in a closed form. In particular, it is possible to use the results of the dissertation work [360], where such integrals are calculated. Since the above equations are obtained for dynamic calculations of a plate, a particular case of the static equilibrium of a plate on the Winklerian (Winkler) foundation presents a great interest. In this case, in (4.3.7) it is possible to assume that $m = 0$ and having the continuity of moments and shear forces on the contour of plate $\Gamma$, we write the following:

$$W(\xi, \eta) = \frac{F_1(\xi, \eta) + Q(\xi, \eta)}{D(\xi^2 + \eta^2) + k}. \tag{4.4.11}$$

Hence,

$$w(x, y) = \frac{1}{2\pi} \iint_{-\infty}^{\infty} e^{-i(\xi x + \eta y)} \frac{F_1(\xi, \eta) + Q(\xi, \eta)}{D(\xi^2 + \eta^2) + k} d\xi \, d\eta. \tag{4.4.12}$$

As a result of the satisfaction of the boundary conditions we obtain:

$$\frac{1}{2\pi} \iint_{-\infty}^{\infty} \frac{F_1(\xi, \eta) e^{-i(\xi x + \eta y)}}{D(\xi^2 + \eta^2) + k} d\xi \, d\eta = w_q(x, y), \tag{4.4.13}$$

$$\frac{i n_x}{2\pi} \iint_{-\infty}^{\infty} \frac{F_1(\xi, \eta) e^{-i(\xi x + \eta y)}}{D(\xi^2 + \eta^2) + k} d\xi \, d\eta = -n_x \frac{\partial}{\partial x} w_q(x, y),$$

$$(x, y \in \Gamma),$$

where

$$w_q(x, y) = \frac{1}{2\pi} \iint_{-\infty}^{\infty} e^{-i(\xi x + \eta y)} \frac{Q(\xi, \eta)}{D(\xi^2 + \eta^2) + k} d\xi \, d\eta.$$

# Chapter 5

# The Propagation of Elastic Waves and Their Interaction with the Engineering Structures

## 5.1 The Propagation of Seismic Waves in the Laminar Inhomogeneous Medium

A number of scientific works [17, 51, 111, 245, 252, 317, 329] have been dedicated to the question of exploring propagation of seismic waves in homogeneous and laminar mediums. In [252] a solution to a problem of oscillation of multilayer medium with dynamic characteristics constant for each layer is given. The propagation of waves in an inhomogeneous medium is studied in [246]. For determining the laws of variation in the shear modulus and density, some approximate and exact solutions are obtained. New solutions to this problem are found below, to be used for the study of the oscillations of laminar medium. The harmonic oscillations of elastic inhomogeneous medium are described by the equation:

$$G(x)\frac{d^2w}{dx^2} + \frac{dw}{dx}\frac{dG}{dx} + \omega^2\rho(x)w = 0 \tag{5.1.1}$$

*Static and Dynamic Analysis of Engineering Structures: Incorporating the Boundary Element Method,*
First Edition. Levon G. Petrosian and Vladimir A. Ambartsumian.
© 2020 John Wiley & Sons Ltd. Published 2020 by John Wiley & Sons Ltd.

where $w$ – the displacement of medium; $\omega$ – the angular frequency of oscilla-tions; and $G(x)$, $\rho(x)$ – the shear modulus and medium density respectively. Let us examine a case, when $G(x)$ and $\rho(x)$ change exponentially:

$$G(x) = G_0 e^{ax}, \quad \rho(x) = \rho_0 e^{bx} \tag{5.1.2}$$

here $G_0$, $\rho_0$, $a$, $b$ – constants.

Substituting (5.1.2) into (5.1.1), we obtain:

$$e^{ax} G_0 \left[ \frac{d^2w}{dx^2} + a\frac{dw}{dx} + a^2 e^{(b-a)x} w \right] = 0. \tag{5.1.3}$$

Equation (5.1.3) is solved in the Bessel functions, if $b - a \neq 1$ [161, 405]. In this case the solution of Eq. (5.1.3) is represented in the form of:

(a) If $a$ is integer, then

$$w = A_1 e^{-a\frac{x}{2}} I_a \left( 2\alpha_0 e^{\frac{x}{2}} \right) + B_1 e^{-a\frac{x}{2}} H_a \left( 2\alpha_0 e^{\frac{x}{2}} \right), \tag{5.1.4}$$

(b) If $a$ is noninteger (frictional, non-integral) then

$$w = A_1 e^{-a\frac{x}{2}} I_a \left( 2\alpha_0 e^{\frac{x}{2}} \right) + B_1 e^{-a\frac{x}{2}} I_{-a} \left( 2\alpha_0 e^{\frac{x}{2}} \right), \tag{5.1.5}$$

where $\alpha_0 = \omega^2 \frac{\rho_0}{G_0}$, $A_1$, $B_1$ – arbitrary constants; $I_a(x)$, $I_{-a}(x)$ – the functions of Bessel of the first kind; and $H_a(x)$ – Hankel function of the first kind.

If $b - a = 1$, Eq. (5.1.3), is solved in the elementary functions [160]: With $4\alpha_0^2 - 1 > 0$

$$w = e^{-\frac{x}{2}} \left[ A_1 \cos \frac{x}{2}\sqrt{4\alpha_0^2 - 1} + B_1 \sin \frac{x}{2}\sqrt{4\alpha_0^2 - 1} \right] \tag{5.1.6}$$

With $1 - 4\alpha_0^2 > 0$

$$w = A_1 e^{\left[ -\frac{1}{2} + \frac{1}{2}\sqrt{1-4\alpha_0^2} \right]x} + B_1 e^{\left[ -\frac{1}{2} + \frac{1}{2}\sqrt{1-4\alpha_0^2} \right]x} \tag{5.1.7}$$

The solution of Eq. (5.1.1) in special functions was found when the vari-ables $G(x)$ and $\rho(x)$, are given in the form of:

$$G(x) = G_0 \left( 1 + k\frac{x}{l} \right)^m; \quad \rho(x) = \rho_0 \left( 1 + k\frac{x}{l} \right)^n \tag{5.1.8}$$

where $G_0$, $\rho_0$, $k$, $l$, $m$, and $n$ – are constants.

In the case of a layer of final thickness, $l$ – is the thickness of the layer. Substituting (5.1.8) into (5.1.1), we obtain:

$$G_0\left(1 + k\frac{x}{l}\right)^m \frac{d^2w}{dx^2} + G_0 m\frac{k}{l}\left(1 + k\frac{x}{l}\right)^{m-1}\frac{dw}{dx} + \omega^2\rho_0\left(1 + k\frac{x}{l}\right)^n w = 0.$$

$$(5.1.9)$$

Making the replacement of the independent variable according to the formula $1 + k\frac{x}{l} = z$, and converting (5.1.9), we obtain:

$$G_0 z^m \left[z^2\frac{d^2w}{dz^2} + mz\frac{dw}{dz} + \alpha^2 z^{n-m+2}w\right] = 0,\qquad (5.1.10)$$

where

$$\alpha^2 = \omega^2\rho_0\frac{l^2}{G_0 k^2}$$

The solution of Eq. (5.1.10) is expressed in Bessel functions [161]: if $\nu = \dfrac{1-m}{n-m+2}$ is a noninteger (frictional) then,

$$w = \left(1 + k\frac{x}{l}\right)^{\frac{1-m}{2}} [A_1 I_\nu(b) + B_1 I_{-\nu}(b)]\qquad (5.1.11)$$

if $\nu = \dfrac{1-m}{n-m+2}$ is an integer, then

$$w = \left(1 + k\frac{x}{l}\right)^{\frac{1-m}{2}} [A_1 I_\nu(b) + B_1 H_\nu(b)],\qquad (5.1.12)$$

where

$$b = \frac{2\alpha}{n-m+2}\left(1 + k\frac{x}{l}\right)^{\frac{n-m+2}{2}}.$$

Let us examine the oscillations of a laminar half-space under the excitation of a horizontal seismic shear wave; Figure 5.1. We assume that the displacement of the wave approaching the boundary of $n$ and $n+1$ layers is given, and the determination of the displacements in the top layers is required. The displacement in the $i$th layer $w_i$ is sought in the form:

$$w_i(x) = A_i X_i(x_i) + B_i Z_i(x_i);\qquad 0 \le x_i \le l_i,\qquad (5.1.13)$$

where $X_i(x_i)$, $Z_i(x_i)$ – are solutions of Eq. (5.1.10), determined by the expressions (5.1.11) or (5.1.12); $A_i$, $B_i$ – are arbitrary constants; and $l_i$ – height of

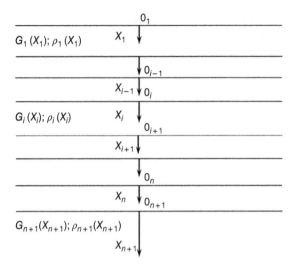

**Figure 5.1**   Diagram of laminar half-space.

$i$th layer. Satisfying the conditions of continuity and stresses on the contact surface of $i$th and $(i+1)$th of layers, we obtain:

$$A_{i+1}X_{i+1}(0) + B_{i+1}Z_{i+1}(0) = A_iX_i(l_i) + B_iZ_i(l_i), \qquad (5.1.14)$$

$$G_{i+1}(0)[A_{i+1}X'_{i+1}(0) + B_{i+1}Z'_{i+1}(0)] = G_i(l_i)[A_iX'_i(l_i) + B_iZ'_i(l_i)],$$

where through $G_i(l_i)$ and $G_{i+1}(0)$ – are designated as values of the variable of the height of the shear modulus of the $i$th and $i+1$ layers, in the sections $X_i = l_i$ and $X_{i+1} = 0$ respectively.

Solving Eqs. (5.1.14) relative to $A_i$ and $B_i$ we obtain:

$$A_i = A_{i+1}a_{i+1} + B_{i+1}b_{i+1}; \quad i = 1, 2, \dots n$$

$$B_i = A_{i+1}c_{i+1} + B_{i+1}d_{i+1}; \qquad (5.1.15)$$

$$a_{i+1} = \frac{\Delta_{1i}}{\Delta_{0i}}, \qquad b_{i+1} = \frac{\Delta_{2i}}{\Delta_{0i}}, \qquad c_{i+1} = \frac{\Delta_{3i}}{\Delta_{0i}}, \qquad d_{i+1} = \frac{\Delta_{4i}}{\Delta_{0i}}$$

$$\Delta_{0i} = G_i(l_i)[X_i(l_i)Z'_i(l_i) - Z_i(l_i)X'_i(l_i)],$$

$$\Delta_{1i} = G_i(l_i)X_{i+1}(0)Z'_i(l_i) - G_{i+1}(0)Z_i(l_i)X'_{i+1}(0),$$

$$\Delta_{2i} = G_i(l_i)Z_{i+1}(0)Z'_i(l_i) - G_{i+1}(0)Z_i(l_i)Z'_{i+1}(0),$$

$$\Delta_{3i} = G_{i+1}(0)X_i(l_i)X'_{i+1}(0) - G_i(l_i)X_{i+1}(0)X'_i(l_i),$$

$$\Delta_{4i} = G_{i+1}(0)X_i(l_i)Z'_{i+1}(0) - G_i(l_i)Z_{i+1}(0)X'_i(l_i).$$

Furthermore, the condition of the absence of stresses on the free surface must be satisfied:

$$G_1(0)[A_1X'_1(0) + B_1Z'_1(0)] = 0 \qquad (5.1.16)$$

The value of the displacement of the incident wave in the section of $X_{n+1} = 0$, is also given:

$$G_{n+1}(0)[A_{n+1}X_{n+1}(0) + B_{n+1}Z_{n+1}(0)] = C \qquad (5.1.17)$$

Using conditions (5.1.15), (5.1.16), and–(5.1.17), all coefficients $A_i$ and $B_i$ ($i = 1, 2, \dots n+1$) can be determined. To do this the values $A_i$ and $B_i$ [they are determined from (5.1.15) with ($i = 2, 3, \dots n$)] are substituted into the expression of $A_1$ and $B_1$ [they are determined from (5.1.15) with ($i = 1$)], and using (5.1.16), and (5.1.17) we obtain two equations, from which $A_1$ and $A_{n+1}$ are determined. From (5.1.16) and (5.1.17) we calculate $B_1$ and $B_{n+1}$, while from (5.1.15) we calculate all the remaining coefficients that are expressed through $C$. Let us examine the oscillations of a non-homogeneous medium in the form of a single-layer base. For this case we can use the formulas derived above with $n = 1$. We assume that in the upper layer, the module of elasticity $G$ increases in depth according to the linear law ($m = 1$), and the density $\rho$ is constant ($n = 0$). In this case we assume $k = 1/2, 2$, which correspond to an increase of the shear modulus within the limits of the upper layer by 1.5 and 3 times. For comparison, we also examine a case of constant $G$ and ($m = 0, n = 0$). We determine the relative displacement $W_1(0)/W_1(l_1)$:

$$\frac{W_1(0)}{W_1(l_1)} = \frac{X_1(0)Z'_1(0) - Z_1(0)X'_1(0)}{X_1(l_1)Z'_1(0) - Z_1(l_1)X'_1(0)}, \qquad (5.1.18)$$

where

$$X_1(x) = I_0\left[2\alpha_1 l_1\left(1 + k\frac{x}{l}\right)^{\frac{1}{2}}\right],$$

$$Z_1(x) = Y_0 \left[ 2\alpha_1 l_1 \left( 1 + k\frac{x}{l} \right)^{\frac{1}{2}} \right], \quad \alpha_1 = \frac{\omega}{c_1}$$

$Y_0(x)$ – the function of Bessel of the second kind of zero order; $c_1$ – the velocity of propagation of shear waves in a layer with a constant modulus of elasticity; and $l_1$ – the height of layer. The dependences of the relative displacement of free surface layer $W_1(0)/W_1(l_1)$ on the wave number $\alpha_1 l_1$ are calculated using formula (5.1.18), and built according to Figure 5.2.

For these curves the values of $\alpha_1 l_1$ exist where the resonance phenomenon takes place. In the case of a layer with a constant module of elasticity $\alpha_1 l_1 = 1.57$. With variable modulus of elasticity, the top layer is more rigid and the resonance frequencies are equal to $\alpha_1 l_1 = 1.775$ and $\alpha_1 l_1 = 2.36$, i.e. it is more than in the case of a layer with constant characteristics. The displacements of a free surface with the fixed frequency $\alpha_1 l_1$ in the layer with constant shear modulus are more than in the layer with the variable shear modulus.

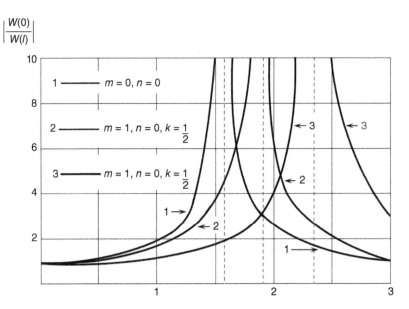

**Figure 5.2** Dependence of the relative displacement of free surface of the layer on the wave number.

## 5.2 Diffraction of Horizontal Waves on the Semi-cylindrical Base of Structure

Let us examine the stationary diffraction of a horizontal shear waves (SH waves) in a half-plane on a semi-cylindrical base and on a semi-cylindrical rigid foundation of a structure. This formulation of the problem has practical interest from the point of view of the requirements of seismic resistance [21, 172, 176, 252], especially since the obtained solution gives the opportunity to estimate the influence of soil conditions of a base on the values of the reactions of a structure. The contemporary state of the question of diffraction of the elastic waves on different kinds of inclusions and cavities, and in homogeneities is stated in [130]. The problems about the wave diffraction in a half-plane on a semi-cylindrical inclusion and on a semi-circular cylindrical groove (recess, seizure) are examined in [114, 251]. Under the conditions of anti-plane deformation let us examine the harmonic steady-state oscillations of a half-plane under the action of a plane shear wave falling at an angle $\gamma$ to the axis of $ox$ (Figure 5.3). The displacements $W_k$, $(k = 0, 1)$ of the half-plane satisfy the wave equation:

$$\Delta^2 W_k(r, \theta) + \alpha_k^2 W_k(r, \theta) = 0; \quad \alpha_k = \frac{\omega}{c_k}; \quad c_k = \sqrt{\frac{\mu_k}{\rho_k}}; \quad k = 0, 1. \quad (5.2.1)$$

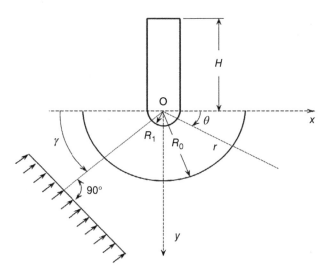

**Figure 5.3** Scheme of a structure with semi-cylindrical foundation and base.

where:

$r$, $\theta$ – polar coordinates with the center at a point of 0, $W_k$, $c_k$, $\mu_k$; and
$\rho_k$ – displacement, the wave propagation velocity of shift, the shear
modulus and the mass of the unit volume of medium in the domains
respectively:

with $k = 0$, $R_0 \leq r \leq \infty$; $0 \leq \theta \leq \pi$;

with $k = 1$, $R_0 \leq r \leq R_1$; $0 \leq \theta \leq \pi$;

and $\omega$ – angular frequency of the steady oscillations.

The design schema of the construction is accepted in the form of a
cantilever with the occupying domain: $0 \geq y \geq -H$; $-R_1 \leq X \leq R_1$. The
semi-cylindrical foundation of the construction is accepted as absolutely
rigid. The displacements of the structure $W_2(y)$ satisfy the one-dimensional
wave equation:

$$\frac{\partial^2 W_2}{\partial y^2} + \alpha_2^2 W_2 = 0; \quad \alpha_2 = \frac{\omega}{c_2}; \quad c_2 = \sqrt{\frac{\mu_2}{\rho_2}} \tag{5.2.2}$$

where $c_2$, $\mu_2$, and $\rho_2$ – a velocity of shear waves, the shear modulus and the
mass of the unit volume of the material of the structure, respectively. The
incident wave is represented in the form:

$$W_0^{(1)} = f e^{i[\alpha_0(y\sin\gamma + x\cos\gamma) - \omega t]} = f e^{i[\alpha_0 r\cos(\theta - \gamma) - \omega t]}, \tag{5.2.3}$$

$f$ – the value of the amplitude of the incident wave. Henceforth, the multiplier $e^{-i\omega t}$ will be omitted. The displacement $W_0$ we search for in the form
of:

$$W_0 = W_0^{(1)} + W_0^{(2)} + W_0^{(3)} \tag{5.2.4}$$

where $W_0^{(2)}$ – reflection from the free surface of the homogeneous half-plane
of the plane wave; and $W_0^{(3)}$ – the wave scattered from the semi-cylindrical
surface. Bearing in mind the well-known decomposition of the plane wave
through the solutions of Eq. (5.2.1) [130], we obtain:

$$W_0^{(1)} + W_0^{(2)} = f[e^{i\alpha_0 r\cos(\theta - \gamma)} + e^{i\alpha_0 r\cos(\theta + \gamma)}]$$

$$= 2f \sum_{n=0}^{\infty} \varepsilon_n i^n I_n(\alpha_0^r) \cos n\gamma \cos n\theta; \quad \varepsilon_n = \begin{cases} 2, & n \geq 1 \\ 1, & n = 1 \end{cases} \quad ; i = \sqrt{-1},$$

$$\tag{5.2.5}$$

where $I_n(\alpha_0^r)$ – Bessel function of the first kind of the $n$th order. $W_0^{(3)}$ is accepted in the form of:

$$W_0^{(3)} = \sum_{n=0}^{\infty} A_n H_n(\alpha_0^r) \cos n\theta, \qquad (5.2.6)$$

here $H_n(\alpha_0^r)$ – Hankel function of the first kind of the $n$th order; and $A_n$ – arbitrary constants. The displacements $W_1$ and $W_2$ are represented in the form of:

$$W_1 = \sum_{n=0}^{\infty} [B_n I_n(\alpha_1^r) + C_n H_n(\alpha_1^r)] \cos n\theta; \qquad (5.2.7)$$

$$W_2 = D \cos \alpha_2 y + E \sin \alpha_2 y,$$

where $B_n$, $C_n$, $D$, and $E$ – arbitrary constants. The accepted expressions $W_0$, $W_1$ satisfy boundary conditions on a free surface which have the form:

$$\tau_{\theta z}(r, 0) = \frac{\mu}{r} \frac{\partial W}{\partial \theta}(r, 0) = 0, \quad \tau_{\theta z}(r, \pi) = \frac{\mu}{r} \frac{\partial W}{\partial \theta}(r, \pi) = 0 \qquad (5.2.8)$$

here $\tau_{\theta z}$ – shear stress, acting on the area $\theta =$ constant and parallel to the coordinate axis of $0Z$. The continuity conditions of displacements $W_0$, $W_1$ and stresses $\tau_{rz}^{(0)}$ and $\tau_{rz}^{(1)}$ on the contact circle $r = R_0$ must be satisfied:

$$W_0(R_0, \theta) = W_1(R_0, \theta); \qquad (5.2.9)$$

$$\mu_0 \frac{\partial W}{\partial \theta}(R_0, \theta) = \mu_1 \frac{\partial W_1}{\partial \theta}(R_0, \theta).$$

The condition of the dynamic equilibrium of the semi-cylindrical rigid body has the form:

$$-\frac{\pi R_1^2}{2} \rho_3 \omega^2 W_1(R_1 \theta) = \int_0^{\pi} \tau_{rz}^{(1)}(R_1 \theta) R_1 d\theta + 2R_1 \tau_{rz}^{(2)}(y = 0), \qquad (5.2.10)$$

where $\rho_3$ – the mass of the unit volume of the material of foundation; and $\tau_{rz}^{(2)}(y = 0)$ – shear stress in the section $y = 0$ of the structure. Furthermore, the boundary conditions for the structure must be satisfied:

$$W_1(R_1 \theta) = W_2(y = 0); \quad \mu_2 \frac{dW_2}{dy}(y = -H) = 0. \qquad (5.2.11)$$

Let us proceed to the determination of the unknown coefficients, which determine the stress-strained state of a medium and structure. Using (5.2.4), (5.2.5), (5.2.6), (5.2.7) from (5.2.8), (5.2.10), and (5.2.11), we obtain:

$$\sum_{n=0}^{\infty}[2f\varepsilon_n i^n I_n(\alpha_0 R_0)\cos n\gamma + A_n H_n(\alpha_0 R_0)]\cos n\theta$$

$$= \sum_{n=0}^{\infty}[B_n I_1(\alpha_1 R_0)\cos n\gamma + C_n H_n(\alpha_1 R_0)]\cos n\theta; \qquad (5.2.12)$$

$$\sum_{n=0}^{\infty}\mu_0[2f\varepsilon_n i^n I_n(\alpha_0 R_0)\cos n\gamma + A_n H_n'(\alpha_0 R_0)]\cos n\theta$$

$$= \sum_{n=0}^{\infty}\mu_1[B_n I_n'(\alpha_1 R_0)\cos n\gamma + C_n H_n'(\alpha_1 R_0)]\cos n\theta; \qquad (5.2.13)$$

$$-\frac{\pi R_1^2}{2}\alpha_1^2\mu_1\frac{\rho_3}{\rho_1}\sum_{n=0}^{\infty}[B_n I_n(\alpha_1 R_1) + C_n(\alpha_1 R_1)]\cos n\theta$$

$$= \mu_1[B_0 I_0'(\alpha_1 R_1) + C_0 H_0'(\alpha_1 R_1)]R_1\pi + 2R_1\mu_2\alpha_2 E; \qquad (5.2.14)$$

$$\sum_{n=0}^{\infty}[B_n I_n(\alpha_1 R_1) + C_n H_n(\alpha_1 R_1)]\cos n\theta = D; \qquad (5.2.15)$$

$$E = -D\tan\alpha_2 H. \qquad (5.2.16)$$

From (5.2.15) it follows that:

$$B_n I_n(\alpha_1 R_1) + C_n H_n(\alpha_1 R_1) = 0 \quad \text{with } n \geq 1 \qquad (5.2.17)$$

$$B_n I_n(\alpha_1 R_1) + C_0 H_0(\alpha_1 R_1) = D \qquad (5.2.18)$$

Keeping in mind (5.2.17), from (5.2.14) and (5.2.18) we determine the coefficients $B_0$, $C_0$ expressed through $E$:

$$B_0 = E\frac{\Delta_1}{\Delta}; \quad C_0 = E\frac{\Delta_2}{\Delta}; \qquad (5.2.19)$$

where

$$\Delta = 2i$$

$$\Delta_1 = sH_0(\alpha_1 R_1) + PH_1(\alpha_1 R_1),$$

$$\Delta_2 = -sI_0(\alpha_1 R_1) + PI_1(\alpha_1 R_1),$$

$$s = 2R_1\alpha_2\frac{\mu_2}{\mu_1} - \frac{\pi}{2}(R_1\alpha_1)^2 \cdot \frac{\rho_3}{\rho_1}\cot\alpha_2 H,$$

$$P = \pi\alpha_1 R_1\cot\alpha_2 H.$$

Taking into account the orthogonality of $\cos n\theta$ in the interval $0 \leq \theta \leq \pi$ from (5.2.12) and (5.2.13) we will obtain two algebraic equations, connecting the coefficients $A_n$, $B_n$, and $C_n$. Excluding from these equations $C_n$ with the aid of (5.2.17), we obtain the system of two equations for determining $A_n$ and $B_n$ $(n \geq 1)$. Solving these equations, we obtain the explicit expressions for determining these coefficients:

$$A_n = \frac{\delta_1}{\delta}; B_n = \frac{\delta_2}{\delta}; n \geq 1, \tag{5.2.20}$$

Where

$$\delta = \mu_1\alpha_1 H_n(\alpha_0 R_0)\left\{\frac{I_n(\alpha_1 R_1)}{H_n(\alpha_1 R_1)}\left[-\frac{n}{\alpha_1 R_0}H_n(\alpha_1 R_0) + H_{n-1}(\alpha_1 R_0)\right]\right.$$

$$+ \frac{n}{\alpha_1 R_0}I_n(\alpha_1 R_0) - I_{n-1}(\alpha_1 R_0)\bigg\} + \mu_0\alpha_0\left[-\frac{n}{\alpha_0 R_0}H_n(\alpha_0 R_0) + H_{n-1}(\alpha_0 R_0)\right]$$

$$\times\left[I_n(\alpha_1 R_0) - \frac{I_n(\alpha_1 R_1)}{H_n(\alpha_1 R_1)} \cdot H_n(\alpha_1 R_0)\right];$$

$$\delta_1 = \mu_1\alpha_1 K_n I_n(\alpha_0 R_0)\left\{-\frac{I_n(\alpha_1 R_1)}{H_n(\alpha_1 R_1)}\left[-\frac{n}{\alpha_1 R_0}H_n(\alpha_1 R_0) + H_{n-1}(\alpha_1 R_0)\right]\right.$$

$$- \frac{n}{\alpha_1 R_0}I_n(\alpha_1 R_0) + I_{n-1}(\alpha_1 R_0)\bigg\} + \mu_0\alpha_0 K_n\left[-\frac{n}{\alpha_0 R_0}I_n(\alpha_0 R_0) + I_{n-1}(\alpha_0 R_0)\right]$$

$$\times\left[-I_n(\alpha_1 R_0) + \frac{I_n(\alpha_1 R_1)}{H_n(\alpha_1 R_1)} \cdot H_n(\alpha_1 R_0)\right];$$

$$\delta = \mu_1\alpha_1 H_n(\alpha_0 R_0)\left\{\frac{I_n(\alpha_1 R_1)}{H_n(\alpha_1 R_1)}\left[-\frac{n}{\alpha_1 R_0}H_n(\alpha_1 R_0) + H_{n-1}(\alpha_1 R_0)\right]\right.$$

$$+ \frac{n}{\alpha_1 R_0} I_n(\alpha_1 R_0) - I_{n-1}(\alpha_1 R_0) \Big\} + \mu_0 \alpha_0 \left[ -\frac{n}{\alpha_0 R_0} H_n(\alpha_0 R_0) + H_{n-1}(\alpha_0 R_0) \right]$$

$$\times \left[ I_n(\alpha_1 R_0) - \frac{I_n(\alpha_1 R_1)}{H_n(\alpha_1 R_1)} H_n(\alpha_1 R_0) \right];$$

$$\delta_1 = \mu_1 \alpha_1 K_n I_n(\alpha_0 R_0) \Big\{ -\frac{I_n(\alpha_1 R_1)}{H_n(\alpha_1 R_1)} \left[ -\frac{n}{\alpha_1 R_0} H_n(\alpha_1 R_0) + H_{n-1}(\alpha_1 R_0) \right]$$

$$- \frac{n}{\alpha_1 R_0} I_n(\alpha_1 R_0) + I_{n-1}(\alpha_1 R_0) \Big\} + \mu_0 \alpha_0 K_n \left[ -\frac{n}{\alpha_0 R_0} I_n(\alpha_0 R_0) + I_{n-1}(\alpha_0 R_0) \right]$$

$$\times \left[ -I_n(\alpha_1 R_0) + \frac{I_n(\alpha_1 R_1)}{H_n(\alpha_1 R_1)} \cdot H_n(\alpha_1 R_0) \right];$$

$$K_n = 2f \varepsilon_n i^n \cos n\gamma;$$

$$\delta_2 = \mu_0 K_n \frac{n}{R_0} [I_n(\alpha_0 R_0) H_n(\alpha_0 R_0) - H_n(\alpha_0 R_0) I_n(\alpha_0 R_0)]$$

$$+ \mu_0 \alpha_0 K_n [I_n(\alpha_0 R_0) H_{n-1}(\alpha_0 R_0) - H_n(\alpha_0 R_0) I_{n-1}(\alpha_0 R_0)].$$

Using (5.2.13) and (5.2.14), let us find two equations which connect $A_0$, $B_0$, and $C_0$. Substituting (5.2.19), instead of $C_0$, we obtain two equations relative to $A_0$ and $E$. The expressions for their determination take the form:

$$E = \frac{\xi_1}{\xi}; \quad A_0 = \frac{\xi_2}{\xi}; \xi_1 = \frac{2K_0 \mu_0}{i\pi R_0}; \qquad (5.2.21)$$

$$\xi = \frac{1}{2i}[sH_0(\alpha_1 R_1) + PH_1(\alpha_1 R_1)][\mu_0 \alpha_0 I_0(\alpha_1 R_0) H_1(\alpha_0 R_0)$$

$$- \mu_1 \alpha_1 I_1(\alpha_1 R_0) H_0(\alpha_0 R_0)] + \frac{1}{2i}[sI_0(\alpha_1 R_1) + PI_1(\alpha_1 R_1)]$$

$$\times [-\mu_0 \alpha_0 H_0(\alpha_1 R_0) H_1(\alpha_0 R_0) + \mu_1 \alpha_1 H_0(\alpha_0 R_0) H_1(\alpha_1 R_0)];$$

$$\xi_2 = \frac{K_0}{2i}[sH_0(\alpha_1 R_1) + PH_1(\alpha_1 R_1)][-\mu_0 \alpha_0 I_1(\alpha_0 R_0) I_0(\alpha_1 R_0)$$

$$+ I_0(\alpha_0 R_0) I_1(\alpha_1 R_0)] + \frac{K_0}{2i}[sI_0(\alpha_1 R_1) + PI_1(\alpha_1 R_1)]$$

$$\times [\mu_0 \alpha_0 I_1(\alpha_0 R_0) H_0(\alpha_1 R_0) - I_0(\alpha_0 R_0) H_1(\alpha_1 R_0)]$$

The coefficients are determined by formulas (5.2.16), (5.2.17), and (5.2.19). Thus, the problem is completely solved. Additional calculations are carried out showing the degree of the influence of the characteristics of a semi-cylindrical base on the displacement of the upper end of the structure. The value of the modulus of this displacement was determined by the formula:

$$|W_2(-H)| = \frac{8f}{\pi \sin \alpha_2 H} \frac{1}{(a^2 + b^2)^{1/2}}, \qquad (5.2.22)$$

$$a = [sI_0(\alpha_1 R_1) + PI_1(\alpha_1 R_1)][\alpha_0 R_0 Y_0(\alpha_1 R_0) Y_1(\alpha_0 R_0)$$

$$- \frac{\mu_1}{\mu_0} \alpha_1 R_0 Y_0(\alpha_0 R_0) Y_1(\alpha_1 R_0)\Big] + [sY_0(\alpha_1 R_1) + PY_1(\alpha_1 R_1)]$$

$$\times \Big[-\alpha_0 R_0 I_0(\alpha_1 R_0) Y_1(\alpha_0 R_0) + \frac{\mu_1}{\mu_0} \alpha_1 R_0 I_1(\alpha_1 R_0) Y_0(\alpha_0 R_0)\Big],$$

$$b = [sI_0(\alpha_1 R_1) + PI_1(\alpha_1 R_1)][-\alpha_0 R_0 Y_0(\alpha_1 R_0) I_1(\alpha_0 R_0)$$

$$+ \frac{\mu_1}{\mu_0} \alpha_1 R_0 Y_1(\alpha_1 R_0) I_0(\alpha_0 R_0)\Big] + [sY_0(\alpha_1 R_1) + PY_1(\alpha_1 R_1)]$$

$$\times \Big[\alpha_0 R_0 I_0(\alpha_1 R_0) I_1(\alpha_0 R_0) - \frac{\mu_1}{\mu_0} \alpha_1 R_0 I_1(\alpha_1 R_0) I_0(\alpha_0 R_0)\Big].$$

where $Y_0$ and $Y_1$ – the functions of Bessel of the second kind of zero and first order. The tables of Bessel functions listed in [187, 409] were used for the calculation. The frequency of an external action was taken as equal to $\omega = 15.7\ \text{s}^{-1}\ T = \frac{2\pi}{\omega} = 0.4\text{s}$. This approximately corresponds to the predominant frequencies of the oscillations of ground during strong earthquakes. The values of the shear wave propagation velocities in the half-plane and in the construction were accepted as the following:

$$C_0 = 600\ \text{ms}^{-1}, \quad C_1 = 300,600, 900\ \text{ms}^{-1}, \quad C_2 = 50 \div 1000\ \text{ms}^{-1}.$$

Experiments show that the values of $C_2$ for framework, large-panel, and stone buildings vary within the limits of $200 \div 900\ \text{ms}^{-1}$ [171]. In our case the interval of the variation of $C_2$ is taken somewhat larger for the purposes of accounting for other types of buildings and constructions, for which there is no experimental data thus far. The other parameters necessary for the calculation were accepted as follows:

$$\rho_0 = \rho_1 = 0.16\ T^2\ \text{m}^{-4}, \qquad \rho_2 = 0.038 \div 0.042\ T^2\ \text{m}^{-4},$$

$$R_1 = 6\ \text{m}, \qquad R_0 = 40\ \text{m}, \qquad H = 30\ \text{m}.$$

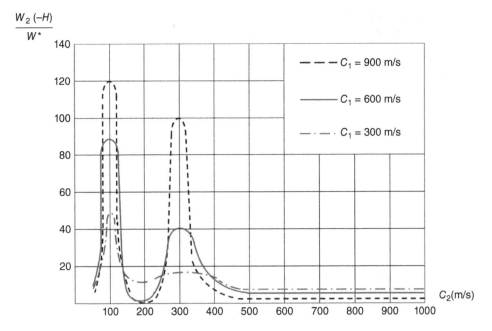

**Figure 5.4**   Dependencies of the relative displacement $W_2(-H)/W^*$ of the structure on the velocity of propagation of a shear wave.

Figure 5.4 shows the dependencies of the relative displacement $W_2(-H)/W^*$ ($W^*$ – displacement of a free surface of a homogeneous half-plane) from a shear wave propagation velocity in a construction on three types of ground of a semi-cylindrical base. The value of $C_0$ in this case was taken as fixed. The graphics have the characteristic maximums, which correspond to values 100, 300 ms$^{-1}$. With these values the phenomenon of resonance occurs. The given structure, the design schema of which is represented in the form of a shifted cantilever with the build-in lower end, has the following natural frequencies:

$$\omega_j = \frac{2j-1}{2}\pi\frac{C_2}{H}, \quad j = 1,2\cdots$$

A structure for which $C_2 = 300$ ms$^{-1}$, has the first natural frequency equal to $\omega_1 = 15.7$ s$^{-1}$, which coincides with the frequency of the external action. For a structure with the value of $C_2 = 100$ ms$^{-1}$, the second natural frequency is equal to $\omega_2 = 15.7$ s$^{-1}$, i.e. occurs in the case of resonance with this frequency. Figure 5.4 shows another characteristic of regularity – a reduction of the maximum displacements of a construction in a resonance zone with the decrease of the value of the wave propagation velocity in a semi-cylindrical base. For example, with $C_1 = 300$ ms$^{-1}$ the displacement of a structure decreases 3.44 times in comparison with the case of $C_1 = 900$ ms$^{-1}$. This means that the presence of a semi-cylindrical pliable base can lead to the reduction of the reaction of structures.

## 5.3    Method of Calculation of the Lining of Tunnels to Seismic Resistance

The method of calculation of the lining of tunnels to seismic resistance is based on the application of accelerograms from strong earthquakes. The works [16, 87, 88, 130, 174, 191, 223, 254] are dedicated to the question of interaction of longitudinal or transverse elastic plane waves with the circular openings or inclusions. The calculations of lining of tunnels to seismic resistance is given in the research works [14, 93, 107, 227]. Below we present the methodology of calculations of linings of tunnels on seismic resistance, based on the application of the accelerogram of strong earthquakes. Let us examine a thin-walled ring, which is located in the limitless space; Figure 5.5. It is assumed that the contact between the ring and elastic medium under the seismic exposure (influences) is not disrupted.

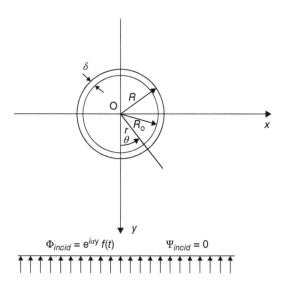

**Figure 5.5**    Scheme of the action of a seismic wave on the lining of a tunnel.

The seismic action is represented in the form of a plane non-stationary wave. The strength of the ring is possible to check in two limiting (extreme) cases. In the first case the strength of the ring is checked against the effect on a plane longitudinal wave, moving parallel to the axis of ox. In this first case, predominantly vertical oscillations occur on the free surface. In the second case the strength is checked against the excitation on a transverse wave, having predominantly horizontal oscillations occurring on the free surface. In both cases, it is additionally assumed that the kinematic parameters of the ground particle movement, located on the front of a wave approaching the underground structure, are known and assigned in the form of an accelerogram, seismogram, and velisogram, obtained during actual strong earthquakes. The methodology based on the application of the accelerograms of strong earthquakes, is successfully applied during calculations of above-ground structures [12, 18, 172]. In this case this methodology is applied to the calculation of underground structures. Let us examine the action of a longitudinal wave on a thin-walled ring. In the case of a plane strain the equations of the theory of elasticity lead to two wave equations relative to the potentials of longitudinal and transverse waves $\Phi$ and $\Psi$:

$$\Delta^2\Phi - \frac{1}{c_1^2}\frac{\partial^2\Phi}{\partial t^2} = 0; \tag{5.3.1}$$

$$\Delta^2\Psi - \frac{1}{c_1^2}\frac{\partial^2\Psi}{\partial t^2} = 0,$$

where $\Delta$ – the two-dimensional operator of Laplace; and $c_1, c_2$ – the velocity of propagation of longitudinal and transverse waves, respectively. The solution of the problem is represented in the polar system of coordinates $r, \theta$. The external action is characterized by the potential of a plane longitudinal wave [254]:

$$\Phi = \sum_{P=1}^{\infty}\sum_{n=1}^{\infty} \varepsilon_n i^n b_p I_n(\alpha_p r)\cos n\theta \sin \omega_p t; \tag{5.3.2}$$

where

$$\varepsilon_n = \begin{cases} 1, & n \geq 1 \\ 2, & n = 0 \end{cases}; \quad i = \sqrt{-1}, \quad \alpha_p = \frac{\omega_p}{c_1}, \quad \omega_p = \frac{P\pi}{T},$$

$I_n(\alpha_p r)$ – Bessel function of the first kind of the $n$th order; $b_p$ – coefficients of decomposition of a non-stationary action in Fourier series in terms of the sines (sins); and $T$ – period of external action. As in the work [116], in this case we accept that the external action (excitation, effect, exposure, influence) is periodic. If the duration of accelerograms are designated through $\Delta$, then the interval between two actions will be equal $T - \Delta$. In practice, the value of $T$ is selected in such a way that prior to the beginning of any non-periodic external action, the conditions of equating the displacements and the speed of the lining of tunnel to zero are complete. Thus, even though we assume the periodic excitation instead of the transient one, the obtained results are completely reliable because the chosen interval is such that the actual initial conditions of the problem are satisfied [100]. The solution to the wave equations is sought in the form of a double series in the following form:

$$\Phi_{incident} = \sum_{P=1}^{\infty} \sum_{n=1}^{\infty} A_{np} H_n(\alpha_p r) \cos n\theta \sin \omega_p t; \qquad (5.3.3)$$

$$\Psi = \sum_{P=1}^{\infty} \sum_{n=1}^{\infty} B_{np} H_n(\beta_p r) \sin n\theta \sin \omega_p t; \quad \beta = \frac{\omega_p}{c_2},$$

here $H_n(\alpha_p r)$, $H_n(\beta_p r)$ – Hankel functions of the first kind of $n$th order; and $A_{np}$, $B_{np}$ – complex constants, determined by satisfying the boundary conditions. The displacements $U_r$ and $U_\theta$, and stresses $\sigma_{rr}$, $\sigma_{\theta\theta}$, $\tau_{r\theta}$ are expressed as potentials in the following formulas:

$$U_r = \frac{\partial \Phi}{\partial r} + \frac{1}{r}\frac{\partial \Psi}{\partial \theta}; \quad U_\theta = \frac{1}{r}\frac{\partial \Phi}{\partial \theta} \frac{\partial \Psi}{\partial r};$$

$$\sigma_{rr} = \lambda \left[ \frac{\partial^2 \Phi}{\partial r^2} + \frac{1}{r}\frac{\partial \Phi}{\partial r} + \frac{1}{r^2}\frac{\partial^2 \Phi}{\partial \theta^2} \right] + 2\mu \left[ \frac{\partial^2 \Phi}{\partial r^2} + \frac{1}{r}\frac{\partial^2 \Psi}{\partial r \partial \theta} - \frac{1}{r^2}\frac{\partial \Psi}{\partial \theta} \right], \qquad (5.3.4)$$

$$\sigma_{\theta\theta} = \lambda \left[ \frac{\partial^2 \Phi}{\partial r^2} + \frac{1}{r}\frac{\partial \Phi}{\partial r} + \frac{1}{r^2}\frac{\partial^2 \Phi}{\partial \theta^2} \right] + 2\mu \left[ \frac{1}{r}\frac{\partial \Phi}{\partial r} + \frac{1}{r^2}\frac{\partial^2 \Phi}{\partial \theta^2} + \frac{1}{r^2}\frac{\partial \Psi}{\partial \theta} - \frac{1}{r}\frac{\partial^2 \Psi}{\partial r \partial \theta} \right]$$

$$\tau_{r\theta} = \mu \left[ \frac{2}{r}\frac{\partial^2 \Phi}{\partial r \partial \theta} - \frac{2}{r^2}\frac{\partial \Phi}{\partial \theta} + \frac{1}{r^2}\frac{\partial^2 \Psi}{\partial \theta^2} - \frac{\partial^2 \Psi}{\partial r^2} + \frac{1}{r}\frac{\partial \Psi}{\partial r} \right]$$

Where $\lambda$, $\mu$ – the elastic Lame's coefficients. Substituting (5.3.3) into (5.3.4), we will obtain:

$$U_r = \sum_{P=1}^{\infty} \sum_{n=1}^{\infty} N_{np} \cos n\theta \sin \omega_p t;$$

$$U_\theta = \sum_{P=1}^{\infty} \sum_{n=1}^{\infty} Q_{np} \sin n\theta \sin \omega_p t;$$

$$\sigma_{rr} = \sum_{P=1}^{\infty} \sum_{n=1}^{\infty} K_{np} \cos n\theta \sin \omega_p t; \qquad (5.3.5)$$

$$\tau_{r\theta} = \sum_{P=1}^{\infty} \sum_{n=1}^{\infty} M_{np} \sin n\theta \sin \omega_p t;$$

$$N_{np} = A_{np}c_{1p} + B_{np}c_{2p} + \overline{A}_{np}\overline{c}_{1p}$$

$$Q_{np} = A_{np}d_{1p} + B_{np}d_{2p} + \overline{A}_{np}\overline{d}_{1p}$$

$$K_{np} = A_{np}a_{1p} + B_{np}a_{2p} + \overline{A}_{np}\overline{a}_{1p}$$

$$M_{np} = A_{np}b_{1p} + B_{np}b_{2p} + \overline{A}_{np}\overline{b}_{1p},$$

where:

$$c_{1p} = \frac{1}{r}[-nH_n(\alpha_p r) + \alpha_p r H_{n-1}(\alpha_p r)],$$

$$c_{2p} = \frac{n}{r}H_n(\beta_p r),$$

$$d_{1p} = -\frac{n}{r}H_n(\alpha_p r), \qquad (5.3.6)$$

$$d_{2p} = \frac{1}{r}[nH_n(\beta_p r) - \beta_p r H_{n-1}(\beta_p r)],$$

$$a_{1p} = \frac{2\mu}{r^2}[H_n(\alpha_p r)(n^2 + n - 0.5r^2\beta_p^2) - \alpha_p r H_{n-1}(\alpha_p r)],$$

$$a_{2p} = \frac{2\mu}{r^2}[-(n^2 + n)H_n(\beta_p r) + n\beta_p r H_{n-1}(\alpha_p r)],$$

$$b_{1p} = \frac{2\mu}{r^2}[(n^2 + n)H_n(\alpha_p r) - n\alpha_p r H_{n-1}(\alpha_p r)],$$

$$b_{2p} = \frac{2\mu}{r^2}[-(n^2 + n - 0.5r^2\beta_p^2)H_n(\beta_p r) + \beta_p r H_{n-1}(\beta_p r)],$$

$$\overline{A}_{np} = \varepsilon_n i^n b_p.$$

$\overline{a}_{1p}, \overline{b}_{1p}, \overline{c}_{1p}$, and $\overline{d}_{1p}$ are obtained from $a_{1p}, b_{1p}, c_{1p}$, and $d_{1p}$ by replacing functions $H_n(z)$ with functions $I_n(z)$. Thus, the parameters of a stress-strained state of an elastic medium are determined by expressions (5.3.5) and (5.3.6), where the constants $A_{np}$, $B_{np}$ are to be determined from the boundary conditions.

The lining of a tunnel is represented as a thin-walled ring, predominantly bent. The equation of the forced oscillations (vibrations) of the thin-walled ring takes the form [357, 358]:

$$\frac{\partial^6 W_r}{\partial \theta^6} + 2\frac{\partial^4 W_r}{\partial \theta^4} + \frac{\partial^2 W_r}{\partial \theta^2} + \frac{mR^2}{EI}\frac{\partial^2}{\partial t^2}\left(\frac{\partial^2 W_r}{\partial \theta^2} - W_r\right)$$

$$= \frac{R^4}{EI}\left(\frac{\partial^2 \sigma_{rr}}{\partial \theta^2} - \frac{\partial \tau_{r\theta}}{\partial \theta}\right), \tag{5.3.7}$$

where $W_r$ – displacement of the ring in the radial direction; $EI$ – flexural rigidity (bending stiffness) of the ring; $R$ – the radius of the ring; and $m$ – the mass of the unit of length of the ring. The displacements of the ring in the radial and tangential directions $W_r$, and $W_\theta$ are connected as follows:

$$W_r = \frac{\partial W_\theta}{\partial \theta} \tag{5.3.8}$$

The displacements of the ring are sought in the form:

$$W_r = \sum_{P=1}^{\infty}\sum_{n=1}^{\infty}[C_{np}\cos n\theta + D_{np}\sin n\theta]\sin \omega_p t, \tag{5.3.9}$$

$$W_\theta = \sum_{P=1}^{\infty}\sum_{n=1}^{\infty}\frac{1}{n}[C_{np}\cos n\theta - D_{np}\cos n\theta]\sin \omega_p t.$$

Substituting (5.3.9) and the values of $\sigma_{rr}$ and $\sigma_{r\theta}$, determined by expressions (5.3.5) into (5.3.7), we obtain the following dependencies for the coefficients $C_{np}$ and $D_{np}$:

$$D_{np} = 0;$$

$$C_{np} = -\frac{\lambda_p^2}{m\omega_p^2}\frac{(K_{np}n^2 + M_{np}n)}{[-n^2(n^2-1)^2 + \lambda_p^4(1+n^2)]}. \tag{5.3.10}$$

here:

$$\lambda_p^4 = \frac{mR^4}{EI}\omega_p^2.$$

The boundary conditions of this problem are expressed in the requirement of the equality of the radial and tangential displacements of the medium and ring. They take the form:

$$U_r(R) = W_r(R); \quad U_\theta(R) = W_\theta(R). \tag{5.3.11}$$

In the open form the conditions in (5.3.11) are written in the form of:

$$A_{np}c_1 + B_{np}c_2 + \overline{A}_{np}\overline{c}_1 + \frac{1}{\Delta}(K_{np}n^2 + nM_{np}) = 0 \tag{5.3.12}$$

$$A_{np}d_1 + B_{np}d_2 + \overline{A}_{np}\overline{d}_1 + \frac{1}{\Delta}(K_{np}n + M_{np}) = 0$$

where,

$$\Delta = -\frac{EI}{R^4}n^2(n^2 - 1)^2 + m\omega_p^2(1 + n^2)$$

Substituting the expressions $K_{np}$ and $M_{np}$ from (5.3.5) into (5.3.12) we will obtain the equations for determining the coefficients $A_{np}$ and $B_{np}$.

$$A_{np}\delta_{11} + B_{np}\delta_{12} = \overline{A}_{np}\delta_{10},$$

$$A_{np}\delta_{21} + B_{np}\delta_{22} = \overline{A}_{np}\delta_{20} \tag{5.3.13}$$

$$n = 0, 2, 3 \cdots, \quad p = 1, 2, 3 \cdots$$

where

$$\delta_{11} = \Delta c_1 + n^2 a_1 + nb_1,$$

$$\delta_{12} = \Delta c_2 + n^2 a_2 + nb_2,$$

$$\delta_{21} = \Delta d_1 + na_1 + b_1,$$

$$\delta_{22} = \Delta d_2 + na_2 + b_2,$$

$$\delta_{10} = -\overline{c}_1\Delta - n^2\overline{a}_1 - n\overline{b}_1,$$

$$\delta_{20} = -\overline{d}_1\Delta - n\overline{a}_1 - \overline{b}_1.$$

The equations are valid when $\Delta \neq 0$; $n \neq 1$. The case of $n = 1$ corresponds to the motion of the ring as an absolutely rigid body. Coefficients with (at) $n = 1$ are determined using the conditions:

$$U_r(R) = U\cos\theta; \quad U_\theta(R) = -U\sin\theta, \tag{5.3.14}$$

Here $U$ – displacement of the ring in the direction $y$, which is determined with the use of a condition of the dynamic equilibrium of a rigid non-deformable ring:

$$-\pi(R^2 - R_0^2)\rho_1\omega_p^2 U = \int_0^{2\pi} [\sigma_{rr}\cos\theta - \tau_{r\theta}\sin\theta]Rd\theta. \qquad (5.3.15)$$

As a result, conditions (5.3.14) take the form:

$$A_{1p}c_1 + B_{1p}c_2 + \overline{A}_{1p}\overline{c}_1 = \frac{m}{R}[A_{1p}H_1(\alpha R) + B_{1p}H_1(\beta R) + \overline{A}_{1p}I_1(\alpha R)] \qquad (5.3.16)$$

$$A_{1p}d_1 + B_{1p}d_2 + \overline{A}_{1p}\overline{d}_1 = -\frac{m}{R}[A_{1p}H_1(\alpha R) + B_{1p}H_1(\beta R) + \overline{A}_{1p}I_1(\alpha R)]$$

In the case of $\Delta = 0$, the conditions for determining $A_{np}, B_{np}$ have another form. Meanwhile, in this case the resonance takes place, and the conditions (5.3.11) take the form:

$$\sigma_{rr}(R) = 0; \quad U_r(R) = \frac{\partial U_\theta}{\partial\theta}(R). \qquad (5.3.17)$$

The first of the conditions (5.3.17) means that with the resonance oscillations the ring and the surrounding medium oscillate synchronously, and therefore, they do not press on each other. The second condition shows the connection between the normal and tangential displacements [324]. These conditions are valid for the ring, and in the case of the resonance they are also valid for the boundary points of the surrounding medium. The conditions (5.3.17) in the open form are written as follows:

$$A_{np}(n^2a_1 + nb_1) + B_{np}(n^2a_2 + nb_2) = -\overline{A}_n(n^2\overline{a}_1 + n\overline{b}_1), \qquad (5.3.18)$$

$$A_{np}(c_1 - nd_1) + B_{np}(c_2 - nd_2) = \overline{A}_n(n\overline{d}_1 - \overline{c}_1).$$

$$n = 2, 3\cdots, \quad p = 1, 2, \cdots$$

Thus from (5.3.13), (5.3.16), and (5.3.18) the coefficients $A_{np}, B_{np}$ are determined, and with that using the formulas (5.3.5) all parameters of the stress-strain condition (state) of the thin-walled ring are determined as well. Here, one should bear in mind that these coefficients are complex and, therefore, in practical calculations the values of $A_{np}, B_{np}, a_1, a_2, b_1, b_2, c_1, c_2, d_1$, and $d_2$ should be substituted in a complex form. Further, it is necessary to equate the real and imaginary parts of the equation, and instead of two equations obtain four equations to determine the four coefficients, characterizing the real and imaginary parts of the coefficients $A_{np}$, $B_{np}$. The external excitation in the proposed methodology is given as the decomposition potential of an incident longitudinal wave in the form of a number of the series (5.3.2) on the time coordinate, and is characterized by the coefficient $b_p$. In actuality, with the aid of the measuring devices we can measure accelerations, velocities, or displacements of ground at depth corresponding to the location of an underground structure or on the free surface. Let us assume that we have obtained the accelerogram of the vertical motion of the ground at the level of underground structure. We decompose the accelerogram in to the Fourier series in the interval of $2T$.

$$\frac{\partial^2 U_y}{\partial t^2} = \sum_{P=1}^{\infty} C_p \cdot \sin p \frac{\pi}{T} t, \tag{5.3.19}$$

where

$$C_p = \frac{2}{\pi} \int_{t_0}^{t_0+\Delta} \frac{\partial^2 U_y}{\partial t^2} \cdot \sin p \frac{\pi}{T} dt$$

$\Delta$ – duration of accelerogram; and $2t_0$ – interval between the periodic actions.

The potential of the incident wave in the Cartesian coordinates is represented in the form of:

$$\Phi_{incident} = \sum_{P=1}^{\infty} e^{i\alpha_p y} \cdot b_p \sin \omega_p t. \tag{5.3.20}$$

The acceleration, which corresponds to this potential will be determined by the expression:

$$\frac{\partial^2 U_y}{\partial t^2} = -\sum_{P=1}^{\infty} e^{i\alpha_p y} \cdot i\alpha_p \omega_p^2 b_p \cdot \sin \omega_p t. \tag{5.3.21}$$

With known accelerogram, which corresponds to the level of $y = 0$, the connection (link) between the coefficients $b_p$ and $C_p$ is found by equating expressions (5.3.19) and (5.3.21):

$$b_p = -\frac{C_p}{i\alpha_p \omega_p^2} \tag{5.3.22}$$

In order to use the expressions to determine the seismic pressure on the lining of the tunnel with the given accelerogram, we substitute $b_p$ in expressions (5.3.5) and (5.3.6) with expression (5.3.22). Additionally, to calculate the lining of a tunnel when the accelerogram of ground oscillation is not available, we can use the velociogram or seismogram recorded during a strong earthquake at the level of the projected underground structure. In these cases also we use formulas (5.3.5) and (5.3.6), substituting $b_p$ with the following expressions:

$$b_p = -\frac{C_p^*}{i\alpha_p\omega_p},$$ (5.3.23)

$$b_p = -\frac{C_p^{**}}{i\alpha_p},$$ (5.3.24)

Here $C_p^*$, $C_p^{**}$ – are the coefficients of decomposition of velociogram or seismogram in Fourier series in terms of the sines in the form (5.3.19). Let us examine a concrete example of determination of the seismic pressure on the lining of a tunnel of a round cross-section. The external excitation is presented in the form of a vertical component of an accelerogram of a real 8-magnitude earthquake in San Jose, California on April 9, 1955. Since at present, accelerogram records obtained at different depths from free surface are very rare, we therefore considered it possible to use the above-mentioned accelerogram, obtained on the free surface. It should be noted that this assumption will be true for shallow underground structures. However, with significant depths of bedding, the accelerograms recorded at the appropriate depth must be used for the construction or alternatively analytical methods, calculating the ordinates of accelerograms at any depth, similar records on free surface can be used. In particular, the dependency of the acceleration of oscillations of ground on the depths, obtained in natural conditions could be used (see [93]). Having the value of the coefficient, characterizing the degree of the decrease of acceleration at the given depth, we multiply all ordinates of the accelerogram by this coefficient and obtain the reduced ordinates of the accelerogram, which correspond to the given depth. The dimensions of the lining are: $R = 2.75$ m, $R_0 = 2.55$ m. The thickness of the lining is: 0.2 m. The material of lining is reinforced concrete with the following characteristics:

$$E = 315 \cdot \frac{10^4 T}{m^2}; \quad \mu = 126 \cdot \frac{10^4 T}{m^2}; \quad \gamma = 2.5T/m^3,$$

where $E$ – modulus of elasticity; $\mu$ – shear modulus; and $\gamma$ – volumetric weight. The values of the parameters characteristics of lining, in (5.3.7) are:

$$m = \frac{0.05Ts^2}{m}, \quad I = 1 \cdot \frac{0.2^3}{12m^4}, \quad EI/R_4 = 42.6T/m^2.$$

Three types of characteristics of the ground medium were examined. The values of the speeds of transverse waves were taken; $C_2 = 275, \ 550, \ 825$ m/s, and values of the speeds of longitudinal waves corresponding to them were $C_1 = 470.2, \ 940.5, 1410.75$ m/s. The volume weight of ground; $\gamma = 1.6, \ 2.0, \ 2.5 \ T/m^3$. The accelerogram had the duration $\Delta = 3$ s, $2t_0 = 0.15$ s. The values of the normal and tangential stresses acting on the lining of a tunnel $\sigma_{r\theta}(R)$, $\sigma_{rr}(R)$ and the value of normal and tangential displacements of lining in all tabulated points were calculated. The step of tabulation of the accelero-gram is 0.02 s. It seems that the stresses and displacements over time were at their maximum values at $t = 0.68$ s. Tables 5.1 and 5.2. Figures 5.6–5.8 show the diagrams for $\sigma_{rr}(R), \ \tau_{r\theta}(R), \ U_r(R)$, and $U_\theta(R)$ (Tables 5.3 and 5.4). In Table 5.3 the values of contour stresses are presented.

**Table 5.1**  The values of contour stresses ($t = 0.68$ s).

|  | $\sigma_{rr}(\text{kg/cm}^2)$ | | | $\tau_{r\theta}(\text{kg/cm}^2)$ | | |
|---|---|---|---|---|---|---|
| $\theta$ radian | $C_1 = 470$ m/s | $C_1 = 940$ m/s | $C_1 = 1410$ m/s | $C_1 = 470$ m/s | $C_1 = 940$ m/s | $C_1 = 1410$ m/s |
| 0 | 2.38 | 15.33 | 32.22 | 0 | 0 | 0 |
| 1/16 | 2.33 | 15.03 | 31.60 | 0.46 | 2.98 | 6.28 |
| 2/16 | 2.20 | 14.16 | 29.77 | 0.90 | 5.84 | 12.32 |
| 3/16 | 1.98 | 12.75 | 26.79 | 1.31 | 8.49 | 17.88 |
| 4/16 | 1.63 | 10.84 | 22.79 | 1.67 | 10.8 | 22.77 |
| 5/16 | 1.33 | 8.52 | 17.90 | 1.97 | 12.72 | 26.77 |
| 6/16 | 0.91 | 5.87 | 12.33 | 2.19 | 14.15 | 29.76 |
| 7/16 | 0.47 | 2.99 | 6.29 | 2.33 | 15.02 | 31.59 |
| 8/16 | 0.38 | 0 | 3.98 | 2.37 | 15.32 | 32.21 |

**Table 5.2**  The values of contour displacements ($t = 0.68$ s).

|  | $U_r(\text{cm})$ | | | $U_\theta(\text{cm})$ | | |
|---|---|---|---|---|---|---|
| $\theta$ radian | $C_1 = 470$ m/s | $C_1 = 940$ m/s | $C_1 = 1410$ m/s | $C_1 = 470$ m/s | $C_1 = 940$ m/s | $C_1 = 1410$ m/s |
| 0 | −0.152 | −0.246 | −0.220 | 0 | 0 | 0 |
| 1/16 | −0.149 | −0.242 | −0.225 | 0.022 | 0.004 | 0.008 |
| 2/16 | −0.140 | −0.227 | −0.212 | 0.043 | 0.007 | 0.002 |
| 3/16 | −0.126 | −0.204 | −0.190 | 0.063 | 0.010 | 0.002 |
| 4/16 | −0.107 | −0.174 | −0.162 | 0.080 | 0.013 | 0.003 |
| 5/16 | −0.084 | −0.137 | −0.127 | 0.094 | 0.015 | 0.004 |
| 6/16 | −0.058 | −0.094 | −0.087 | 0.105 | 0.017 | 0.004 |
| 7/16 | −0.030 | −0.048 | −0.045 | 0.112 | 0.018 | 0.004 |
| 8/16 | 0 | 0 | 0 | 0.114 | 0.018 | 0.004 |

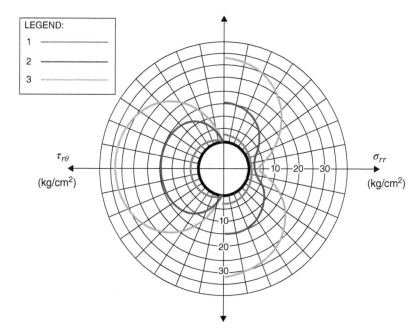

**Figure 5.6**  Distribution is at the maximum over time with ($t = 0.68$ s) normal and shearing stresses acting on the tunnel lining with different types of ground: $1 - C_1 = \frac{470.2\ \text{m}}{\text{s}}$; $2 - C_1 = 940.5$ m/s; $3 - C_1 = 1410.7\frac{\text{m}}{\text{s}}$.

**Table 5.3**  The values of contour stresses ($t = 2.46$ s).

| $\theta$ radian | $\sigma_{rr}$(kg/cm²) | | | $\tau_{r\theta}$(kg/cm²) | | |
|---|---|---|---|---|---|---|
| | $C_1 = 470$ m/s | $C_1 = 940$ m/s | $C_1 = 1410$ m/s | $C_1 = 470$ m/s | $C_1 = 940$ m/s | $C_1 = 1410$ m/s |
| 0 | −1.16 | −7.59 | −15.82 | 0 | 0 | 0 |
| 1/16 | −1.14 | −7.45 | −15.51 | −0.23 | −1.49 | −3.09 |
| 2/16 | −1.07 | −7.02 | −14.61 | −0.45 | −2.92 | −6.06 |
| 3/16 | −0.96 | −6.31 | −13.15 | −0.65 | −4.24 | −8.80 |
| 4/16 | −0.82 | −5.37 | −11.18 | −0.83 | −5.39 | −11.20 |
| 5/16 | −0.64 | −4.21 | −8.78 | −0.97 | −6.33 | −13.16 |
| 6/16 | −0.44 | −2.90 | −6.05 | −1.08 | −7.03 | −14.62 |
| 7/16 | −0.22 | −1.48 | −3.08 | −1.14 | −7.46 | 31.59 |
| 8/16 | −0.38 | −0.90 | −1.74 | −1.16 | −7.60 | 32.21 |

In Table 5.4 the values of contour displacements are presented. The given data, showing the changing values of normal and tangential displacements at the contour of a thin-walled ring, illustrates that the ring is practically not deformed and moves as a rigid body. This allows for a considerable simplification in calculations and in the formulas determining the displacements of

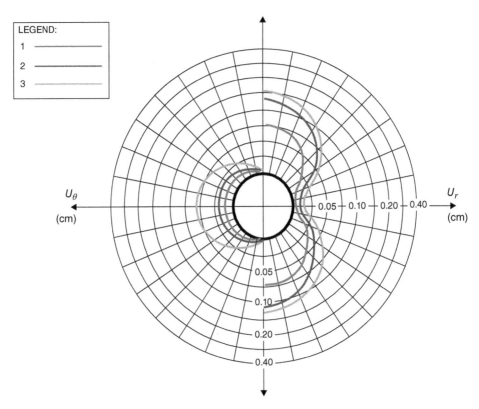

**Figure 5.7** Distribution is at the maximum over time with ($t = 0.68$ s) normal and tangential displacements acting on the tunnel lining with different types of ground: $1 - C_1 = \frac{470.2 \text{ m}}{\text{s}}$; $2 - C_1 = 940.5$ m/s; $3 - C_1 = 1410.7$ m/s.

**Table 5.4** The values of contour displacements ($t = 2.46$ s).

| | $U_r$(cm) | | | $U_\theta$(cm) | | |
|---|---|---|---|---|---|---|
| $\theta$ radian | $C_1 = 470$ m/s | $C_1 = 940$ m/s | $C_1 = 1410$ m/s | $C_1 = 470$ m/s | $C_1 = 940$ m/s | $C_1 = 1410$ m/s |
| 0 | 0.074 | 0.119 | 0.111 | 0 | 0 | 0 |
| 1/16 | 0.072 | 0.117 | 0.109 | −0.011 | −0.002 | −0.004 |
| 2/16 | 0.068 | 0.110 | 0.103 | −0.021 | −0.003 | −0.007 |
| 3/16 | 0.061 | 0.099 | 0.093 | −0.031 | −0.047 | −0.001 |
| 4/16 | 0.052 | 0.085 | 0.079 | −0.039 | −0.006 | −0.001 |
| 5/16 | 0.041 | 0.066 | 0.062 | −0.046 | −0.007 | −0.002 |
| 6/16 | 0.028 | 0.046 | 0.043 | −0.051 | −0.007 | −0.002 |
| 7/16 | 0.014 | 0.023 | 0.023 | −0.054 | −0.008 | −0.002 |
| 8/16 | 0 | 0 | 0 | −0.055 | −0.008 | −0.002 |

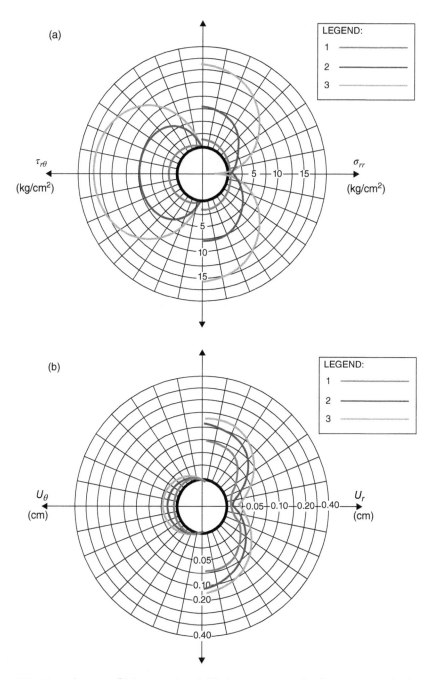

**Figure 5.8**  Distribution of (a) normal and (b) shearing stress displacements in the lining of the tunnel with ($t = 2.46$ s) with different types of ground: $1 - C_1 = \dfrac{470.2 \text{ m}}{\text{s}}$; $2 - C_1 = \dfrac{940.5 \text{ m}}{\text{s}}$; $3 - C_1 = 1410.7$ m/s.

the ring (5.3.9) to take into account only the addend, which corresponds to the value $n = 1$. Another regularity is visible from the figure, i.e. the change in the contour stresses and displacements of the ring depending on the type of ground.

## 5.4   A Study of the Action of Seismic Wave on the Rigid Ring Located in the Half-plane

Let us examine a thin ring located in a half-plane, Figure 5.9. The seismic action is represented in the form of the harmonic wave. Having a solution for a case of harmonic oscillations, and using the methods of harmonic analysis, it is possible to obtain the expressions for displacements and stresses of a medium, and for arbitrary-in-time oscillations as well. We will consider the impact of a longitudinal wave. We have a case of a plane strain (deformation). In this case the equations of the theory of elasticity relative to the displacements are reduced to two equations relative to the potentials of the longitudinal $\phi$ and transverse waves $\psi$.

$$\Delta\phi + \alpha^2\phi = 0, \quad \Delta\psi + \beta^2\psi = 0 \tag{5.4.1}$$

where

$$\alpha = \frac{\omega}{C_1}; \quad \beta = \frac{\omega}{C_2}.$$

The external excitation is characterized by the potential of a longitudinal wave

$$\phi_0^{(1)} = e^{i(\alpha y - \omega t)} \tag{5.4.2}$$

If the ring is absent, then the wave is reflected from the free surface:

$$\phi_0^{(2)} = e^{-i(\alpha y - \omega t)} \tag{5.4.3}$$

The solution of the problem is sought in the form of:

$$\phi = \phi_0^{(1)} + \phi_0^{(2)} + \phi_1;$$

$$\phi_1 = \sum_{n=-\infty}^{\infty} A_n H_n(\alpha r_1)e^{in\theta_1} + B_n H_n(\alpha r_2)e^{in\theta_2}; \tag{5.4.4}$$

$$\psi = \sum_{n=-\infty}^{\infty} C_n H_n(\beta r_1)e^{in\theta_1} + D_n H_n(\beta r_2)e^{in\theta_2}.$$

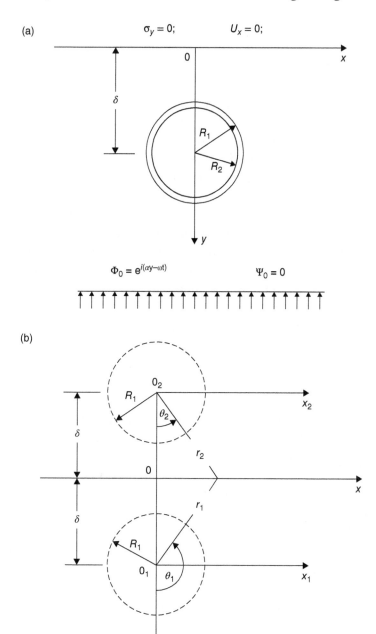

**Figure 5.9**   Schemas used with solutions of problems of the impact of seismic waves on an underground structure: (a) the scheme of the underground structure; (b) the system of the polar coordinate.

All components of the stress-strained condition (state) contain the multiplier $e^{-i\omega t}$, which henceforth will be omitted. The solution for (5.4.4) contains the products $H_n(\alpha r_1)e^{in\theta_1}$, $H_n(\beta r_1)e^{in\theta_2}$, $i = 1, 2$, which are the solutions of Eqs. (5.4.1) in the polar coordinate with centers at points $O_1$ and $O_2$, Figure 5.9.

$A_n$, $B_n$, $C_n$, and $D_n$ – are arbitrary constants, which must be determined from the boundary conditions. The algorithm of the solution to a problem of the type (5.4.4) earlier was used in the work [130] when solving problem of acoustic medium. The solution of the type in Eq. (5.4.4) gives the opportunity to solve the stated problem with a certain approximation. The proximity is that the boundary conditions on the cylindrical plane are satisfied accurately while on the free surface, mixed boundary conditions are satisfied. Under the influence of the longitudinal wave: $\sigma_y = 0$; $U_x = 0$, with $y = 0$. Under the influence of transverse wave: $\tau_{xy} = 0$; $U_y = 0$, with $y = 0$, (5.4.5) where $\sigma_y$; $\tau_{xy}$ – normal and shear (tangential) stresses respectively; and $U_x$; $U_y$ – displacements of medium in the directions of $x$ and $y$. From the physical point of view the conditions (5.4.5), it means that in one case on the earth's surface, only vertical oscillations are present while horizontal oscillations are equal to zero, and in the second case the opposite is true. Under the seismic influences such conditions frequently occur. If the epicenter of an earthquake is found under the construction, then predominantly vertical oscillations on the earth's surface are present. With a distant epicenter the reverse picture is observed. If during an actual earthquake both displacements with approximately close amplitudes arise on free surface, then the adoption of the boundary conditions in (5.4.5) can be seen as a separate calculation of the underground structure by vertical and then horizontal impacts, whose added value gives the stressed state of the given impact in the structure. The stresses, acting on areas $r = const.$, $\theta = const.$; $\sigma_{rr}, \sigma_{\theta\theta}, \tau_{r\theta}$ are expressed as potentials $\phi$ and $\psi$ by formulas (5.4.4). Substituting the expressions $\phi_1$ and $\psi$ into (5.4.4), we obtain:

$$\sigma_{rr} = \sum_{n=-\infty}^{\infty} [A_n\Delta_1(r_1) + C_n\Delta_2(r_1)]e^{in\theta_1} + [B_n\Delta_1(r_2) + D_n\Delta_2(r_2)]e^{in\theta_2}; \quad (5.4.5)$$

$$\tau_{r\theta} = \sum_{n=-\infty}^{\infty} [A_n\Delta_3(r_1) + C_n\Delta_4(r_1)]e^{in\theta_1} + [B_n\Delta_3(r_2) + D_n\Delta_4(r_2)]e^{in\theta_2};$$

$$\sigma_{\theta\theta} = \sum_{n=-\infty}^{\infty} [A_n\Delta_5(r_1) + C_n\Delta_6(r_1)]e^{in\theta_1} + [B_n\Delta_5(r_2) + D_n\Delta_6(r_2)]e^{in\theta_2};$$

where:

$$\Delta_1(r_i) = \frac{2\mu}{r_i^2}[(n^2 + n - 0.5r_i^2\beta^2)H_n(\alpha r_i) - \alpha r_i H_{n-1}(\alpha r_i)],$$

$$\Delta_2(r_i) = -\frac{2\mu}{r_i^2}i[n(n+1)H_n(\beta r_i) - n\beta r_i H_{n-1}(\beta r_i)],$$

$$\Delta_3(r_i) = -\frac{2\mu}{r_i^2}i[n(n+1)H_n(\alpha r_i) - n\alpha r_i H_{n-1}(\alpha r_i)],$$

$$\Delta_4(r_i) = -\frac{2\mu}{r_i^2}[(n^2 + n - 0.5r_i^2\beta^2)H_n(\beta r_i) - \beta r_i H_{n-1}(\beta r_i)] \qquad (5.4.6)$$

$$\Delta_5(r_i) = -\frac{2\mu}{r_i^2}[(n^2 + n - \alpha^2 r_i^2 + 0.5\beta^2 r_i^2)H_n(\alpha r_i) - \alpha r_i H_{n-1}(\alpha r_i)],$$

$$\Delta_6(r_i) = \frac{2\mu}{r_i^2}i[n(n+1)H_n(\beta r_i) - n\beta r_i H_{n-1}(\beta r_i)],$$

$$i = 1, 2.$$

In order to satisfy the boundary conditions (5.4.5) on the surface of $y = 0$ we will use the formulas, which express the connection of stresses in the Cartesian coordinates through stresses in the polar coordinates [130]:

$$\sigma_{yy} = \sigma_{rr}\cos^2\theta + \sigma_{\theta\theta}\sin^2\theta + 2r_{r\theta}\sin\theta\cos\theta,$$

$$\tau_{xy} = (\sigma_{rr} - \sigma_{\theta\theta})\sin\theta\cos\theta + r_{r\theta}(\sin^2\theta - \cos^2\theta), \qquad (5.4.7)$$

$$\sigma_{xx} = \sigma_{rr}\sin^2\theta + \sigma_{\theta\theta}\cos^2\theta - 2r_{r\theta}\sin\theta\cos\theta.$$

The relationship of the corresponding displacements is as follows:

$$U_y = U_r\cos\theta - U_\theta\sin\theta \qquad (5.4.8)$$

$$U_x = U_r\sin\theta + U_\theta\cos\theta$$

where $U_r$, $U_\theta$ – displacements in the directions of the polar coordinates $r$ and $\theta$. These displacements are expressed through the functions $\phi$ and $\psi$ as follows:

$$U_r = \frac{\partial\phi}{\partial r} + \frac{1}{r}\frac{\partial\psi}{\partial\theta}; U_\theta = \frac{1}{r}\frac{\partial\phi}{\partial\theta} + \frac{\partial\psi}{\partial r}. \qquad (5.4.9)$$

After substituting the expressions $\phi$ and $\psi$ in the (5.4.9), we will obtain:

$$
U_r = \sum_{n=-\infty}^{\infty} \left[ A_n H_n'(\alpha r_1) + C_n \frac{in}{r_1} H_n(\beta r_1) \right] e^{in\theta_1}
$$
$$
+ \left[ B_n H_n'(\alpha r_2) + D_n \frac{in}{r_2} H_n(\beta r_2) \right] e^{in\theta_2}; \tag{5.4.10}
$$

$$
U_\theta = \sum_{n=-\infty}^{\infty} \left[ A_n \frac{in}{r_1} H_n(\alpha r_1) - C_n H_n'(\beta r_1) \right] e^{in\theta_1}
$$
$$
+ \left[ B_n \frac{in}{r_2} H_n(\alpha r_2) - D_n H_n'(\beta r_2) \right] e^{in\theta_2}.
$$

After substituting expressions (5.4.5) and (5.4.10) into (5.4.7) and (5.5.8) respectively, it is possible to ensure that the boundary conditions (5.4.5) are satisfied, if:

$$
A_n = -B_{-n}; C_n = D_{-n}. \tag{5.4.11}
$$

For satisfaction of the boundary conditions in a cylindrical cavity we will modify expressions (5.4.4). The functions containing the polar coordinates $r_2$ and $\theta_2$ are expressed through coordinates $r_1$ and $\theta_1$. Then we will take advantage of the formulas [130]:

$$
H_n(\alpha r_2) e^{in\theta_2} = \sum_{p=-\infty}^{\infty} [H_{n-p}(2\delta\alpha) I_p(\alpha r_1)] e^{in\theta_1}, \tag{5.4.12}
$$

$$
H_n(\beta r_2) e^{in\theta_2} = \sum_{p=-\infty}^{\infty} [H_{n-p}(2\delta\beta) I_p(\beta r_1)] e^{in\theta_1}, \quad r_1 < 2\delta
$$

As a result, taking into account (5.4.11) the second addend in the expressions of $\phi_1$ and $\psi$ will take the form:

$$
\sum_{n=-\infty}^{\infty} B_n H_n(\alpha r_2) e^{in\theta_2} = - \sum_{n=-\infty}^{\infty} A_{-n} \sum_{p=-\infty}^{\infty} H_{n-p}(2\delta\alpha) I_p(\alpha r_1) e^{in\theta_1}
$$

$$
= \sum_{n=-\infty}^{\infty} S_{np} I_n(\alpha r_1) e^{in\theta_1}; \quad \text{where} \quad S_{np} = - \sum_{p=-\infty}^{\infty} A_p (-1)^{n+p} H_{n+p}(2\delta\alpha),
$$

$$
\tag{5.4.13}
$$

$$\sum_{n=-\infty}^{\infty} D_n H_n(\beta r_2) e^{in\theta_2} = \sum_{n=-\infty}^{\infty} Q_{np} I_n(\beta r_1) e^{in\theta_1},$$

where,

$$Q_{np} = - \sum_{p=-\infty}^{\infty} C_p (-1)^{n+p} H_{n+p}(2\delta\beta).$$

Using the representation in the form of an infinite series of the functions (5.4.2) and (5.4.3) through the products of Bessel and trigonometric functions we obtain:

$$\phi_0^{(1)} + \phi_0^{(2)} = \sum_{n=-\infty}^{\infty} I_n(\alpha r_1) \alpha_n, \tag{5.4.14}$$

$$\alpha_n = 2i \sin\left(\delta\alpha + n\frac{\pi}{2}\right)$$

The potentials $\phi$ and $\psi$ taking into account (5.4.13) and (5.4.14) take the form:

$$\phi = \sum_{n=-\infty}^{\infty} [A_n H_n(\alpha r_1) + (\alpha_n + S_{np}) I_n(\alpha r_1)] e^{in\theta_1}; \tag{5.4.15}$$

$$\psi = \sum_{n=-\infty}^{\infty} [C_n H_n(\beta r_1) + Q_{np} I_n(\beta r_1)] e^{in\theta_1}.$$

Stresses $\sigma_{rr}$ and $\tau_{r\theta}$ on the ring contour ($r_1 = R_1$) are determined by the expression:

$$\sigma_{rr} = \sum_{n=-\infty}^{\infty} [A_n \Delta_1(R_1) + C_n \Delta_2(R_1) + (\alpha_n + S_{np}) \overline{\Delta}_1(R_1) + Q_{np} \overline{\Delta}_2(R_1)] e^{in\theta_1},$$

$$\tag{5.4.16}$$

$$\tau_{r\theta} = \sum_{n=-\infty}^{\infty} [A_n \Delta_3(R_1) + C_n \Delta_4(R_1) + (\alpha_n + S_{np}) \overline{\Delta}_3(R_1) + Q_{np} \overline{\Delta}_4(R_1)] e^{in\theta_1},$$

where $\Delta_1(R_1)$, $\Delta_2(R_1)$, $\Delta_3(R_1)$, and $\Delta_4(R_1)$ – are determined by the formulas (5.4.6), into which $r_i = R_1$, $\overline{\Delta}_1(R_1)$, $\overline{\Delta}_2(R_1)$, $\overline{\Delta}_3(R_1)$, are substituted; and $\overline{\Delta}_4(R_1)$ – are determined by the formulas (5.4.6), into which, instead of $H_n(\alpha R_1)$, $H_n(\beta R_1)$, we substitute $I_n(\alpha R_1)$, and $I_n(\beta R_1)$. Further, let as build the equation of the dynamic equilibrium for the ring. Projecting all forces, acting on the ring in the direction $y$, we obtain:

$$\pi \rho_1 (R_1^2 - R_2^2) \frac{\partial^2 U}{\partial t^2} = \int_0^{2\pi} (\sigma_{rr} \cos\theta_1 - \tau_{r\theta} \sin\theta_1) R_1 d\theta_1, \tag{5.4.17}$$

where $R_1$, $R_2$ – the external and internal diameters of a ring, respectively; $\rho_1$ – the mass of the unit volume of the material of the ring; and $U$ – the displacement of the ring in the direction $y$. The displacement $U$ is connected with the displacemnts $U_r$ and $U_\theta$ by the formulas:

$$U_r = U \cos \theta_1; \quad U_\theta = -U \sin \theta_1 \qquad (5.4.18)$$

Using the expressions for $\phi$ and $\psi$ (5.4.15), we will obtain the values of the radial and tangential displacements $U_r$ and $U_\theta$:

$$U_r = \sum_{n=-\infty}^{\infty} [A_n H_n(\alpha r_1) + (\alpha_n + S_{np}) I_n(\alpha r_1) - C_n H_n'(\beta r_1) + Q_{np} I_n'(\beta r_1)] e^{in\theta_1};$$

$$(5.4.19)$$

$$U_\theta = \sum_{n=-\infty}^{\infty} \left[ A_n \frac{in}{r_1} H_n(\alpha r_1) + \frac{in}{r_1}(\alpha_n + S_{np}) I_n(\alpha r_1) - C_n H_n'(\beta r_1) + Q_{np} I_n'(\beta r_1)] e^{in\theta_1}.$$

Bearing in mind that all displacements and stresses change according to the harmonic law, and using (5.4.16) and (5.4.20), from (5.4.17) we find $U$:

$$U = K[(A_1 - A_{-1})a + (C_1 + C_{-1})b + (\alpha_1 + S_1 - \alpha_{-1} - S_{-1})\bar{a} + (Q_1 + Q_{-1})\bar{b}],$$

where,

$$a = -\Delta_1(R_1) + i\Delta_3(R_1), \qquad b = -\Delta_2(R_1) + i\Delta_4(R_1) \qquad (5.4.20)$$

$$\bar{a} = -\bar{\Delta}_1(R_1) + i\bar{\Delta}_3(R_1), \qquad \bar{b} = -\bar{\Delta}_2(R_1) + i\bar{\Delta}_4(R_1)$$

$$K = \frac{\eta}{R_1 \gamma \mu \beta^2}; \quad \gamma = 1 - \frac{R_2^2}{R_1^2}.$$

where $\eta = \rho/\rho_1$, $\rho$ – the mass of the unit volume of the medium. Substituting the value of $U$ into (5.4.18) we obtain an infinite system of equations, from which, after some conversions (transformations), it is possible to determine the unknown coefficients $A_n$ and $C_n$ ($n = 0, \pm 1, \pm 2, \dots$)

$$\sum_{n=-\infty}^{\infty} \left[ A_n H_n'(\alpha R_1) + C_n \frac{in}{R_1} H_n(\beta R_1) + (\alpha_n + S_{np}) I_n'(\alpha R_1) \right.$$

$$\left. + \frac{in}{R_1} Q_{np} I_n(\beta R_1) \right] e^{in\theta_1} = \frac{U}{2}(e^{i\theta_1} + e^{-i\theta_1}), \qquad (5.4.21)$$

$$\sum_{n=-\infty}^{\infty} \left[ A_n \frac{in}{R_1} H_n(\beta R_1) - C_n H_n'(\beta R_1) + \frac{in}{R_1}(\alpha_n + S_{np})I_n(\alpha R_1) \right.$$

$$\left. - Q_{np}I_n'(\beta R_1) \right] e^{in\theta_1} = \frac{U}{2} i(e^{i\theta_1} - e^{-i\theta_1}) \cdot \quad (p = 0, \pm 1, \pm 2, \ldots)$$

Substituting the value of $U$ into (5.4.21), and equating the expressions which have identical multipliers $e^{in\theta_1}$, we obtain:

$$\sum_{p=-\infty}^{\infty} A_p\{-(-1)^{n+p}H_{n+p}(2\delta\alpha)I_n'(\alpha R_1)$$

$$+ \frac{1}{2}(f_{1n} + f_{-1,n})K[(-1)^p \bar{a}H_{p+1}(2\delta\alpha) - (-1)^p \bar{a}H_{p-1}(2\delta\alpha)$$

$$+ a(f_{1p} - f_{-1,p})]\} + C_p\{(-1)^{n+p}H_{n+p}(2\delta\beta)\frac{in}{R_1}I_n(\beta R_1)$$

$$+ f_{np}\frac{in}{R_1}H_n(\beta R_1) + \frac{1}{2}(f_{1n} + f_{-1,n})K[(-1)^p \bar{b}H_{p+1}(2\delta\beta)$$

$$- (-1)^p \bar{b}H_{p-1}(2\delta\beta) + b(f_{1p} + f_{-1,p})]\} = -\alpha_n I_n'(\alpha R_1), \quad (5.4.22)$$

$$\sum_{p=-\infty}^{\infty} A_p\left\{-(-1)^{n+p}H_{n+p}(2\delta\alpha)\frac{in}{R_1}I_n(\alpha R_1)\right.$$

$$+ f_{np}\frac{in}{R_1}H_n(\alpha R_1) + \frac{i}{2}(f_{1n} - f_{-1,n})K[\bar{a}(-1)^p\bar{a}H_{p+1}(2\delta\alpha)$$

$$- \bar{a}(-1)^pH_{p-1}(2\delta\alpha) + a(f_{1p} - f_{-1,p})]\}$$

$$+ C_p\{-(-1)^{n+p}H_{n+p}(2\delta\beta)I_n'(\beta R_1)$$

$$- f_{np}H_n'(\beta R_1) + \frac{i}{2}(f_{1n} - f_{-1,n})K[-(-1)^p\bar{b}H_{p+1}(2\delta\beta)$$

$$- (-1)^p\bar{b}H_{p-1}(2\delta\beta) + b(f_{1p} + f_{-1,p})]\} = -\frac{in}{R_1}\alpha_n I_n(\alpha R_1);$$

$$n = 0, \pm 1, \pm 2, \cdots$$

where

$$f_{np} = \begin{cases} 1; & p = n \\ 0; & p \neq n. \end{cases}$$

The infinite system of Eqs. (5.4.22) is completely regular. For the general case of a multi-coherent body, the regularity of such systems is proven by the infinite equations in [130]. It is possible to determine the values of the coefficients $A_p$ and $C_p$, which are complex numbers, by solving the final system of Eqs. (5.4.22). Thus, all components of the stress-strained state can be determined. In particular, normal and shear stresses, acting on the ring are determined by the formulas (5.4.16). The boundary conditions (5.4.17) and (5.4.18) correspond to the case of an absolutely rigid ring. There is no difficulty whatsoever to solve the problem in a similar setting of transferring, and the same is true with cases of deformable flexible rings. The boundary conditions for this latter case are determined by equating of the radial and tangential displacements of the ring and medium. The simpler solution can certainly be obtained in the case of an absolutely rigid non-deformable ring. In that case, the normal stresses are greater than in the case of a deformable ring. With an increase in the flexibility of the ring, these stresses decrease and in the extreme case, of an absolutely flexible ring they approach zero.

## 5.5   Calculations of Underground Structures with Arbitrary Cross-section under Seismic Action Impact

Let us examine an underground structure with arbitrary cross-section, which is located in an infinite (limitless, boundless) space under a plane strain conditions (Figure 5.10).

The seismic action is represented in the form of a plane (flat) longitudinal or transverse wave. The calculations of tunnels with arbitrary cross-sections to seismic resistance are done using the method of the point satisfaction of boundary conditions. This method previously was applied at torsion of bars at a bent stability and oscillations of thin plates of complex configurations.[1] As seen below we propose to use this method to solve problems of diffraction of elastic waves around obstacles of complex form. During the application of this method, a set of functions is used, which accurately solves the problem of diffraction of elastic waves in a circular cavity or inclusion (dirt). The parameters of the stress-strained state are sought in the form of the following series:

$$\sigma_{rr} = \sum_{n=0}^{\infty} [A_n a_{1n} + B_n a_{2n} + \overline{A}_n \overline{a}_{1n}] \cos n\theta,$$

---

[1] Conway H.D. The bending, buckling, and flexural vibration of simply supported polygonal plates by point matching. *Trans. ASME, Ser. E*, 1961, Vol. **28**, No. 2, 288.

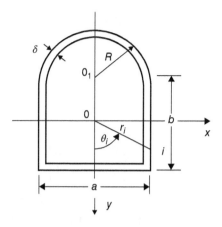

**Figure 5.10** Scheme of an underground tunnel structure.

$$\sigma_{\theta\theta} = \sum_{n=0}^{\infty} [A_n e_{1n} + B_n e_{2n} + \overline{A}_n \overline{e}_{1n}] \cos n\theta,$$

$$\tau_{r\theta} = \sum_{n=0}^{\infty} [A_n b_{1n} + B_n b_{2n} + \overline{A}_n \overline{b}_{1n}] \sin n\theta, \qquad (5.5.1)$$

$$U_r = \sum_{n=0}^{\infty} [A_n c_{1n} + B_n c_{2n} + \overline{A}_n \overline{c}_{1n}] \cos n\theta,$$

$$U_\theta = \sum_{n=0}^{\infty} [A_n d_{1n} + B_n d_{2n} + \overline{A}_n \overline{d}_{1n}] \sin n\theta,$$

where the coefficients $a_{1n}$, $a_{2n}$, $b_{1n}$, $b_{2n}$, $e_{1n}$, and $e_{2n}$ are determined from expressions (5.4.6) with the aid of the relationships (using ratios):

$$a_{1n} = \Delta_1(r); \qquad a_{2n} = \frac{\Delta_2(r)}{i};$$

$$b_{1n} = -\frac{\Delta_3(r)}{i}; \qquad b_{2n} = \Delta_4(r);$$

$$e_{1n} = \Delta_5(r); \qquad e_{2n} = \frac{\Delta_6(r)}{i}.$$

The coefficients $C_{1n}$, $C_{2n}$, $d_{1n}$, and $d_{2n}$ are determined with the aid of the expressions:

$$C_{1n} = \frac{1}{r}[-nH_n(\alpha r) + \alpha r H_{n-1}(\alpha r)]; \qquad C_{2n} = \frac{n}{r}H_n(\beta r);$$

$$d_{1n} = -\frac{n}{r}H_n(\alpha r); \qquad d_{2n} = \frac{1}{r}[nH_n(\beta r) - \beta r H_{n-1}(\beta r)];$$

$$A_n = \varepsilon_n i^n; \qquad \varepsilon_n = \begin{cases} 1; & n = 0 \\ 2; & n \geq 1 \end{cases} \qquad i = \sqrt{-1}$$

Formulas (5.5.1) determine the stresses, acting on the areas $r = const.$, $\theta = const.$ The stresses which act on mutually-perpendicular areas $n = const.$ and $t = const.$ can be expressed through the stresses, acting on the areas $r = const.$, $\theta = const.$:

$$\sigma_{nn} = \sigma_{rr}\cos^2\alpha + \sigma_{\theta\theta}\sin^2\alpha + 2\tau_{r\theta}\sin\alpha\cos\alpha;$$

$$\sigma_{tt} = \sigma_{rr}\sin^2\alpha + \sigma_{\theta\theta}\cos^2\alpha - 2\tau_{r\theta}\sin\alpha\cos\alpha; \qquad (5.5.2)$$

$$\tau_{nt} = (\sigma_{\theta\theta} - \sigma_{rr})\sin\alpha\cos\alpha + \tau_{r\theta}(\cos^2\alpha - \sin^2\alpha).$$

The corresponding relationships for the displacements take the form:

$$U_{nn} = U_r\cos\alpha - U_{\theta r}\sin\alpha; \qquad (5.5.3)$$

$$U_{tt} = U_r\sin\alpha + U_\theta\cos\alpha,$$

where $\alpha$ – the angle between the straight line $\theta = const.$, presenting the direction of a normal at the given point in Figure 5.11. The angle $\alpha$ can be determined by the formulas:

$$\tan\alpha = \frac{k_2 - k_1}{1 + k_1 k_2}; \cos\alpha = \frac{1 + k_1 k_2}{\sqrt{1 + k_1^2}\sqrt{1 + k_2^2}}; \sin\alpha = \frac{k_2 - k_1}{\sqrt{1 + k_1^2}\sqrt{1 + k_2^2}}$$

$$(5.5.4)$$

$k_1$, $k_2$ – angular coefficients of the corresponding straight lines Figure 5.11. If the contour of the underground structure is described by the function $y = f(x)$, then the angular coefficients, which correspond to the point on the curve with the coordinates $x_i$ and $y_i$, can be determined as follows:

$$k_1 = \frac{y(x_i)}{x_i}; \qquad k_2 = -\frac{1}{y'(x_i)} \qquad (5.5.5)$$

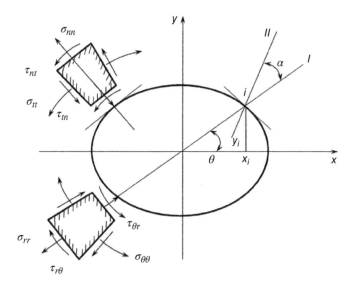

**Figure 5.11**  Scheme of a tunnel cross-section and designations of stresses in different coordinate systems.

The boundary conditions are formulated as follows. If the given problem is about an impact of a seismic wave on a non-reinforced tunnel of an arbitrary cross-section, then the satisfaction of the following conditions is necessary:

$$\sigma_{nn}(r_i, \theta_i) = 0; \tau_{nt}(r_i, \theta_i) = 0; \quad i = 1, 2, \dots K \tag{5.5.6}$$

$K$ – number of the considered contour points. The proximity of this method consists of the fact that the boundary conditions are satisfied at certain points. The accuracy of the solution can be increased by increasing the numbers of points considered.

In the open form the conditions (5.5.6), have the form:

$$\sigma_{nn}(r_i, \theta_i) = \sum_{n=0}^{K}[A_n a_{1n}(r_i) + B_n a_{2n}(r_i) + \overline{A}_n \overline{a}_{1n}(r_i)] \cos n\theta_i \cos^2 \alpha_i$$

$$+ [A_n e_{1n}(r_i) + B_n e_{2n}(r_i) + \overline{A}_n \overline{e}_{1n}(r_i)] \sin n\theta_i \sin^2 \alpha_i$$

$$+ [A_n b_{1n}(r_i) + B_n b_{2n}(r_i) + \overline{A}_n \overline{b}_{1n}(r_i)]2 \sin \alpha_i \cos \alpha_i \sin n\theta_i = 0 \tag{5.5.7}$$

$$\tau_{nt}(r_i, \theta_i) = \sum_{n=0}^{K}[A_n e_{1n}(r_i) + B_n e_{2n}(r_i) + \overline{A}_n \overline{e}_{1n}(r_i)]\cos n\theta_i \sin \alpha_i \cos \alpha_i$$

$$- [A_n a_{1n}(r_i) + B_n a_{2n}(r_i) + \overline{A}_n \overline{a}_{1n}(r_i)]\cos n\theta_i \sin \alpha_i \cos \alpha_i$$

$$+ [A_n b_{1n}(r_i) + B_n b_{2n}(r_i) + \overline{A}_n \overline{b}_{1n}(r_i)]\sin n\theta_i(\cos^2\alpha_i - \sin^2\alpha_i) = 0$$

$$i = 1, 2 \cdots K$$

In (5.5.7) the functions $a_{1n}(r_i)$, $a_{2n}(r_i)$, $e_{1n}(r_i)$, $e_{2n}(r_i)$, ... are determined by expressions (5.5.1) in which $r$ is substituted with $r_i$, representing the polar coordinates of the considered point $i$. The number of the unknowns in (5.5.7) is equal to the number of the equations $2K$. Since the assigned coefficients determined by expressions (5.5.1) are complex quantities, the unknown coefficients $A_n$ and $B_n$ are also complex. Therefore, to solve the equations we equate the real and complex expressions, and instead of two equations of the form (5.5.7) we obtain four such equations. For evaluating the accuracy of the proposed method the calculations of an unsupported tunnel with elliptical cross-section under the action (influence) of an elastic longitudinal wave, propagating in the $Ox$ and $Oy$ directions were conducted (Figure 5.11). The reason for picking such method of calculations is due to the fact that for this case numerical results were obtained by another method – the perturbation forms of boundary method [195]. The equation of the ellipse in the Cartesian coordinates can be written in the form of:

$$y = b\sqrt{1 - (x/a)^2} \tag{5.5.8}$$

where $a, b$ – semi-axis (half axis) of ellipsis. The angle $\alpha$ is determined from the expression:

$$\cos \alpha = \left(\frac{x}{a}\right)^{-2}\left\{1 + \left(\frac{b}{a}\right)^2\left[1 - \left(\frac{x}{a}\right)^2\right]\left(\frac{x}{a}\right)^{-2}\right\}^{-1/2}$$

$$\times \left\{1 + \left(\frac{x}{a}\right)^{-2}\left(\frac{b}{a}\right)^{-2}\left[1 - \left(\frac{x}{a}\right)^2\right]\right\}^{-1/2} \tag{5.5.9}$$

The polar coordinates $r_i, \theta_i$ are determined from the expression:

$$r_i = \sqrt{x_i^2 + y_i^2}; \quad \theta_i = arc\tan\frac{y_i}{x_i}. \tag{5.5.10}$$

Using "Nairi-4" software, the system of algebraic equations (5.5.7) was solved with $K = 11$. Due to the symmetry of the problem only half of the ellipse was considered. Thus, the boundary conditions were satisfied at 11 points of the semi-contour (half contour) of the ellipsis. The solutions of the system of equations were determined to be the coefficients $A_n$ and $B_n$ ($n = 0$, 1, 2 ... 10). Using these coefficients we determined the contour stresses $\sigma_{tt}(r_i,$ $\theta_i)$ to be different from zero. While calculating the values of dynamic coefficient, representing the relation of the contour stresses to the stress of the incident wave $\sigma_{tt}(r_i, \theta_i)$ at different points on the contour presents interest to us. Since stresses $\sigma_{tt}$ determined by formula (5.5.2) are complex values, we first determined the real and complex values of these stresses $\mathrm{Re}\,\sigma_{tt}$ and $\mathrm{Im}\,\sigma_{tt}$ and then their maximum values according to the formula:

$$|\sigma_{tt}| = \sqrt{(\mathrm{Re}\,\sigma_{tt})^2 + (\mathrm{Im}\,\sigma_{tt})^2} \tag{5.5.11}$$

The formula (5.5.11) represents the modules of complex numbers. The results of the calculations, represented in the form of dynamic coefficients, are given in Tables 5.5–5.8. The tabulated data in Tables 5.5–5.7 relate to the case, when the front of the incident longitudinal wave is parallel to the semi-major axis of the ellipsis. The ratio of the semi-axes of the ellipsis is taken equal to $a/b = 3/2$. The values of the parameter $\alpha_a$, characterizing seismic impact, are accepted as $\alpha_a = 0.2, 0.4$, and $0.6$. The calculations were performed for two values of the Poisson's ratio: $\nu = 0.25$ and $\nu = 0.30$. The maximum dynamic coefficients with $\nu = 0.25$, and $\alpha_a = 0.2, 0.4, 0.6$ came

Table 5.5 Values of stresses on the contour of a tunnel with elliptical cross-section when $\alpha_a = 0.2$, $a/b = 3/2$.

| $\theta$ radian | $\nu = 0.25$ | | | $\nu = 0.30$ | | |
|---|---|---|---|---|---|---|
| | $\dfrac{\mathrm{Re}\,\sigma_{tt}}{\sigma_{incid.}}$ | $\dfrac{\mathrm{Im}\,\sigma_{tt}}{\sigma_{incid.}}$ | $\dfrac{|\sigma_{tt}|}{\sigma_{incid.}}$ | $\dfrac{\mathrm{Re}\,\sigma_{tt}}{\sigma_{incid.}}$ | $\dfrac{\mathrm{Im}\,\sigma_{tt}}{\sigma_{incid.}}$ | $\dfrac{|\sigma_{tt}|}{\sigma_{incid.}}$ |
| 0 | 0.29 | 0.05 | 0.30 | 0.30 | 0.08 | 0.31 |
| 20 | 0.17 | 0.06 | 0.18 | 0.19 | 0.06 | 0.20 |
| 40 | 0.09 | 0.04 | 0.10 | 0.10 | 0.05 | 0.11 |
| 60 | 1.18 | 0.35 | 1.23 | 1.22 | 0.39 | 1.28 |
| 80 | 3.01 | 0.82 | 3.12 | 3.12 | 0.80 | 3.22 |
| 90 | 3.85 | 1.01 | 3.98 | 3.96 | 0.98 | 4.08 |
| 100 | 2.95 | 0.77 | 3.05 | 3.00 | 0.74 | 3.09 |
| 120 | 1.11 | 0.30 | 1.15 | 1.18 | 0.27 | 1.21 |
| 140 | 0.24 | 0.07 | 0.25 | 0.25 | 0.10 | 0.27 |
| 160 | 0.33 | 0.12 | 0.35 | 0.35 | 0.08 | 0.36 |
| 180 | 0.40 | 0.09 | 0.41 | 0.40 | 0.13 | 0.42 |

**Table 5.6**   Values of stresses on the contour of a tunnel with elliptical cross-section when $\alpha_a = 0.4$, $a/b = 3/2$.

| $\theta$ radian | $v = 0.25$ | | | $v = 0.30$ | | |
|---|---|---|---|---|---|---|
| | $\dfrac{\mathrm{Re}\,\sigma_{tt}}{\sigma_{incid.}}$ | $\dfrac{\mathrm{Im}\,\sigma_{tt}}{\sigma_{incid.}}$ | $\dfrac{\|\sigma_{tt}\|}{\sigma_{incid.}}$ | $\dfrac{\mathrm{Re}\,\sigma_{tt}}{\sigma_{incid.}}$ | $\dfrac{\mathrm{Im}\,\sigma_{tt}}{\sigma_{incid.}}$ | $\dfrac{\|\sigma_{tt}\|}{\sigma_{incid.}}$ |
| 0 | 0.34 | 0.06 | 0.35 | 0.36 | 0.08 | 0.37 |
| 20 | 0.24 | 0.07 | 0.025 | 0.27 | 0.11 | 0.29 |
| 40 | 0.23 | 0.10 | 0.25 | 0.30 | 0.08 | 0.31 |
| 60 | 1.27 | 0.32 | 1.31 | 1.27 | 0.50 | 1.36 |
| 80 | 3.12 | 0.84 | 3.23 | 3.20 | 0.88 | 3.32 |
| 90 | 4.02 | 1.03 | 4.15 | 4.16 | 0.77 | 4.23 |
| 100 | 3.05 | 0.83 | 3.16 | 3.11 | 0.83 | 3.22 |
| 120 | 1.20 | 0.35 | 1.25 | 1.25 | 0.36 | 1.30 |
| 140 | 0.25 | 0.07 | 0.26 | 0.26 | 0.10 | 0.27 |
| 160 | 0.36 | 0.12 | 0.38 | 0.37 | 0.12 | 0.39 |
| 180 | 0.44 | 0.09 | 0.45 | 0.45 | 0.09 | 0.46 |

**Table 5.7**   Values of stresses on the contour of a tunnel with elliptical cross-section when $\alpha_a = 0.6$, $a/b = 3/2$.

| $\theta$ radian | $v = 0.25$ | | | $v = 0.30$ | | |
|---|---|---|---|---|---|---|
| | $\dfrac{\mathrm{Re}\,\sigma_{tt}}{\sigma_{incid.}}$ | $\dfrac{\mathrm{Im}\,\sigma_{tt}}{\sigma_{incid.}}$ | $\dfrac{\|\sigma_{tt}\|}{\sigma_{incid.}}$ | $\dfrac{\mathrm{Re}\,\sigma_{tt}}{\sigma_{incid.}}$ | $\dfrac{\mathrm{Im}\,\sigma_{tt}}{\sigma_{incid.}}$ | $\dfrac{\|\sigma_{tt}\|}{\sigma_{incid.}}$ |
| 0 | 0.24 | 0.07 | 0.25 | 0.26 | 0.10 | 0.27 |
| 20 | 0.09 | 0.04 | 0.10 | 0.11 | 0.07 | 0.13 |
| 40 | 0.57 | 0.19 | 0.60 | 0.61 | 0.20 | 0.64 |
| 60 | 1.28 | 0.36 | 1.33 | 1.31 | 0.40 | 1.37 |
| 80 | 1.86 | 0.42 | 1.91 | 1.93 | 0.63 | 2.03 |
| 90 | 3.45 | 1.06 | 3.61 | 3.53 | 1.11 | 3.70 |
| 100 | 1.77 | 0.63 | 1.88 | 1.85 | 0.55 | 1.93 |
| 120 | 1.21 | 0.31 | 1.25 | 1.20 | 0.42 | 1.27 |
| 140 | 0.53 | 0.15 | 0.55 | 0.54 | 0.15 | 0.56 |
| 160 | 0.04 | 0.03 | 0.05 | 0.04 | 0.05 | 0.06 |
| 180 | 0.27 | 0.13 | 0.30 | 0.29 | 0.13 | 0.32 |

out equal: 3.93, 4.16, 3.61. With $v = 0.30$ the values of the same coefficients are as follows: 4.08, 4.23, 3.7. The maximum dynamic coefficient in all cases examined occurs at the point with the polar coordinates $r_i = a$, $\theta_i = \pi/2$ . Let as recall that under the static impact of the given load the dynamic coefficient is equal to 3.72. Hence it follows that under the dynamic action of the load the maximum stresses are 11% higher than

**Table 5.8** Values of stresses on the contour of a tunnel with elliptical cross-section when $\alpha_a = 0.2$, $a/b = 2/3$.

| $\theta$ radian | $v = 0.25$ | | | $v = 0.30$ | | |
|---|---|---|---|---|---|---|
| | $\dfrac{\text{Re } \sigma_{tt}}{\sigma_{incid.}}$ | $\dfrac{\text{Im } \sigma_{tt}}{\sigma_{incid.}}$ | $\dfrac{\lvert \sigma_{tt} \rvert}{\sigma_{incid.}}$ | $\dfrac{\text{Re } \sigma_{tt}}{\sigma_{incid.}}$ | $\dfrac{\text{Im } \sigma_{tt}}{\sigma_{incid.}}$ | $\dfrac{\lvert \sigma_{tt} \rvert}{\sigma_{incid.}}$ |
| 0 | 0.54 | 0.10 | 0.55 | 0.56 | 0.15 | 0.58 |
| 20 | 0.40 | 0.13 | 0.42 | 0.43 | 0.13 | 0.56 |
| 40 | 0.38 | 0.12 | 0.40 | 0.44 | 0.13 | 0.46 |
| 60 | 1.76 | 0.42 | 1.81 | 1.80 | 0.43 | 1.85 |
| 80 | 2.01 | 0.61 | 2.10 | 2.10 | 0.69 | 2.21 |
| 90 | 2.46 | 8.84 | 2.60 | 2.56 | 0.86 | 2.70 |
| 100 | 1.92 | 0.56 | 2.00 | 2.02 | 0.64 | 2.12 |
| 120 | 1.69 | 0.37 | 1.73 | 1.73 | 0.50 | 1.80 |
| 140 | 0.36 | 0.12 | 00.38 | 0.38 | 0.12 | 0.40 |
| 160 | 0.34 | 0.08 | 0.35 | 0.35 | 0.12 | 0.37 |
| 180 | 0.38 | 0.12 | 0.40 | 0.40 | 0.13 | 0.42 |

under the static action of the load. Let us compare the results obtained by the proposed method with the corresponding data obtained using the "perturbation forms of boundary" method [195]. The maximum dynamic coefficient in this case came out to be equal to 4.106 with $v = 0.28$. In our case, the corresponding coefficients came out equal to 4.16 with $v = 0.25$ and 4.23 with $v = 0.30$. As can be seen from the comparison of the maximum values of dynamic coefficients, the results the obtained by us are very close to the results of the work [195]. Approximately the same ratio occurs when comparing the dynamic coefficients corresponding to other contour points. Further, we considered a case of the impact of an elastic longitudinal wave on a tunnel with elliptical cross-section when the front of the incident wave is parallel to the semi-minor axis of the ellipsis. The ratio of the semi-axes of the ellipsis is taken equal to $a/b = 2/3$. Calculations were performed for the following values of the input parameters: $\alpha_a = 0.2$, $v = 0.25$, $v = 0.30$. The results of the calculations are given in Table 5.8. The maximum dynamic coefficient occurs at the point with polar coordinates $r_i = a$, $\theta_i = \pi/2$ and is equal to 2.6 with $v = 0.25$. Due to the static action of this load, the dynamic coefficient is equal to 2.33. Thus, it is possible to state that the proposed methodology of solving problems of diffraction of elastic waves gives completely satisfactory results and can be applied to seismic calculation of underground structures. The boundary conditions for when a tunnel lining is a thin-walled shell with an arbitrary curvilinear outline, is built as follows.

We determine the displacements of shell from the stresses acting on the contour. We equate the normal and tangential displacements of the shell to the corresponding displacements of the surrounding elastic medium. Building such conditions for individual points of the contour, we obtain a system of equations, the solutions of which are the unknown coefficients. In certain cases a shell can be considered as the non-deformable body, and the boundary conditions for that case will be different. The solution of a problem without accounting for deformability of a shell is considerably simpler. Total forces acting on the lining of a tunnel in the directions of $Oy$ and $Ox$ (Figure 5.11) respectively will be:

$$F_y = \int_L [\sigma_{nn} \cos(\theta_i - \alpha_i) + \tau_{nt} \sin(\theta_i - \alpha_i)]ds; \qquad (5.5.12)$$

$$F_x = \int_L [\sigma_{nn} \sin(\theta_i - \alpha_i) + \tau_{nt} \cos(\theta_i - \alpha_i)]ds.$$

where $L$ – length of the contour.

If a longitudinal wave acts on the underground structure, then at any point $i$ of the counter the following conditions must be satisfied:

$$U_{nn,i} = U \cos(\theta_i - \alpha_i); \quad U_{tt,i} = U \sin(\theta_i - \alpha_i) \qquad (5.5.13)$$

where $U_{nn,i}$, $U_{tt,i}$ – normal and tangential displacements of the contour at the point $i$ respectively; and $U$ – displacement of the entire structure in the direction of $Oy$. $U$ – is determined by the equation:

$$-m\omega^2 U = F_y \qquad (5.5.14)$$

where $m$ – it is the mass of the entire structure.

# Chapter 6

# The Special Features of the Solution of Dynamic Problems by the Boundary Element Methods (BEM)

## 6.1 One Method of Calculation: The Hilbert Transform and its Applications to the Analysis of Dynamic System

The solution of dynamic problems using the Boundary Element Method (BEM) requires building of the Green's functions in their spectral (frequency) or temporary representation (view). Accounting for real properties of materials of structures or bases, in particular the internal friction and dependences of elastic characteristics on the frequency perturbations, is an extremely important question to be analyzed. Accounting for internal friction not only provides a more adequate description of the dynamic processes in the most important and dangerous resonance modes (regimes) of oscillations, but it also considerably facilitates the solution of dynamic problems, since it eliminates the need for calculating improper integrals with the special features on the real axis during the construction of the

*Static and Dynamic Analysis of Engineering Structures: Incorporating the Boundary Element Method,*
First Edition. Levon G. Petrosian and Vladimir A. Ambartsumian.
© 2020 John Wiley & Sons Ltd. Published 2020 by John Wiley & Sons Ltd.

corresponding Green's functions. It also eliminates the need to satisfy the additional conditions of emission or their analogs, which ensure the dispersion of energy at infinity. In the theory of structures on an elastic foundation, the dynamic problems, as a rule, invoke taking into account the internal friction in the material of a base. The damping properties in structural materials are expressed substantially less than in ground, and therefore, they can be neglected. With bases of coherent type the elastic isotropic half-space is usually used. The damping of oscillations due to internal friction is rarely considered, because dissipation of energy in infinity ensures damping of free oscillations and the limitedness of amplitudes at steady processes [75, 76]. With the use of the discrete bases such as a Winkler-type elastic support, the internal friction in the base in the majority of research is accounted for with the aid of the Voight model or the frequency-independent model of E.S. Sorokin. For example, the equation of stationary (steady-state) oscillations of a two dimensional structure on an elastic foundation is written in the form of:

$$Lw - m\omega^2 w + k(\omega)w = qe^{i\omega t}$$

where $L$ – operator of the elastic forces of structure; $m$ – the mass per unit area; $w(x, y, t)$ – vertical displacement; $q(x, y)$ – the function of the amplitudes of external action; $\omega$ – circular frequency; and $k(\omega)$ – the complex coefficient of bed. In the case of the Voight's model:

$$k(\omega) = k_0 + ic,$$

where $k_0$ – real common coefficient of bed, for a frequency independent model of E.S. Sorokin:

$$k(\omega) = k_0(a + ib),$$

where $a$ and $b$ – parameters of the internal friction, introduced by E.S. Sorokin [345].

In recent years, numerous studies of oscillations of dissipative systems were devoted mostly to the construction of different models of internal friction and their use for the dynamic design of structures. In the scientific literature and applications the preference is given to the linear models, used in practical calculations and in scientific research, as they do not cause computational difficulties and allow to correctly estimate the influence of internal friction on the oscillating process. The hypothesis of Voight, elementary frequency-independent models (hypothesis E.S.

Sorokin, frequency-independent, elastic-viscous resistance) and hered-itary type models, most prevalently described by integral operators in the linear models, have gained the widest application. For construction materials, the most essential experimentally established fact influencing the selection of models of internal friction is the independence of frequency parameters from damping – coefficient of losses $\gamma$ and logarithmic decre-ment $\delta$ in common, real actions, and range of frequency of perturbances (impacts, influences). Therefore, in practical calculations the models of frequency-independent internal friction must be used. With the use of this widespread model, such as the Voight's hypothesis, which frequently occurs when solving dynamic problems of the theory of elasticity, the obtained results can be considerably distorted in comparison with the real results as a result of the frequency dependence of dissipative forces and their increasing suppression of the high-frequency components of motion.

When building linear models of frequency-independent internal fric-tion it becomes necessary to observe the fundamental principle of mechan-ics – the causality principle, requiring a reaction of a physically realizable system to a certain action occurs only after such action is applied to the sys-tem, i.e. the present output depends only on the past and present inputs, not on future inputs. The transfer function of a system completely defines its input – output relationship in a steady-state. The impulse transfer func-tion $y_\delta(t)$ of a causal system, being a Fourier transform of the transmission function $\Phi(\omega)$, is other than zero only on the positive semi-axis of time

$$y_\delta(t) = \frac{1}{2\pi} \int_{-\infty}^{\infty} \Phi(\omega)e^{i\omega t}\,d\omega = 0, \qquad (t < 0).$$

These constraints are associated with the interdependence of real and imaginary parts of the transmission function. Meanwhile in the dynamic calculations of construction and machine-building designs, a priori repre-sentations of structural behavior with harmonic loads are frequently used, assigning different forms of complex rigidity or pliability. In particular, this approach is used in the well-known hypothesis of internal friction of Bock-Schlippe and E.S. Sorokin, where the complex stiffness (rigidity) is introduced in various modes into the equations of harmonic oscillations of a system, without analyzing its conformity to the principle of causality. From the theory of dynamic systems, it follows that between the real and imaginary parts of the transmission function of a causal system,

$$\Phi(\omega) = P(\omega) + iQ(\omega) \tag{6.1.1}$$

must exist the following dependencies [38]

$$P(\omega) = -\frac{1}{\pi} \int_{-\infty}^{\infty} \frac{Q(\alpha)d\alpha}{\alpha - \omega}, \tag{6.1.2}$$

$$Q(\omega) = \frac{1}{\pi} \int_{-\infty}^{\infty} \frac{P(\alpha)d\alpha}{\alpha - \omega},$$

where the integrals are understood to be the principal values. The formulas (6.1.2) are defined by the Hilbert transformation, the calculation of which presents certain difficulties. Therefore, a direct verification of the single-valued (unambiguous) relationship between the frequency characteristics $P(\omega)$ and $Q(\omega)$ is difficult for real systems. Below we present a new way of calculating the transformation of Hilbert using Fourier transformation of the functions $P(\omega)$ and $Q(\omega)$ [273], which simplifies the task as there are extensive tables for Fourier transforms. The Hilbert transformation follows directly from the real modes of the Fourier transform and, therefore, the original $s(\tau)$ and its Hilbert transformation

$$S(t) = \frac{1}{\pi} \int_{-\infty}^{\infty} \frac{s(\tau)d\tau}{\tau - t} \tag{6.1.3}$$

are bound by the following formulas:

$$F_s(\alpha) = iF_S(\alpha)sign\alpha, \quad F_S(\alpha) = -iF_s(\alpha)sign\alpha, \tag{6.1.4}$$

where

$$F_s(\alpha) = \frac{1}{\sqrt{2\pi}} \int_{-\infty}^{\infty} s(\tau)e^{i\alpha\tau}d\tau, \tag{6.1.5}$$

$$F_S(\alpha) = \frac{1}{\sqrt{2\pi}} \int_{-\infty}^{\infty} S(\tau)e^{i\alpha\tau}d\tau,$$

are the Fourier transformations of the functions $s(\tau)$ and $S(t)$, respectively. The formula (6.1.5) makes it possible to calculate the integrals (6.1.2) in such cases when one of the transformations (6.1.5) is known. Actually, let the integral $F_S(\alpha)$ be known, then according to the formula (6.1.4) we can calculate $F_s(\alpha)$, where according to the conversion (inversion) formula

$$s(\tau) = \frac{1}{\sqrt{2\pi}} \int_{-\infty}^{\infty} F_s(\alpha)e^{-i\alpha\tau}d\alpha, \tag{6.1.6}$$

we can determine the function $s(\tau)$. In many instances calculations of the integral (6.1.6) can be simpler than calculations of the transformations (6.1.2). Let us examine several examples of calculating the Hilbert's transformations.

(a) Let $S(t) = \delta(t)$, where $\delta(t)$ – Dirac-delta function. Then

$$F_S(\alpha) = \frac{1}{\sqrt{2\pi}} \ , \quad F_s(\alpha) = i\frac{sign\alpha}{\sqrt{2\pi}}.$$

According to (6.1.6), we obtain

$$s(\tau) = \frac{i}{\sqrt{2\pi}} \int_{-\infty}^{\infty} sign\alpha\, e^{-i\alpha\tau}\, d\alpha = \frac{1}{2} \int_{0}^{\infty} \sin\alpha\tau d\alpha = \frac{1}{\pi\tau}.$$

Here we used the formula, known from the theory of the generalized functions [115]

$$\int_{0}^{\infty} e^{i\xi x}\, d\xi = ix^{-1} + \pi\delta(x),$$

where it follows:

$$\int_{0}^{\infty} \sin\xi x d\xi = \frac{1}{x}, \quad \int_{0}^{\infty} \cos\xi x d\xi = \pi\delta(x).$$

Substituting the obtained expression for $s(\tau)$ in (6.1.3), it is possible to obtain an integral representation of the Dirac-delta-function in the form of the convolution of two functions $1/\pi t$

$$\delta(t) = \frac{1}{\pi^2} \int_{-\infty}^{\infty} \frac{d\tau}{\tau(t - \tau)}.$$

It is also not difficult to notice that the formula obtained for $s(\tau)$ directly follows from the basic property of the delta-function. It follows from (6.1.3) that

$$s(\tau) = -\frac{1}{\pi} \int_{-\infty}^{\infty} \frac{S(t)dt}{t - \tau},$$

and with $S(t) = \delta(t)$ immediately we will obtain

$$s(\tau) = \frac{1}{\pi\tau}.$$

(b) If the function $S(t)$ has the mode (form) of the unit function of Heaviside: $S(t) = H(t)$, where

$$H(t) = \begin{cases} 1 & (t \geq 0) \\ 0 & (t < 0) \end{cases},$$

then

$$F_S(\alpha) = \frac{1}{\sqrt{2\pi}} \int_0^\infty e^{i\alpha t} dt = \sqrt{\frac{\pi}{2}} \delta(\alpha) + i \frac{1}{\alpha\sqrt{2\pi}}. \tag{6.1.7}$$

Hence

$$s(\tau) = \frac{1}{\sqrt{2\pi}} \int_{-\infty}^\infty \left[ \sqrt{\frac{\pi}{2}} \delta(\alpha) + i \frac{1}{\alpha\sqrt{2\pi}} \right] sign\alpha e^{-i\tau\alpha} d\alpha = -\frac{1}{\pi} \int_0^\infty \frac{\cos\tau\alpha}{\alpha} d\alpha. \tag{6.1.8}$$

The obtained improper integral diverges, and from the standpoint of classical analysis, does not exist. In order to give this integral meaning, consider the Fourier transformation of the function: $y = |x|^{-2m-1}$. According to ([115] page 227), in the designations given above

$$F_y(\alpha) = F_{|x|^{-2m-1}}(\alpha) = \frac{\alpha^{2m}}{\sqrt{2\pi}} [C_0^{(2m+1)} - C_1^{(2m+1)} \ln|\alpha|], \tag{6.1.9}$$

where

$$C_0^{(2m+1)} = \frac{2 \cdot (-1)^m}{(2m)!} \left[ 1 + \frac{1}{2} + ... + \frac{1}{2m} + \Gamma'(1) \right], \tag{6.1.10}$$

$$C_1^{(2m+1)} = \frac{2 \cdot (-1)^m}{(2m)!},$$

$\Gamma'(1)$ – derivative of the gamma function, the value of which can be determined through the $\psi$ function, which is the logarithmic derivative of the gamma function

$$\psi = \frac{\Gamma'(z)}{\Gamma(z)}.$$

Thus $\Gamma'(1) = \psi(1)\Gamma(1)$, and therefore (I.S. Gradstein, I.M. Ridjick [123]), $\Gamma'(1) = -C$, where $C = 0.5772...$ – Euler's constant. At the same time from (6.1.9) it follows:

$$F_{|x|^{-2m-1}}(\alpha) = \sqrt{\frac{2}{\pi}} \int_0^\infty \frac{\cos\alpha x dx}{x^{2m+1}}, \tag{6.1.11}$$

where when $m = 0$ we obtain the integral (6.1.8). Hence,

$$s(\tau) = -\frac{1}{2\pi^2}[1 - C - \ln|\alpha|]. \tag{6.1.12}$$

Now let us examine the case $S(t) = t^n$ $(n = 1, 2, 3, \dots)$. We have:

$$F_S(\alpha) = \sqrt{\frac{2}{\pi}} \int_0^\infty t^n \cos \alpha t \, dt. \tag{6.1.13}$$

With even $n$, and

$$F_S(\alpha) = i\sqrt{\frac{2}{\pi}} \int_0^\infty t^n \sin \alpha t \, dt, \tag{6.1.14}$$

with odd $n$.

Calculating the integrals (6.1.13) and (6.1.14), when $n$ is even, we find [92, 295]:

$$F_S(\alpha) = \sqrt{\frac{2}{\pi}} \frac{\Gamma(n+1)}{\alpha^{n+1}} \cos \frac{(n+1)\pi}{2},$$

and with odd $n$

$$F_S(\alpha) = i\sqrt{\frac{2}{\pi}} \frac{\Gamma(n+1)}{\alpha^{n+1}} \sin \frac{(n+1)\pi}{2}.$$

Consequently and respectively

$$F_s(\alpha) = i\sqrt{\frac{2}{\pi}} \, signa \frac{\Gamma(n+1)}{\alpha^{n+1}} \cos \frac{(n+1)\pi}{2},$$

$$F_s(\alpha) = -\sqrt{\frac{2}{\pi}} \, signa \frac{\Gamma(n+1)}{\alpha^{n+1}} \sin \frac{(n+1)\pi}{2}.$$

Further when $n$ is even we obtain

$$s(\tau) = \frac{2i\Gamma(n+1)\cos \frac{(n+1)\pi}{2}}{\pi} \int_0^\infty \frac{\cos \alpha \tau \, d\alpha}{\alpha^{n+1}}. \tag{6.1.15}$$

The integral in (6.1.15) can be calculated according to the formula (6.1.9). So, taking into account (6.1.10) and designating $n = 2m$, we find:

$$\int_0^\infty \frac{\cos \alpha \tau \, d\alpha}{\alpha^{2m+1}} = \frac{(-1)^m \tau^{2m}}{(2m)!} \left[ 1 + \frac{1}{2} + \dots + \frac{1}{2m} - C + \ln|\tau| \right].$$

With odd $n$ it is necessary to use the following formula ([115] page 218) with introduction of an additional multiplier through a different designation of the Fourier transform:

$$F_{x-2m}(\alpha) = \frac{(-1)^m}{\sqrt{2\pi}} |\alpha|^{2m-1} \frac{\pi}{(2m-1)!}.$$

Since,

$$F_{x-2m}(\alpha) = \sqrt{\frac{2}{\pi}} \int_0^\infty \frac{\cos \alpha x \, dx}{x^{2m}} .$$

then

$$\int_0^\infty \frac{\cos \alpha x \, dx}{x^{2m}} = \frac{(-1)^m \pi}{2(2m-1)!} |\alpha|^{2m-1}.$$

Let us further use the aforementioned method to calculate the Hilbert's transformation to analyze dynamic models of a base, the internal friction of which is described by hypothesis of Voight, E.S. Sorokin, and Bock-Schlippe [345].

Let us first examine an elasto-viscous Winkler foundation with complex rigidity (stiffness)

$$K(\omega) = k + ic\omega.$$

The real and imaginary parts of the transfer function of the base take the form:

$$P(\omega) = \text{Re}K^{-1}(\omega) = \frac{k}{k^2 + c^2\omega^2}, \tag{6.1.16}$$

$$Q(\omega) = \text{Im}K^{-1}(\omega) = -\frac{c\omega}{k^2 + c^2\omega^2}.$$

Let us verify whether the dependencies in (6.1.2) comply with the characteristics in (6.1.16). We will calculate the Fourier transform of $P(\omega)$ function:

$$F_P(\alpha) = \frac{k}{\sqrt{2\pi}} \int_{-\infty}^\infty \frac{e^{i\omega\alpha} d\omega}{k^2 + c^2 \cdot \omega^2} = \sqrt{\frac{\pi}{2}} C^{-1} e^{-|\alpha|kC^{-1}}.$$

Hence,

$$F_Q(\alpha) = -i\text{sign}\alpha \sqrt{\frac{\pi}{2}} C^{-1} e^{-|\alpha|kC^{-1}},$$

$$Q(\omega) = -\frac{i}{2c} \int_{-\infty}^\infty \text{sign}\alpha e^{-|\alpha|kC^{-1} - i\alpha\omega} d\alpha = -\frac{c\omega}{k^2 + \omega^2 c^2},$$

i.e. we come to the function $Q(\omega)$, given by the formula (6.1.16). Thus, the Voight's coefficient of the bed satisfies the conditions of the physically realizable dynamic model. The inertia-less base with the internal friction of E.S. Sorokin is described by the differential equation of harmonic oscillation

$$k(u + iv)y = qe^{i\omega t}, \qquad (6.1.17)$$

Where the correctly determined parameters of complex rigidity (stiffness) take the form [369, 374]

$$u = \frac{1}{\sqrt{1 + \gamma^2}}, \quad v = \frac{\gamma}{\sqrt{1 + \gamma^2}} \ ,$$

The parameters $u$ and $v$ do not depend on the frequency of disturbance (perturbance, excitation). Here $\gamma$ - coefficient of losses. Since $u^2 + v^2 = 1$, then for this model

$$P(\omega) = uk^{-1}, \quad Q(\omega) = -vk^{-1}. \qquad (6.1.18)$$

It should be noted that the complex rigidity (stiffness) of the corresponding hypothesis of Bock-Schlippe has a similar form according to which the equation of harmonic oscillations of the base takes the form (mode)

$$\frac{k \cdot \gamma}{\omega} y + k\dot{y} = qe^{i\omega t}$$

The complex rigidity (stiffness) in this case

$$K(\omega) = k(1 + i\gamma),$$

coincides with the complex rigidity for the hypothesis of E.S. Sorokin, which commonly uses values of parameters $u = 1$ and $v = \gamma$ in dynamic calculations. Let us verify that the dependencies in (6.1.2) comply with the characteristics in (6.1.16). We will determine Fourier the transform of $P(\omega)$ function

$$F_P(\alpha) = \frac{u}{k\sqrt{2\pi}} \int_{-\infty}^{\infty} e^{i\omega\alpha} d\omega = \frac{u\sqrt{2\pi}}{k} \delta(\alpha), \qquad (6.1.19)$$

hence

$$F_Q(\alpha) = -iuk^{-1}\sqrt{2\pi} \ sign \ \alpha\delta(\alpha),$$

and

$$Q(\omega) = -\frac{iu}{k} \int_{-\infty}^{\infty} sign \ \alpha e^{-i\omega\alpha} \delta(\alpha) d\alpha = 0.$$

So as we can see, for these models the dependencies in (6.1.2) are not executed, and consequently these models are not physically realizable. Thus, the presented method of calculating the Hilbert transformation allows us to comparatively easily analyze the aprioristic models of dynamic systems based on creation of complex rigidity or complex pliability of systems. Additionally, this method makes it possible to analytically determine one of the components of complex rigidity according to the experimentally defined other component.

## 6.2    Construction of Green's Function for Bases Having Frequency-Dependent Internal Friction

With respect to the constructions made from traditional materials like steel, concrete, reinforced concrete, wood, brick and so on, there is already sufficiently convincing work by many authors that proves the need for describing the internal friction using frequently-independent type of models. On the other hand, for ground conditions, the experiments have shown that there are more adequate models with significant dependence of parameters of complex rigidity (stiffness) on frequency.

The questions of dynamic calculations of structures and bases in non-stationary systems for materials with a sufficiently arbitrary dependency of complex rigidity on the frequency have only been touched upon in the literature.

One of the main methods of research of linear dynamic dissipative systems is construction and analysis of their characteristics in a frequency domain. For example, in the case of transfer functions, their parameters in many instances can be determined experimentally with sufficiently high accuracy, with subsequent transition into the time domain. A similar approach is used during construction of certain models of elastic-viscous media and of internal friction with the introduction of complex modules or complex rigidities, based on aprioristic mechanical representations (views). For researching such dynamic systems in the time domain it is necessary to move to their transient characteristics (features) using the Fourier transformation. This procedure is not simple in all cases. Furthermore, the procedure assumes that the transfer function possesses certain properties ensuring observance of the principle of causality. Addressing the elastic basis-foundations, specifically in the case when its material possesses elasto-viscous characteristics dependent on the frequency of the cyclic deformation, in order to the constructing Green's function it is necessary to solve quasi-static task. Please remember that construction of Green's function is the base of BEM. First we have to solve the problem of steady oscillations and then substitute the complex values instead of the modulus of elasticity $E$ or coefficient of bed $k$ accordingly

$$E(\omega) = E'(\omega) + iE''(\omega) \quad \text{or} \quad k(\omega) = k'(\omega) + ik''(\omega).$$

Let us examine for example, the construction (building) of Green's function for a plate lying on a Winklerian type elasto-viscous base with the arbitrary dependence of the parameters of the complex coefficient of bed on the frequency [372, 374, 377]. We write the equation for steady (established) oscillations of a plate in complex amplitudes, determining Green's function for a stationary problem

$$D\nabla^2\nabla^2 w(x,y) + [k(\omega) - m\omega^2]w(x,y) = \delta(x - x_1)\delta(y - y_1), \qquad (6.2.1)$$

where, as well as above,

$$k(\omega) = \int_0^\infty R(z)e^{-i\omega z}dz$$

$R(z)$ – the kernel of the relaxation of the Winklerian model of the base. We will designate

$$k'_0(\omega) = \int_0^\infty \cos \omega z R(z)\, dz,$$

$$k''(\omega) = \int_0^\infty \sin \omega z R(z)\, dz$$

and in (6.2.1) will obtain the Fourier transform. Considering that the displacement of the plate together with all derivatives disappears at infinity, which is ensured by the presence of damping, we will obtain:

$$[D(\xi^2 + \eta^2)^2 + k(\omega) - m\omega^2]\overline{W}(\xi,\eta) = (2\pi)^{-1}. \qquad (6.2.2)$$

The transfer function of a system in the transforms space is determined by the formula (let us designate it by index "e")

$$W_e(\xi,\eta,\omega) = \frac{1}{2\pi}[D(\xi^2 + \eta^2)^2 + k(\omega) - m\omega^2]^{-1}. \qquad (6.2.3)$$

The transfer function, determined by the formula (6.2.3), coincides with the transfer function of a system with one degree of freedom, which has complex rigidity $2\pi k(\omega) = 2\pi[D(\xi^2 + \eta^2)^2 + k(\omega)]$ and mass $2\pi m$. It is sufficient to take Fourier transform from the corresponding transfer function to obtain the impulse transient function in the space of transforms,

$$\overline{W}_\delta(\xi, \eta, t) = \frac{1}{2\pi} \int_0^\infty \overline{W}_e(\xi, \eta, \omega) e^{i\omega t} d\omega \qquad (6.2.4)$$

Then using the double inversion of Fourier from (6.2.4), we find the impulse transient function in the space of the originals, i.e. the reaction of the plate to the action of a unit instantaneous impulse.

It should be noted, however, that practical calculations of the Fourier transform using (6.2.4) in a closed mode is not feasible, and the use of numerical integration complicates an already complex calculations of the double Fourier integrals for obtaining the originals. Therefore, calculations of the impulse transient function (6.2.4) in a closed mode is important from the practical point of view.

Thus, knowing the parameters of the complex coefficient of bed we can obtain the impulse transient function of a plate, and then using the Duhamel's formula, solve for any dynamic load $q(x, y, t)$. However, according to the theory of dynamic systems the parameters of the complex coefficient of bed must possess the properties described above ensuring physical realization of the system. In other words it is necessary to observe the principle of causality. Functions $k_0'(\omega)$ and $k_0''(\omega)$ must be positive with $\omega > 0$, moreover $k_0'(\omega)$ must be even, and $k_0''(\omega)$ an odd function of frequency $\omega$. The coefficient of losses:

$$\gamma(\omega) = \frac{k_0''(\omega)}{k_0'(\omega)}. \qquad (6.2.5)$$

The following representation for complex rigidity can be written for systems having unlimited creep

$$k_0(\omega) = |\omega|^\alpha k_*(\omega), \quad (0 < \alpha < 1, k_*(0) \neq 0, k_*'(0) > 0) \qquad (6.2.6)$$

In order to observe (comply with) the one-to-one (single-validness) relationship between its real and imaginary parts the transfer function of a causal system should not have poles in the lower half-plane or on the real axis of a complex plane. In addition, because the actual constructions possess limited creep, they belong to the minimum-phase systems, i.e. systems realizing a minimum shift of a phase at any frequency at the same amplitude. Thus, the transfer function must have not only poles, but also zeroes on the entire lower half-plane, including the real axis. In general, for transfer functions we will assume the existence of at least one pole at the beginning of the coordinates (with $\omega = 0$), corresponding to the systems with unlimited creep under static load. For studying free oscillations of a system it is sufficient to build an impulse transient function (6.2.4).

$$\overline{W}_\delta(\xi, \eta, t) = \frac{1}{4\pi^2} \int_{-\infty}^{\infty} \frac{e^{i\omega t}}{-m\omega^2 + k(\omega)} d\omega \qquad (6.2.7)$$

This can also be represented in the form of the following Mellin's integral by means of the substituting $=i\omega$.

$$\overline{W}_\delta(\xi, \eta, t) = \frac{1}{4\pi^2 i} \int_{-i\infty}^{i\infty} \frac{e^{pt}}{mp^2 + \overline{k}_0'(p) + \overline{k}_o''(p)} dp. \qquad (6.2.8)$$

Here,

$$\overline{k}_0'(p) = k_0'(-ip), \qquad \overline{k}_o''(p) = ik_0''(-ip) \qquad (6.2.9)$$

According to experimental data, for all construction materials $k_0'(\omega)$–smooth function with the small changes in a range of frequencies, which represents an interest in the dynamic calculations, and $k''(\omega) \ll k'(\omega)$. Therefore, the denominator of the integrant expression (6.2.8) has a degree of freedom not more than the second, and consequently, not more than two roots different from zero. Furthermore, as was indicated above, it is possible to have a root of $p = 0$. For the analysis of the poles of the integrant in (6.2.8) let us write:

$$p = \pm i\overline{\lambda}(p)\sqrt{1 + \overline{\gamma}(p)}, \qquad (6.2.10)$$

where

$$\lambda(p) = \sqrt{\frac{\overline{k}_0'(p)}{m}}, \qquad \overline{\gamma}(p) = \frac{\overline{k}_o''(p)}{\overline{k}_0'(p)} = i\gamma(\omega). \qquad (6.2.11)$$

Taking into account that $\overline{\gamma}(p) \ll 1$, it is possible to assume

$$\sqrt{1 + \overline{\gamma}(p)} = 1 + \frac{\overline{\gamma}(p)}{2}.$$

Then

$$p = \pm i\overline{\lambda}(p)\left[1 + \frac{\overline{\gamma}(p)}{2}\right], \qquad (6.2.12)$$

The assumptions adopted above, as applied to the weak damping and the presence of two zero roots in a frequency equation, make it possible to obtain the approximate dependencies from the expression (6.2.12)

$$p_1 = i\bar{\lambda}(p_0^+)\left[1 + \frac{\overline{\gamma}(p_0^+)}{2}\right], \tag{6.2.13}$$

$$p_2 = -i\bar{\lambda}(p_0^-)\left[1 + \frac{\overline{\gamma}(p_0^-)}{2}\right],$$

where $i|p_0| = |\omega_0|$ – the natural frequency of conservative system with the rigidity (stiffness) $k_0'(\omega)$; and $\omega_0^+$, $\omega_0^-$ – the solution of equation

$$-m\omega_0^2 + k_0'(\omega_0) = 0 \tag{6.2.14}$$

or

$$mp^2 + \overline{k}_0'(p_0) = 0 \tag{6.2.15}$$

The roots of $p_0^+$ and $p_0^-$ are purely imaginary, and also satisfy the equations

$$p_0^+ = i\bar{\lambda}(p_0^+), \quad p_0^- = -i\bar{\lambda}(p_0^-). \tag{6.2.16}$$

Taking into account that $\bar{\lambda}(p)$ is an even function and $\overline{\gamma}(p)$ is an odd function, we obtain:

$$p_1 = i\bar{\lambda}(p_0^+)\left[1 + \frac{\overline{\gamma}(p_0^+)}{2}\right], \tag{6.2.17}$$

$$p_2 = -i\bar{\lambda}(p_0^-)\left[1 - \frac{\overline{\gamma}(p_0^+)}{2}\right].$$

Turning to $|\omega_0|$, we will obtain two complex-conjugate roots

$$p_1 = -\frac{|\omega_0|\gamma(|\omega_0|)}{2} + i|\omega_0|, \tag{6.2.18}$$

$$p_2 = -\frac{|\omega_0|\gamma(|\omega_0|)}{2} - i|\omega_0|.$$

Thus, the integrand expression in (6.2.8) is a multi-valued function which has the branch point of $p = 0$, and two poles that lie on the left half-plane. The integral (6.2.8) can be examined on the first sheet of

Riemann (Riemannian) surface with the section (a cut) along the negative semi-axis $\text{Re} p < 0$ as

$$\int_{-i\infty}^{i\infty} \overline{\Phi}(p) e^{pt} dp = \lim_{\rho \to 0} \left( \int_{-i\infty}^{-i\rho} + \int_{C_\rho} + \int_{i\rho}^{i\infty} \right) \Phi(p) e^{pt} dp. \qquad (6.2.19)$$

Here $\overline{\Phi}(p) = [mp^2 + \overline{k}_0'(p) + \overline{k}_0''(p)]^{-1}$; $C_\rho$ – the semicircle of the radius $\rho$, which envelopes to the right of the branching point $p = 0$. Let us select the contour (loop), as shown in Figure 6.1.

We will consider the integral in the circle $C_\rho$. The existent assessment:

$$\left| \int_{C_\rho} \frac{e^{pt} dp}{mp^2 + \overline{k}_0'(p) + \overline{k}_0''(p)} \right| = \left| \int_{C_\rho} \frac{p^{-\alpha} e^{pt} dp}{S(p)} \right| \leq 2\pi \rho^{1-\alpha} max \left| \frac{e^{pt}}{S(\rho)} \right|, \qquad (6.2.20)$$

Where,

$$S(\rho) = m|\rho^{2-\alpha}| + \overline{k}_*(\rho). \qquad (6.2.21)$$

The right part of the last inequality with $\rho \to 0$ approaches zero, therefore the left part also approaches zero. On the other hand, when $R \to \infty$ the integrals along the arcs $C_R$ approach zero due to Jordan's lemma (theorem).

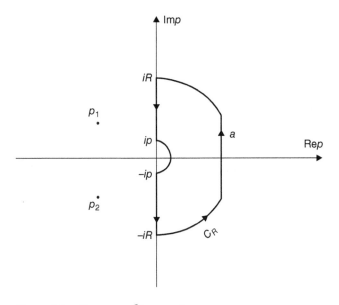

**Figure 6.1**   Contour of integration.

Since inside the contour, represented in Figure 6.1, the integrand function does not have singular points, the following is true:

$$\overline{W}_\delta(\xi,\eta,t) = \frac{1}{4\pi^2 i} \int_{a-i\infty}^{a+i\infty} \frac{e^{pt}}{mp^2 + \overline{k}_0'(p) + i\overline{k}_o''(p)} dp. \quad (a > 0) \quad (6.2.22)$$

For calculations of the integral (6.2.22), following the works [288, 413] let us compose another contour (Figure 6.2). We will present the integral (6.2.22) in the form:

$$\lim_{\substack{R \to \infty \\ \rho \to 0}} \frac{1}{4\pi^2 i} \int_{a-iR}^{a+iR} \overline{\Phi}(p)e^{pt}dp = \sum_{j=1}^{2} res[e^{p_j t}\overline{\Phi}(p_j)] \quad (6.2.23)$$

$$+ \lim_{\substack{R \to \infty \\ \rho \to 0}} \frac{1}{4\pi^2 i}\left[\left(\int_{l_2} + \int_{l_3} + \int_{C_\rho} + \int_{C_R}\right)\overline{\Phi}(p)e^{pt}dp\right].$$

As has already been indicated above, integrals in small and large circles tend to approach zero with $\rho \to 0$ and $R \to \infty$ respectively. Hence it follows

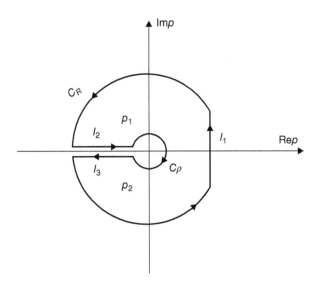

**Figure 6.2**  Contour of integration.

that

$$\overline{W}_e(\xi, \eta, t) = \frac{1}{4\pi^2 i} \int_0^\infty [\overline{\Phi}(re^{i\pi}) - \overline{\Phi}(re^{-i\pi})]e^{-rt} dr \qquad (6.2.24)$$

$$+ \sum_{j=1}^2 res[e^{p_j t} \overline{\Phi}(p_j)].$$

Taking into account the conjugality of real functions from conjugated arguments, we finally write

$$\overline{W}_\delta(\xi, \eta, t) = \Omega^{-1} exp[-0, 5|\omega_0| \gamma (|\omega_0|)t] \cos(|\omega_0|t - \varepsilon) + A(t), \qquad (6.2.25)$$

Where,

$$\Omega = \frac{\pi}{2} \left\{ \left[ -m|\omega_0| \gamma + \text{Re}\frac{\overline{dk}_0'(p_1)}{dp} + \text{Re}\frac{\overline{dk}_0''(p_1)}{dp} \right]^2 \right.$$

$$+ \left. \left[ 2m|\omega_0| + \text{Im}\frac{\overline{dk}_0'(p_1)}{dp} + \text{Im}\frac{\overline{dk}_0''(p_1)}{dp} \right]^2 \right\}^{\frac{1}{2}}, \qquad (6.2.26)$$

$$\varepsilon = \tan^{-1} \left[ \frac{2m|\omega_0| + \text{Im}\frac{\overline{dk}_0'(p_1)}{dp} + \text{Im}\frac{\overline{dk}_0''(p_1)}{dp}}{-m|\omega_0| \gamma + \text{Re}\frac{\overline{dk}_0'(p_1)}{dp} + \text{Re}\frac{\overline{dk}_0''(p_1)}{dp}} \right] \qquad (6.2.27)$$

Let us examine in more detail the function $A(t)$. Since

$$\overline{k}_0(re^{-i\varphi}) = \overline{k}_0^*(re^{i\varphi}), \qquad (6.2.28)$$

where * – indicates a complex conjugate value, and assuming

$$\overline{k}_0(re^{i\pi}) = r^\alpha[a(r) + ib(r)],$$

where $a(r)$ – nondecreasing function, and $a(0) \neq 0$, we obtain

$$A(t) = \frac{1}{2\pi^2} \int_0^\infty \frac{r^{-\alpha} b(r) e^{-rt} dr}{[mr^{2-\alpha} + a(r)]^2 + b^2(r)}. \qquad (6.2.29)$$

According to (6.2.29) $A(t)$ is a monotonically decreasing function of time, with the initial value

$$A(0) = \frac{1}{2\pi^2} \int_0^\infty \frac{r^{-\alpha}b(r)dr}{[mr^{2-\alpha} + a(r)]^2 + b^2(r)},$$

for which the following approximation can be obtained;

$$A(0) \approx \frac{b_{max}}{2\pi^2(2-\alpha)a^2(0)} \left[\frac{a(0)}{m}\right]^z \frac{\Gamma(z)\Gamma(2-z)}{\Gamma(2)}, \quad z = \frac{1-\alpha}{2-\alpha}.$$

Taking into account the properties of the gamma-functions, we find that

$$A(0) \approx \frac{b_{max}}{2\pi(2-\alpha)^2 a^2(0)} \left[\frac{a(0)}{m}\right]^z \sec\frac{\pi\alpha}{4(2-\alpha)},$$

or with $\alpha \ll 1$,

$$A(0) \approx \frac{b_{max}}{8\pi a^{3/2}(0)m^{1/2}}.$$

Thus, the system performs free oscillations with the frequency $|\omega_0|$ and logarithmic decrement

$$\delta(|\omega_0|) = \pi\gamma(|\omega_0|), \tag{6.2.30}$$

relative to the position of the quasi-static equilibrium, described by the function $A(t)$, and the corresponding small deformations of aftereffect. Thus, a closed solution is obtained for the transform of impulse transient function $\overline{W}_\delta(\xi, \eta, t)$. The impulse transient response function of the plate will be determined by the formula

$$w_\delta(x, y, t) = \frac{1}{2\pi} \iint_{-\infty}^{\infty} \overline{W}_\delta(\xi, \eta, t) e^{-i(\xi x + \eta y)} d\xi\, d\eta.$$

Further it is possible to obtain a solution of a problem for an arbitrary load $q(x, y, t)$ by convolution

In addition, we studied the behavior of a dynamic system made of materials, characteristics of which have an arbitrary dependency on the frequency of the cycles of deformation.

## 6.3   The Green's Functions of Systems with the Frequency-Independent Internal Friction

Let us examine the Green's functions for one-dimensional dynamic systems with damping, described by different models of frequency-independent and frequency-dependent internal friction. The Green functions in the case of harmonic steady oscillations of a dynamic problem based on the elasto-viscous analogy can be expressed through a Green's function of the corresponding static problem with the replacement of the elastic characteristics with complex modules. For nonstationary problems the construction of Green's function can be realized (carried out) on the basis of applying integral transformations in the spatial coordinate and reducing the equations of motion to ordinary differential equations with respect to, similar to the equation of motion of dissipative system with one degree of freedom. To solve a system with an arbitrary linear damping, primary attention will be given to the frequency-independent internal frictions as the most relevant to the real properties of traditional building structures. As it relates to the present book describing problems for calculating structures on an elastic foundation, these models are of interest to us for describing the dissipative properties of structures themselves, while the description of the dissipative properties in general, as was noted above, is given more adequately with frequency-dependent models. In dynamics and seismic resistance of structures, a well-known method is used to describe the internal friction. This method is based on the introduction into the differential equations of motion, divided into the generalized coordinates of dissipative terms (members) with such coefficients which ensure the frequency independence of the decrement of oscillations for each mode of oscillations. In the works [271, 278, 362, 369, 381, 382], the linear models of internal friction are built, one of which described by the integral operator with Abel's kernel, ensures strict frequency independence of decrements of oscillations and coefficients of losses. The other is based on the introduction of fractional operators of elastic and inertial forces, and leads exactly to the same results as the above-mentioned artificial technique. The last model, named the model of the elasto-viscous frequency-independent resistance, possesses within it a certain conditionality (the coefficients with the dissipative terms depend on the inertia and elastic properties of the system), sufficient simplicity and convenience for solving practical problems, and it allows to take into account the frequency-independent energy losses strictly at the level of initial equations of motion. First let us examine a stationary one-dimensional wave problem based on the example of the longitudinal oscillations of an isotropic uniform unlimited bar, in the central section of which acts a single harmonic load. Alternatively we will consider a problem based on the example of the lateral oscillations of a string. The equation of oscillations we write in the form

$$-(ic)^n \int_{-\infty}^{t} R(t-\tau)\frac{\partial^n u}{\partial x^n}(x,\tau)d\tau - c^2\frac{\partial^2 u}{\partial x^2}(x,t) + \frac{\partial^2 u}{\partial t^2}(x,t)$$

$$= \rho^{-1}\delta(x)e^{i\omega t}. \tag{6.3.1}$$

With boundary conditions:

$$\lim_{|x|\to\infty} u(x,t) = 0. \tag{6.3.2}$$

Here $c$ – wave propagation velocity, equal at the lateral oscillations of the string to $c = \sqrt{Sm^{-1}}$ and at the longitudinal oscillations of a bar $c = \sqrt{E_0/\rho}$, where $S$, $m$ – tension of the string and its linear mass; $E_0$, $\rho$ – modulus of the elasticity of the bar and its density; and $R(t)$ – relaxation kernel. Equation (6.3.1) also describes the propagation of plane waves in the isotropic elastic medium; in this case $c = \sqrt{\mu\rho^{-1}}$ for the transverse waves, and $c = \sqrt{(\lambda + 2\mu)\rho^{-1}}$ for the waves of expansion, where $\lambda, \mu$ – Lamé parameters.

The following basic models of a medium are examined: (a) general (common) linear model; (b) the elastic-viscous model of Voight; and (c) the model of the frequency-independent elasto-viscous resistance [371, 377]. For these models respectively we have:

1. $n = 2$;

2. $n = 2$;  $R(t-\tau) = \varepsilon\delta'(t-\tau)E_0^{-1}$;

3. $n = 1$;  $R(t-\tau) = \gamma\delta'(t-\tau)$,

where $\gamma$ – coefficient of losses; and $\varepsilon$ – damping coefficient (factor) according to Voight. Let us proceed in the left and right sides of Eq. (6.3.1) to the Fourier transforms for the coordinate $x$, and we will taking into account (6.3.2)

$$-(ic)^n(-i\xi)^n \int_{-\infty}^{t} R(t-\tau)U(\xi,\tau)d\tau + c^2\xi^2 U(\xi,t) + \frac{\partial^2 U}{\partial t^2}(\xi,t)$$

$$= \frac{e^{i\omega t}}{\rho\sqrt{2\pi}}. \tag{6.3.3}$$

Here $U(\xi,\tau)$ – the transform of displacement

$$U(\xi,\tau) = \frac{1}{\sqrt{2\pi}} \int_{-\infty}^{\infty} u(x,t)e^{-i\xi x}dx$$

We will search for the solution of Eq. (6.3.3) in the form of [297]

$$U(\xi, \tau) = A_0(\xi)e^{i[\omega t + \varphi(\xi)]}. \tag{6.3.4}$$

Substituting (6.3.4) into (6.3.3) and replacing the variable $t - \tau = z$ in the integrant, we obtain

$$A_0(\xi)[-\omega^2 + c^2\xi^2 + (-ic)^n(-i\xi)^n(R_c - iR_s)] = \frac{e^{-i\varphi(\xi)}}{\rho\sqrt{2\pi}},$$

where

$$R_c(\omega) = \int_0^\infty R(z)\cos\omega z dz,$$

$$R_s(\omega) = \int_0^\infty R(z)\sin\omega z dz.$$

From here

$$\begin{cases} A_0(\xi) = \dfrac{1}{\sqrt{2\pi}\,\rho c^2\xi^2}\left[\left(1 + R_c - \dfrac{\omega^2}{c^2\xi^2}\right)^2 + R_s^2\right]^{-1/2} \\[2ex] \varphi(\xi) = \tan^{-1}\dfrac{R_s}{\left(1+R_c-\frac{\omega^2}{c^2\xi^2}\right)} \end{cases} \quad , \quad (n = 2) \tag{6.3.5}$$

and

$$\begin{cases} A_0(\xi) = \dfrac{1}{\sqrt{2\pi}\,\rho}[(c\xi R_c + \omega^2 - c^2\xi^2)^2 + c^2\xi^2 R_s^2]^{-1/2} \\[2ex] \varphi(\xi) = \tan^{-1}\dfrac{c\xi R_s}{(c\xi R_c + \omega^2 - c^2\xi^2)} \end{cases} \quad , \quad (n = 1). \tag{6.3.6}$$

For the models mentioned above we have:

$$(b) \qquad R_c = 0, \ R_s = -\frac{\varepsilon\omega}{E_0};$$

$$(c) \qquad R_c = 0, \ R_s = -\gamma\omega.$$

Therefore,

$$A_0(\xi) = \frac{1}{\sqrt{2\pi}\,\rho c^2\xi^2}\left[\left(1 - \frac{\omega^2}{c^2\xi^2}\right)^2 + \frac{\varepsilon^2\omega^2}{E_0^2}\right]^{-1/2},$$

$$\varphi(\xi) = \tan^{-1} \frac{\varepsilon\omega}{E_0 \left( \frac{\omega^2}{c^2\xi^2} - 1 \right)} \, ,$$

and

$$A_0(\xi) = \frac{1}{\sqrt{2\pi}\,\rho} [(\omega^2 - c^2\xi^2)^2 + \gamma^2 c^2 \xi^2 \omega^2]^{-1/2},$$

$$\varphi(\xi) = \tan^{-1} \frac{\gamma\omega c\xi}{(c^2\xi^2 - \omega^2)}.$$

Thus

$$u(x, t) = \frac{1}{\sqrt{2\pi}} \int_{-\infty}^{\infty} A_0(\xi) e^{i[\omega t + \varphi(\xi) - \xi x]} d\xi \qquad (6.3.7)$$

The amplitude of the displacement

$$\overline{U}(x, \omega) = \sqrt{\frac{2}{\pi}} \; (B_c^2 + B_s^2),$$

$$B_c = \int_0^{\infty} A_0(\xi) \cos \xi x \cos \varphi(\xi) d\xi,$$

$$B_s = \int_0^{\infty} A_0(\xi) \sin \xi x \sin \varphi(\xi) d\xi.$$

When solving nonstationary boundary problems it is necessary to determine the nonstationary Green's function solution of the problem of free oscillations of a system under the impact of an instantaneous impulse. In this case the equation of motion (6.3.3) transformed by Fourier takes the form:

$$(-i\xi)^n \int_{-\infty}^{t} R(t - \tau) U(\xi, \tau) d\tau + c^2 \xi^2 U(\xi, t) + \frac{\partial^2 U}{\partial t^2}(\xi, t) = \frac{\delta(t)}{\rho\sqrt{2\pi}}. \qquad (6.3.8)$$

To solve (6.3.8) we can use the standard methods of the theory of viscoelasticity [70], in particular the Laplace transform. However, we can use the results for the arbitrary linear model, obtained in this paragraph, taking into account that the solutions of Eqs. (6.3.8) and (6.3.3) are connected by Fourier transformations the same way as the transfer function is connected with the impulse transient function [(formula (6.2.7)]. Let us write the transfer function of the system, described by Eq. (6.3.3) with the fixed value of the parameter $\xi$

$$\Phi(\omega, \xi) = \frac{1}{\rho\sqrt{2\pi}\,[c^2\xi^2 - \omega^2 + (-c\xi)^n(R_c - iR_s)]}. \qquad (6.3.9)$$

By analogy with (6.2.7) (with an accuracy to a multiplier) we can accept that:

$$m = \rho$$

$$k'(\omega) = \rho c^2\xi^2 + (-1)^n\rho c^n\xi^n R_c^n(\omega), \quad k''(\omega) = (-1)^{n+1}\rho c^n\xi^n R_s(\omega).$$

Now the solution of Eq. (6.3.8), considered as an impulse transient function of the same system, can be written as (6.2.25). The general solution of the problem is obtained by the inverse Fourier transform:

$$u(x, t) = \frac{1}{\sqrt{2\pi}} \int_{-\infty}^{\infty} e^{i\xi x} \left\{ \Omega^{-1}(\xi)\, exp\left[ -\frac{|\omega_0|\gamma|\omega_0|t}{2} \right] \right.$$

$$\times \cos[|\omega_0|t - \varepsilon(\xi)] + A(\xi, t)\}d\xi, \qquad (6.3.10)$$

where $\Omega(\xi)$, $\varepsilon(\xi)$, and $A(\xi, t)$ are determined by formulas (6.2.26), (6.2.27), and (6.2.29) respectively, with the fixed value of the parameter $\xi$.

For basic models (Voight's model and the elasto-viscous frequency-independent resistance) the construction of the nonstationary function of Green can be easily realized from the examination of the equations of motion in the differential form. In particular, the differential equation of longitudinal oscillation of a bar, taking into account the frequency-independent elasto-viscous resistance, can be written in the form:

$$E_0 Lu(x, t) + \gamma\sqrt{E_0\rho}\, L^{1/2}\, \dot{u}(x, t) + \rho\ddot{u}(x, t)$$

$$= \delta(x)\delta(t), \qquad (|x| < \infty, t > 0). \qquad (6.3.11)$$

where $L = -\frac{\partial^2}{\partial x^2}$ is the operator of elastic forces. We will search for the solution of Eq. (6.3.11), which approaches zero at infinity, together with the first derivative of the spatial coordinate. Let us accept the initial conditions as zero. To interpret the operator $L^{1/2}$ we will consider that the domain of definition of operators is the entire real (material) axis. Therefore,

$$L^{1/2} = \pm i\frac{\partial}{\partial x^2},$$

so, that Eq. (6.3.11) takes the form

$$-E_0 u'' \pm i\gamma\sqrt{E_0\rho}\ \dot{u}' + \rho\ddot{u} = \delta(x)\delta(t), \tag{6.3.12}$$

where the prime indicates the differentiation with respect to $x$, and the sign before the dissipative term is selected as the opposite to the sign of the spectral parameter. To solve Eq. (6.3.12) we will use Fourier transformation. Moving (6.3.12) to Fourier's transforms, we obtain:

$$E_0 \xi^2 U + \gamma|\xi|\sqrt{E_0\rho}\ \dot{U} + \rho\ddot{U} = \frac{\delta(t)}{\sqrt{2\pi}}, \tag{6.3.13}$$

where

$$U(\xi, t) = \frac{1}{\sqrt{2\pi}} \int_{-\infty}^{\infty} u(x, t) e^{i\xi x} dx. \tag{6.3.14}$$

The solution of Eq. (6.3.14) at zero initial conditions takes the form:

$$U(\xi, t) = \frac{\sin p_1 \xi t}{\sqrt{2\pi}\ \rho p_1 \xi} \exp\left(-\frac{\gamma c|\xi|t}{2}\right), \tag{6.3.15}$$

where

$$p_1 = c\sqrt{1 - \gamma^2/4}. \tag{6.3.16}$$

Converting (6.3.15) according to Fourier, we will find

$$u(x, t) = \frac{1}{\pi\rho p_1} \int_0^{\infty} \frac{\cos\xi x \sin p_1 \xi t \exp\left(\frac{\gamma c|\xi|t}{2}\right)}{\xi} d\xi. \tag{6.3.17}$$

The integral (6.3.17) is calculated [123]

$$u(x, t) = \frac{1}{2\pi\rho p_1}\left\{\tan^{-1}\frac{\gamma c p_1 t^2}{x^2 - p_1^2 t^2 + \frac{\gamma^2 c^2 t^2}{4}} + \begin{array}{l}\pi\left(x^2 - p_1^2 t^2 + \frac{\gamma^2 c^2 t^2}{4} < 0\right) \\ 0\left(x^2 - p_1^2 t^2 + \frac{\gamma^2 c^2 t^2}{4} \geq 0\right)\end{array}\right\}. \tag{6.3.18}$$

By assuming that $v_1 = \sqrt{1 - \frac{\gamma^2}{2}}, v_2 = \sqrt{\gamma}\sqrt{1 - \frac{\gamma^2}{2}}$, the solution to (6.3.18) leads to the form

$$u(x, t) = \frac{1}{2\pi\rho p_1}\left\{\tan^{-1}\frac{c^2 t^2 v_2^2}{x^2 - c^2 t^2 v_1^2} + \pi[1 - H(|x| - ctv_1)]\right\}, \tag{6.3.19}$$

where $H(|x| - ctv_1)$ – the Heaviside unit step function. The first derivative of the displacement, which determines the stress with the longitudinal oscillations of a bar or the angle of rotation of the section of a string will be:

$$\frac{\partial u}{\partial x}(x,t) = \frac{1}{2\pi\rho p_1} \left[ \frac{(x^2 - c^2 t^2 v_1^2)^2}{(x^2 - c^2 t^2 v_1^2) + c^4 t^4 v_2^2} - \pi\delta(|x| - ctv_1) \right]. \qquad (6.3.20)$$

In the conclusion of this Section let us examine problems of construction of Green's function for a nonstationary problem with transverse oscillations of a bar and plate, taking into account the frequency-independent elasto-viscous resistance. The differential equation of oscillations of a bar can be written in the following form:

$$EIy^{IV} - \gamma\sqrt{EIm}\,\dot{y}^{II} + m\ddot{y} = \delta(x)\delta(t), \qquad (6.3.21)$$

$$(|x| < \infty, \quad t > 0).$$

Assuming that with $|x| \to \infty$, the displacement of the bar $y(x, t)$ and its derivatives over the coordinate approach zero, and by using the Fourier transform on $x$ in (6.3.21) we obtain

$$\ddot{Y}(\xi,t) + \gamma p \xi^2 \dot{Y}(\xi,t) + p^2 \xi^4 Y(\xi,t) = \frac{\delta(t)}{m\sqrt{2\pi}}, \qquad (6.3.22)$$

where

$$Y(\xi,t) = \frac{1}{\sqrt{2\pi}} \int_{-\infty}^{\infty} y(x,t) e^{i\xi x} dx,$$

$$p = \sqrt{EIm^{-1}}.$$

The solution of Eq. (6.3.22) with zero initial conditions takes the form (with lower-order accuracy [second-order])

$$Y(\xi,t) = \frac{1}{\sqrt{2\pi}} \cdot \frac{\sin p\xi^2 t}{mp\xi^2} \exp\left(-\frac{\gamma p \xi^2 t}{2}\right). \qquad (6.3.23)$$

Converting (6.3.23), we find the displacement under the impact of the instant impulse

$$y(x,t) = \frac{1}{\pi mp} \int_0^\infty \frac{\cos\xi x \sin p\xi^2 t \exp\left(-\frac{\gamma p t \xi^2}{2}\right)}{\xi^2} d\xi. \qquad (6.3.24)$$

As a result of (6.3.24) for Green's function we have an expression

$$G(x - x_1, t - t_1) = \frac{1}{mp\pi} \int_0^\infty \frac{\cos \xi(x - x_1) \sin p\xi^2(t - t_1) \exp\left[-\frac{\gamma p\xi(t-t_1)}{2}\right]}{\xi^2} d\xi.$$

In the initial section of $x = 0$ the displacement will be

$$y(0, t) = \frac{1}{\pi mp} \int_0^\infty \frac{\sin p\xi^2 t \exp\left(-\frac{\gamma pt\xi^2}{2}\right)}{\xi^2} d\xi.$$

Let us switch over to the new variable of integration $z = p\xi^2 t$, we obtain

$$y(0, t) = \frac{\sqrt{t}}{2\pi m\sqrt{p}} \int_0^\infty \frac{e^{-\frac{\gamma z}{2}} \sin z}{z\sqrt{z}} dz.$$

The last integral is calculated [123]

$$y(0, t) = \frac{\left(1 + \frac{\gamma^2}{4}\right)\sqrt{t}}{m\sqrt{\pi p}} \sin\left(\frac{1}{2}\tan^{-1}\frac{2}{\gamma}\right).$$

For the real constructions $\gamma \ll 0.1$, therefore by disregarding $\gamma^2$ in comparison with the unit ("1") and by performing a simple conversion, we will obtain

$$y(0, t) = \frac{\sqrt{(1 - \gamma/2)t}}{m\sqrt{2mp}}.$$

As we can see, the influence of internal friction in the case of free oscillations of the unlimited beam somewhat decreases the displacements, although their total increase, related to the absence of supporting links occurs just as well without taking into account damping and is proportional to $\sqrt{t}$. The Green's function of a nonstationary problem with lateral oscillations of an elastic plate with frequency-independent elasto-viscous resistance is determined from the equation

$$D\nabla^2\nabla^2 w(x, y, t) + \gamma\sqrt{Dm}\, \nabla^2\dot{w}(x, y, t) + m\ddot{w}(x, y, t) = \delta(x)\delta(y)\delta(t), \quad (6.3.25)$$

$$(|x|, \ |y| < \infty, \quad t > 0).$$

Assuming as we did above, that at infinity the displacement of the plate and its derivatives disappear, and performing a dual Fourier transformation in (6.3.25) we obtain an equation similar to (6.3.22)

$$\ddot{W}(\xi,\eta,t) + \gamma p(\xi^2 + \eta^2)\dot{W}(\xi,\eta,t) + p^2(\xi^2 + \eta^2)^2 W(\xi,\eta,t) = \frac{\delta(t)}{2\pi m}, \quad (6.3.26)$$

where

$$W(\xi,\eta,t) = \frac{1}{2\pi} \iint_{-\infty}^{\infty} w(x,y,t)e^{i(\xi x + \eta t)}, \quad p = \sqrt{Dm^{-1}}.$$

With the accuracy to the small of the second order

$$W(\xi,\eta,t) = \frac{1}{2\pi} \cdot \frac{\sin p(\xi^2 + \eta^2)t}{mp(\xi^2 + \eta^2)} \exp\left(-\frac{\gamma p(\xi^2 + \eta^2)t}{2}\right). \quad (6.3.27)$$

Converting (6.3.27) we obtain

$$w(x,y,t) = \frac{1}{mp\pi^2} \iint_0^{\infty} \cos\xi x \cos\eta y \frac{\sin(\xi^2 + \eta^2)t}{(\xi^2 + \eta^2)} e^{\left(-\frac{\gamma p(\xi^2+\eta^2)t}{2}\right)} d\xi d\eta. \quad (6.3.28)$$

Green's function, which is the function of difference arguments $x - x_1$, $y - y_1$, $t - t_1$, can be represented in the form

$$G(x - x_1, y - y_1, t - t_1)$$

$$= \frac{1}{mp\pi^2} \iint_0^{\infty} \cos\xi(x - x_1)\cos\eta(y - y_1)\frac{\sin(\xi^2 + \eta^2)t}{(\xi^2 + \eta^2)}$$

$$\times \exp\left(-\frac{\gamma p(\xi^2 + \eta^2)t}{2}\right) d\xi d\eta \quad (6.3.29)$$

Further calculation of the Green's function can be performed by the decomposition of cosines in exponential series.

Thus, in the case of the dynamic calculation of structures on the elastic foundation, the results of the previous paragraph make it possible to take into account the dependency of the parameters of stiffness (rigidity) and damping on frequency both for the structures and for the base. These dependencies are reflected in the construction of the corresponding functions of Green, separately for the structures and bases, or for the general Green function of the entire system: "structure – base/foundation."

Summarizing the special features of the solutions of dynamic problems by the boundary equations method, we come to the following conclusions:

A new method of the analysis of physical realization of dynamic systems, based on calculations of the conversions (transformations, transforms) of Hilbert of real and imaginary frequency characteristics through their Fourier transformations is proposed. The method is illustrated based on the example of the elasto-viscous base, which possesses the internal friction, described by the models of Voigt and E.S. Sorokin.

The solution to a dynamic problem for a base with frequency-dependent elasto-viscous characteristics is obtained for the first time. Based on the example of construction of Green's function for a plate, which lies on Winklerian-type elasto-viscous base, the opportunity of obtaining a close solution of the equation of oscillations for the generalized coordinate in the time domain is shown, if experimentally determined parameters of complex rigidity of the base are known.

We obtained closed expressions for Green's functions for a number of one-
dimensional and two-dimensional dynamic problems taking into account the frequent-independent internal friction.

## 6.4   The Numerical Realization of Boundary Element Method (BEM)

Here we examine the numerical realization of the direct BEM (See Table 2.1 – DBEM) for solving plane stationary problems of the linear theory of elasticity, taking into account the internal friction (A.G. Sarkisian [315]). Let us write the boundary Eq. (2.2.23) of Direct Boundary Element Method (DBEM) for the domain with piecewise-smooth boundary in the following form:

$$\gamma u_i(p) = \int_\Gamma G_{ij}(q - p; \omega) t_j(q) d\Gamma_q - \int_\Gamma T_{ij}(q, p; \omega) u_j(q) d\Gamma_q, \qquad (6.4.1)$$

$$(i, j = 1, 2),$$

where $\gamma = 0.5$, if the boundary is smooth or point $p$ lies in the section of the piecewise-smooth boundary; $u_j$, $t_j$ – the component of the complex vectors of displacements and stresses respectively; and $G_{ij}$, $T_{ij}$ – directed along $x_j$ are components of complex vectors of displacements and stresses at the point $q$ as a result of the action of unit load in the direction $x_i$ at the point $p$, representing the elements of matrices $G$ and $G_q$

respectively. The second integral is understood to be the principal (main) value.

Henceforth the independent summation over the repetitive lower indexes is accepted. The most common approach to solve singular integral equations consists of reducing them to a system of algebraic equations and the numerical solution of the latter. Following this approach, we divide the boundary $\Gamma$ into $N$ elements in such a way that the latter would not contain within themselves points of non-smoothness (rough, roughness, no smoothness) of the boundary, and the discrete analogue of Eqs. (6.4.1) we write as:

$$\gamma u_i(p) = \sum_{m=1}^{N} \int_{\Gamma_m} G_{ij}(q^m - p; \omega) t_j(q^m) d\Gamma_{q^m}$$

$$- \sum_{m=1}^{N} \int_{\Gamma_m} T_{ij}(q^m, p; \omega) u_j(q^m) d\Gamma_{q^m}, \qquad (6.4.2)$$

where $q^m$ – points of $m$th boundary element. We approximate $u_j$ and $t_j$ in the limits of each boundary element according to the interpolation formula of Lagrange of the second kind (order), and assume that elements themselves vary linearly between their limited (geometrically) nodes.

$$u_j(q^m) = \frac{1}{2}\eta(\eta - 1)U_{j1}^m + (1 - \eta^2)U_{j2}^m + \frac{1}{2}\eta(1 + \eta)U_{j3}^m \equiv N_\alpha(\eta)U_{j\alpha}^m,$$

$$t_j(q^m) = \frac{1}{2}\eta(\eta - 1)P_{j1}^m + (1 - \eta^2)P_{j2}^m + \frac{1}{2}\eta(1 + \eta)P_{j3}^m \equiv$$

$$\equiv N_\alpha(\eta)P_{j\alpha}^m, \qquad (6.4.3)$$

$$X_j^m(\eta) = \frac{1}{2}(1 - \eta)X_{j1}^m + \frac{1}{2}(1 + \eta)X_{j2}^m \equiv M_\beta(\eta)X_{j\beta}^m,$$

where $U_{j\alpha}^m$, $P_{j\alpha}^m$ – the corresponding displacements and stresses in the $\alpha$th node in $j$th direction of the $m$th element; $X_j^m(\eta)$, $X_{j\beta}^m$ – the global coordinates of arbitrary internal point and $\beta$th of geometric node of $m$th element respectively; and $\eta$ – is normalized, i.e. converted to (transformed to) the interval $[-1, 1]$, local coordinate of the element. The direction of the local coordinate system coincides with the given direction of the traversal (bypass, to go around) of the boundary, according to which the considered domain at the bypass remains at the left at all times.

The interpolation (functional) nodes within the element can be positioned arbitrarily, however it is more expedient to have the interpolation nodes at central and edge points. The total number of these nodes is $2N$. Let us assume, that the boundary of the considered interval is smooth. Then the number of the unknown boundary values of displacements and stresses in a correctly stated problem will be equal to $4N$. We will get the same number of equations if in each boundary element at two points we compose the relationship (6.4.2). These points, i.e. the points of the application of a unit external load, can also be positioned arbitrarily in the limits of an element, but to avoid excessive complexities in calculating singular integrals it is expedient not to combine the point of the application of the unit external load with the edge nodes. We place one of them in the center of the left half of the boundary element, i.e. at the point with the local coordinate $\eta_1 = -0.5$, and we combine the other with the central functional node ($\eta_2 = 0.0$), then we call them the main points of application of the unit external influence or the major points of observation. Thus, for the $k$th boundary element we obtain

$$0.5 N_\alpha(\eta_\theta) U_{i\alpha}^k + \sum_{m=1}^{N} \left\{ \int_{-1}^{1} T_{ij}[q^m(\eta), p^k(\eta_\theta); \omega] N_\alpha(\eta) J^m(\eta) d\eta \right\} U_{j\alpha}^m$$

$$= \sum_{m=1}^{N} \left\{ \int_{-1}^{1} G_{ij}[q^m(\eta) - p^k(\eta_\theta); \omega] N_\alpha(\eta) J^m(\eta) d\eta \right\} P_{j\alpha}^m, \qquad (6.4.4)$$

$$(\theta = 1, 2),$$

$$J^m(\eta) = \left( \frac{dx_j(\eta)}{d\eta} \frac{dx_j(\eta)}{d\eta} \right)^{1/2} = \frac{1}{2}[(X_{j2}^m - X_{j1}^m)(X_{j2}^m - X_{j1}^m)] = \frac{1}{2} R^m,$$

where $J^m$ – is the Jacobian of the transformation of the element of the line upon the transfer from the given system to a local normalized system of coordinates $\eta$; and $R^m$ – the length of the $m$ boundary element. Further, composing similar equations for all boundary elements, we will obtain the system of $4N$ algebraic equations, which in a matrix format we write as:

$$\sum_{m=1}^{N} H_\alpha^{km} U_\alpha^m = \sum_{m=1}^{N} G_\alpha^{km} P_\alpha^m, \qquad (6.4.5)$$

where

$$U_\alpha^m = \{u_{j\alpha}^m\}_{j=1,2}, \quad P_\alpha^m = \{P_{j\alpha}^m\}_{j=1,2}, \quad H_\alpha^{km} = [h_{\alpha ij}^{km}]_{i,j=1,2},$$

$$h_{\alpha ij}^{km} = 0.5 N_\alpha(\eta_\theta)\delta_{ij}\delta_{km} + J^m \int_{-1}^{1} T_{ij}(q^m(\eta), p^k(\eta_\theta); \omega) N_\alpha(\eta) d\eta,$$

$$G_\alpha^{km} = [g_{\alpha ij}^{km}]_{i,j=1,2},$$

$$g_{\alpha ij}^{km} = J^m \int_{-1}^{1} G_{ij}[q^m(\eta) - p^k(\eta_\theta); \omega] N_\alpha(\eta) d\eta.$$

Additional complexities can arise when a boundary is not smooth (in our case it has corners) where the surficial forces are determined not unambiguously but only by their limiting values from left to right in the direction of the bypass, Figure 6.3. In this case, with certain boundary conditions at corner points (for example, when only displacements are assigned) in the system (6.4.5), extra unknowns will appear making solving it impossible.

Well known methods (P.K. Banerjee, R. Butterfield [23–29]) overcoming these difficulties are based on the approximate approaches, which consist in one case in the distortion of the geometry of a corner with its rounding; in the second in the artificial equating of the unknown multivalued surface stresses in the corner; and in the third ("the concept of multiple nods") in calculations of unknown functions only at a certain distance from the corner point. The last concept, with modifications, leads to the most accurate results. In the works of S.E. Mikhaylov [228] and V.P. Klepikov [178], the singularity of densities of integral equations in the vicinity of the corner point is approximated by a known singular weight function. The degree of singularity of the weight function is determined based on the value of the interior angle at this corner point.

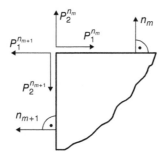

**Figure 6.3**   Corner point (6.4.1).

Below we present another approach to solving boundary equations for the domains with corner points that allows us to avoid these difficulties (A.G. Sarkisian [315]. The essence of this approach consists of the following: depending on the quantity of extra unknowns of the system (6.4.5), on the boundary of the domain we select a corresponding number of additional points of application of a unit external load (additional observation points), which are non-coincident with the corner nodes. These points do not coincide with the corner nodes. In each of these points the relationship of the form (6.4.4) is constructed, so that the number of equations becomes equal to the number of unknowns.

The placement of additional points of application of a unit external load outside of corner points frees us from the superfluous difficulties when calculating breaking and singular components of particular integrals. At the same time numerical experiments show that for well-conditionality of the system (6.4.5) these points should be located at the distance 0.01 – 0.05 of length of the element from the corner point. This is sufficient for a linear independence of equations of the form (6.4.4), written at the selected points on both sides of the corner (angle).

Now the main difficulty is the calculation of the matrix elements $G_\alpha^{km}$ and $H_\alpha^{km}$, i.e. the integrals of the form:

$$J_{ij}^1 = \int_{-1}^{1} G_{ij}[q^m(\eta) - p^k(\eta_\theta); \omega] N_\alpha(\eta) d\eta, \qquad (6.4.6)$$

$$J_{ij}^2 = \int_{-1}^{1} T_{ij}[q^m(\eta), p^k(\eta_\theta); \omega] N_\alpha(\eta) d\eta, \qquad (6.4.7)$$

where

$$G_{ij}[q^m(\eta) - p^k(\eta_\theta); \omega] = \frac{1}{2\pi\rho c_2^2}(\Psi\delta_{ij} - \chi r_{,i} r_{,j}) \qquad (6.4.8)$$

$$T_{ij}[q^m(\eta), p^k(\eta_\theta); \omega] = \sigma_{jk}^i[q^m(\eta) - p^k(\eta_\theta); \omega] n_k(q)$$

$$= \frac{1}{2\pi}\left[\left(\frac{d\Psi}{dr} - \frac{1}{r}\chi\right)\left(\delta_{ij}\frac{\partial r}{\partial n} + r_j n_i\right) - 2\frac{d\chi}{dr} r_{,i} r_{,j} \frac{dr}{dn}\right. \qquad (6.4.9)$$

$$\left. - \frac{2}{r}\chi\left(n_j r_{,i} - 2r_{,i} r_{,j}\frac{\partial r}{\partial n}\right) + \left(\frac{c_1^2}{c_2^2} - 2\right)\left(\frac{d\Psi}{dr} - \frac{d\chi}{dr} - \frac{1}{r}\chi\right) r_{,i} n_j\right],$$

$$\Psi = \frac{i\pi}{2}\left\{H_0^{(1)}(k_2 r) + \frac{c_2}{\omega r}\left[\frac{c_2}{c_1}H_1^{(1)}(k_1 r) - H_1^{(1)}(k_2 r)\right]\right\}, \qquad (6.4.10)$$

$$\chi = \frac{i\pi}{2}\left[\frac{c_2^2}{c_1^2}H_2^{(1)}(k_1 r) - H_2^{(1)}(k_2 r)\right],$$

$$\frac{d\Psi}{dr} = -\frac{i\pi\omega}{2c_2}H_1^{(1)}(k_2 r) - \frac{1}{r}\chi$$

$$\frac{d\chi}{dr} = \frac{i\pi}{2}\left(\frac{c_2}{c_1}\right)^2\frac{\omega}{c_1}H_1^{(1)}(k_1 r) + \frac{d\Psi}{dr} - \frac{1}{r}\chi$$

$$r_{,k} \equiv \frac{\partial r}{\partial x_k} = \frac{x_k(q) - x_k(p)}{r}, \qquad (k = i, j) \qquad (6.4.11)$$

$$k_1 = \frac{\omega}{c_1}, \qquad k_2 = \frac{\omega}{c_2},$$

$$c_1^2 = \frac{\lambda + 2\mu}{\rho}, \quad c_2^2 = \frac{\mu}{\rho}.$$

Here $\lambda$ and $\mu$ are determined from the formulas:

$$\lambda = \frac{E\sigma}{(1 + \sigma)(1 - 2\sigma)}, \qquad \mu = \frac{E}{2(1 + \sigma)}$$

moreover, $E$ – the complex module of elasticity accepted in accordance with the model of the frequency-independent relative damping in the formula

$$E(\omega) = E_1|\omega|^\alpha\left(\cos\frac{\pi\alpha}{2} + i\,\mathrm{sign}\,\omega\,\sin\frac{\pi\alpha}{2}\right),$$

index $(\cdot, i)$ indicates the differentiation of the given function along the coordinate $x_i$; $\delta_{ij}$ – Kronecker's symbol; $n_k(q)$ – the directed cosines of the external normal $\boldsymbol{n}$ to the boundary of the domain at the point $q$; $H_n^{(1)}(\cdot)$–the function of Hankel of the first kind of order $n$th, in calculations of which we will use its decomposition in series; $r$ – the distance between the point of application of a unit external load and point $q$, $x_k(q)$; and $x_k(p)$ – coordinates of the points $q$ and $p$ respectively. Analytical calculations of these integrals are connected with serious technical difficulties. When the observation points are not located in the boundary element $(m \neq k)$, the integrands in (6.4.6) and (6.4.7) remain limited, and for their integration we use the quadrature formula of Gauss with a unit weighting function. In the case when the observation points are located in the boundary element $(m = k)$, both of these integrals have a singularity, i.e. logarithmic and the first order, respectively. In the first case the singularity is weak (S.G. Mikhlin [229, 230, 232]), the integral exists in the usual sense and for its calculation

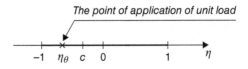

**Figure 6.4**  The symmetrical intervals relative to a singulatity point.

we use special quadrature formula with the logarithmic weighting function (P.K. Banerjee, R. Butterfield [27]). In the second case the integral exists only in the sense of the principal value (*p. v.*) Cauchy. For the calculation of this integral we transform $J_{ij}^2$ into the following form (argument in $T_{ij}$ is omitted):

$$J_{ij}^2 = p.v. \int_{-1}^{1} T_{ij}\{N_\alpha(\eta) - N_\alpha(\eta_\theta)\}d\eta + N_\alpha(\eta_\theta)p.v. \int_{-1}^{1} T_{ij}d\eta. \qquad (6.4.12)$$

The first of these integrals exists in the usual sense, since $N_\alpha(\eta) - N_\alpha(\eta_\theta)$ turns into zero at a singular point and the integrant expression remains limited everywhere. This integral is computed according to the quadrature formula of Gauss with the unit weighting function. The second addend when $\eta_\theta = 0$, i.e. when the singular point is in the center of the interval of integration equal to zero. The same integral, when $\eta_\theta \neq 0$, we divide (see Figure 6.4) into symmetrical sections in relation to the singular point $[-1, c]$ and the remaining $[c, 1]$ sections; in the first section, as it was noted, the integral equals to zero, and in the second it is regular, and can be calculated with the aid of the Gauss procedures.

Now, when the matrix elements $G_\alpha^{km}$ and $H_\alpha^{km}$ are known from (6.4.5), together with additional equations (if there are any) and given boundary conditions, we can determine the unknowns on the border of the displacements and stresses. For this, we consider a continuous pass-through numeration of nodes, assigning different numbers to the left and right for nodes of breaking points of stresses. Taking into account the conditions $U_3^m = U_1^{m+1}$, and (in the nodes of continuity) $P_3^m = P_1^{m+1}$, we will transform Eq. (6.4.5) to the following form:

$$[H_{ef}]_{(4N+B)\times 4N}\{U_f\}_{4N\times 1} = [G_{ek}]_{(4N+B)\times(4N+2Q+2M)}$$

$$\times [P_k]_{(4N+2Q+2M)\times 1} \qquad (6.4.13)$$

$$N = \sum_{k=1}^{n} N^{(k)}; \quad B = \sum_{k=1}^{n} B^{(k)}$$

where $B$ – the number of additional equations, which, depending on boundary conditions in the corner nodes is less than or equal to $2Q$; $Q$ – the number of angular points; and $M$ – the number of break-points of stresses in the smooth section of the boundary. Further, taking into account the boundary conditions and transposing (rearranging) the columns in (6.4.13), we finally will obtain a system of algebraic equations with complex coefficients for complex vectors of unknown displacements and stresses

$$[A_{en}]_{(4N+B)\times(4N+B)}\{X_n\}_{(4N+B)\times1} = \{F_n\}_{(4N+B)\times1}, \tag{6.4.14}$$

Equating the corresponding real and imaginary parts of the right and left sides of Eqs. (6.4.14), and solving the obtained system $2(4N+B)$ of the equations with the aid of the standard programs, we obtain the unknown complex vector $\{X_n\}$. After determining the unknowns at the boundary, the calculation of displacements and stresses at the internal points of the domain is done according to the formulas:

$$u_i(p) = \sum_{m=1}^{N} \left\{ J^m \int_{-1}^{1} G_{ij}[q^m(\eta) - p; \omega]N_\alpha(\eta)d\eta \right\} P_{j\alpha}^m$$

$$- \sum_{m=1}^{N} \left\{ J^m \int_{-1}^{1} T_{ij}[q^m(\eta) - p; \omega]N_\alpha(\eta)d\eta \right\} U_{j\alpha}^m, \tag{6.4.15}$$

$$\sigma_{ik}(p) = \sum_{m=1}^{N} \left\{ J^m \int_{-1}^{1} D_{ikj}(q^m(\eta), p; \omega)N_\alpha(\eta)d\eta \right\} P_{j\alpha}^m$$

$$- \sum_{m=1}^{N} \left\{ J^m \int_{-1}^{1} S_{ikj}(q^m(\eta), p; \omega)N_\alpha(\eta)d\eta \right\} U_{j\alpha}^m, \tag{6.4.16}$$

where

$$D_{ikj}(q^m(\eta), p; \omega) = [\sigma_{jk}^i(q^m - p; \omega)]^T = -\sigma_{jk}^i(q^m - p; \omega), \tag{6.4.17}$$

$$S_{ikj}(q^m, p; \omega) = \rho c_2^2 \left[ \left( \frac{c_1^2}{c_2^2} - 2 \right) \times T_{rj,r}\delta_{ik} + T_{ij,k} + T_{kj,i} \right]. \tag{6.4.18}$$

As a numerical example, let us consider a problem of propagation of elastic waves in plane with circular cavity from a uniformly distributed radial stationary harmonic load. Assume that on the contour of the circular opening in the infinite plane with the density $\rho = 0.24 \cdot 10^{-5}$ Kg/cm$^3$, complex

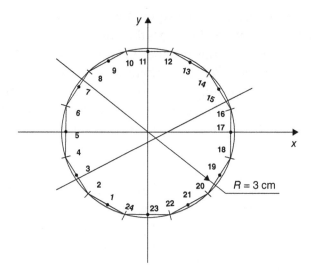

**Figure 6.5**   The partitioning of the boundary with the numeration of functional nodes.

**Table 6.1**   The numerical results of a problem of propagation of elastic waves in plane with the circular opening.

| Nodes# | Amplitude of displacements | | Phase shift | |
|---|---|---|---|---|
| | on x | on y | on x | on y |
| 1 | 0.77362 − 0.3 | 0.13696 − 0.2 | 0.31411 + 0.1 | 0.31411 + 0.1 |
| 2 | 0.11320 − 0.2 | 0.11211 − 0.2 | 0.31411 + 0.1 | 0.31411 + 0.1 |
| 3 | 0.13546 − 02 | 0.77931 − 0.3 | 0.31411 + 0.1 | 0.31411 + 0.1 |
| 4 | 0.15062 − 0.2 | 0.34501 − 03 | 0.31411 + 0.1 | 0.31411 + 0.1 |
| 5 | 0.15727 − 0.2 | 0.14755 − 0.4 | 0.31411 + 0.1 | 0.31413 + 0.1 |
| 6 | 0.14968 − 0.2 | 0.35446 − 0.3 | 0.31411 + 0.1 | 0.62827 + 0.1 |
| 7 | 0.13694 − 0.2 | 0.77367 − 0.3 | 0.31411 + 0.1 | 0.62827 + 0.1 |
| 8 | 0.11219 − 0.2 | 0.113218 − 0.2 | 0.31411 + 0.1 | 0.62827 + 0.1 |
| 9 | 0.77917 − 0.3 | 0.13548 − 0.2 | 0.31411 + 0.1 | 0.62827 + 0.1 |
| 10 | 0.34493 − 0.3 | 0.15065 − 0.2 | 0.31411 + 0.1 | 0.62827 + 0.1 |
| 11 | 0.14743 − 0.4 | 0.15730 − 0.2 | 0.31408 + 0.1 | 0.62827 + 0.1 |
| 12 | 0.35442 − 0.3 | 0.15071 − 0.2 | 0.62827 + 0.1 | 0.62827 + 0.1 |
| 13 | 0.77364 − 0.3 | 0.13696 − 0.2 | 0.62827 + 0.1 | 0.62827 + 0.1 |
| 14 | 0.11320 − 0.2 | 0.11220 − 0.2 | 0.62827 + 0.1 | 0.62827 + 0.1 |
| 15 | 0.13547 − 0.2 | 0.78930 − 0.3 | 0.62827 + 0.1 | 0.62827 + 0.1 |
| 16 | 0.15062 − 0.2 | 0.34496 − 0.3 | 0.62827 + 0.1 | 0.62827 + 0.1 |
| 17 | 0.15728 − 0.2 | 0.14753 − 0.4 | 0.62827 + 0.1 | 0.62825 + 0.1 |
| 18 | 0.15068 − 0.2 | 0.35449 − 0.3 | 0.62827 + 0.1 | 0.31411 + 0.1 |
| 19 | 0.13695 − 0.2 | 0.77377 − 0.3 | 0.62827 + 0.1 | 0.31411 + 0.1 |
| 20 | 0.11419 − 0.2 | 0.11321 − 0.2 | 0.62827 + 0.1 | 0.31411 + 0.1 |
| 21 | 0.77925 − 0.3 | 0.13549 − 0.2 | 0.62827 + 0.1 | 0.31411 + 0.1 |

Table 6.1   (*continued*)

| Nodes# | Amplitude of displacements | | Phase shift | |
| --- | --- | --- | --- | --- |
| | on $x$ | on $y$ | on $x$ | on $y$ |
| 22 | 0.34502 – 0.3 | 0.15064 – 0.2 | 0.62827 + 0.1 | 0.31411 + 0.1 |
| 23 | 0.14788 – 0.4 | 0.15730 – 0.2 | 0.62826 + 0.1 | 0.31411 + 0.1 |
| 24 | 0.35431 – 0.3 | 0.15071 – 0.2 | 0.31411 + 0.1 | 0.31411 + 0.1 |

modulus of elasticity $E = (200000 + 100\text{i})$ Kg/cm$^2$, and with Poisson's coefficient $\nu = 0.1$ a uniformly distributed radial harmonic load an an amplitude of $P = 100$ Kg/cm and angular frequency of $\omega = 20$ radian/c is applied. The partitioning of the boundary with the numeration of functional nodes is shown in Figure 6.5. The results of the calculations are given in Table 6.1

## 6.5   The Construction of the Green's Function of the Dynamic Stationary Problem for the Elasto-Viscous Half-Plane

Construction of the Green's function is the most important stage when solving boundary value problems using the methods of boundary integral equations. Boundary equations are built on the basis of the Green's Function. Usually Green's functions are used for unlimited domains, since they take the simplest form and can be obtained in a closed form. However, during the solution of many applied problems, the use of Green's function for an unlimited domain can lead to superfluous operations. In particular, for problems of dynamics of structure and the theory of seismic resistance, it is necessary to examine limited spaces, which are a part of or an addition to a half-space (half-plane). Hence, the Green's function is more effective in a case of a semi-infinite domain, which frees us from the need for solving additional integral equations on the boundary of the half-space (half-plane). In the present paragraph we offer a solution to a problem of steady oscillations of an isotropic homogeneous elasto-viscous half-plane. These oscillations are caused by the action of an arbitrarily oriented concentrated force, changing in time according to the harmonic law (A.I. Tseitlin, D.R. Atadzhanov, and A.G. Sarkisyan [372], A.I. Tseitlin, L.G .Petrosian, D.R. Atadzhanov [379]). A similar problem for the elastic half-space and elastic half-plane was examined in the works of K.M. Case, R.D. Hazeltine [59], and V.A. Ilichev, O.Y. Shakhter [149]. It is assumed that the material of the medium possesses internal friction, which is described

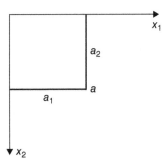

**Figure 6.6**   Half-plane in
Cartesian coordinates.

by the model of the frequency-independent and relative damping, A.I.
Tseitlin [369]. In this case the complex module takes the form of:

$$E(\omega) = E_1|\omega|^\alpha \left(\cos\frac{\pi\alpha}{2} + i\,\mathrm{sign}\omega \sin\frac{\pi\alpha}{2}\right), \qquad (6.5.1)$$

and Poisson's ratio $\sigma$ is taken as real and does not depend on the frequencies
of harmonic oscillations $\omega$. Here $E_1 = E(1)$; $\alpha = 2\pi^{-1}\tan^{-1}\gamma$; $\gamma$ – loss factors
(coefficient of loses). This model provides the ideal frequency independence
of energy loss on the entire frequency axis making it possible to investigate
any dynamic processes; however, this excludes the consideration of a static
load as a result of an unlimited creep with $\omega = 0$. Let us examine a half-plane
$x_2 > 0$, Figure 6.6, in the Cartesian coordinates $x(x_1, x_2)$. Assume that at a
certain point of half-plane $a(a_1, a_2)$ a unit concentrated force, changing in
time according to the harmonic law, is applied.

  The solution of the problem of steady oscillations of a half-plane under
the action of this load, after the exclusion of time, is described by the matrix
function of Green $g = [g_{ij}]$, where $g_{ij}(i, j = 1, 2)$ – displacement along the $i$th
coordinate from the action of the unit force, directed along $j$th coordinate.
The Green's function satisfies the equations

$$\nabla^*g(x; a) + \rho\omega^2 g(x; a) + \delta(x; a) = 0, \qquad (6.5.2)$$

where $\nabla^* = [\nabla^*_{ij}]-$ is the matrix differential operator

$$\nabla^*_{ij} = \mu\delta_{ij}\nabla + (\lambda + \mu)\frac{\partial^2}{\partial x_i \partial x_j}; \qquad (i = 1, 2),$$

$$\nabla = \frac{\partial^2}{\partial x_1^2} + \frac{\partial^2}{\partial x_2^2};$$

$\delta_{ij}$ – Kronecker's delta; $\delta(x; a)$ – two dimensional Dirac delta function, concentrated at point $a$

$$\delta(x; a) = \begin{bmatrix} \delta(x_1 - a_1)\delta(x_2 - a_2) & 0 \\ 0 & \delta(x_1 - a_1)\delta(x_2 - a_2) \end{bmatrix};$$

$\rho$ – material density of half-plane; and $\lambda$ – Lame's constants.

$$\lambda = \frac{E\sigma}{(1+\sigma)(1-2\sigma)}, \qquad \mu = \frac{E}{2(1+\sigma)}.$$

As a result of (6.5.1), we assume:

$$\lambda = \lambda^*(E' + iE''), \qquad \mu = \mu^*(E' + iE'')$$

where

$$\lambda^* = \frac{\sigma}{(1+\sigma)(1-2\sigma)}, \qquad \mu^* = \frac{1}{2(1+\sigma)},$$

$$E'(\omega) = E_1|\omega|^\alpha \cos\frac{\pi\alpha}{2}, \quad E''(\omega) = E_1|\omega|^\alpha sign\omega \sin\frac{\pi\alpha}{2}.$$

Boundary of the half-plane is free of stresses, and has the following boundary conditions:

$$T_n g(x_1 + 0, a) = 0 \qquad (n_1 = 0, n_2 = 1) \qquad (6.5.3)$$

where

$T_n = [T_{nij}]$ – matrix differential operator of stresses with elements:

$$T_{nij} = \lambda n_i \frac{\partial}{\partial x_j} + \mu n_j \frac{\partial}{\partial x_i} + \mu \delta_{ij} \frac{\partial}{\partial n},$$

$$\frac{\partial}{\partial n} = \sum_i^2 n_i \frac{\partial}{\partial x_i};$$

$n$ – the direction of the internal normal with directing cosines $n_1$ and $n_2$. With $n_1 = 0$ and $n_2 = 1$, we obtain:

$$T_{nij} = \lambda \delta_{2i} \frac{\partial}{\partial x_j} + \mu \left( \delta_{2i} \frac{\partial}{\partial x_i} + \delta_{ij} \frac{\partial}{\partial x_2} \right). \qquad (6.5.4)$$

Furthermore, we assume that as a result of the internal friction, the components of the vector of displacements and stresses tensor become zero at infinity. Let us somehow prolong $g(x, a)$ into the extended domain $\Omega$. As the extended domain we take the entire plane. Now let us apply a delta-transformation to Eq. (6.5.2), by multiplying on the left side by the two-dimensional delta-function $\delta(x, x^*)$, concentrated at the point $x^*(x_1^*, x_2^*)$, and integrating over the entire domain $\Omega$. Using the Green's function in the domains of continuity, i.e. in both half-planes $x_2 > 0$ and $x_2 < 0$, and taking into account the properties of a delta-function, we will obtain:

$$\nabla^* g(x^*; a) + \rho \omega^2 g(x^*; a) + \delta(x^*; a) = \int_{-\infty}^{\infty} [\delta(x_0, x^*) \Delta T_n g(x_0, a)$$

$$- [T_n \delta(x_0, x^*)]^T \, \Delta g(x_0, a)] dx_1, \quad (6.5.5)$$

where $x_0 \in \Gamma$ – the point on the border of the half-plane; $x^* \in \Omega$; and $\Delta$ – indicates the difference in values of corresponding vector-functions on the inner $\Gamma^+$ and outer $\Gamma^-$ boundary of the half-plane. For solving Eq. (6.5.5) we use the known Green's function for the entire plane $G(x^*, x)$, which satisfies equation (6.5.2), with $-\infty < x, x^* < \infty$. Then from (6.5.5) we find:

$$g(x^*, a) = G(x^*, a) - \int_{-\infty}^{\infty} [G(x^*, x_0) \Delta T_n g(x_0, a)$$

$$- [T_n G(x_0, x^*)]^T \, \Delta g(x_0, a)] dx_1 \quad (6.5.6)$$

Thus, we presented the function $g(x, a)$ in the entire plane. As it was mentioned above, there are various ways to select jumps of unknown functions on a boundary of a half-plane. Let us pick the following option: $\Delta g(x_0; a) \neq 0$; $\Delta T_n g(x_0, a) \neq 0$. This case corresponds to prolongation of the function $g(x, a)$ by zero into the extended domain $\Omega$. Assuming that the outer boundary of the half-plane $\Gamma^-$ remains undeformed, we obtain the relationships:

$$g(x_0, a) = T_n g(x_0, a) = 0, \quad (x_0 \in \Gamma^-). \quad (6.5.7)$$

Consequently, $\Delta g(x_0, a) = g(x_0; a)$, $\Delta T_n g(x_0, a) = T_n g(x_0, a)$, $(x_0 \in \Gamma^+)$. Then (6.5.6) will take the form:

$$g(x^*, a) = G(x^*, a) - \int_{-\infty}^{\infty} \{ G(x^*, x_0) T_n g(x_0, a)$$

$$- [T_n G(x_0, x^*)]^T g(x_0, a) \} dx_1 \quad (6.5.8)$$

Using known properties of the Green's function for a plane:

$$G(x, x^*) = G(x^*, x),$$

$$T_n G(x, x^*) = - T_n G(x^*, x),$$

as well as introducing designations

$$G_q(x, x^*) = -[T_n G(x^*, x)]^T \tag{6.5.9}$$

and using the boundary conditions (6.5.3), we obtain:

$$\frac{1}{2} g(x_0^*, a) = G(x_0^*, a) - \int_{-\infty}^{\infty} G_q(x_0^*, x_0) g(x_0, a) dx_1. \tag{6.5.10}$$

Here the integral on the right side is understood as having the principal value. For solving the obtained integral in Eq. (6.5.10), let us apply Fourier transformation to both parts on the coordinate $x_1^*$

$$\frac{1}{2} \bar{g}(\xi_1, 0; a) = \bar{G}(\xi_1, 0; a) - \iint_{-\infty}^{\infty} G_q(x_0^*, x_0) g(x_0^*, a) e^{i\xi_1 x_1^*} dx_1^* dx_1. \tag{6.5.11}$$

where the feature indicates Fourier transform on the coordinate $x_1^*$, for example

$$\bar{g}(\xi_1, 0; a) = \int_{-\infty}^{\infty} g(x_0^*, a) e^{i\xi_1 x_1^*} dx_1^*.$$

Since $G_q(x_0^*, x_0)$ is the function of the difference of the arguments $x_0^* - x_0$, the dual integral on the right side of (6.5.11) can be presented in the form of (K.M. Case, R.D. Hazeltine [59])

$$\iint_{-\infty}^{\infty} G_q(x_0^*, x_0) g(x_0^*, a) e^{i\xi_1 x_1^*} dx_1^* dx_1 = \tilde{G}_q(\xi_1) \bar{g}(\xi_1, 0; a), \tag{6.5.12}$$

where,

$$\tilde{G}_q(\xi_1) = \int_{-\infty}^{\infty} G_q(R) e^{i\xi_1 R} dR, \tag{6.5.13}$$

$$(R = x_1^* - x_1).$$

Thus, from (6.5.10) we will obtain:

$$\left[ \frac{1}{2} E + \tilde{G}_q(\xi_1) \right] \bar{g}(\xi_1, 0; a) = \bar{G}(\xi_1, 0; a), \tag{6.5.14}$$

where $E$ – unit matrix. Designating $A$ as the matrix

$$A = \left[ \frac{1}{2} E + \tilde{G}_q(\xi_1) \right], \tag{6.5.15}$$

from (6.5.14) we find

$$\bar{g}(\xi_1, 0; a) = A^{-1} \bar{G}(\xi_1, 0; a). \tag{6.5.16}$$

Using the Green's function $G$ for a plane, and generated by its tensor of stresses $G_q$ in (6.5.15) and (6.5.16), leads to cumbersome calculations. By doing as follows we can bypass these difficulties. We have

$$\overline{G}(\xi_1, 0; a) = \frac{1}{2\pi} \int_{-\infty}^{\infty} \overline{\overline{G}}(\xi_1, \xi_2; a) e^{-i\xi_2 x_2^*} d\xi_2, \tag{6.5.17}$$

$$\overline{G}_q(\xi_1, x_2^*) = \frac{1}{2\pi} \int_{-\infty}^{\infty} \widetilde{\overline{G}}_q(\xi_1, \xi_2) e^{-i\xi_2 x_2^*} d\xi_2,$$

where $\overline{\overline{G}}$, $\widetilde{\overline{G}}_q -$ are dual transforms of Fourier from $G$ and $G_q$. For determining $\widetilde{\overline{G}}_q$ we will use formula (6.5.9). Taking into account that $G_q(x^*, x) = -G_q(x, x^*)$, and by applying Fourier transform to its both parts, we obtain

$$\widetilde{\overline{G}}_q(\xi_1, \xi_2) = \iint_{-\infty}^{\infty} [T_n G(x^*, x)]^T e^{i(\xi_1 x_1^* + \xi_2 x_2^*)} dx_1^* dx_2^* \tag{6.5.18}$$

The elements of the matrix $\widetilde{\overline{G}}_q$ will be:

$$[\widetilde{\overline{G}}_q]_{jk} = \iint_{-\infty}^{\infty} \left[ \lambda \delta_{2k} \frac{\partial}{\partial x_m^*} + \mu \left( \delta_{2m} \frac{\partial}{\partial x_k^*} + \delta_{km} \frac{\partial}{\partial x_2^*} \right) \right]$$

$$\times G_{mj}(x^*, x) e^{i(\xi_1 x_1^* + \xi_2 x_2^*)} dx_1^* dx_2^* \tag{6.5.19}$$

Integrating by parts the right side of (6.5.19), we obtain:

$$[\widetilde{\overline{G}}_q]_{jk} = -i[\lambda \delta_{2k} \xi_m + \mu \xi_k \delta_{2m} + \mu \xi_2 \delta_{km}] \overline{\overline{G}}_{mj}(\xi_1, \xi_2) \tag{6.5.20}$$

Let us determine $\overline{\overline{G}}_{mj}$ using the dynamic Lame's equations. Let the unit load act in the horizontal (on $x_1^*$) direction. Lame's equations in the expanded form are taken in the form:

$$(\lambda + \mu) \left( \frac{\partial^2 G_{11}}{\partial x_1^{*2}} + \frac{\partial^2 G_{21}}{\partial x_1^* \partial x_2^*} \right) + \mu \left( \frac{\partial^2 G_{11}}{\partial x_1^{*2}} + \frac{\partial^2 G_{11}}{\partial x_2^{*2}} \right)$$

$$+ \delta(x_1^* - x_1)\delta(x_2^* - x_2) + \rho\omega^2 G_{11} = 0, \tag{6.5.21}$$

$$(\lambda + \mu) \left( \frac{\partial^2 G_{21}}{\partial x_2^{*2}} + \frac{\partial^2 G_{11}}{\partial x_1^* \partial x_2^*} \right) + \mu \left( \frac{\partial^2 G_{21}}{\partial x_1^{*2}} + \frac{\partial^2 G_{21}}{\partial x_2^{*2}} \right) + \rho\omega^2 G_{11} = 0.$$

Applying Fourier transform on the coordinate $x_1^*$ to Eqs. (6.5.21) and having previously designated $x_1^* - x_1 = R_1$, $x_2^* - x_2 = R_2$, we will obtain:

$$(\lambda + 2\mu)\xi_1^2\widetilde{\widetilde{G}}_{11} + (\lambda + \mu)\xi_1\xi_2\widetilde{\widetilde{G}}_{21} + \mu\xi_2^2\widetilde{\widetilde{G}}_{11} - 1 - \rho\omega^2\widetilde{\widetilde{G}}_{11} = 0 \qquad (6.5.22)$$

$$(\lambda + 2\mu)\xi_2^2\widetilde{\widetilde{G}}_{21} + (\lambda + \mu)\xi_1\xi_2\widetilde{\widetilde{G}}_{11} + \mu\xi_1^2\widetilde{\widetilde{G}}_{21} - 1 - \rho\omega^2\widetilde{\widetilde{G}}_{21} = 0$$

where,

$$\widetilde{\widetilde{G}}_{ij}(\xi_1, \xi_2) = \iint_{-\infty}^{\infty} G_{ij}(R_1, R_2)e^{i(\xi_1 R_1 + \xi_2 R_2)}dR_1\,dR_2$$

The multiplier $e^{i(\xi_1 x_1)}e^{i(\xi_2 x_2)}$ in (6.5.22) was omitted for the simplicity. Using the following designation:

$$k_e^2 = \frac{\rho\omega^2}{(\lambda + 2\mu)}; \qquad k_t^2 = \frac{\rho\omega^2}{\mu};$$

$$\frac{\xi_1^2}{k_e^2} + \frac{\xi_2^2}{k_t^2} - 1 = a_{11}; \qquad (6.5.23)$$

$$\left(\frac{1}{k_e^2} - \frac{1}{k_t^2}\right)\xi_1\xi_2 = a_{12} = a_{21};$$

$$\frac{\xi_2^2}{k_e^2} + \frac{\xi_1^2}{k_t^2} - 1 = a_{22};$$

from (6.5.22) we will obtain the following system of algebraic equations:

$$a_{11}\widetilde{\widetilde{G}}_{11} + a_{12}\widetilde{\widetilde{G}}_{21} = \frac{1}{\rho\omega^2} \qquad (6.5.24)$$

$$a_{21}\widetilde{\widetilde{G}}_{11} + a_{22}\widetilde{\widetilde{G}}_{21} = 0$$

Hence we easily find:

$$\widetilde{\widetilde{G}}_{11} = \frac{1}{\rho\omega^2}\left(\frac{a_{22}}{a_{11}a_{22} - a_{12}a_{21}}\right) = \frac{1}{\rho\omega^2}\left(\frac{\xi_1^2}{\xi_2^2 - k_e^2} + \frac{k_t^2 - \xi_1^2}{\xi_1^2 - k_t^2}\right),$$

$$\widetilde{\widetilde{G}}_{21} = -\frac{1}{\rho\omega^2}\left(\frac{a_{21}}{a_{11}a_{22} - a_{12}a_{21}}\right) = \frac{1}{\rho\omega^2}\left(\frac{\xi_1\xi_2}{\xi^2 - k_e^2} - \frac{\xi_1\xi_2}{\xi^2 - k_t^2}\right),$$

where,

$$\xi^2 = \xi_1^2 + \xi_2^2.$$

Under the action of a vertical unit load along the axis $x_2^*$, analogously by doing likewise, we obtain:

$$\widetilde{\widetilde{G}}_{22} = \frac{1}{\rho\omega^2}\left(\frac{\xi_2^2}{\xi^2 - k_e^2} + \frac{k_t^2 - \xi_2^2}{\xi^2 - k_t^2}\right);$$

$$\widetilde{\widetilde{G}}_{12} = \frac{1}{\rho\omega^2}\left(\frac{\xi_1\xi_2}{\xi^2 - k_e^2} - \frac{\xi_1\xi_2}{\xi^2 - k_t^2}\right).$$

Thus, the formula for the general term, taking into account the omitted multipliers, takes the form

$$\widetilde{\widetilde{G}}_{ij}(\xi_1, \xi_2; x) = \frac{e^{i(\xi_1 x_1 + \xi_2 x_2)}}{\rho\omega^2}\left(\frac{\xi_m\xi_j}{\xi^2 - k_e^2} + \frac{\delta_{mj}k_t^2 - \xi_m\xi_j}{\xi^2 - k_t^2}\right) \tag{6.5.25}$$

Substituting (6.5.20) into (6.5.17) and taking into account (6.5.25) and the relationships

$$\widetilde{G}_q(\xi_1, x_2^*) = e^{-i\xi_1 x_1}\widetilde{\overline{G}}_q(\xi_1, x_2^*), \tag{6.5.26}$$

we obtain:

$$2\pi i\rho\omega^2\widetilde{G}_{qik}(\xi_1, x_2^*) = \int_{-\infty}^{\infty}\left(\lambda\delta_{2k}\sum_m^2 \frac{\xi_m^2\xi_j}{\xi^2 - k_e^2} + \lambda k_t^2\delta_{2k}\frac{\xi_j}{\xi^2 - k_t^2}\right.$$

$$-\frac{\lambda\delta_{2k}\sum_m^2 \xi_m^2\xi_j}{\xi^2 - k_t^2} + \frac{2\mu\xi_2\xi_j\xi_k}{\xi^2 - k_e^2} + \frac{\mu\delta_{2j}k_t^2\xi_j}{\xi^2 - k_t^2} - \frac{2\mu\xi_2\xi_j\xi_k}{\xi^2 - k_e^2}$$

$$\tag{6.5.27}$$

$$+ \left.\frac{\mu k_t^2\delta_{kj}\xi_{j2}}{\xi^2 - k_t^2}\right)e^{-i\xi_2 x_2^*}d\xi_2, \quad (j, k = 1, 2)$$

After calculating integrals (I.S. Gradstein, I.M. Ryzhik [123]) let us find:

$$2\pi i\rho\omega^2[\widetilde{G}_q(\xi_1, x_2^*)]_{ik} = \lambda\xi_1^2\delta_{2k}[(\xi_j - \delta_{2j}\xi_2)I_0 + \delta_{2j}I_2]$$

$$+ \lambda\delta_{2k}[(\xi_j - \delta_{2j}\xi_2)I_3 + \delta_{2j}I_4] + \lambda k_t^2\delta_{2k}[(\xi_j - \delta_{2j}\xi_2)I_{0t} + \delta_{2j}I_{1t}] \tag{6.5.28}$$

$$+ 2\mu[(\xi_k - \delta_{2k}\xi_2)(\xi_j - \delta_{2j}\xi_2)I_2 + (\xi_j - \delta_{2j}\xi_2)\delta_{2k}I_\xi + (\xi_k - \delta_{2k}\xi_2)\delta_{2j}I_3$$

$$+ \delta_{2k}\delta_{2j}I_4] + \mu k_t^2[\delta_{2j}(\xi_k - \delta_{2k}\xi_2)I_{0t} + \delta_{2j}\delta_{2k}I_{1t} + \delta_{kj}I_{1t}],$$

where,

$$I_{0e} = \frac{\pi}{R_e}e^{-R_ex_2^*}, \qquad I_{0t} = \frac{\pi}{R_t}e^{-R_tx_2^*},$$

$$I_0 = I_{0e} - I_{0t},$$

$$I_{1e} = -i\pi e^{-R_ex_2^*}, \qquad I_{1t} = -i\pi e^{-R_tx_2^*},$$

$$I_2 = I_{1e} - I_{1t} = i\pi(e^{-R_tx_2^*} - e^{-R_ex_2^*}),$$

$$I_{3e} = -\pi R_e e^{-R_ex_2^*}, \qquad I_{3t} = -\pi R_t e^{-R_tx_2^*}, \tag{6.5.29}$$

$$I_3 = I_{3e} - I_{3t} = \pi(R_t e^{-R_tx_2^*} - R_e e^{-R_ex_2^*}),$$

$$I_{4e} = i\pi R_e^2 e^{-R_ex_2^*}, \qquad I_{4t} = i\pi R_t^2 e^{-R_tx_2^*},$$

$$I_4 = I_{4e} - I_{4t} = i\pi(R_e^2 e^{-R_ex_2^*} - R_t^2 e^{-R_tx_2^*}),$$

$$R_e = \sqrt{\xi_1^2 - k_e^2} \qquad R_t^2 = \sqrt{\xi_1^2 - k_t^2} \qquad x_2^* > 0.$$

Accomplishing a limited transition to $x_2^* \to +0$, we will obtain the boundary value of $\widetilde{G}_q(\xi_1, x_2^*)$

$$2\pi i\rho\omega^2\widetilde{G}_{qik}(\xi_1, +0) = (\lambda + 2\mu)\delta_{2k}(\xi_j - \delta_{2j}\xi_2)I_3^0 + \lambda\xi_1^2\delta_{2k}(\xi_j - \delta_{2j}\xi_2)I_{0e}^0$$

$$+ \lambda k_t^2\delta_{2k}(\xi_j - \delta_{2j}\xi_2)I_{0t}^0 + 2\mu\delta_{2j}(\xi_k - \delta_{2k}\xi_2)I_3^0 \tag{6.5.30}$$

$$+ \mu k_t^2[\delta_{2j}(\xi_k - \delta_{2k}\xi_2)I_{0t}^0 + \delta_{kj}I_{1t}^0],$$

where

$$I_{0e}^0 = \frac{\pi}{R_e}; \; I_{0t}^0 = \frac{\pi}{R_t};$$

$$I_{1e}^0 = I_{1t}^0 = -i\pi; \; I_3^0 = \pi(R_t - R_e).$$

Taking into account the relationship between the boundary and direct singular value of the integral of the stress tensor $G_q$ (V.D. Kupradze [199])

$$\widetilde{G}_q(\xi_1) = \widetilde{G}_q(\xi_1, +0) + \frac{1}{2}E, \tag{6.5.31}$$

After simple computations from (6.5.30) we will find:

$$[\widetilde{G}_q(\xi_1)]_{jk} = \frac{1}{2}[\delta_{2k}(\xi_j - \delta_{2j}\xi_2)\alpha(\xi_1) + \delta_{2j}(\xi_k - \delta_{2k}\xi_2)\beta(\xi_1)], \tag{6.5.32}$$

where,

$$\alpha(\xi_1) = i\frac{2(R_t - R_e)R_t + k_t^2}{R_e k_t^2};$$ (6.5.33)

$$\beta(\xi_1) = -i\frac{2(R_t - R_e)R_t + k_t^2}{R_e k_t^2}.$$

Now, according to (6.5.15) we obtain:

$$A = \frac{1}{2}\begin{bmatrix} 1 & \xi_1 \ \alpha(\xi_1) \\ \xi_1 & \beta(\xi_1) & 1 \end{bmatrix}.$$

The reciprocal (inverse) matrix $[A]^{-1}$ takes the form:

$$[A]^{-1} = \frac{2}{1 - \xi_1^2\alpha(\xi_1)\beta(\xi_1)}\begin{bmatrix} 1 - \xi_1 & \beta(\xi_1) \\ -\xi_1 & \alpha(\xi_1) & 1 \end{bmatrix}^T.$$ (6.5.34)

Now from (6.5.8) taking into account (6.5.16) we determine the Green's function for the half-plane

$$g(x^*, a) = G(x^*, a) + B^*(x^*, a),$$ (6.5.35)

where,

$$B^*(x^*, a) = -\frac{1}{2\pi}\int_{-\infty}^{\infty}\tilde{G}_q(\xi_1, x_2^*)[A]^{-1}G(\xi_1, 0; a)e^{-i\xi_1 x_1^*}d\xi_1.$$ (6.5.36)

In particular, let the unit load act at the point $a = (0, a_2)$. For this case we write the elements $\overline{G}(\xi_1, 0; 0, a_2)$

$$\overline{\overline{G}}_{11}(\xi_1, 0; 0, a_2) = \frac{1}{2\rho\omega^2}\left(e^{-R_e a_2}\frac{\xi_1^2}{R_e} - R_t e^{-R_t a_2}\right),$$

$$\overline{\overline{G}}_{12}(\xi_1, 0; 0, a_2) = \overline{\overline{G}}_{21}(\xi_1, 0; 0, a_2)$$

$$= -\frac{i}{2\rho\omega^2}(\xi_1 e^{-R_t a_2} - \xi_1 e^{-R_e a_2}),$$ (6.5.37)

$$\overline{\overline{G}}_{22}(\xi_1, 0; 0, a_2) = \frac{1}{2\rho\omega^2}\left(\frac{\xi_1^2}{R_t}e^{-R_t a_2} - R_e e^{-R_e a_2}\right),$$

And the elements of $\widetilde{G}_q(\xi_1, x_2^*)$ will be:

$$\widetilde{G}_{q11}(\xi_1, x_2^*) = \frac{\xi_1^2}{k_t^2}(e^{-R_t x_2^*} - e^{-R_e x_2^*}) - \frac{1}{2}e^{-R_t x_2^*},$$

$$\widetilde{G}_{q12}(\xi_1, x_2^*) = \frac{\xi_1}{i}\left[\frac{R_t}{k_t^2}e^{-R_t x_2^*} - \left(\frac{1}{2R_e} + \frac{R_t^2}{k_t^2 R_e}\right)e^{-R_e x_2^*}\right], \qquad (6.5.38)$$

$$\widetilde{G}_{q21}(\xi_1, x_2^*) = \frac{\xi_1}{i}\left[\left(\frac{1}{2R_t} + \frac{R_t}{k_t^2}\right)e^{-R_t x_2^*} - \frac{R_e}{k_t^2}e^{-R_e x_2^*}\right],$$

$$\widetilde{G}_{q22}(\xi_1, x_2^*) = \frac{\xi_1^2}{k_t^2}(e^{-R_e x_2^*} - e^{-R_t x_2^*}) - \frac{1}{2}e^{-R_e x_2^*}.$$

Let us introduce the designations:

$$\widetilde{G}_q(\xi_1, x_2^*)[A]^{-1} = C(\xi_1, x_2^*), \qquad (6.5.39)$$

$$B(\xi_1, x_2^*; 0, a_2) = C(\xi_1, x_2^*)\overline{G}(\xi_1, 0; 0, a_2). \qquad (6.5.40)$$

Thus, we can define:

$$g(x^*, a; 0, a_2) = G(x^*; 0, a_2) - \frac{1}{2\pi}\int_{-\infty}^{\infty} B(\xi_1, x_2^*; 0, a_2)e^{-i\xi_1 x_1^*}d\xi_1, \qquad (6.5.41)$$

where the matrix elements $B(\xi_1, x_2^*; 0, a_2)$ have the form:

$$B_{11} = \frac{\xi_1^2}{N(\xi_1)\mu k_t^2 R_e}\left\{[(2\xi_1^2 - k_t^2)^2 + 4\xi_1^2 R_e R_t]\left(\frac{R_t R_e}{2\xi_1^2}e^{-R_t(x_2^*+a_2)} + \frac{1}{2}e^{-R_t(x_2^*+a_2)}\right)\right.$$

$$\left. -2(2\xi_1^2 - k_t^2)R_e R_t(e^{-R_t x_2^*}e^{-R_e a_2} + e^{-R_e x_2^*}e^{-R_t a_2})\right\},$$

$$B_{12} = \frac{i\xi_1}{N(\xi_1)\mu k_t^2}\left\{\frac{1}{2}[(2\xi_1^2 - k_t^2)^2 + 4\xi_1^2 R_e R_t](e^{-R_t(x_2^*+a_2)} + e^{-R_e(x_2^*+a_2)})\right.$$

$$\left. -2(2\xi_1^2 - k_t^2)R_e R_t e^{-R_t x_2^*}e^{-R_e a_2} - 2(2\xi_1^2 - k_t^2)R_e R_t e^{-R_t x_2^*}e^{-R_e a_2}\right\},$$

$$B_{21} = \frac{-i\xi_1}{N(\xi_1)\mu k_t^2} \left\{ \frac{1}{2}[(2\xi_1^2 - k_t^2)^2 + 4\xi_1^2 R_e R_t](e^{-R_t(x_2^* + a_2)} + e^{-R_e(x_2^* + a_2)}) \right.$$

$$\left. - 2(2\xi_1^2 - k_t^2)R_e R_t e^{-R_e x_2^*} e^{-R_t a_2} - 2\xi_1^2(2\xi_1^2 - k_t^2)e^{-R_t x_2^*} e^{-R_e a_2} \right\},$$

$$B_{22} = \frac{\xi_1^2}{N(\xi_1)\mu k_t^2 R_t} \left\{ [(2\xi_1^2 - k_t^2)^2 + 4\xi_1^2 R_e R_t] \left( \frac{R_e R_t}{2\xi_1^2} e^{-R_e(x_2^* + a_2)} + \frac{1}{2} e^{-R_t(x_2^* + a_2)} \right) \right.$$

$$\left. - 2(2\xi_1^2 - k_t^2)R_e R_t(e^{-R_t x_2^*} e^{-R_e a_2} + e^{-R_e x_2^*} e^{-R_t a_2}) \right\}$$

where

$N(\xi_1) = 2(2\xi_1^2 - k_t^2)^2 - 4\xi_1^2 R_e R_t$ – Rayleigh function.

Substituting $a_2 = 0$, into the obtained solution we obtain the solution of the Lamb problem of the harmonic force acting on the half-plane and applied to its boundary

$$g(x_1^*, x_2^*; 0, 0) = G(x_1^*, x_2^*; 0, 0) - \frac{1}{2\pi} \int_{-\infty}^{\infty} B(\xi_1, x_2^*; 0, 0)e^{-i\xi_1 x_1^*} d\xi_1, \quad (6.5.42)$$

where

$$G(x_1^*, x_2^*; 0, 0) = \frac{1}{2\pi} \int_{-\infty}^{\infty} \overline{G}(\xi_1, x_2^*; 0, 0)e^{-i\xi_1 x_1^*} d\xi_1 \quad (6.5.43)$$

Substituting (6.5.42) into (6.5.43), we will obtain:

$$g(x_1^*, x_2^*; 0, 0) = \frac{1}{2\pi} \int_{-\infty}^{\infty} [\overline{G}(\xi_1, x_2^*; 0, 0) - B(\xi_1, x_2^*; 0, 0)]e^{-i\xi_1 x_1^*} d\xi_1. \quad (6.5.44)$$

In particular, if a vertical force is acting on the half-plane, then

$$g_i(x_1^*, x_2^*; 0, 0) = \frac{1}{2\pi} \int_{-\infty}^{\infty} [G_{i2}(\xi_1, x_2^*; 0, 0) - B_{i2}(\xi_1, x_2^*; 0, 0)]e^{-i\xi_1 x_1^*} d\xi_1.$$

$$(6.5.45)$$

Determining the elements of $G_{i2}(\xi_1, x_2^*; 0, 0)$, from (6.5.43) after simple transformations let us find:

$$g_{12}(x^*;0) = \frac{i}{2\pi\mu} \int_{-\infty}^{\infty} \frac{\xi_1[(2\xi_1^2 - k_t^2)e^{-R_e x_2^*} - 2R_t R_e e^{-R_t x_2^*}]e^{-i\xi_1 x_1^*}}{N(\xi_1)} d\xi_1, \quad (6.5.46)$$

$$g_{22}(x^*;0) = \frac{1}{2\pi\mu} \int_{-\infty}^{\infty} \frac{R_e[(2\xi_1^2 - k_t^2)e^{-R_e x_2^*} - 2\xi_1^2 e^{-R_t x_2^*}]e^{-i\xi_1 x_1^*}}{N(\xi_1)} d\xi_1. \quad (6.5.47)$$

The solutions (6.5.46) and (6.5.47) exactly coincide with the solution of the Lamb's problem.

Figure 6.7 shows graphs of the amplitudes of oscillations on a surface of a half-plane under the action of a unit force, located at the depth of 10 m

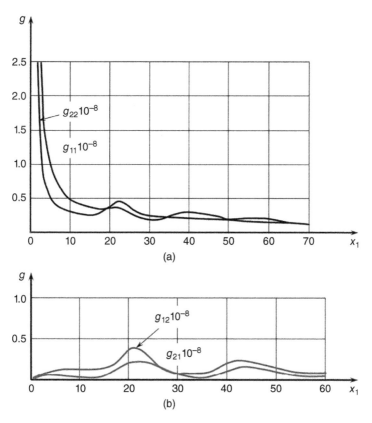

**Figure 6.7**   (a) Graphs of the amplitudes of the oscillations of the surface of a half-plane under the action of a unit force. (b) Graphs of the amplitudes of the oscillations of the surface of a half-plane under the action of a unit force.

(along the vertical coordinate axis) from the surface of ground, and changing in time according to the harmonic law. The angular (circular) frequency of the force change $\omega = 40$ radian/s, the modulus of elasticity of the material of half-plane, $E = (1) = 20000 k\Gamma/cm^2$, and the coefficient of loss (losses) $\gamma = 0.2$. The calculations were carried out according to the given above formulas on the mainframe YeS-1080.

# Chapter 7

# The Questions of the Static and Dynamic Analysis of Structures on an Elastic Foundation

Structures on elastic foundations are widespread in industrial, civil, transportation, and energy construction. Structures on elastic foundations have received considerable attention due to their wide applicability in many engineering disciplines. Since interaction between structural foundations and supporting soil has great importance in many engineering applications, a considerable amount of research has been conducted on perfecting the methods of their calculation. The improvement of these methods is an important task, since decisions are often made during structural design processes which either result in significant unnecessary reserves of strength and rigidity, or at the other extreme, can lead to reduction in their resource or damages of constructions under the action of external loads. The reason these structural errors occur is largely explained by the difficulty of modeling the elastic properties of soil. The following works are dedicated to the study

*Static and Dynamic Analysis of Engineering Structures: Incorporating the Boundary Element Method*,
First Edition. Levon G. Petrosian and Vladimir A. Ambartsumian.
© 2020 John Wiley & Sons Ltd. Published 2020 by John Wiley & Sons Ltd.

of structures on elastic foundations: S.N. Akour [2], S.M. Aleynikov [6], R.K. Anant, K. Man-Gi [20], M.A. Biot [35], N.M. Borodachev [43], S.V. Bosakov [44], W.D. Carrier, J.T. Christian [58], Z. Celep, D. Turham, R.Z. Al-Zaid [60], R. Chang, E.J. Hang, K. Rim [62], C. Chen [63], F. Cheng, C. Pantelides [65], Y.K. Cheung, O.C. Zienkievich [68], S.W. Choi, T.S. Jang [69], T.F. Conry, A. Seireg [74], D. Dinev [90], M. Eisenberger, J. Clastornik [99], C. Floris, F.P. Lamacchia [105, 106], M. Di Paola, M. Zingales,[1] K.E. Egorov, V.A. Barvashov [96], K.E. Egorov, I.A. Simvulidi [97], N.M. Gersevanov [117], M.I. Gorbunov-Pasadov [121, 122], S. Guenfoud, S.V. Bosakov, D.F. Laefer [129], M. Hetenyi [142], C. Hu, G.A. Hartley [145], T.S. Jang [155], F.S. Kadish [158], V. Kolar, I. Nemic [180], V.M. Kulakov, B.M. Tolkatchev [197], B.Y. Lashchenikov [205], A.D. Kerr [166–168], G.K. Klein [177], A.N. Krylov [192], Q. Liu, J. Ma [211], B.P Makarov, B.E. Kochetkov [219], L.I. Manvelov, E.S. Bartoshevich [221], E.S. Melerski [226], D. Milovic [234], P.L. Pasternak [261], M.M. Filonenko-Borodich [103, 104], V.Z. Vlassov, N.N. Leontiev,[2] B.N. Zhemochkin, A.P. Sinitsyn [414], Y.T. Chernov [66, 67], V.A. Ilichev [148, 150], A.G. Ishkova [154], S.N. Klepikov [178], B.G. Korenev [186–189], E.A. Palatnikov [253], J. Park, H. Bai, T.S. Jang [255], G. J. Popov [291, 292], L.G. Petrosian [274–277], E.J. Sapountakis, J.T. Katsikadelis [313], O.I. Schechter [318], A.P.S. Selvadurai [320], V.M. Seymov [322], G.S. Shapiro [323][3], I.Y. Shtaerman [328], I.A. Simvulidi [331, 332], S.N. Sinha [333], S.N. Sirosh, A. Ghali, A.G. Razaqpur [336], F.R. Skott [338], C. Skroll [339], D.N. Sobolev [344], I.B. Teodoru [352–354], D. Thambiratnam, Y. Zhuge [355], W.T. Straughan [346], O.J. Svec, R.M. Hardy [349], S.P. Timoshenko [357, 358], G.R. Tomlin, R. Butterfield [359], K. Tiwari, R. Kuppa [386], V.I. Travush [360–364], A.I. Tseitlin [366, 367, 373], N.A. Tsitovich,[4] E. Tsudik [383, 384], A.S. Veletsos, B. Verbic [392], T.M. Wang, J.E. Stephens [402], Y Wang, B. Zhang, S. Shepard [403], P. Zissimos, Z.P. Mourelatos, M.G. Parsons [417], J.H. Yin,[5]and many others. Many studies, have pursued finding a convenient representation of physical behavior of a real structural component supported on a foundation. There are several practical foundation models as well as their proper mathematical formulations. A broad

---

[1] Di Paola M., Zingales M. (2008). Long-range cohesive interactions of non-local continuum faced by fractional calculus. *International Journal of Solids and Structures*, **45** p. 5642–5659.

[2] Vlassov V.Z., Leontiev N.N. Beams, plates, and shells on an elastic foundation. M.Phizmatgiz, 1960, 492p.

[3] Selvadurai A.P.S. A contact problem for a Reissner plate and an isotropic elastic half space. *J. Mech. Theor. Appl.* **3**, 181–196.

[4] Tsitovich, N.A. 1973. *Soil Mechanics.* Moscow: Visshaya Shkola. 560p.

[5] Yin J.H. Closed Form Solution for Reinforced Timoshenko Beam on Elastic Foundation, *J. of Engin. Mech.*, 2000, Vol. **126**, 868–874.

range of engineering problems, such as beam and plates, have been solved numerically with finite element and boundary element methods (BEM).

The methods of analysis of beams on an elastic foundation were produced by Karmvir Tiwari and Ramakrishna Kuppa [386]. The response of a beam on an elastic foundation under static and dynamic loads was studied. The foundation model is based on the Winkler hypothesis. Winkler assumed the foundation model consists of a closely spaced independent linear spring system. According to Winkler's idealization, the deformation of the base due to applied load is confined to loaded regions only. The fundamental problem with using this model is its failure to account for any soil coherence. The paper presents various methods for treating the analysis of beams on elastic foundation subjected to point load in the transverse direction. An example problem is solved using finite element package ANSYS, where the methods of analysis of beams on elastic foundation are studied, and the finite element method for static and dynamic analysis of beams is provided. An example is taken and solved for both static and dynamic loading and the results show the behavior of beams on elastic foundation when subjected to loads with spatial and temporal variation. This provides good groundwork to study structures on continuous foundations.

A numerical procedure proposed by Jinsoo Park, Hyeree Bai, and T.S. Jang [255] is applied for the numerical analysis of static deflection of an infinite beam on a nonlinear elastic foundation. A one-way spring model is used for the modeling of fully nonlinear elastic foundation. The nonlinear procedure involves Green's function technique and an iterative method using the pseudo spring coefficient. The workability of the numerical procedure is demonstrated through showing the validity of the solution and the convergence test with some external loads.

Plates and tanks on elastic foundations as an application of the finite element method have been examined by Y.K. Cheung and O.C. Zienkiewicz [68]. The authors explained that in the case of a Boussinesq medium the solution was pursued mathematically, and in the case of a Winkler's medium the solution was found analytically. The problems of slabs and tanks resting on semi-infinite elastic continuum (Boussinesq's type) or on individual springs (Winkler's type) were solved using the finite element method. The paper "Dynamic Timoshenko Beam Columns on Elastic Media" has been presented by F. Cheng and C. Pantelides [65] in the Journal of Structural Engineering. The Timoshenko model of elastic foundation is based on Timoshenko beam theory, and considers both the bending and shear deformations. The plane sections still remain plane after bending but are no longer normal to the longitudinal axis. Differential equations, stiffness coefficients, and fixed end forces are formulated for a Timoshenko beam-column on an elastic foundation subjected to lateral time-dependent excitations and static axial loads. The theoretical formulation includes shear and bending deformations as well as rotatory inertia, with emphasis on two approaches. These two approaches differ in terms of the assumed shear component of the static axial load on the cross section. The first approach is based on the assumption that the shear component of the axial load is calculated from the total slope; in the second approach, the shear component of the axial load is calculated only from the bending slope. The dynamic stiffness coefficients and fixed-end forces are expressed in terms of non-dimensional parameters associated with the effects of transverse and rotatory inertia, axial force, elastic media, and shear and bending deformations. When the individual effect is not considered, the associated parameter can be dropped. The significance of the individual parameters on natural frequencies and dynamic response of typical beams is then extensively examined. These two approaches are also studied by comparing the response behavior, and it is found that they differ appreciably with increasing axial loads and decreasing slenderness ratios. Comparison of the natural frequencies shows that the second approach gives higher values than the first.

An analytical formulation using the principle of minimum potential energy is presented by R.K. Anant and K. Man-Gi [20] to study the dynamic behavior of a rectangular plate resting on an elastic half-space and subjected to a uniformly distributed load. In this work, the compatibility at the interface of the plate and soil medium is satisfied by integrating the Boussinesq's formula that relates the contact stress and the soil surface deformation.

The response of a Bernoulli-Navier beam having visco-elastic behavior and resting on elastic medium was studied by C. Floris and F.P. Lamacchia [105, 106, 401]. In these works, the authors presented an algorithm for the compatibility of the displacements between the beam and the elastic medium.

The dynamic analysis of a plate resting on elastic half-space with distributive properties has been studied by Salah Guenfoud, Sergey V. Bosakov, and Debra F. Laefer [59, 64, 129]. The study provides a semi-analytical approach to determine the natural frequencies and natural shapes of a rectangular plate resting on an elastic foundation with distributive properties (Boussinesq's type) and its response to an external vertical harmonic excitation. The following aspects are neglected: damping, inertia of the elastic foundation, and friction in the contact zone between the plate and the surface of the elastic foundation. The approach is based on the mixed method known as Zhemochkin's method [414], wherein the rectangular plate is divided into a finite number of identical elements, and in the center of each element is placed a rigid link through which the contact between the plate and the surface of elastic foundation is accomplished. The approach assumes that contact between the structure and the surface of an elastic foundation is replaced by a finite contact in rigid links and that the mass of each element is concentrated in its center.

This work gives a semi-analytical approach for the dynamic analysis of a plate resting on an elastic half-space with distributive properties. Such calculations have been associated with significant mathematical challenges, often leading to unrealizable computing processes. Therefore, the dynamic analysis of beams and plates interacting with the surfaces of elastic foundations has, to date, not been adequately solved. To advance this work, the deflections of the plate are determined by the Ritz method, and the displacements of the surface of elastic foundation are determined by studying Green's function. The coupling of these two studies is achieved by a mixed method, known in the theory of elasticity as Zhemochkin's method [414], which allows determination of reactive forces in the contact zone and, hence, the determination of other physical magnitudes. The obtained solutions can be applied to study the dynamic interaction between soils and structures and to assess numerical computations through various numerical methods programs. Natural frequencies, natural shapes, and the dynamic response of a plate due to external harmonic excitation are determined. Validation with a Winkler problem illustrates the distributive property effects on the results of the dynamic analysis.

The dynamic behavior of nonlinear beams on an elastic foundation has been studied by S.N. Akour [2]. A simply supported nonlinear beam resting on a linear elastic foundation and subjected to harmonic loading is investigated. Parametric study is carried out in the view of the linear model of the problem, and Hamilton's principle is utilized in deriving the governing equations. The well-known forced doffing oscillator equation is obtained and the equation is analyzed numerically using the Runk-Kutta technique. Three main parameters are investigated; the damping coefficient, the natural frequency, and the coefficient of the nonlinear term. Stability regions are unveiled.

The evaluation of large deflections of plates on elastic foundations was examined by S.N. Sinha [333]. The evaluation of moderately large deflections of uniformly loaded plates on elastic foundations of the Winkler type is presented using Berger's approximation, which neglects the strain energy due to the second invariant of the middle surface strains. Because of this approximation, the problem may be formulated in terms of two decoupled nonlinear equations. Series solutions are obtained for circular and rectangular plates under various support conditions, and are numerically evaluated and presented in the form of graphs for plates of various aspect ratios with simply supported edges. Results obtained are compared with some known solutions, and it is concluded that Berger's approximate method yields results of sufficient accuracy for practical purposes.

The basic response functions for elastic foundations was studied by Anestis S. Veletsos and Branislav Verbic [392]. Approximate, close-form expressions are presented for the response to both harmonic and impulsive excitations of a massless, circular, rigid foundation which is supported at the surface of an elastic half-space and is excited horizontally, vertically, or in rocking motion. The impulse response functions considered include: (i) functions for the displacements of the disk due to unit impulsive forces; and (ii) functions for the forces necessary to produce unit impulsive displacements. The functions of the first type are also evaluated numerically, and the results are used to assess the accuracy of the analytical approximations proposed. Some comparisons are also made with results obtained for a viscously-damped, simple oscillator with or without mass. With the impulse response functions presented, analyses of the transient response of foundations and foundation-structure systems can be carried out directly in the time domain. This approach may be more efficient or of wider applicability in specific applications than the corresponding frequency-domain approach.

The analysis of thick plates on elastic foundations by finite elements is conducted by Otto J. Svec, and R.M. Hardy [349]. The plate is subdivided into an assemblage of triangular elements. The triangular plate-bending element developed earlier for contact problems is here enriched by shear degrees-of-freedom to include the plate thickness effect. The vertical displacement field of the plate element is approximated by a seventh-degree polynomial, resulting in 33 degrees-of-freedom, while shear is represented by independent linear or cubic polynomials. The total number of degrees-of-freedom for these two types of elements is 39 and 53. The analysis is performed for two different foundation models: (i) a homogeneous isotropic elastic half-space; and (ii) a Winkler spring model. The contact between the plate and its support is assumed to be frictionless. The contact pressure is approximated by cubic polynomials under each triangular element. A parametric study is performed by varying plate thickness and elastic moduli of the plate and foundation. Results, deformations, and contact stress are presented and analyzed.

In the work of Q. Liu and J. Ma [211] an analytical method is introduced to deal with the coupling problem for Euler-Bernoulli beams on elastic bi-dimensional foundations by considering the horizontal and vertical displacements of the beam-function system. The approach is an extension of the modified V.Z. Vlasov model [400]. With separation of the variables, the horizontal and vertical displacements are expressed as the displacements function at the ground surface and the attenuation function along the depth of the foundations, respectively. The governing equations and the corresponding boundary conditions of the model are obtained via the variational principle. Then, the differential operator method is used to uncouple the governing equations and boundary conditions. An iterative procedure is executed to accomplish the numerical implementation. A parametric study is conducted to illustrate the effects of the applied loadings and the physical and geometry properties on the static responses of the beam and foundations. The numerical results show that the coupling elasticity should be taken into account as cases of flexible beams and high soil Poisson ratios. Moreover, the horizontal loads on the beam significantly affect the response of the beam-foundation system.

The analysis of inhomogeneous elastic half-space has been presented by W. David Carrier and John T. Christian [58]. The solutions for various loads applied to homogeneous and inhomogeneous elastic half-spaces have played an important role in the development of foundation engineering, particularly in conjunction with the design of rafts and mats. In the present paper, a finite element computer program was used to solve for the stresses and settlements of a circular load resting on a half-space. The computer results agree quite well with the theoretical answers in those cases for which analytical solutions exist. These results give confidence to the finite element analyses for those cases for which no analytical solutions have been found.

In this study conducted by M. Di Paola, F. Marino, and M. Zingales [89] a different generalization of the Winkler model has been introduced, representing the interactions between foundation elements as distance-decaying body forces that each column applies on adjacent and non-adjacent surrounding foundation elements. The common models of elastic foundations are provided by supposing that they are composed by elastic columns with some interactions between them, such as contact forces, that yield a differential equation involving gradients of the displacement field. In this paper, a new model of elastic foundation is proposed introducing into the constitutive equation of the foundation body forces depending on the relative vertical displacements and on a distance-decaying function ruling the amount of interactions. Different choices of the distance-decaying function correspond to different kind of interactions and foundation behavior. The use of an exponential distance-decaying function yields an integro-differential model, while a fractional power-law decay of the distance-decaying function yields a fractional model of elastic foundation ruled by a fractional differential equation. It is shown that in the case of exponential-decaying function the integral equation represents a model in which all the gradients of the displacement function appear, while the fractional model is an intermediate model between integral and gradient approaches. A fully equivalent discrete point-spring model of long-range interactions that may be used for the numerical solution of both integral and fractional differential equation is also introduced. Some Green's functions of the proposed model have been included in the paper and several numerical results have also been reported to highlight the effects of long-range forces and the governing parameters of the linear elastic foundation proposed.

In the work of D. Dinev [90] an analytical solution of beams on elastic foundations by singularity functions deals with a new manner of obtaining a closed-form analytical solution of the problem of bending of a beam on elastic foundation. The basic equations are obtained by a variational formulation based on the minimum of the total potential energy functional. The basic methods for solving the governing equations are considered and their advantages and disadvantages are analyzed. The author proposes a felicitous approach for solving the equilibrium equation and applying the boundary conditions by transformation of the loading using singularity functions. This approach, combined with the resources of the modern computational algebra systems, allows a reliable and effective analysis of beams on an elastic foundation. The numerical examples show the applicability of the approach for the solution of some problems of soil-structure interaction.

The approximated and the numerical methods such as Ritz and Galerkin methods, finite differences, finite elements, and differential equation quadrature's methods are widely used for solving basic equations by many scientists; I.B. Teodoru [352–354], C. Chen [63], V. Kolar [180], E.S. Melerski [226]. All of these methods are superior in comparison with the analytical methods for solving complex problems with various boundary conditions or loads. The numerical methods also have disadvantages such as difficulty studying the influence of the problem's parameters on the solution, and the need of a special computer program to obtain a solution of the problem. The implementation of a special finite element or the development of a computer code requires particular skills from the user [45].

In the books of E. Tsudik [383, 384], the author provides practical methods of analysis and describes in detail various foundations, including simple beams on elastic foundations, and complex foundations such as supported on piles. Most of the methods are for three soil models; Winkler foundation, elastic half-space, and elastic layer. Numerical examples throughout the book illustrate the applications of the methods. Analysis of beams on elastic foundation is performed currently by using special computer programs based on numerical methods, such as Finite Difference Method and Finite Element Method. However, these programs are not always available, and their application is limited. They cannot be used for other soil models such as elastic half-space or elastic layer and others. The research also includes practical recommendations for analysis of beams and frames on Winkler foundation, by replacing the soil with elastic spring supports and individual foundations with line elements that allows using programs for analysis of statically indeterminate systems to analyze various beams and frames supported on elastic foundation. A significant part of [383] is dedicated to the Method of Initial Parameters and its application in analysis of beams and frames on elastic foundation. The proposed method of analysis takes into account the absolutely rigid elements located at the area of beam intersection. A computer program based on this method allows performing computer analysis of beams on elastic foundations. Several numerical examples illustrate application of the method to practical analysis. This book also includes tables developed by I.A. Simvulidi [331, 332] that allow analyzing beams supported on elastic half-space.

In the theory of analysis of structures on elastic foundations and soil mechanics, several models of stress-strained foundations allowing with some degree of certainty to describe the stress-strained condition of the ground with its interaction with coatings or above-ground constructions were developed. The most widely used models of elastic homogeneous and heterogeneous half-space, and elastic layer, enable taking into account the distributive properties of soil. Discrete models of Winklerian type do not take into account these properties, and therefore are not exact; they are however, simple to use in practical computing. The Winklerian model is based on the pure bending beam theory, and the deformations outside the loaded area are neglected and taken as zero. No interaction between springs is considered, and the spring constant may depend on a number of parameters, such as stiffness of structure, geometry of structure, soil profile, and behavior. In recent decades the practice has started to use combined models (Repnikov's model). Historically, the first and simplest model with one coefficient of bed was the Fuss-Winkler model, which has come under criticism in different periods of time as insufficiently experimentally substantiated. However, at present it is more frequently used in practical calculations than other models.

The significant number of experimental works performed in recent years by F.S. Kadysh [158], L.I. Manvelov, E.S. Bartoshevich [221], and E.A. Palatnikov [253] showed that the model with one coefficient of bed gives satisfactory results in the analysis of beams and plates lying on a sandy base or thin enough layer of a deformable soil. With the dynamic calculation of plates on elastic foundation the model with one coefficient of bed bears a more conditional nature, since it does not account for the inertia properties of the base. In a number of research works, however, it is noted that the account of the inertness of base has relatively little influence on the magnitudes of displacements in dynamic computing of extended beams and plates; see Z.M. Gershunov, A.S. Yakovlev, or N.A. Nikolaenko [248]. In the work of V.A. Ilichev and V.G. Taranov [150] the transfer functions of system ground base, obtained experimentally, were approximated, in particular by the transfer functions of systems with one degree of freedom. The formulas for determining the equivalent rigidity and joined mass of ground are presented, and it is shown that in the frequency range of 16–24 Hz these values weakly depend on frequency of induced (forced) oscillations. Experiments also showed that the amplitudes of the oscillations of the surface of ground rapidly decrease with the distance from the source. This, apparently, gives reason to take into account corrections according to the results of experiments modeled with one coefficient of bed and with the dynamic calculations of extent plates. Confirmation of this can be found in the results of experimental studies of the oscillations of the extent plates (Z.I. Berodze, Y.T. Chernov). From the works dedicated to the static computation of extent structures on the elastic foundation, we first should note the scientific works of B.G. Korenev [186, 187], in which models for calculation of beams and plates were considered, according to the design schemas, as unlimited. The influence functions or Green functions are constructed and a significant number of problems calculating beams and plates on general or linear-deformed base are solved. The detail calculation of isotropic elastic plates based on those modeled with one coefficient of bed, is examined. The method of the compensating loads is also well developed in the works of B.G. Korenev [186], and in combination with the solution for the unlimited domains, it is the effective method of calculation of the structures of finite dimensions. In recent years this method underwent further development in the works E.S. Ventsel [393], for solving some problems of structural mechanics. From the research works dedicated to the static calculation of the unlimited beams and plates with different models of linear-deformed base let us also note works by I.G. Alperin, Y.P. Ziukin, G.S. Tselikov, G.A. Malikova, Y.K. Tkachev, A.I. Tseitlin, and O.Y. Shakhter. In those cases when the location of the load is near the edge of the extent plate, the influence of two edges can be neglected, and displacements and stresses are determined from the calculation of semi-infinite plate. In the work of G.S. Shapiro, the Fourier transformations were used in the construction of the function of influence for the semi-infinite plate, lying on the Winklerian base. The Wiener-Hopf method is used in the work of R.V. Serebriany [321] with the solution of the problem about the calculation of the boundary section of the plate resting on the elastic layer of finite thickness.

In the work of B.G. Korenev and E.I. Chernigovskaya [189] the solution is given for the semi-infinite plate with the aid of the method of compensating loads, and the tables needed for calculation are presented. The scientific works of M.I. Gorbunov-Pasadov [121, 122], G.J. Popov [291, 292], V.I. Travush [360, 361], and A.I. Tseitlin [371], L.G. Petrosian [278, 377] are dedicated to the static computation of semi-infinite plates. A number of problems of calculating non-isolated structures on an elastic foundation, namely hinged plates or unconnected aligned plates are examined by B.G. Korenev [186, 188] and R.V. Serebriany [321].

As an alternative to the Winkler foundation, the foundation was considered as semi-infinite and on isotropic elastic continuum (M.I. Gorbunov-Pasadov [121, 122]), increasing the mathematical complexity of the model. However, despite its completeness, it has been shown that for solid-like materials the displacement field away from the loaded region decays faster than their counterpart, as predicted with the use of elastic half-space model. The idea to represent foam rubber-like material with high numbers of voids with the continuum mechanics theory of isotropic solids may also be questionable.

A.G. Ishkova [154] examined circle plates under a uniform load. For that purpose she used reaction pressure in the form of a sum of addends, one of which was determined by the Buscinesk solution, and another of which was determined by the power series. It was shown for the first time in this work that theoretically the reaction pressures are equal to infinity at the edge of the flexible structures. For another task with the axial symmetry of the plate of unlimited sizes and loaded concentrated force, O.I. Schechter [318] obtained exact solutions in the integrals, which contain Bessel functions. This solution is similar to the Hertz solution for the unlimited plate by Winkler, which was widely used in practice of the design of continuous plates under the grid of columns. V.I. Travush [363, 364] proposed and used the method of the generalized solutions during calculation of the uninsulated structures on the linear-deformed base of common form. Let us point out that the solutions obtained in the calculation of the uninsulated structures can automatically be applied in the calculation of semi-infinite plates on the Winklerian base. In his research works, Y.T. Chernov [66, 67] developed the method of calculation of the extent bending structures (beam and plates) on the elastic foundations with the areas of heterogeneity and anisotropy under the action of static and dynamic loads, while also taking into account the physical nonlinearity and creep of the material of plate and base under the action of static loads. The proposed method is based on the reduction of deflection equations and physical dependencies for anisotropic bending elements to integral equations of the second kind in the domain relative to the fictitious self-balanced load of the type of the distributed angular deflection (potentials of special type). The kernels of integral equations are derivatives of Green's function of the operators, which correspond to the equations of bending of elastic elements with constant rigidity. In the calculation of a structure of finite dimensions by the methods of boundary integral equations (BIEM), the system of resolving integral equations along with (BIEM) will also contain integral equation of the second kind for the domain.

The works of P.K. Banerjee, R. Butterfield [27], C.A. Brebbia, C. Walker [50], C.A. Brebbia, Zh. Telles, L. Wrobel [49], Y.V. Veryuzhsky [394–396], Y.D. Kopeikin [183], V.D. Kupradze [199], and V.M. Kulakov, B.M. Tolkachev [197] obtain the methods based on the BIEM and methods of the theory of the potential recently gaining widespread acceptance in the calculations of the structures occupying finite domains. The calculation of structures lying on the elastic foundation of the coherent type is the area of structural mechanics where the standard calculation methods are difficult to apply directly. There are also well-known defects of such models as elastic half-plane, caused by the unlimited displacements of surface with an unbalanced load. The coherency of the base when using linear models, based on various solutions of the problems of the theory of elasticity, also causes unsurmountable obstacles during the design of some types of structures, in particular uninsulated structures according to the terminology of B.G. Korenev [188] of beams and plates with free ends (edges). Uninsulated structures are widely used in the construction of roadway and airport pavements, foundations of buildings, coating slopes of hydraulic structures, and so on. Designing them for static and dynamic actions is very important; however, due to the above-mentioned difficulties associated with the advent of nonintegrated singularities in the points corresponding to the joints of the uninsulated structures, their calculation on a coherent basis was not developed free from strong assumptions until now. Actually, for uninsulated beams and plates with free ends (edges), lying on the coherent base, strict design (calculation) methods are not available. Another shortcoming of the existing models is their non-universality; different models are required for different types of grounds. Of course, the most convenient way would be to describe the properties of different soils (grounds) within the framework of a united model, changing only the necessary values of the physical parameters of this model. In the present work the aim is to develop a universal model of base, which would be deprived of the above-indicated deficiencies. In particular, this model must satisfy the following conditions:

1. Provide limited displacement of the base surface under any real actions (impacts, influences).
2. Safeguard the distributive properties of the base and offer the ability to calculate the jumps of displacements on its surface, i.e. offer the special features of both discrete and continuous models.
3. Possess universality and the ability of obtaining parameters of the most common models of the base with the limiting values, such as homogeneous elastic half-space (semi-continuum), half-space (semi-continuum) with the modulus of elasticity changing with the depth according to the power law, and Winklerian base.

Satisfaction of the first condition makes it possible to obtain absolute sediments of base with all load cases, in particular with the plane strain (deformation). Due to the second property the model provides the ability to solve any problems for uninsulated structures. Finally, the third condition is necessary for the unity of approaches to the calculation of structures on ground basis of different geological structure and the comparability of the results, obtained for different types of ground conditions.

## 7.1   The Kernel of the Generalized Model of Elastic Foundation (Base)

Simultaneously with the development of the theory and methods of calculation according to the hypothesis of elastic half-space, work was conducted on the use of the models of ground, the distributive ability of these models would fall somewhere in between Winkler elastic half-space models. The most important question in the theory of analysis and design of structures on ground base is the choice of the base model. In research works of E. Winkler, V.Z. Vlasov, N.N. Leontiev [400], N.M. Gersevanov [117], K.E. Egorov [94, 95], B.N. Zhemochkin [414], G.K. Klein, [177], B.G. Korenev [188], K. Marguerre, P.L. Pasternak [261], G.Y. Popov, [291,292], A.R. Rzhanitsin [312], A.P. Sinitsin [335], D.N. Sobolev [344], M.M. Filonenko-Borodich [103, 104] and others, a significant number of mechanical and mathematical models were proposed, each of which has its advantages and disadvantages, as well as a specific area of application, in which it gives the most accurate results.

One of the first intermediate hypotheses was proposed by K. Wieghardt,[6] who examined the possibility of using the theory of elasticity for computing structures on elastic foundations, according to which the relationship between the pressure and settlement is expressed through diminishing exponential function. M.M. Filonenko-Borodich proposed to use a membrane and a laminar model of ground (soil). The distributive ability of the model is achieved by introducing non-extendable thread with constant horizontal projection of tension to the top of a set of springs. This model requires continuity between the individual spring elements in the Winkler's model by connecting them to a thin elastic membrane under constant tension. In spatial model the thread is replaced by a membrane. V.Z. Vlasov proposed a theory of calculation of beams on elastic foundation, using a variational method of bringing two-dimensional problems of the theory of elasticity, described by partial differential equations, to simpler one-dimensional problems. Further development of this theory extended into the works of N.N. Leontiev, where the model of the base was taken in the form of vertical plates on the upper rectilinear edge on which a beam rests. The problem is considered under

---

[6] Wieghardt K. Uber den Balken auf nachgiebiger Unterlage "Zeitchrift fur Angew. Mathematik und Mechanik." Bd. 2, H. 3, 1922.

conditions of plane stress state (planar stressed state, two-dimensional stress state). The plate (disc) has limited height and rests on the uncompressible base. It is assumed that the deformations of transverse elongation remain constant on the width of base, and longitudinal displacements are absent. In addition, another member with a second derivative is introduced, therefore making it possible to consider the influence of shearing stresses in the elastic base Hetenyi's model of elastic foundation suggested in literature, can be regarded as a fair compromise between two extreme approaches: Winkler foundation and isotropic continuum. In this model, the interaction among the discrete springs is accomplished by incorporating an elastic beam or an elastic plate, which undergoes flexural deformation only. The method of superposition is presented in [142]. The method uses solutions of simple problems of infinity long beams with different simple loads to construct the final solution of an arbitrary beam, loads, and supports.

The model of elastic foundation proposed by P.L. Pasternak [261], is also characterized by two parameters. The springs are connected to an isotropic layer of incompressible vertical element which deform in transverse shear only. The model assumes the existence of shear interaction between the spring elements. The first of these, $c_1$, connects (links) the intensity of the vertical rebuff (resistance) of ground $\sigma$ with its settlement $w$ by Winklerian dependence $\sigma = c_1 w$. The independent from $c_1$ second parameter – shear coefficient, makes it possible to determine the intensity of vertical shearing force $t$ as a product of $c_2$ on the derivative of settlement in the corresponding direction $t = c_2 \partial w / \partial x$. The continuity in this model is characterized by the consideration of the shear layer. A comparison of this model with that of M.M. Filonenko-Borodich [104] implies their physical equivalency. It should be noted that the two-parameter models mentioned above have the same deficiency; during computing, concentrated forces are detected at the ends of the beams and on the edges of plates, even in the case when the edges of structures are free. The authors of two-parameter models call these forces fictitious, however, the interpretation of a disabled entity cannot be correct, because the ground cannot take the load in the form of concentrated forces. When using two-parameter models we are freed from one of the shortcomings of calculating in from the hypothesis of elastic half-space (semi-continuum), where damping of the settlements either is absent or it occurs more slowly than the real settlements of the surface of the ground. Another deficiency is the appearance of fictitious forces on the edges of structures, and some difficulties also arise when determining the numerical values of the model parameters. In addition to the two-parameter models in technical literature, we also encounter the three-parameter model of V.A. Barvashov and V.G. Fedorovsky, in which two-bed coefficients are covered with the layer of Winkler's spring with stiffness $c_3$. This model satisfies the conditions of monotone decreasing settlements as the distance decreases from the point of application of the load, i.e. at infinity, the task may have a limited solution and presence of the break of settlement in the places of breaking load. The universal formula for determining the low of distribution of stiffness coefficient (the model of the variable coefficient of bed S.N. Klepikov[7]) under the rectangular foundations has been proposed by S.A. Rivkin.[8]

The model of stiffness coefficient is adapted works of many authors including P.P. Shagin; B.A. Kositsin [190]; V.I. Lishak; D.N. Sobolev [344]; Ch. Gracgoff,[9] and others. The method of M. Kany[10] adjoins to calculation using a variable coefficient of bed, where in essence, the stress is defined taking into account the model of homogeneous elastic half-space, but the deformations at each point of surface are considered as the sum of compressions of separate layers in soil column under this point from the corresponding stresses. It is known that settlement of the surface of the base can depend on the bearing surface of the foundation. The greater the area of footing , the more deeper, condensed, and compacted layers enter into work, i.e. scale factor in calculation of structures on elastic deformable base becomes relevant. G.K. Klein [177] addressed this phenomenon by introducing a module that varied with depth, as, for example, linear or square parabola. K.E. Egorov [94] introduces not the module of elastic half-space, but the compressible layer, below which the base is taken as absolutely rigid. The thickness of the layer is selected so that with the modules of deformation obtained by the conventional stamp testing, the calculated settlement of the surface layer would correspond to real actual settlement. K.E. Egorov [95] also developed the method of calculation of round plates of finite stiffness on the compressible layer. E.F. Vinokurov developed the iteration technique of calculating the bases with the application of a model of the layer of finite thickness in combination with the finite-differences method, which was further developed in the works of M.S. Gritsuk. The value of internal forces in the structure interacting with the deformable base is determined by the law of distribution of reactive pressure. A flat

---

[7] Klepikov, S.N. (1967). *Design Structure on Elastic Foundation*. Kiev: Budivelnik.
[8] Rivkin, S.A. (1969). *Design Foundations*. Kiev: Budivelnik.
[9] Graszhoff, H. (August, 1951). *Ein einfaches Naherungvefahren fur Berechnung elastisch gehelttetes Balken*. Die Bautechnik.
[10] Kany, M. (1972). Berechnung von Flachen grundungen. Berlin, 1959.

sole is almost always non-optimal, since it causes significant bending moments in the plate or beam. The application of a convex surface of support can achieve the elastic working base under the entire sole, and also reduce internal efforts in the structure, due to the rational distribution of contact stresses. Problems of control of the distribution of reactive pressure attracted the attention of many authors, including of N.M. Borodachev, P.D. Yevdokimov, S.A. Rivkin, F. Simonsen, E.A. Sorochan, A.N. Tetior, K. Hirschfeld, and others. Two approaches to the solution of this problem are known. One of them consists of the creation of a heterogeneous base in the plan, the deformation of which under the middle part of the foundation (place of the application of load from the above ground construction) is less than at other points. Such a condition is possible to attain either by the method of the unequal compaction of base or by replacement of its specific sections by less deformable materials (sand, gravel, by concrete layer). The second approach consists of the application of foundations with the nonplanar sole, the shape of which may be formed either in the process of construction loading or in the manufacture of the foundation plate (slab), which is apparently the most appropriate method of regulating the distribution of the reactions of the base. In particular, N.M. Borodachev [43] determines the required shape of the sole of the die, after assigning any desired diagram of reaction pressure. Accounting for distribution properties of the elastic foundation in the analysis of structures using BEM required not only development of methods of solution integro-differential systems, but also the specific improvements of the models of elastic foundations themselves. Two ways are possible. The survey of literature shown above is given partially from the works of M.I. Gorbunov-Pasadov, T.A. Malikova,[11] B.I. Solomin,[12] and Y.T. Chernov [66], and relates to the directions associated with the construction of models, based on their internal mechanical content. Since the main disadvantages in the most common models of the elastic foundations with distributive properties elastic half-space and elastic half-plane are the unlimited stress concentration at the boundary points of contact and unlimited displacements with the unbalanced load (for the half-plane), then improving these models would be accomplished by the introduction of the mechanical corrections of a geometric (rounding of angles by I.Y. Shtaerman [328]) or physical (introduction of zones of plastic

---

[11] Gorbunov-Pasadov, M.I. and Malikova, T.A. (1973). *Calculation of Structures on Elastic Foundation.* Moscow: Stroyizdat, 627.
[12] Solomin, V.I. (1973). *To the Calculation of Foundation Plates with the Load Applied Near the Corner.* Moscow: SMCC, 2.

deformations) character. However, this way leads to extremely complicated and cumbersome models, the practical use of which is not feasible in real structures. A different way of constructing the models, proposed by B.G. Korenev [188] and developed by G.J. Popov [291], consists of determining the external reactions of base without the penetration into the mechanical scheme of load. Such external reaction is Green's function of base, called also the kernel of the model of base. By definition the kernel of base, which enters into the integro-differential system of equilibrium or motion of structure on elastic foundation, are actually postulated properties and behavior under various types of loading. Therefore, for improving the models of coherent elastic foundation we choose the ground as the most simple and not requiring a priori mechanical constructions for such a complex, heterogeneous, and varied medium. Furthermore, it should be noted that the determination of the mechanical properties of the base through its kernel has another important advantage over the purely mechanical models. This advantage consists of the opportunity of the simple and economic determination of the characteristics of the kernel directly from the results of soil tests, while the experimental determination of mechanical parameters of the model can become the independent task, the solution of which involves considerable technical difficulties. The most convenient method of describing the complete class of the linear-deformed models of elastic foundations through the kernel of base (function of influence), which is to coordinate time dependence of the displacement of the surface of base due to action of unit force, was developed in detail by B.G. Korenev [188], and then integrated by G.J. Popov [291, 292].

The kernel of any isotropic linear-deformed base, i.e. the vertical displacement of the surface of base at a point $(x, y)$ due to action of vertical unit force applied at the point $(x_1, y_1)$, is the function of the distance between these points $\theta$

$$K(x, x_1; y, y_1) = K_0(r), \tag{7.1.1}$$

where $\theta$ – is a certain physical constant, determined experimentally or on the basis of solution of the corresponding mechanical problems. Using a kernel of the model of elastic foundation (i.e. actually Green's function of the boundary-value problem which describes the stress-strained state of medium of base), it is possible to write the integral relations connecting the displacement surface of base $w(x, y)$ with the contact stress $p(x_1, y_1)$

$$\iint_{-\infty}^{\infty} K\left(\sqrt{(x - x_1)^2 + (y - y_1)^2}\right) p(x_1, y_1) dx_1 dy_1 = w(x, y),$$

and thereby, to obtain a sufficiently simple formulation of the contact conditions necessary in the solution of any contact problem or the problem of loading by the external load surface of the base. Our task is to generalize three basic models of the elastic foundation within the framework of united universal model of base. The kernel takes the following form (with accuracy up to constant):

- Elastic homogeneous isotropic half-space

$$K_0(r) = \frac{1}{2\pi r} \tag{7.1.2}$$

- Elastic half-space with the modulus of elasticity, which is changed with the depth, according to the power law (model G.K. Klein)

$$K_0(r) = \frac{1}{(2\pi r)^{1+\nu}} \tag{7.1.3}$$

where $\nu > 0$ the parameter, which determines the dependence of the modulus of elasticity on the depth.

For the Winklerian base the kernel of model can be written as

$$K_0(x - x_1; y - y_1) = \delta(x - x_1)(y - y_1) \tag{7.1.4}$$

It is necessary to conduct complicated transformations of generalized functions in order to come to variable $r = \sqrt{(x - x_1)^2 + (y - y_1)^2}$. Therefore let us defer receiving representations for the kernel $K_0(r)$ in the case of the Winkler foundation. Naturally, it is possible to come to it after consideration of properties of integral transformations of kernels of linearly-deformable foundation. The Fourier and Hankel transforms for the kernel of base can be built determining the following integral representations [188, 292].

$$K(x, y) = \frac{\theta}{4\pi^2} \iint_{-\infty}^{\infty} h\left(\sqrt{\alpha^2 + \beta^2}\right) \frac{\exp[-i(\alpha x + \beta y)]}{\sqrt{\alpha^2 + \beta^2}} d\alpha d\beta \tag{7.1.5}$$

$$K(r) = \frac{\theta}{2\pi} \int_0^{\infty} h(t) I_0(tr) dt \tag{7.1.6}$$

Here $I_0(tr)$ – is the Bessel function; and $h(t)$ – the density of the kernel. Using the inversion formula for the Hankel transformation, we obtain from (7.1.6)

$$h(t) = 2\pi t \int_0^{\infty} r K_0(r) I_0(tr) dr \tag{7.1.7}$$

Let us note that the density is connected with the function $c(t)$ by dependence $h(t) = tc(t)$ introduced by B.G. Korenev. The function $h(t)$ for the three above-mentioned models takes the form:

- Homogeneous elastic isotropic half-space

$$h(t) \equiv 1 \qquad (7.1.8)$$

- Elastic half-space with the modulus of elasticity varies with depth according to a power law [292]

$$h(t) = \frac{\pi t^{\nu} \sec \frac{\nu \pi}{2}}{\Gamma(1 + \nu)} \qquad (7.1.9)$$

- Winklerian base

$$h(t) \equiv t \qquad (7.1.10)$$

Now it is not difficult to determine kernel $K_0(r)$ for the Winkler's foundation model. Using function (7.1.10) in (7.1.6), we find

$$K_0(r) = \frac{1}{2\pi} \int_0^{\infty} t I_0(tr) dt.$$

The last integral diverges. To give it meaning in the class of generalized functions let us write the condition of the orthogonality of the Bessel function zero index on the infinite interval.

$$r \int_0^{\infty} t I_0(rt) I_0(r_1 t) dt = \delta(r - r_1)$$

Taking into account that $\lim_{r_1 \to 0} I_0(r_1 t) = 1$, we obtain

$$\int_0^{\infty} t I_0(rt) dt = r^{-1} \delta(r),$$

and consequently

$$K_0(r) = \frac{\delta(r)}{2\pi r}, \qquad (7.1.11)$$

is a well-known representation of delta-function in polar coordinates. For generalizing the kernels of elastic foundations models (7.2)–(7.4), let us introduce into the examination the following axisymmetric kernel

$$K(r) = \theta K_0(r) = \frac{\theta}{\left(2\pi \sqrt{R^2 + \varepsilon^2}\right)^{1+\nu}} \qquad (7.1.12)$$

It is not difficult to see that in the limiting cases the kernel (7.1.12) becomes the kernels for the homogeneous isotropic half-space ($\varepsilon = 0$, $\nu = 0$) and for the model G.K. Klein ($\varepsilon = 0$), defined by formulas (7.1.2) and (7.1.3) respectively. At the same time the introduction of the parameters $\varepsilon$ and $\nu$ allows us to obtain, with the solution of different problems, the convergent integrals and take into account a certain extent due to the smoothing of solutions such phenomena as local plastic deformation, natural rounding at the corners, and so on. It is necessary to note that the kernel (7.1.12) is essentially a generalization of the kernel used by B.G. Korenev [188]

$$K(r) = \frac{\theta}{\sqrt{r^2 + \varepsilon^2}}$$

and the kernel (7.1.3). Now we should justify the transition from the kernel (7.1.12) to (7.1.11). Let us for this purpose represent physical constant $\theta$ in the form of

$$\theta = 4\pi^2 k_0 \varepsilon^{\nu/2}$$

and fix the value $k_0$. Let us calculate the limit

$$\lim_{\varepsilon \to 0, \nu \to 2} \frac{4\pi^2 k_0 \varepsilon^{\nu/2}}{\left(2\pi \sqrt{R^2 + \varepsilon^2}\right)^{1+\nu}} = \lim_{\varepsilon \to 0} \frac{k_0 \varepsilon}{2\pi r(r^2 + \varepsilon^2)}.$$

From the theory of generalized function [115] it is known that

$$\lim_{\varepsilon \to 0} \frac{\varepsilon}{r^2 + \varepsilon^2} = \delta(r) \tag{7.1.13}$$

Thus

$$K(r) = k_0 \frac{\delta(r)}{2\pi r},$$

which corresponds to (7.1.11)

We calculate the density $h(t)$ of the kernel of the proposed foundation model.

Using (7.1.12) in (7.1.7), we find

$$h(t) = \frac{t^{\alpha+\nu} \varepsilon^{\alpha}}{2^{\nu-\alpha} \pi^{\alpha} \Gamma(\alpha + \nu)} K_{\alpha}(\varepsilon t), \tag{7.1.14}$$

where $K_{\alpha}(\varepsilon t)$ – is MacDonald's function; and $\alpha = \frac{1-\nu}{2}$. The limit transitions from (7.1.14) give: with $\nu \to 0$

$$h(t) = e^{-\varepsilon t}; \tag{7.1.15}$$

with $\nu \to 0$, $\varepsilon \to 0$ we come to the density of the kernel of homogeneous elastic half-space

$$h(t) \equiv 1. \tag{7.1.16}$$

If we accept $\nu = 2$, then

$$\theta h(t) = \frac{t\theta}{4\pi^2 \varepsilon} e^{-\varepsilon t}.$$

Assuming as above $\theta = 4\pi^2 \varepsilon k_0$, we obtain with $\varepsilon = 0$

$$\theta h(t) = t k_0. \tag{7.1.17}$$

Expression (7.1.17) corresponds to the density of the kernel of Winklerian base. In the case of the two-dimensional problem the kernel of the model of elastic foundation $\overline{K}(x - x_1) = \overline{K}(|x - x_1|) = \overline{K}(s)$ can be represented by density $h(t)$ according to formula [292]

$$\overline{K}(s) = \theta \int_0^\infty \frac{h(t)}{t} \cos st\, dt$$

Assuming in this formula $h(t)$ is equal to (7.1.14), we find

$$\overline{K}(s) = \frac{\theta \varepsilon^\alpha}{2^{\nu-\alpha} \pi^\nu \Gamma(\alpha + \nu)} \int_0^\infty t^{-\alpha} K(\varepsilon t) \cos st\, dt. \tag{7.1.18}$$

The last integral is calculated using formula 6.699.12 [123], just with $\text{Re}\,\varepsilon > 0$, $s > 0$, $\alpha < 1/2$. If we consider that by the definition of MacDonald's function

$$K_\mu(z) = \frac{\pi i}{2} e^{\frac{\pi \mu i}{2}} H_\mu^{(1)}(iz) = \frac{\pi i}{2} e^{-\frac{\pi \mu i}{2}} H_{-\mu}^{(1)}(iz),$$

where $H_\mu^{(1)}(iz)$— the Hankel function of the first kind

$$K_\mu(z) = K_{-\mu}(z).$$

So, we have

$$\overline{K}(s) = \frac{\theta_\nu}{(s^2 + \varepsilon^2)^{\nu/2}}, \tag{7.1.19}$$

where

$$\theta_\nu = \frac{\theta \Gamma(\nu/2)}{2^{\nu+1} \pi^{\nu-1/2} \, \Gamma\left(\frac{1+\nu}{2}\right)}.$$

Let us note that the condition $\alpha < 1/2$ corresponds to the accepted value $\nu > 0$.

As we see, the kernel (7.1.18) has sufficiently simple form and with finite $\nu$, and $\varepsilon$ does not have singularities on the entire semiaxis $0 \le y < \infty$. The function of influence determinates by formula (7.1.19) from a mechanical point of view is sufficiently substantiated. It has a finite value

$$\overline{K}(0) = \theta_y \varepsilon^{-\nu}$$

at the point of application of force and tends to zero at infinity, which corresponds to the engineering notions of real elastic foundations; which is different for example from the model of the elastic homogeneous half-plane, the kernel of which in the case of a plane problem takes the form

$$\overline{K}(s) = \frac{\theta}{\pi} \ln \frac{1}{s} + c, \tag{7.1.20}$$

where $c$ – is an arbitrary constant, with the singularities in zero and at infinity. Assuming that $\nu \to 0$, $\nu \to 2$, and $\varepsilon \to 0$, we can obtain four above-mentioned kernels: G.K. Klein ($\varepsilon \to 0$); B.G. Korenev ($\nu \to 0$); elastic homogeneous half-plane ($\nu \to 0$, $\varepsilon \to 0$); and Winklerian base ($\nu \to 2$, $\varepsilon \to 0$). However, it is impossible to accomplish limit transitions in formula (7.1.19) directly, since in this case the conditions of the convergence of integral (7.1.18) are disrupted. In order to get around this difficulty let us find the derivative of kernel – angle of the surface inclination of the base

$$\overline{K}'(s) = -\frac{\theta \varepsilon^{\alpha}}{2^{\nu-\alpha} \pi^{\nu} \Gamma(\alpha + \nu)} \int_0^{\infty} t^{1-\alpha} K_{\alpha}(\varepsilon t) \sin s t \, dt.$$

This integral is calculated with the aid of formula 6.699.11 [123]

$$\int_0^{\infty} x^{1+\mu} K_{\mu}(ax) \sin bx \, dx = \sqrt{\pi} \, (2a)^{\mu} \, \Gamma\left(\frac{3}{2} + \mu\right) b(b^2 + a^2)^{-\frac{3}{2}-\mu},$$

if we assume that $\mu = -\alpha$, $a = \varepsilon$, and $b = s$. We will obtain

$$\overline{K}'(s) = -\frac{\theta \Gamma\left(\frac{3}{2} - \alpha\right) s(s^2 + \varepsilon^2)^{\alpha-3/2}}{2^{\nu} \pi^{\nu-1/2} \Gamma(\alpha + \nu)} = -\frac{\nu \theta \Gamma\left(\frac{\nu}{2}\right) s(s^2 + \varepsilon^2)^{\alpha-3/2}}{2^{\nu+1} \pi^{\nu-1/2} \Gamma(\alpha + \nu)}. \tag{7.1.21}$$

Now it is possible to carry out limiting transitions; with $\nu \to 0$ we have

$$\overline{K}'(s) = -\frac{\theta s}{\pi(s^2 + \varepsilon^2)}. \tag{7.1.22}$$

Integrating (7.1.22), we find the corresponding plane kernel

$$\overline{K}(s) = -\frac{\theta}{2} \ln\left(\frac{1}{s^2 + \varepsilon^2}\right) + c, \tag{7.1.23}$$

and with $\varepsilon \to 0$ we come to kernel (7.1.20) for the elastic homogeneous half-plane with accuracy up to a physical constant. On the other hand, from (7.1.21) with $\varepsilon \to 0$ we have

$$\overline{K}(s) = -\frac{\theta\Gamma\left(1 + \frac{v}{2}\right) s^{-1-v/2}}{2^v \pi^{v-1/2}\Gamma\left(\frac{1+v}{2}\right)}$$

and after integration we have

$$\overline{K}(s) = \frac{\theta\Gamma\left(1 + \frac{v}{2}\right)}{2^v \pi^{v-1/2}v\Gamma\left(\frac{1+v}{2}\right) s^{v/2}} + c. \tag{7.1.24}$$

From the condition (with $v \neq 0$) we obtain $c = 0$. The kernel (7.1.24) corresponds to elastic half-plane with the modulus of elasticity, which is changed with the depth according to the power law. The limiting transitions from (7.1.24) to (7.1.20) is accomplished with the aid of formula [292]

$$\lim_{v \to 0}\left[\frac{1}{v}\left(\frac{1}{|s|^v} - 1\right)\right] = \ln\frac{1}{|s|}.$$

Thus from (7.1.12) with the aid of indicated limiting operations, we can obtain the kernels (7.1.20), (7.1.23) and (7.1.24) that correspond to known models or kernels. Let us note that proposed kernel (7.1.23) was not considered in the scientific literature previously. Figure 7.1 shows the graphs of the influence functions $K_0(r)$ for the proposed model with the different values of $v$ and $\varepsilon$. The question about the selection of these parameters must be solved, taking into account the real properties of ground base. An important special feature of the model is the possibility of the adequate description of the properties of the ground with the rapidly damped sediments out of the loaded area, which are usually described by Winklerian type models.

Assuming in (7.1.21) that $v = 2$, then we will obtain

$$\overline{K}'(s) = -\frac{\theta s}{2\pi^3(s^2 + \varepsilon^2)},$$

from where by integration we find

$$\overline{K}(s) = -\frac{\theta}{4\pi^3(s^2 + \varepsilon^2)}.$$

With $\theta = 4\pi^2 k_0\varepsilon$, where $k_0$ – is a fixed coefficient of bed, we will have

$$\overline{K}(s) = \frac{k_0\varepsilon}{(s^2 + \varepsilon^2)}. \tag{7.1.25}$$

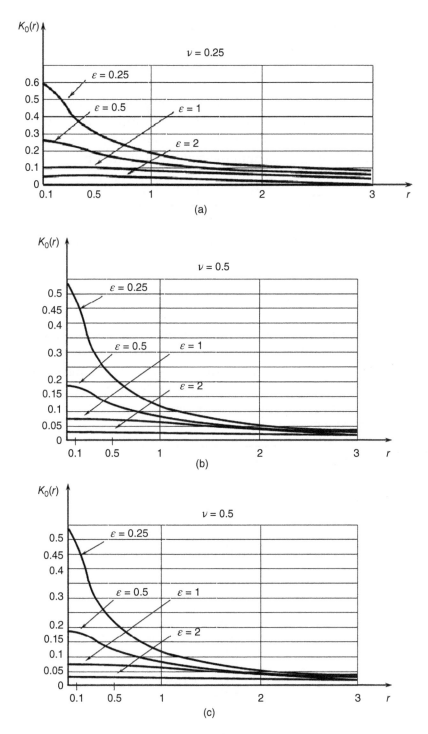

**Figure 7.1** (a) Influence function $K_0(r)$; (b) influence function $K_0(r)$; (c) influence function $K_0(r)$.

The limiting transitions with $\varepsilon \to 0$ is accomplished (carried out, implemented) according to formula (7.1.13)

$$\overline{K}(s) = k_0 \delta(s).$$

The kernel (7.1.25) can be used for describing the stress-strain state for soils that are only a little cohesive. The general form of the kernel of base with the plane strained state, and with $v \neq 0$, $\varepsilon \neq 0$ can be obtained by integration (7.1.21) taking into account the condition at infinity

$$\overline{K}(s) = -\frac{\theta \Gamma\left(\frac{3}{2} - \alpha\right)(s^2 + \varepsilon^2)^{-v/2}}{v 2^v \pi^{v-1/2} \Gamma(\alpha + v)} \tag{7.1.26}$$

and using well known formula from the theory of gamma-functions

$$\Gamma(1 + z) = z\,\Gamma(z). \tag{7.1.27}$$

Figure 7.2 shows the graphs of the kernel (7.1.26) with the different values of the parameters $v$ and $\varepsilon$.

Along with the kernels foundation for the spatial and planar (two-dimensional) problems considered in the Cartesian coordinate system, it is also of interest to obtain the corresponding expressions for the tasks, formulated in polar coordinates. As it is well known, the distance between two points in the plan with the polar coordinates $(r, \varphi)$ and $(r_1, \varphi_1)$ is determined by the formula:

$$R = \sqrt{r^2 + r_1^2 - 2rr_1 \cos(\varphi - \varphi_1)}.$$

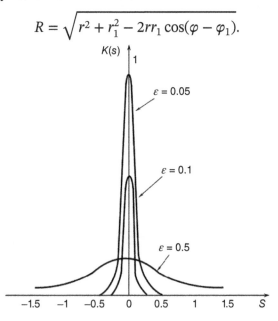

Figure 7.2   Graph of function $K(s)$.

Therefore, the relationship between the load on the base $p(r_1, \varphi_1)$ and displacement of its surface $w(r, \varphi)$ can be represented in the form of:

$$\int_0^{2\pi} \int_0^\infty K\left(\sqrt{r^2 + r_1^2 - 2rr_1 \cos(\varphi - \varphi_1)}\right) p(r_1, \varphi_1) r_1 d\varphi_1 dr_1 = w(r, \varphi)$$

Using the integral representation (7.1.6) for the kernel $K(R)$, we obtain

$$\int_0^{2\pi} \int_0^\infty \int_0^\infty p(r_1, \varphi_1) I_0(tR) h(t) r_1 d\varphi_1 dr_1 dt = w(r, t) \qquad (7.1.28)$$

Bearing in mind the formula for summation of Bessel functions

$$I_0\left(t\sqrt{r^2 + r_1^2 - 2rr_1 \cos(\varphi - \varphi_1)}\right) = 2\sum_{n=0}^{\infty}{}' I_n(tr) I_n(tr_1) \cos n(\varphi - \varphi_1)$$

Let us rewrite relationship (7.1.28) in the following form:

$$\frac{\theta}{2\pi} \sum_{n=0}^{\infty}{}' \int_0^\infty \int_0^{2\pi} h(t)\overline{p}_n(\varphi, t) I_n(tr) \cos n(\varphi - \varphi_1) dt d\varphi_1 = w(r, \varphi), \quad (7.1.29)$$

where

$$\overline{p}_n(\varphi, t) = \int_0^\infty p(r_1, \varphi_1) I_n(tr_1) r_1 dr_1.$$

Here, prime in the sign of sum means that the multiplier $\frac{1}{2}$ with the zero member (term) is introduced. The greatest interest is the case of the axisymmetric loading of the base. In this case $\overline{p}_n(\varphi, t)$ does not depend on angular coordinate, and all terms except zero in the sum of (7.1.29) will turn to zero:

$$\theta \int_0^\infty h(t)\overline{p}_0(t) I_0(tr) dt = w(r) \qquad (7.1.30)$$

where

$$\overline{p}_0(t) = \int_0^\infty p(r_1) I_0(tr_1) r_1 dr_1.$$

Returning to the original kernel, i.e. by substituting value $\overline{p}_0(t)$ in (7.1.30) we obtain

$$\theta \int_0^\infty K(r, r_1) p(r_1) r_1 dr_1 = w(r), \qquad (7.1.31)$$

where the kernel of the axisymmetric problem has the form:

$$K_0(r, r_1) = \int_0^\infty h(t)I_0(r_1 t)I_0(rt)dt. \tag{7.1.32}$$

For the considered model, i.e. with the kernel density $h(t)$ expressed by the formula (7.1.14), which for simplification we will represent in the form

$$h(t) = B_\nu t^{\alpha+\nu} K_\alpha(\varepsilon t), \tag{7.1.33}$$

where

$$B_\nu = \frac{\varepsilon^\alpha}{2^{\nu-\alpha}\pi^\nu\Gamma(\alpha + \nu)}.$$

The integral of (7.1.32) is calculated using the formula 6.578.2 [123]. As a result of calculations we obtain

$$K_0(r, r_1) = 2^{\alpha+\nu-1}B_\nu(\varepsilon)^{-\alpha-\nu-1}\Gamma\left(\frac{\nu+1}{2}\right)\Gamma\left(\frac{2\alpha+\nu+1}{2}\right)$$

$$\times F_4\left(\frac{\nu+1}{2}, \frac{2\alpha+\nu+1}{2}, 1, 1; -\frac{r^2}{\varepsilon^2}, -\frac{r_1^2}{\varepsilon_1^2}\right). \tag{7.1.34}$$

Here, $F_4$ – hypergeometric function of two variables, is determined by double series [123]

$$F_4(\alpha, \beta, \gamma, \gamma'; x, y) = \sum_{m=0}^\infty \sum_{n=0}^\infty \frac{(\alpha)_{m+n}(\beta)_{m+n}}{(\gamma)_m(\gamma')_n m! n!}x^m y^n,$$

where

$$(\gamma)_m = \frac{\Gamma(\gamma + m)}{\Gamma(\gamma)}.$$

The area of convergence of the hypergeometric function $F_4$ is determined by the ratio $|\sqrt{x}| + |\sqrt{y}| < 1$, i.e. for formula (7.1.34) to be satisfied the conditions $r + r_1 < \varepsilon$. This condition is a substantial limitation and allows calculating only for those points of the surface of the base which are located near the local load. In other cases it is necessary to calculate numerically the integral (7.1.32). In the case of $h(t) = e^{-\varepsilon t}$ ($\nu = 0$) the expression for the axisymmetric kernel is considerably simple (see integral 6.612.3 [123]):

$$K_0(r, r_1) = \frac{1}{\pi\sqrt{rr_1}}Q_{-1/2}\left(\frac{\varepsilon^2 + r^2 + r_1^2}{\sqrt{rr_1}}\right) \tag{7.1.35}$$

where $Q_{-1/2}$ – is the Legendre functions of the second kind.

When $\nu = 2$, $h(t) = t(4\pi^2\varepsilon)^{-1}e^{-\varepsilon t}$, and according to 6.626.1 [123]

$$K_0(r, r_1) = \frac{\varepsilon}{4\pi^2} \sum_{m=0}^{\infty} \frac{\Gamma(2 + 2m)}{m!\Gamma(m + 1)} F\left(-m, -m; 1, \frac{r}{r_1}\right) \tag{7.1.36}$$

Hypergeometric series in (7.1.36) is broken. However, the calculation of this kernel is still quite a cumbersome computational procedure, and in this case it is easier to calculate the directly well-convergent integral (7.1.32)

## 7.2   The Determination of the Characteristics of the Generalized (Unified, Integrated) Model of the Elastic Foundation (Base)

Let us examine now the most important question about the determination of the characteristics of the proposed model of base, i.e. the parameters of kernel (7.1.12) based on the tests of real ground. Suppose loading of a rigid (hard) circular stamp with a small footprint results in a measured vertical displacement of the stamp $z_0$ and the distance of $r_1$ and $r_2$ from the stamp to the multiple circular points, for which the displacements are $z_1$ and $z_2$, respectively (Figure 7.3). The selection of radii $r_1$ and $r_2$ must be such that the displacements $z_0$, $z_1$, and $z_2$ would not be close to each other.

The displacement of stamp $z_0$ with a small radius due to unit force can be identified with the $K(0)$. According to (7.1.12)

$$z_0 = K(0) = \frac{\theta}{(2\pi\varepsilon)^{1+\nu}} \tag{7.2.1}$$

Analogously for the displacements $z_1$ and $z_2$ we have

$$z_1 = \frac{\theta}{\left(2\pi\sqrt{r_1^2 + \varepsilon^2}\right)^{1+\nu}} \tag{7.2.2}$$

$$z_2 = \frac{\theta}{\left(2\pi\sqrt{r_2^2 + \varepsilon^2}\right)^{1+\nu}}.$$

We convert the last expressions to the form

$$\alpha_1^\beta = \frac{\varepsilon}{\sqrt{r_1^2 + \varepsilon^2}}, \qquad \alpha_2^\beta = \frac{\varepsilon}{\sqrt{r_2^2 + \varepsilon^2}}, \tag{7.2.3}$$

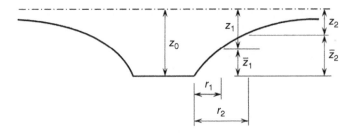

**Figure 7.3**   Schema of the hole during of die testing.

where

$$\alpha_1 = \frac{z_1}{z_0} \ , \quad \alpha_2 = \frac{z_2}{z_0} \ , \quad \beta = \frac{1}{1+\nu} \ .$$

Raising the left and right parts of (7.2.3) into the square and excluding the parameter $\nu$ by taking the logarithm, we find

$$\ln\left(\frac{\varepsilon^2}{r_1^2 + \varepsilon^2}\right)\ln \alpha_2 - \ln\left(\frac{\varepsilon^2}{r_2^2 + \varepsilon^2}\right)\ln \alpha_1 = 0 \qquad (7.2.4)$$

The solution of this equation in the elementary functions requires some assumptions. For example, we cannot arbitrarily choose the values $\alpha_1$ and $\alpha_2$; we have to choose the value of parameters within a certain selection. In particular, it is convenient to assume that $\alpha_1 = \alpha_2^2$. Then Eq. (7.2.4) will have the following solution

$$\varepsilon = \sqrt{\frac{r_2^4}{r_1^2 - 2r_2^2}} \qquad (7.2.5)$$

The parameter $\nu$ is determined by the formula

$$\nu = \frac{\ln \alpha_1}{\ln \dfrac{\varepsilon}{\sqrt{r_1^2 + \varepsilon^2}}} - 1 \qquad (7.2.6)$$

Obtaining the solution of Eq. (7.2.4) in the final form required the assumption $\alpha_1 = \alpha_2^2$,which contributes some complications into the treatment of experimental data. However, these complications are insignificant since they require the construction of several points of an experimental line of influence and the selection of two points on this curve, displacements of which are connected with the indicated dependence. The last constant, $\theta$, will be determined from formula (7.2.1)

$$\theta = z_0(2\pi\varepsilon)^{1+\nu}. \qquad (7.2.7)$$

If we now express the kernel of the base through (7.2.7), we can obtain a convenient formula

$$K(r) = \frac{z_0}{(r^2\varepsilon^{-2} + 1)^{\frac{1+\nu}{2}}} \qquad (7.2.8)$$

If the base has small coherence, then for its kernel, as it was indicated above, it is necessary to adopt the following expression

$$K(r) = \frac{4\pi^2 k_0 \varepsilon^{\frac{\nu}{2}}}{(2\pi\sqrt{r^2 + \varepsilon^2})^{1+\nu}}, \qquad (7.2.9)$$

where $k_0$ – is a some constant, which is similar in mining to the coefficient of bed of Winklerian base. In this case, all the assumptions made above remain valid. The parameters $\nu$ and $\varepsilon$ are determined by formulas (7.2.5) and (7.2.6), and for constant $k_0$ we have

$$k_0 = \frac{z_0 \varepsilon^{1-\nu/2}}{(2\pi)^{1-\nu}}. \qquad (7.2.10)$$

After expressing the kernel of the foundation through $k_0$ we will come again to formula (7.2.8). Thus, it is possible to calculate the values of the two parameters, $\varepsilon$ and physical constant $\theta$, determining three displacements on an experimental curve, which is sufficiently accurate for the engineering practice and can be taken as the influence function (kernel) of the base. It is of interest to determine the kernel base on the results of field soil tests. We will try to use the data from the extensive experiments of L.I. Manvelov and E.S. Bartoshevich [221]. Figure 7.4 shows the experimental curve of the surface of the base out of a circular, rigid die obtained with testing of silty (dusty) loams (solid line).

Let us accept, according to the graph,

$$z_0 = 1.0\text{mm}, \quad z_1 = 0.09\text{mm}, \quad z_3 = 0.3\text{mm}.$$

The corresponding distances will be:

$$r_0 = 0, \quad r_1 = 210\text{mm}, \quad r_2 = 80\text{mm}.$$

We have

$$\alpha_1 = 0.09; \quad \alpha_2 = 0.3.$$

The calculations by formulas (7.2.5) and (7.2.6) give:

$$\varepsilon = 36.2; \quad \nu = 0.356.$$

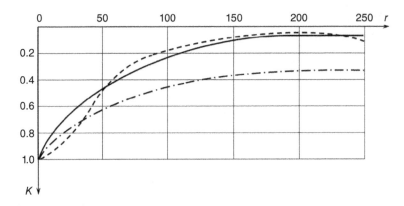

**Figure 7.4**   Kernel $K(r)$ for silty (dusty) loams.

For the physical constant $\theta$ we obtain:

$$\theta = 1 \cdot (2\pi \cdot 36.2)^{1.356} = 1570$$

The kernel of elastic foundation model according to (7.2.8) takes the form:

$$K_0(r) = \frac{1}{(\,0.000763r^2 + 1\,)^{0.678}}.$$

Figure 7.4 shows the graphs of the influence functions on the experiments of L.I. Manvelov and E.S. Bartoshevich [221], formula (7.2.8) (dashed line), and isotropic elastic half-space model (dot-and-dash line). As we can see, the proposed model gives results quite close to the experiment.

It is of interest to also obtain an expression for the plane kernel $\overline{K}(s)$. According to (7.1.19) we have, in this case

$$\overline{K}(s) = \frac{1570\ \Gamma(0.178)\pi^{0.144}(s^2 + 1310)^{-0.178}}{2^{1.356}\ \Gamma(0.678)}$$

For calculation of the value of $\Gamma(0.178)$, let us use formula (7.1.27). From the tables of Gamma function [409] we obtain:

$$\Gamma(0.178) = \frac{\Gamma(1.178)}{0.178} = \frac{0.924}{0.178} = 5.19$$

$$\Gamma(0.678) = \frac{\Gamma(1.678)}{0.678} = \frac{0.8047}{0.678} = 1.187$$

So, finally we find:

$$\overline{K}(s) = \frac{3162}{(s^2 + 1310)^{0.178}}$$

As we can see, the characteristics of the model for different calculated cases are restored according to the experimental data sufficiently simple.

## 7.3 Contact Problem for the Rigid Die, Lying on the Generalized Elastic Base

The research of properties of proposed model of elastic foundation is most convenient to conduct on the contact tasks for the rigid foundation. These problems not only represent the independent interest, they also allow to reveal the character of contact stresses in the dependence on the parameters of model, thus determining the ranges of values of the parameters for various ground indicated. Great contributions to the development of the theory of contact problems were made by scientists V.M. Abramov, B.L. Abramyan, V.M. Alexsandrov, N.K. Arutynian, L. Ascione, V.A. Babeshko, A.A. Babloyan, H.J.C. Barbosa, N.M. Borodachev, S.V. Bosakov, J.E. Bowles, M.A. Bradford, Z. Celep, Y.K. Cheung, R. Chand, T.F. Cory, L.M. Flitman, Y.C. Fung, L.A. Galin, M.I. Gorbunov-Passadov, V.T. Grinchenko, V.S. Gubenko, V.A. Ilichev, R. Jones, A.I. Kalandia, A.D. Kerr, N.A. Kilchevsky, B.G. Korenev, N.N. Lebedev, M.J. Leonov, K.M. Liew, G. Luko, A.I. Lurie, M.D. Martinenko, V.I. Mosakovsky, N.I. Muskhelishvili, P.D. Panagiotopoulos, G.J. Popov, V.L. Rvachev, A.R. Rzhanitsin, N.A. Rostovtsev, V.M. Seimov, A.S.P. Selvadurai, A.R.D. Silva, R.A.M. Silveira, G.E. Stavroulakis, A.Ph. Ulitko, I.S. Uflyand, S.P. Timoshenko, A.I. Tseitlin, E. Tsudik, G.S. Shapiro, D.I. Sherman, I.J. Shtaerman, O.J. Shakhter, S.T. Smith, C.V.G. Vallabhan, P.M. Varvak, I.I. Vorovich, Y.H. Wang, Y. Weistman, P. Wriggers, K.V. Terzaghi, W.T. Thompson, A.C. Ugural, T. Kobori, T.Y. Yang, and a number of other researchers.

In this paragraph we will examine the plane and axisymmetric contact problems for the rigid die, pressed into the elastic half-space, the properties of which are described by the proposed model. As known [371, 377], the equation of a contact problem can be formulated in the form of a paired system in the case of the axial symmetry

$$\int_0^\infty h(\xi)\widetilde{p}(\xi)I_0(\xi r)d\xi = \frac{w_0 - w(r)}{\theta}; \quad (0 < r < r_0) \tag{7.3.1}$$

$$\int_0^\infty \widetilde{p}(\xi)I_0(\xi r)d\xi = 0, \quad (r_0 < r < \infty)$$

where $\tilde{p}(\xi)$—Hankel transformation of the contact stresses $p(r)$; $w_0$ – the displacement of the central point of die; $w(r)$ ($w(0) = 0$)–the function, which describes the contact surface of die; and $r_0$ – a radius of the die. The function $h(\xi)$ is determined by formula (7.1.14). For solving paired integral equations (7.3.1), one of the methods usually applied in such cases (Cook-Lebedev, B.M. Alexandrov, and A.I. Tseitlin) can be used. These methods require the specific behavior of function at infinity; in particular its performance at infinity must take the form

$$h(\xi) = B\xi^{1-2\mu}[1 + 0(1)], \qquad \frac{1}{2} \geq \mu > 0, \qquad (7.3.2)$$

where $B$ – a certain constant.

Using an asymptotic behavior of MacDonald's function for the considered model we can obtain the following representation with

$$h(\xi) = B\varepsilon^\alpha \xi^{\alpha+\nu-\frac{1}{2}} e^{-\varepsilon\xi}[1 + 0(1)]. \qquad (7.3.3)$$

With $\varepsilon = 0$, $\nu = 0$ the conditions (7.3.2) are satisfied, and the solution of the problem can be reduced to solving the integral equation of Fredholm of the second kind. In this case [371]

$$\tilde{p}_1(\xi) = B^{-1}\rho(\xi) \int_0^{r_0} G_1(\eta)[I_{-\mu}(\xi\eta) + \int_0^{r_0} R(\eta_1, \eta)$$

$$\times \ I_{-\mu}(\xi\eta_1)\eta_1 d\eta_1]d\eta, \qquad (7.3.4)$$

where $\tilde{p}_1(\xi) = \xi^{-\mu}\rho(\xi)\tilde{p}(\xi)$,

$$\rho(\xi) = \sqrt{h(\xi) - 1}, \quad \mu = \frac{0.5 - \alpha - \nu}{2},$$

$$G_1(\eta) = \frac{2^\mu}{\Gamma(1 - \mu)\eta^\mu}\left[w_0 - \eta^{2\mu}\int_0^\eta \frac{w'(r)dr}{(\eta^2 - r^2)^\mu}\right],$$

$R(\eta_1, \eta)$ – the solution of integral equation

$$R(\eta_1, \eta) + \Psi(\eta, \eta_1) + \int_0^{r_0} R(\eta, \eta_2)\Psi(\eta_1, \eta_2)\eta_2 d\eta_2 = 0,$$

with kernel

$$\Psi(\eta, \eta_1) = \int_0^\infty h(\xi)I_{-\mu}(\xi\eta)I_{-\mu}(\xi\eta_1)\xi d\xi.$$

The contact pressure is determined by the formula

$$p(r) = \frac{2^\mu}{\Gamma(1-\mu)} \left[ \frac{G_2(r_0)}{(r_0^2 - r^2)^\mu} - \int_r^{r_0} \frac{G_2'(\eta)d\eta}{(\eta^2 - r^2)^\mu} \right], \tag{7.3.5}$$

where

$$G_2(\eta) = \eta^\mu \int_0^\infty \rho^{-1}(\xi)\tilde{p}_1(\xi)I_{-\mu}(\xi\eta)\xi d\xi.$$

This solution, as we see, is quite cumbersome, but the main obstacle on the way to its use is the constraint (7.3.2). Therefore, to obtain the universal solution, we will use the method proposed by U.G. Plotnikov, which gives the very efficient algorithm to solve contact problems [285–288]. This method reduces to the system of the linear algebraic equations in relation to axisymmetric problem.

$$rA + Rw_0 = 0 \tag{7.3.6}$$

$$R^T A + Q = 0$$

where $r = [r_{ml}]$, $R = [R_m]$, $Q = \frac{P}{2\pi}$; $P$ – force acting on the die; $w_0$ – settlement of the die; $A = [A_m]$; $R_m = \int_0^\infty p_m(r)rdr$; and $A_m$– the coefficients of the expansion of contact pressure into series (in terms) of the coordinate functions.

$$p(r) = \sum_{m=1}^N A_m \, p_m(r) \tag{7.3.7}$$

The coefficients $r_{ml}$ are determined by the formula

$$r_{ml} = \frac{\theta}{2\pi} \int_{-\infty}^\infty h(\xi)\tilde{p}_m(\xi)\tilde{p}_l(\xi)d\xi$$

where $\tilde{p}_m(\xi)$– Hankel transformation of the function $p_m(r)$;

$$\tilde{p}_m = \int_0^\infty rp_m(r)I_0(\xi r)dr$$

Approximation (7.3.7) of contact pressure is accepted in the form:

$$p(r) = \sum_{m=1}^N A_m\Delta_m \left\{ \frac{1}{2\pi r_m} \left[ H\left(r - r_m + \frac{\Delta_m}{2}\right) - H\left(r - r_m - \frac{\Delta_m}{2}\right) \right] \right\}.$$

Here $H$ – is a unit step Heaviside function. Thus the function of contact pressure is approximated by a step function (Figure 7.5), and the value of $A_m$ in the physical sense is the resultant of contact pressure on area $\Delta_m$.

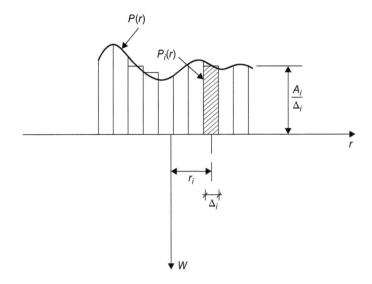

**Figure 7.5**   Approximation of contact pressure.

The plane problem is solved similarly. In this case, in contrast to the axisymmetrical problem in (7.3.6), it is accepted $Q = P$ and

$$r_{ml} = \frac{\theta}{2\pi} \int_0^\infty h(\xi)\tilde{p}_m(\xi)\overline{p_l(\xi)}d\xi,$$

where

$$\tilde{p}_m(\xi) = \frac{1}{2\pi} \int_0^\infty p_m(x)e^{i\xi x}dx,$$

$$\overline{p_l(\xi)} = \frac{1}{2\pi} \int_0^\infty p_l(x)e^{-i\xi x}dx.$$

$\overline{p_l(\xi)}$– A function of the complex-conjugate of $\tilde{p}_l(\xi)$.

The value $R_m$ is determined by the formula

$$R_m = \int_0^\infty p_m(x)dx.$$

The approximation of contact pressure in contrast to (7.3.7) takes the form

$$p(x) = \sum_{i=1}^N A_i \left\{ \frac{1}{\Delta_i} \left[ H\left(x - x_i + \frac{\Delta_i}{2}\right) - H\left(x - x_i - \frac{\Delta_i}{2}\right) \right] \right\} \qquad (7.3.8)$$

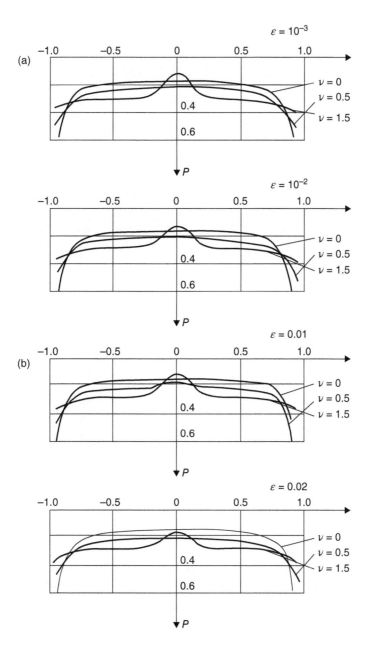

**Figure 7.6**   Diagram of contact stresses.

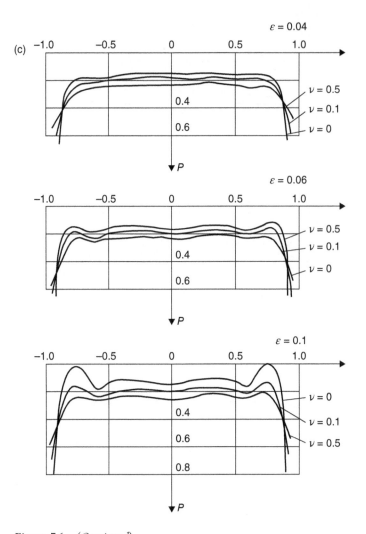

**Figure 7.6**   *(Continued)*

Together with Yu.G. Plotnikov, we have compiled FORTRAN programs of the calculation of contact stresses for axisymmetric and plane contact problems for the numerical solution of the contact problem by the given algorithm, using kernel (7.1.12) as the kernel base proposed in the present work. The results of enumerating the contact stresses, determined with an accuracy to a constant multiplier $(2\pi)^{-1-\nu}$, are shown in Figures 7.6a,b,c and 7.7. Actually, calculations were conducted for the kernel:

$$K(r) = \frac{1}{\left(r^2 + \varepsilon^2\right)^{\frac{1+\nu}{2}}};$$

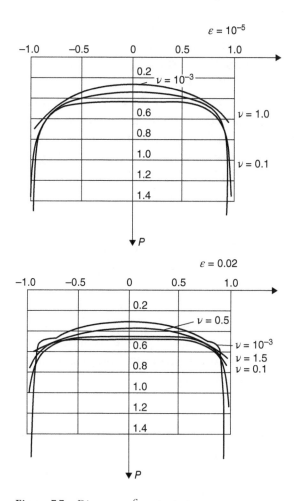

**Figure 7.7**   Diagram of contact stresses.

and in the case of the plane problem, the following kernel was accepted:

$$K(x) = \frac{\varepsilon^{\frac{\nu}{2}}}{(x^2 + \varepsilon^2)^{\frac{\nu}{2}}} \; .$$

For the evaluation of the accuracy of the numerical solution of axisymmetrical and plane contact problems, simultaneously average contact pressure under the die was calculated, which ensured checking the condition of equilibrium.

For this purpose with the limiting values of parameters, we conducted calculations for the ideally elastic half-plane. The analysis of the results of calculations shows that with the different values of the parameters $\nu$ and $\varepsilon$ the proposed model makes it possible to obtain the diagrams of pressure, which are intermediate between the diagrams, characteristic for the Winklerian base and elastic half-space.

In limited cases the diagrams for these two most important models are obtained. Thus the proposed model makes it possible to describe practically any ground bases, examined within the framework of the linear theory, and which possess different cohesiveness. In this case they can have heterogeneous structure in the depth. Such sufficiently common properties are not characteristic for other known models.

## 7.4   On One Method of Calculation of Structures on an Elastic Foundation

The proposed generalized model of the elastic foundation as a result of the regularity of its kernel opens the ways for applying this effective method to the solution of contact problems and problems of design structure on elastic foundations. The calculation of structures on elastic foundations is associated with the different ways of directly reducing the formulated problem to the solution of the integral equation of Fredholm of the second kind. The effectiveness of this method is determined by the opportunity of applying the simple iterative procedure for obtaining the most important characteristic of contact pressure, after which the calculation of structures can be produced, to the combined action of an external load and obtained contact pressure. Let the equation of the equilibrium of structure take the form

$$Lw(x.y) = q(x,y) - p(x,y) \quad (x.y \in D) \qquad (7.4.1)$$

$$\int_D K(x - x_1, y - y_1)p(x_1, y_1)dx_1 d y_1 = w(x,y)$$

where $L$ – the differential operator of the elastic forces of structure; $w(x, y)$ – displacements (deflection); $q(x, y)$ – external load; $p(x, y)$ – contact pressure; and $D$ – the area occupied by the structure. Let us substitute displacement determined from the second Eq. (7.4.1) into the first Eq. (7.4.1). We will obtain

$$p(x,y) + \iint_D R(x, x_1; y, y_1)p(x_1, y_1)dx_1 dy_1 = q(x,y), \qquad (7.4.2)$$

$$R(x, x_1; y, y_1) = LK(x - x_1, y - y_1).$$

Equation (7.4.2) is a two-dimensional linear integral equation of the second kind. Such equations are solvable and the solutions can be built with the aid of the effective numerical procedures, if the kernel of equation is Fredholm's (Fredholmians). For the known models of the coherent elastic foundations the kernel $K(x-x_1, y-y_1)$ is irregular, so $R(x, x_1; y, y_1)$ is also irregular. The proposed generalized model of the elastic foundation is free of this deficiency. Actually the kernel (7.1.12) is regular in the entire plane $x$, $y$ and all its derivatives

$$\frac{\partial^{s+t} K}{\partial x^s \partial y^t} = \frac{\partial^{s+t}}{\partial x^s \partial y^t} \left( \frac{\theta}{\sqrt{\sqrt{(x-x_1)^2 + (y-y_1)^2} + \varepsilon^2}^{1+\nu}} \right)$$

are also regular on the entire plane. At the same time, all derivatives of the kernel $K(x-x_1, y-y_1)$ decrease at infinity more rapidly than $[(x-x_1)^2 + (y-y_1)^2]^{-1/2}$, so that the kernel of Eq. (7.4.2) is integrated (integrable) with the square in any domain $D$ of the plane $x$, $y$, i.e. it is Fredholmian.

$$\iint_D |R(x, x_1; y, y_1)|^2 dx_1 dy_1 < \infty.$$

Due to this fact Eq. (7.4.2) can be easily solved with the aid of the mechanical quadrature's iteration techniques For an illustration of this let us give a solution of the problem about the bend of the beam of the finite length, freely lying on an elastic foundation, described by the generalized model. Let us write the equation of the equilibrium of beam in the form of

$$EIy^{IV}(x) = q(x) - p(x), \qquad (-l < x < l) \tag{7.4.3}$$

$$\theta_y \int_{-l}^{l} [(x-x_1)^2 + \varepsilon^2]^{-\nu/2} p(x_1) dx_1 = y(x).$$

Equation (7.4.3) takes the following form

$$p(x) + \theta_y EI \int_{-l}^{l} R(x, x_1) p(x_1) dx_1 = q(x), \tag{7.4.4}$$

where

$$R(x, x_1) = \frac{2\nu(2+\nu)}{[(x-x_1)^2 + \varepsilon^2]^{2+\nu/2}} - \frac{5\nu(2+\nu)(4+\nu)(x-x_1)^2}{[(x-x_1)^2 + \varepsilon^2]^{3+\nu/2}}$$
$$+ \frac{\nu(2+\nu)(4+\nu)(6+\nu)(x-x_1)^4}{[(x-x_1)^2 + \varepsilon^2]^{4+\nu/2}}$$

It is not difficult to see that when $v > 0$ the integral

$$\int_{-\infty}^{\infty} | R(x,x_1)^2 | \, dx_1 < 0,$$

so for all $l \leq \infty$ Eq. (7.4.4) is an equation of Fredholm of the second kind. For solving this equation let us construct iterative procedure, citing preliminary equation (7.4.4) by substitution $x/l = \bar{x}, \ x_1/l = \bar{x}_1$ to the following form

$$p(\bar{x}) = q(\bar{x}) - \lambda \int_{-1}^{1} R(\bar{x},x_1) p(\bar{x}_1) dx_1,$$

where $\lambda = \frac{\theta_y EI}{l^3}$. This iterative procedure is expressed by the formula

$$p_{n+1}(\bar{x}) = q(\bar{x}) - \lambda \int_{-1}^{1} R(\bar{x},x_1) p_n(\bar{x}_1) dx_1, \qquad (7.4.5)$$

or

$$p_n(\bar{x}) = q(\bar{x}) - \lambda \int_{-1}^{1} R(\bar{x},\bar{x}_1) \left[ q(\bar{x}) - \lambda \int_{-1}^{1} R(\bar{x},\bar{x}_1)[q(\bar{x}) - ... \right] dx_1^n$$

After the determination of $p(x)$, the displacement of beam can be obtained by its calculation for the combined action of load $q - p$. As an example, the beam is loaded by unit uniformly distributed load. The parameters of beam and base are accepted following:

$$\theta = \frac{2(1 - \mu_0^2)}{\pi E_0}; \quad \lambda = \frac{Ebh^3(1 - \mu_0^2)}{6E\pi l^3} \approx \frac{h^3 E}{20l^3 E_0} = \frac{1}{2t}.$$

$E_0, \mu_0$ – the module of elasticity and Poisson ratio of base; $b = 1$ – the width of beam; and $t$ – dimensionless (nondimensional) parameter, which characterizes the rigidity of properties of beam and base. According to M.I. Gorbunov-Pasadov and T.A. Malikova[13] with $\lambda = 0.1$ and $t = 2$ the beam is placed in category of "short," i.e. it can be considered as the beam of finite length. For the parameters of the model of base we take the values, based on the experimental data of L.I. Manvelov and E.S. Bartoshevich [221] and to the data obtained in the paragraph two of the present chapter: $\varepsilon = 36.2$ mm, $v = 0.356$. Upon transfer to the dimensionless

[13] Gorbunov-Pasadov M.I., Malikova T.A. (1973). *Calculation of Structures on Elastic Foundation*. Moscow: Stroyizdat, 627.

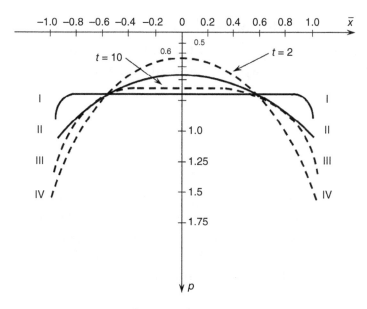

**Figure 7.8**    Diagram of contact stresses.

(non-dimensional) values, in the kernel $R(x, x_1)$ it is necessary to introduce the dimensionless (non-dimensional) parameter $\bar{\varepsilon} = \varepsilon/l$. Let $l = 1000$ mm, then $\bar{\varepsilon} = 0.036$. For $t$ accept two values $t = 2$ and $t = 10$. The results of the calculations have been shown in Figure 7.8.

The diagram of contact stresses for the proposed model is shown by a solid line; the dashed lines show contact stresses for an elastic half-space (M.I. Gorbunov-Pasadov, T.A. Malikova).[14] The second iteration provides a sufficiently precise value of contact stresses. It follows from the Figure 7.8 that when real ground conditions are taken into account, our generalized model produces more accurate results that differ significantly from the output produced by either the model of isotropic elastic half-space and Winklerian grounds model. For isotropic elastic half-space model and Winklerian model we would obtain uniform contact stresses as shown on the graphs (Figure 7.8). As we see, the iteration technique examined is very effective, since it requires the fulfillment of sequential quadratures, in contrast to the solution of the sufficiently complex integral equations of the first kind, which are encountered during calculations of structure on an elastic foundation, described by known models of the coherent type.

[14] Gorbunov-Pasadov M.I., Malikova, T.A. (1973). *Calculation of Structures on Elastic Foundation*. Moscow: Stroyizdat, 627.

## 7.5   The Calculation of the (Non-isolated) Beams and Plates, Lying on an Elastic Foundation, Described by the Generalized Model

Previously it has been noted that the generalized model of the elastic foundation with the kernel

$$K(r) = \frac{\theta}{(2\pi\sqrt{r^2 + \varepsilon^2})^{1+\nu}} \tag{7.5.1}$$

gives the opportunity to solve any problems about the uninsulated plates, which lie on the coherent base. The generalized model of the elastic foundation makes it possible to solve problems about uninsulated plates, in contrast to the classical models, for which it is impossible to obtain such solutions for the plates; for example with the free edge as a result of the presence of non-integrable singularities of the respective kernels. For an illustration of the capabilities of the proposed model let us examine the problem of bending of two semi-infinite uninsulated (non-insulated, non-isolated) plates, which have adjoined edges that are not fixed. The equations of the equilibrium of the plate lying on the elastic foundation, we write in the form:

$$D\nabla^2\nabla^2 w(x,y) = q(x,y) - p(x,y), \qquad (-\infty < x,y < \infty) \tag{7.5.2}$$

$$\iint_{-\infty}^{\infty} p(x_1,y_1)K(x - x_1, y - y_1)dx_1 dy_1 = w(x,y)$$

here $w$ – displacement (deflection) of the plate; $\nabla^2$ – the Laplace operator in Cartesian coordinates; $q$ – the external load; $p$ – the contact pressure; and $D$ – the cylindrical rigidity of the plate. The boundary conditions at the line of separation of the plates:

$$\frac{\partial^2 w}{\partial x^2} + \sigma\frac{\partial^2 w}{\partial y^2} = 0,$$

$$(x = \pm 0, \quad -\infty < y < \infty) \tag{7.5.3}$$

$$\frac{\partial^3 w}{\partial x^3} + (2 - \sigma)\frac{\partial^3 w}{\partial x \partial y^2} = 0,$$

Let us switch over in (7.5.2) to the dimensionless parameters, following works [321] in which we examined the close problem about two hinge connected (jointed) plates on the layer, assuming that:

$$\lambda = \sqrt[3]{D\theta}, \quad x = x_* \lambda, \quad y = y_* \lambda, \quad w_* = w\frac{\lambda}{Q\theta}, \quad p = \frac{Qp_*}{\lambda^2},$$

$$q = \frac{Qq_*}{\lambda^2}, \quad R = R_* \lambda, \quad \varepsilon = \varepsilon_* \lambda, \quad Q = \iint_{-\infty}^{\infty} q(x,y)dxdy.$$

As a result we will obtain integro-differential system

$$\nabla^2 \nabla^2 w_* = q_* - p_* , \tag{7.5.4}$$

$$\iint_{-\infty}^{\infty} K(R_*)p(x_{1*}, y_{1*})dx_{1*}dy_{1*} = w_*.$$

Subsequently, we will use only the dimensionless form of the equations of equilibrium (7.5.4), so an asterisk in the indexes of functions and coordinates can be omitted. Let us implement the Fourier transform in (7.5.4) on the coordinate $y$.

$$\left(\frac{d^2}{dx^2} - \eta^2\right)^2 w_\eta(x) = q_\eta(x) - p_\eta(x),$$

$$\frac{1}{\sqrt{2\pi}} \iint_{-\infty}^{\infty} e^{i\eta y}K(R)p(x_1, y_1)dx_1 dy_1 dy = w_\eta(x).$$

Let us replace the variable $y - y_1 = z$ in the integral equation, so we will obtain:

$$\sqrt{2\pi} \int_{-\infty}^{\infty} K_\eta(R)p_\eta(x_1)dx_1 = w_\eta(x);$$

here the index $\eta$ designated Fourier's transform on the coordinate $y$, for example

$$w_\eta(x) = \frac{1}{\sqrt{2\pi}} \int_{-\infty}^{\infty} w(x,y)e^{i\eta y}dy.$$

The converted boundary conditions (7.5.3) take the following form

$$\frac{\partial^2 w_\eta}{\partial x^2} - \sigma\eta^2 w_\eta = 0,$$

$$(x = \pm 0) \tag{7.5.5}$$

$$\frac{\partial^3 w_\eta}{\partial x^3} - (2 - \sigma)\eta^2 \frac{\partial w_\eta}{\partial x} = 0,$$

Turning further to the Fourier transforms along the coordinate $x$, it is necessary to take into account that along the line $x = 0$ the function $w_\eta$ and its derivatives can have discontinuities

$$\frac{1}{\sqrt{2\pi}} \int_{-\infty}^{\infty} e^{i\xi x} \left(\frac{d^2}{dx^2} - \eta^2\right)^2 w_\eta(x)dx = (\xi^2 + \eta^2)^2 w_{\xi\eta}$$

$$+ \frac{1}{\sqrt{2\pi}} \left(\Delta\frac{d^3 w_\eta}{dx^3}(0) - i\xi\Delta\frac{d^2 w_\eta}{dx^2}(0) - \xi^2\Delta\frac{dw_\eta}{dx}(0) + i\xi^3\Delta w_\eta(0)\right)$$

$$- \frac{2\eta^2}{\sqrt{2\pi}} \left(\Delta\frac{dw_\eta}{dx}(0) - i\xi\Delta w_\eta(0)\right),$$

here through $\Delta$ the jumps of functions are designated, for example

$$\Delta w_\eta(0) = w_\eta(+0) - w_\eta(-0).$$

Taking into account the converted boundary conditions (7.5.5), we obtain

$$(\xi^2 + \eta^2)^2 w_{\xi\eta} + \frac{1}{\sqrt{2\pi}} \left\{ (\xi^2 + \sigma\eta^2)\Delta\frac{dw_\eta}{dx}(0) - i\xi[\xi^2 + (2 - \sigma)\eta^2]w_\eta(0) \right\}$$

$$= q_{\xi\eta} - p_{\xi\eta};$$

$$2\pi K_{\xi\eta}p_{\xi\eta} = w_{\xi\eta}$$

where

$$K_{\xi\eta} = \frac{1}{\sqrt{2\pi}} \iint_{-\infty}^{\infty} e^{i(\xi x + \eta y)} K(\sqrt{x^2 + y^2})dxdy.$$

For the kernel transform of base we can write

$$K_{\xi\eta} = \frac{\theta h\left(\sqrt{\xi^2 + \eta^2}\right)}{2\pi\sqrt{\xi^2 + \eta^2}}.$$

Consequently,

$$w_{\xi\eta} = \frac{q_{\xi\eta} - (2\pi)^{-1/2}\left\{(\xi^2 + \sigma\eta^2)\Delta\frac{dw_\eta}{dx} - \iota\xi[\xi^2 + (2-\sigma)\eta^2]\Delta w_\eta(0)\right\}}{\overline{R}(\xi,\eta)},$$

(7.5.6)

and ,according to the inversion formula of Fourier

$$w(x,y) = w_q(x,y)$$

$$+ \frac{1}{2\pi} \iint_{-\infty}^{\infty} e^{-i(\xi x + \eta y)} \frac{(\xi^2 + \sigma\eta^2)A + \iota\xi[\xi^2 + (2-\sigma)\eta^2]B}{\overline{R}(\xi,\eta)} d\xi d\eta,$$

(7.5.7)

where

$$w_q(x,y) = \frac{1}{2\pi} \iint_{-\infty}^{\infty} \frac{q_{\xi\eta} e^{-i(\xi x + \eta y)}}{\overline{R}(\xi,\eta)} d\xi d\eta.$$

The displacement of the unlimited plate due to internal load

$$\overline{R}(\xi,\eta) = (\xi^2 + \eta^2)^2 + (2\pi K_{\xi\eta})^{-1}$$

$$A(\eta) = -(2\pi)^{-1/2}\Delta\frac{dw_\eta}{dx}(0), \quad B(\eta) = \Delta w_\eta(0)(2\pi)^{-1/2}.$$

Let us switch over to the determination of unknown functions $A(\eta), B(\eta)$. From (7.5.6) it follows

$$w_\eta(x) = \frac{1}{\sqrt{2\pi}} \int_{-\infty}^{\infty} \frac{q_{\xi\eta} + (\xi^2 + \sigma\eta^2)A(\eta) + \iota\xi[\xi^2 + (2-\sigma)\eta^2]B(\eta)}{\overline{R}(\xi,\eta)} e^{-i\xi x} d\xi$$

$$= w_{q\eta}(x) + A(\eta)[C_2'(x,\eta) + \sigma\eta^2 C_1'(x,\eta)]$$

$$+ B(\eta)[C_3(x,\eta) + (2-\sigma)\eta^2 C_2(x,\eta)]$$

where

$$C_1(x,\eta) = \sqrt{\frac{2}{\pi}} \int_0^\infty \frac{\sin \xi x d\xi}{\xi \overline{R}(\xi,\eta)}; \quad C_2(x,\eta) = \sqrt{\frac{2}{\pi}} \int_0^\infty \frac{\xi \sin \xi x d\xi}{\overline{R}(\xi,\eta)}; \quad (7.5.8)$$

$$C_3(x,\eta) = \sqrt{\frac{2}{\pi}} \int_0^\infty \frac{\xi^3 \sin \xi x d\xi}{\overline{R}(\xi,\eta)}.$$

We will further obtain the system of equations for determining the unknowns $A(\eta)$ and $B(\eta)$ satisfying converted boundary conditions (7.5.5).

In this case an essential fact is the convergence of integrals (7.5.7) during the triple $(C_3)$ and fourfold $(C_1, C_2)$ differentiation. With $t \to \infty$, we have

$$K_{\frac{1-\nu}{2}}(\varepsilon t) \sim \sqrt{\frac{\pi}{\varepsilon t}} e^{-\varepsilon t}$$

and consequently, the denominator of the integrand expression in these integrals has exponential growth at infinity. So all integrals can be three times differentiated on parameter $x$. Further, we have

$$C_1'''(0, \eta) = -\sqrt{\frac{2}{\pi}} \int_0^\infty \frac{\xi^2 d\xi}{\overline{R(\xi, \eta)}}; \qquad C_1^{IV}(0, \eta) = 0;$$

$$C_2'''(0, \eta) = -\sqrt{\frac{2}{\pi}} \int_0^\infty \frac{\xi^4 d\xi}{\overline{R(\xi, \eta)}}; \qquad C_2^{IV}(0, \eta) = 0; \qquad (7.5.9)$$

$$C_3''(0, \eta) = 0; \qquad C_3'''(0, \eta) = -\sqrt{\frac{2}{\pi}} \int_0^\infty \frac{\xi^6 d\xi}{\overline{R(\xi, \eta)}}.$$

Hence the system of boundary equations takes the form $(x = +0)$

$$w_{q\eta}''(0) + A(\eta)[C_2'''(0, \eta) + \sigma\eta^2 C_1'''(0, \eta)] - \sigma\eta^2\{w_{q\eta}(0)$$
$$+ A(\eta)[C_2'(0, \eta) + \sigma\eta^2 C_1'(0, \eta)]\} = 0, \qquad (7.5.10)$$

$$w_{q\eta}'''(0) + B(\eta)[C_3'''(0, \eta) + (2 - \sigma)\eta^2 C_1'''(0, \eta)] - (2 - \sigma)\eta^2$$
$$\times \{w'_{q\eta}(0) + B(\eta)[C_3'(0, \eta) + (2 - \sigma)\eta^2 C_2'(0, \eta)]\} = 0,$$

Thus:

$$A(\eta) = \frac{\sigma\eta^2 w_{q\eta}(0) - w_{q\eta}''(0)}{C_3'''(0, \eta) + \sigma\eta^2 C_1'''(0, \eta) - \sigma\eta^2[C_2'(0, \eta) + \sigma\eta^2 C_1(0, \eta)]},$$

$$B(\eta) = \frac{(2 - \sigma)\eta^2 w'_{q\eta}(0) - w_{q\eta}'''(0)}{C_3'''(0, \eta) + (2-\sigma)\eta^2 C_1'''(0, \eta) - (2-\sigma)\eta^2[C_3'(0, \eta) + (2-\sigma)\eta^2 C_2'(0, \eta)]}$$

The displacement of the plate is determined by formula

$$w(x, y) = \frac{1}{\sqrt{2\pi}} \int_{-\infty}^\infty w_\eta(x) e^{-i\eta y} d\eta$$

Analogously we can examine the problem of bending of two free semi-infinite uninsulated beams lying on an elastic foundation, described by the proposed generalized model. The equations of the equilibrium of beams take the form:

$$EIy^{IV}(x) = q(x) - p(x), \quad (-\infty < x < \infty),$$

$$\int_{-\infty}^{\infty} p(x_1)\overline{K}(|x - x_1|)dx_1 = y(x) \tag{7.5.11}$$

The boundary conditions, corresponding to the free ends of the beams are as follows:

$$y''(\pm 0) = y'''(\pm 0) = 0. \tag{7.5.12}$$

Let us add in (7.5.11) to the Fourier transforms along the coordinate $x$, for which we will multiply left and right parts of both equations on $(2\pi)^{-1/2}e^{-i\xi x}$, and integrate over the entire axis. The integration by parts with taking into account (7.5.12) gives:

$$EI\xi^4 y_\xi + \frac{1}{\sqrt{2\pi}}\left[\xi^2\Delta\frac{dy}{dx}(0) - i\xi^3\Delta y(0)\right] = q_\xi - p_\xi, \tag{7.5.13}$$

$$\frac{1}{\sqrt{2\pi}}\iint_{-\infty}^{\infty} p(x_1)\overline{K}(|x - x_1|)e^{i\xi x}dx_1 dx = y_\xi.$$

Assuming in integral Eq. (7.5.12) $x - x_1 = z$, we find

$$\sqrt{2\pi}\, p_\xi \overline{K}_\xi = y_\xi \tag{7.5.14}$$

The calculation of the Fourier transform of the kernel base can be executed using the formula

$$\overline{K}(s) = \theta \int_0^\infty t^{-1}h(t)\cos stdt \tag{7.5.15}$$

Since the kernel is an even function, then

$$\overline{K}_\xi = \sqrt{\frac{2}{\pi}}\int_0^\infty \overline{K}(s)\cos\xi sds$$

Hence, according to the inversion formula of Fourier transform, it follows

$$\overline{K}(s) = \sqrt{\frac{2}{\pi}}\int_0^\infty \overline{K}_\xi \cos\xi sd\xi \tag{7.5.16}$$

Further comparison (7.5.15, 7.5.16) gives:

$$\overline{K}_\xi = \sqrt{\frac{\pi}{2}}\,\theta\xi^{-1}h(\xi).$$

As we see $\overline{K}_\xi$ and $\xi^{-1}h(\xi)$ are also even functions of argument $\xi$. Now we obtain from (7.5.13, 7.5.14)

$$y_\xi = \frac{q_\xi - (2\pi)^{-1/2}\left[\xi^2\Delta\frac{dy}{dx}(0) - i\xi^3\Delta y(0)\right]}{EI\xi^4 + \xi[\pi\theta h(\xi)]^{-1}} \tag{7.5.17}$$

The displacement of beam can be obtained by the formula of inversion for Fourier transform, using (7.5.17) and taking into account that the denominator in (7.5.17) is an even function:

$$y(x) = y_q(x) - \Delta\frac{dy}{dx}(0)F_1(x) + \Delta y(0)F_2(x), \tag{7.5.18}$$

where

$$F_1(x) = \frac{1}{\pi}\int_0^\infty \frac{\xi\cos\xi x\,d\xi}{EI\xi^3 + [\pi\theta h(\xi)]^{-1}},$$

$$F_2(x) = \frac{1}{\pi}\int_0^\infty \frac{\xi^2\sin\xi x\,d\xi}{EI\xi^3 + [\pi\theta h(\xi)]^{-1}},$$

$$y_q(x) = \frac{1}{\sqrt{2\pi}}\int_{-\infty}^\infty \frac{q\exp(-i\xi x)d\xi}{EI\xi^4 + \xi[\pi\theta h(\xi)]^{-1}}.$$

Further, with the displacement (7.5.18) substituting into boundary conditions (7.5.12), we obtain the expression for determining the unknown jumps $\Delta\frac{dy}{dx}(0)$, and $\Delta y(0)$

$$\Delta\frac{dy}{dx}(0) = -\pi EIy_q''(0)F_3^1,$$

$$\Delta y(0) = -\pi EIy_q'''(0)F_5^1,$$

where

$$F_3(x) = \int_0^\infty \frac{\xi^3 d\xi}{\xi^3 + [\theta_1 h(\xi)]^{-1}},$$

$$F_5(x) = \int_0^\infty \frac{\xi^5 d\xi}{\xi^3 + [\theta_1 h(\xi)]^{-1}}, \qquad \theta_1 = \pi EI\theta.$$

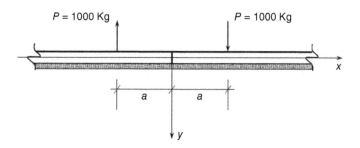

**Figure 7.9** Scheme of loading of the uninsulated beams.

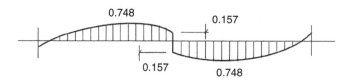

**Figure 7.10** Diagram of the displacements of the uninsulated beams.

Thus, we have obtained the resulting solution of the considered problem. The scheme of the beam is presented in Figure 7.9. In Figure 7.10 we present the results of calculating the deflection of beam in cm, lying on a generalized model of the base with the parameters, determined according to the results of experiments of L.I. Manvelov and E.S. Bartoshevich [221]. For the beam we accept the following; $EI = 2 \cdot 10^9$ Kg cm$^2$, $a = 200$ cm. It is characterized that the end sections of the beams have a jump of displacements that do not allow for other models of the coherent base.

## 7.6 The Forced Oscillations of a Rectangular Plate on an Elastic Foundation

The spectral method of boundary elements (SMBE), closely related to the theory of integral transformations, can be a highly effective means for the solution of boundary-value problems with complex boundary conditions. In the present paragraph the SMBE is used for the solution of the problem of the harmonic oscillations of the rectangular isotropic plate, lying on an elastic foundation of a Winklerian type. This problem is encountered with the dynamic calculation of the pavements of industrial buildings, power (thrust) plates, road and airport paving, when inertia of base, which plays a significant role in the general picture of the dynamic deformation of plates, can be taken into account within the framework of the discrete model. As a special case, from the obtained solution follows the solution of the static problem of the bend of a rectangular plate on an elastic foundation. Similar static problems for the plates with free edges were examined in the works of E.A. Palatnikov [253], and V.I. Travush and V.K. Sangadzhiev [364]. However, in contrast to the works indicated, the following solution gives a sufficiently simple algorithm, and automatically included are unknowns, concentrated at the angular points, that can be obtained with the use of a SMBE. The problem in this case is a specific example of the application of the SMBE to the solution of boundary-value problems, since the area occupied by plate is rectangular, it therefore does not reveal all possibilities of this method in comparison with the domains of complex configuration. The solution to the problem can be obtained by applying the generalized finite integral transformations. Let us write the equation of the harmonic oscillations of the rectangular plate, which lies on the discrete inertia-base under the action of load $q(x, y)e^{i\omega t}$, in the form of:

$$DV^2V^2w(x,y,t) + m\frac{\partial^2 w(x,y,t)}{\partial t^2} + \int_{-\infty}^{t} R(t-\tau)w(x,y,t)d\tau = q(x,y)e^{i\omega t},$$

$$(7.6.1)$$

where $V^2 = \frac{\partial^2}{\partial x^2} + \frac{\partial^2}{\partial y^2}$ — is the Laplace operator in the Cartesian coordinates; $k$ – the coefficient of the bed of the base; $w(x, y, t)$ – the displacement of the plate; $R(t-\tau)$–the kernel of discrete model, which describes the elastic, inertial, and dissipative properties of the base; $q(x, y)$ – the intensity of harmonic load; $\omega$ – the circular frequency of the load; and $D$, $m$ – the cylindrical rigidity and mass per unit area of the plate. Assuming that $w(x, y, t) = w(x, y)e^{i\omega t}$, and excluding time, we will obtain:

$$DV^2V^2w(x,y) + k_0(\omega)w(x,y) = q(x,y), \qquad (7.6.2)$$

where

$$k_0(\omega) = k(\omega) - m\omega^2; \quad k(\omega) = \int_{0}^{\infty} R(z)e^{-i\omega z}dz.$$

For simplicity we will consider $k(\omega)$ as a real function in the future, disregarding the dissipation of energy in the base. Let us introduce designations $|k_0(\omega)|/D = l^{-4}$ and switch over to dimensionless coordinates:

$$x = x_*l, \quad y = y_*l, \quad a = a_*l, \quad b = b_*l.$$

Then Eq. (7.6.2) takes the form

$$[\nabla^2\nabla^2 + signk_0(\omega)]w(x,y) = q_1(x,y), \tag{7.6.3}$$

where

$$q_1(x,y) = D^{-1}l^4q(x_*,y_*).$$

In this case indices with $x_*$ and $y_*$ are omitted for simplicity. The SMBE can be used for calculating the plate with arbitrary boundary conditions. In all cases, together with the distributed boundary densities, the method makes it possible to obtain angular singularity. Subsequently let us limit the examination to the most important case of plates with free edges, which have the greatest applications in civil engineering [380]. The boundary conditions for the free-lying plate with the selected system of coordinates (Figure 7.11) takes the form

$$L_iw(x,y)|_{x=\pm a/2} = 0, \qquad L_{i+2}w(x,y)|_{y=\pm b/2} = 0, \tag{7.6.4}$$

$$(i = 1, 2)$$

where

$$L_1 = \frac{\partial^2}{\partial x^2} + v\frac{\partial^2}{\partial y^2}, \quad L_2 = \frac{\partial^3}{\partial x^3} + (2 - v)\frac{\partial^3}{\partial x\partial y^2},$$

$L_3$ and $L_4$ are obtained by the replacement of $x$ on $y$ in $L_i$ and vice versa.

If an arbitrary loads act on the plate, then it is possible to decompose the load into symmetric and skew-symmetric components (Figure 7.12), and design the plate to the action of each component separately. The general solution is obtained by superposition of the particular solutions. For the

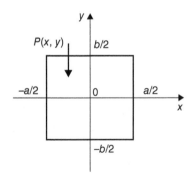

**Figure 7.11**   Design scheme of a plate on an elastic foundation.

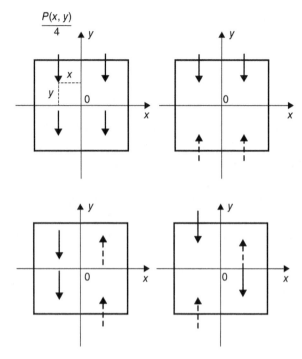

**Figure 7.12** Decomposition of the load on (into) symmetric and skew-symmetric components.

solution of the problem with a symmetric loading, the cosine transformation should be used; with the inversely symmetric loading, we use the sine transformation; and in the mixed case transformation, we use both the cosine and the sine. For the conditions which give the resolving system of boundary equations, we have boundary conditions (7.6.4) and the condition of the equilibrium:

$$\iint_{-\frac{a}{2}-\frac{b}{2}}^{\frac{a}{2}\ \frac{b}{2}} k_0(\omega)\, w\,(x,y)\, dxdy = \iint_{-\frac{a}{2}-\frac{b}{2}}^{\frac{a}{2}\ \frac{b}{2}} q(x,y)\, dxdy. \tag{7.6.5}$$

Let us examine the case of symmetric loading relative to both axes $Ox$ and $Oy$ (Figure 7.12), and we will limit ourselves to considering the domain occupying a quarter of the plate accordingly.

For the solution of Eq. (7.6.3) we will apply the two-dimensional finite cosine transformation. Let us first multiply both parts of the equation on

$\cos \frac{2m\pi x}{a} \cos \frac{2n\pi y}{b}$ and integrate in the limits $\left(0, \frac{a}{2}\right), \left(0, \frac{b}{2}\right)$; we have

$$\int\int_{0\,0}^{a/2\,b/2} (\nabla^2\nabla^2 + signk_o)w(x,y)\cos\alpha_m x \cos\beta_n y dxdy$$

$$= \int\int_{00}^{a/2\,b/2} q_1(x,y)\cos\alpha_m x \ \cos\beta_n y dxdy, \tag{7.6.6}$$

After the integration on the left part of Eq. (7.6.6) we have the form:

$$[(\alpha_m^2 + \beta_n^2)^2 + signk_o]F_c[w(x,y)] = q(m,n) - \sum_{i=1}^{3} Q_i, \tag{7.6.7}$$

where

$$\alpha_m = \frac{2m\pi}{a}, \qquad \beta_n = \frac{2n\pi}{b},$$

$Q_i$ – "compensating" actions (impact, effects) containing unknown densities in the form of some differential expressions from $w(x, y)$ and distributed along the contour of the plate, including angular points as well take the form:

$$Q_1(m, n, a) = (-1)^m W_x'' \left(\frac{a}{2}, n\right) - (-1)^m \alpha_m^2 W_x' \left(\frac{a}{2}, n\right),$$

$$Q_2(m, n, b) = (-1)^n W_y''' \left(\frac{b}{2}, m\right) - (-1)^n \beta_m^2 W_y' \left(\frac{b}{2}, n\right) \tag{7.6.8}$$

$$Q_3(m, n, a, b) = 2(-1)^{m+n} W_{xy}'' \left(\frac{a}{2}, \frac{b}{2}\right) + 2(-1)^m \beta_m^2 W_x' \left(\frac{a}{2}, n\right)$$

$$+ 2(-1)^n \alpha_m^2 W_y' \left(\frac{b}{2}, m\right), \qquad (m, n = 0, 1, 2, \dots, \infty)$$

where

$$W_x^{(j)} \left(\frac{a}{2}, n\right) = \int_0^{b/2} W_x^{(j)} \left(\frac{a}{2}, y\right) \cos\beta_n y dy, \tag{7.6.9}$$

$$W_y^{(j)} \left(\frac{b}{2}, m\right) = \int_0^{a/2} W_y^{(j)} \left(\frac{b}{2}, x\right) \cos\alpha_m x dx.$$

Using the boundary conditions:

$$L_2 w(x,y)\Big|_{x=\frac{a}{2}} = 0, \quad L_4 w(x,y)\Big|_{y=\frac{b}{2}} = 0$$

According to (7.6.4) we have

$$w_x''' \left(\frac{a}{2}, y\right) = (v - 2)\frac{\partial^3 w\left(\frac{a}{2}, y\right)}{\partial x \partial y^2}, \tag{7.6.10}$$

$$w_y''' \left(\frac{b}{2}, x\right) = (v - 2)\frac{\partial^3 w\left(\frac{b}{2}, x\right)}{\partial x^2 \partial y}$$

We multiply both parts of the first Eq. (7.6.10) on $\cos \beta_n y$ and both parts of the second equation on $\cos \alpha_m x$, then integrate respectively in the limits $(0, b/2)$ and $(0, a/2)$. After integrating by parts we obtain

$$\int_0^{b/2} w_x''' \left(\frac{a}{2}, y\right) \cos \beta_n y \, dy = (-1)^n (v - 2)\frac{\partial^2 w\left(\frac{a}{2}, \frac{b}{2}\right)}{\partial x \partial y}$$

$$- (v - 2)\beta_n^2 \int_0^{b/2} w_x' \left(\frac{a}{2}, y\right) \cos \beta_n y \, dy; \tag{7.6.11}$$

$$\int_0^{a/2} w_y''' \left(\frac{b}{2}, x\right) \cos \alpha_m x \, dx = (-1)^m (v - 2)\frac{\partial^2 w\left(\frac{a}{2}, \frac{b}{2}\right)}{\partial x \partial y}$$

$$- (v - 2)\alpha_m^2 \int_0^{b/2} w_y' \left(\frac{b}{2}, x\right) \cos \alpha_m x \, dx.$$

Let us designate according to (7.6.9)

$$A(n) = \int_0^{b/2} w_x' \left(\frac{a}{2}, y\right) \cos \beta_n y \, dy;$$

$$B(m) = \int_0^{a/2} w_y' \left(\frac{b}{2}, x\right) \cos \alpha_m x \, dx; \tag{7.6.12}$$

$$c = \frac{\partial^2 w\left(\frac{a}{2}, \frac{b}{2}\right)}{\partial x \partial y}.$$

Taking into account designations (7.6.12) and equality (7.6.11) in (7.6.8), we obtain:

$$Q_1(m, n, a, b) = (-1)^{m+n}(\nu - 2)c - (-1)^m[(\nu - 2)\beta_n^2 + \alpha_m^2]A(n),$$

$$Q_2(m, n, a, b) = (-1)^{m+n}(\nu - 2)c - (-1)^n[(\nu - 2)\alpha_m^2 + \beta_n^2]B(m), \quad (7.6.13)$$

$$Q_3(m, n, a, b) = (-1)^{m+n} \cdot 2c + 2(-1)^m\beta_n^2 A(n) + 2(-1)^n\alpha_m^2 B(m),$$

$$m, n = 0, 1, 2, \ldots, \infty.$$

As can easily be seen, Eq. (7.6.13) contains three unknown values, which are determined from the boundary conditions and conditions of equilibrium. Let us substitute (7.6.13) into (7.6.7) and after inversion (conversion, transformation), we will obtain the solution of the equation of motion in the form:

$$w(x, y) = \frac{16}{ab} \sum_{m=0}^{\infty}{}' \sum_{n=0}^{\infty}{}' \frac{1}{(\alpha_m^2 + \beta_n^2)^2 + \text{sign } k_0} [q(m, n)$$
$$+ (-1)^m(\nu - 4)(\alpha_m^2 + \beta_n^2)A(n) + (-1)^n(\nu - 4)(\alpha_m^2 + \beta_n^2)B(m)$$
$$+ (-1)^{m+n}(2 - 2\nu)c)] \cos \alpha_m x \cos \beta_n y, \quad (7.6.14)$$

where the prime in the sum symbol means that an additional multiplier, 1/2, is entered (introduced) for the zero member. Inserting (7.6.14) into boundary conditions (7.6.4) with the operators $L_1$ and $L_3$, and into the conditions of equilibrium we will obtain the following system of algebraic equations:

$$\sum_{m=0}^{\infty}{}' \frac{(\alpha_m^2 + \nu\beta_n^2)(-1)^m}{(\alpha_m^2 + \beta_n^2)^2 + \text{sign } k_0}\{q(m, n)$$
$$+ (\nu - 4)(\alpha_m^2 + \beta_n^2)[(-1)^m A(n) + (-1)^n B(m)]\} = 0,$$

$$\sum_{n=0}^{\infty}{}' \frac{(\beta_n^2 + \nu\alpha_m^2)(-1)^n}{(\alpha_m^2 + \beta_n^2)^2 + \text{sign } k_0}\{q(m, n)$$
$$+ (\nu - 4)(\alpha_m^2 + \beta_n^2)[(-1)^m A(n) + (-1)^n B(m)]\} = 0 \quad (7.6.15)$$

$$q(0, 0) + (2 - 2\nu)c = \int\int_{0}^{\frac{a}{2}} {}_{0}^{\frac{b}{2}} q_1(x, y)\, dx dy,$$

$$q(m, n) = \int_0^{\frac{a}{2}} \int_0^{\frac{b}{2}} q_1(x, y) \cos \alpha_m x \cos \beta_n y \, dx dy,$$

$$q_1(x, y) = D^{-1} l^4 q(x, y), \quad \alpha_m = \frac{2m\pi}{a}, \quad \beta_n = \frac{2n\pi}{b}.$$

$$m, n = 0, 1, 2, \dots, \infty.$$

From the third equation of system (7.6.15) we will obtain $c = 0$. Let us designate

$$(\alpha_m^2 + \beta_n^2)^2 + \text{sign } k_0 = A$$

$$a_1(n) = \sum_{m=0}^{\infty} {}' \frac{(\alpha_m^2 + \beta_n^2)(\alpha_m^2 + \nu\beta_n^2)(\nu - 4)}{A},$$

$$b_1(m, n) = \frac{(\alpha_m^2 + \nu\beta_n^2)(\alpha_m^2 + \beta_n^2)(\nu - 4)(-1)^{m+n}}{A}, \qquad (7.6.16)$$

$$c_1(n) = -\sum_{m=0}^{\infty} {}' \frac{(\alpha_m^2 + \nu\beta_n^2)(-1)^m q(m, n)}{A},$$

$$a_2(m, n) = \frac{(\beta_n^2 + \nu\alpha_m^2)(\alpha_m^2 + \beta_n^2)(\nu - 4)(-1)^{m+n}}{A},$$

$$b_2(m) = \sum_{n=0}^{\infty} {}' \frac{(\beta_n^2 + \nu\alpha_m^2)(\alpha_m^2 + \beta_n^2)(\nu - 4)}{A},$$

$$c_2(m) = -\sum_{n=0}^{\infty} {}' \frac{(\beta_n^2 + \nu\alpha_m^2)(-1)^n q(m, n)}{A}.$$

Taking into account the designations of (7.6.16), the system (7.6.15) takes the following form:

$$a_1(n)A(n) + \sum_{m=0}^{\infty} {}' b_1(m, n)B(m) = c_1(n), \qquad (7.6.17)$$

$$\sum_{n=0}^{\infty} {}' a_2(m, n)A(n) + b_2(m)B(m) = c_1(m),$$

$$m, n = 0, 1, \dots, \infty.$$

The coefficients with the unknowns $A(n)$ and $B(m)$ in the system (7.6.17) are represented by the divergent series. Therefore it is necessary to regularize these series after isolating (extracting) from them the generalized functions. Let us examine for an example the expression for $a_1(n)$;

$$a_1(n) = \left[ \sum_{m=0}^{\infty}{}' \frac{(\alpha_m^2 + \beta_n^2)(\alpha_m^2 + \nu\beta_n^2)(\nu - 4)\cos\frac{2m\pi x}{a}(-1)^m}{(\alpha_m^2 + \beta_n^2)^2 + sign\, k_0} \right]_{x=a/2},$$

and let us isolate from its regular and singular parts. For this we will use the formula known from the theory of generalized functions [115]

$$\sum_{m=1}^{\infty} \cos m\alpha = -\frac{1}{2} + \pi \sum_{m=-\infty}^{\infty} \delta(\alpha - 2\pi m).$$

Let us assume in this formula that $\alpha = z - \pi$, then we will obtain

$$\sum_{m=1}^{\infty} (-1)^m \cos mz = -\frac{1}{2} + \pi \sum_{m=-\infty}^{\infty} \delta[z - \pi(2m + 1)].$$

Let us switch over further to a new variable $z = \frac{2\pi x}{a}$ and take into account the known property of delta-function [115]

$$\delta[r(x)] = \sum_{h} \frac{\delta(x - x_h)}{|r'(x_h)|},$$

where $x_h$ – is a simple zero of function $r(x)$. With $r(x) = \frac{2\pi x}{a} - \pi(2m + 1)$, $x_h = \frac{a}{2}(2m + 1)$, and consequently, we have

$$\delta\left[\frac{2\pi x}{a} - \pi(2m + 1)\right] = \frac{a\delta\left[x - \frac{a}{2}(m + 1)\right]}{2\pi}.$$

Hence

$$\sum_{m=1}^{\infty} (-1)^m \cos\frac{2m\pi x}{a} = -\frac{1}{2} + \frac{a}{2} \sum_{m=-\infty}^{\infty} \delta\left[x - (2m + 1)\frac{a}{2}\right], \qquad (7.6.18)$$

where $a = 2\pi$, $x \in (-\pi, \pi)$.

We will obtain

$$
a_1(n) = \Bigg[ (v-4) \sum_{m=1}^{\infty} (-1)^m \cos \frac{2m\pi x}{a} + \frac{v-4}{2}
$$

$$
+ \sum_{m=0}^{\infty}{}' \frac{[(v-1)(\beta_n^4 + \alpha_m^2 \beta_n^2) - 1](v-4) \cos \frac{2m\pi x}{a}(-1)^m}{(\alpha_m^2 + \beta_n^2)^2 + \operatorname{sign} k_0} \Bigg]_{x=\frac{a}{2}}
$$

$$
= \frac{\pi a}{2}(v-4) \lim_{x \to \frac{a}{2}} \sum_{m=-\infty}^{\infty} \delta\Big[x - (2m+1)\frac{a}{2}\Big]
$$

$$
x < a/2
$$

$$
+ \sum_{m=0}^{\infty}{}' \frac{[(v-1)(\beta_n^4 + \alpha_m^2 \beta_n^2) - \operatorname{sign} k_0](v-4)}{(\alpha_m^2 + \beta_n^2)^2 + \operatorname{sign} k_0}
$$

As mentioned above, a strict formulation of boundary conditions requires the determination of the boundary values of the solution of the boundary-value problem on the inner contour of the domain. The value of delta-function in the last expression inside the domain and in the limit on its boundary can be considered equal to zero (although formally the delta-function at this point is not defined). In view of this fact, the singular component can be neglected in the expression for determining the coefficient $a_1(n)$. After discarding the singular part the expression for $a_1(n)$, it takes the form

$$
a_1(n) = (v-4) \sum_{m=0}^{\infty}{}' \frac{(v-1)(\beta_n^4 + \alpha_m^2 \beta_n^2) - \operatorname{sign} k_0}{A} ,
$$

With the aid of similar operations the expression for $b_2(m)$, it takes the form

$$
b_2(m) = (v-4) \sum_{n=0}^{\infty}{}' \frac{(v-1)(\alpha_m^4 + \alpha_m^2 \beta_n^2) - \operatorname{sign} k_0}{(\alpha_m^2 + \beta_n^2)^2 + 1} .
$$

Thus, the series for all coefficients of systems converge more rapidly than

$$
\sum_{m=1}^{\infty} \frac{1}{\alpha_m^2} \quad or \quad \sum_{n=1}^{\infty} \frac{1}{\beta_n^2} ,
$$

Their calculation is expedient to conduct with a computer. The choice of the upper limit of the sum depends on the speed of the decrease of the coefficients $B(m)$, included in the system, and therefore it is necessary to conduct a study of their asymptotic behavior with $m \to \infty$, which is also more conveniently carried out with the aid of the computer. After assigning the upper limit of summation and truncation of the system of equations (7.6.17) to $s$ of order, it brings this system to the form:

$$\sum_{n=0}^{s\,\prime} a_2(m,n)\frac{c_1(n) - \sum\limits_{m=0}^{t\,\prime} b_1(m,n)B(m)}{a_1(n)} + b_2(m)B(m) = c_2(m),$$

or

$$\sum_{m=0}^{t\,\prime} b_1(m,n)\frac{c_2(n) - \sum\limits_{n=0}^{s\,\prime} a_2(m,n)A(n)}{b_2(m)} + a_1(n)A(n) = c_1(n).$$

For square plate $(B(m) = A(m))$ in particular we have:

$$a_1(n)A(n) + \sum_{m=0}^{t\,\prime} b_1(m,n)A(m) = c_1(n), \tag{7.6.19}$$

$$m, n = 0, 1, 2, \dots, t.$$

where $a_1(n)$ and $b_1(m, n)$ are determined from formula (7.6.17), by replacing $\beta_n$ to $\alpha_n$. After the solution of system of equations (7.6.17) the function $w(x, y)$ is determined by the formula (7.6.14). Further, the calculation of internal forces $M_x$, $Q_x$, $M_y$, $Q_y$, and $H_{xy}$ at any point on the plate is achieved according to the known formulas of the theory of plates:

$$M_x(x,y) = -\frac{D}{l^2}\left(\frac{\partial^2}{\partial x^2} + v\frac{\partial^2}{\partial y^2}\right)w(x,y),$$

$$M_y(x,y) = -\frac{D}{l^2}\left(\frac{\partial^2}{\partial y^2} + v\frac{\partial^2}{\partial x^2}\right)w(x,y), \tag{7.6.20}$$

$$Q_x(x,y) = -\frac{D}{l^3}\frac{\partial}{\partial x}\nabla^2 w(x,y),$$

$$Q_y(x,y) = -\frac{D}{l^3}\frac{\partial}{\partial y}\nabla^2 w(x,y),$$

$$H_{xy}(x,y) = -\frac{D}{l^2}(1-v)\frac{\partial^2 w}{\partial x\partial y}\ .$$

The substitution of (7.6.14) into formula (7.6.20) leads to the divergent series, similar to (7.6.15), the regularization of which can be carried out according to the formula (7.6.18). For example, for the square plate the bending moment is described by the expression

$$M(x, y) = \frac{16}{a^2} \sum_{m=0}^{\infty}{}' \sum_{n=0}^{\infty}{}' \{(\alpha_m^2 + \nu\alpha_n^2)q(m, n) + (\nu - 4)(\nu - 1)(\alpha_m^2 + \alpha_n^2)$$

$$\times [(-1)^m \alpha_n^2 A(n) + (-1)^n \alpha_m^2 A(m)] - signk_0 - (\nu - 4)[(-1)^m A(n)$$

$$+ (-1)^n A(m)]\} \frac{\cos \alpha_m x \cos \alpha_n y}{(\alpha_m^2 + \alpha_n^2)^2 + signk_0}. \tag{7.6.21}$$

As an example let us examine a square plate with the given size of $2 \times 2$, with the following types of load:

1. Force concentrated in the center of the plate $P = 1$;
2. Uniform load $q_1 = 100$, distributed along the area of the plate with size $0.1 \times 0.1$ in the center of the plate.

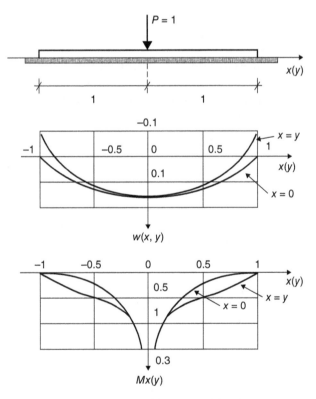

Figure 7.13   Diagram of deflections and bending moments.

The diagrams of the static deflections and bending moments in the sections throughout the diagonal $x = y$ and along the axis $x = 0$ are obtained by taking into account the 25 terms of the series, and are shown in Figures 7.13 and 7.14. Moreover, with the last version of the load of the value of deflection $w = 0.272$ and bending moment $M = 0.2987$ in the center of plate differ from those obtained by an alternative method in the work of V.I. Travush, V.K Sangadzhiev [364] by values less than 1%.

The analysis of the dependence of the deflections of the plate on the frequency of load was conducted for the last version of load. An increase in the frequency of load leads to the decrease of the given rigidity $k_0(\omega)$, and consequently to an increase in the absolute value of reduced length $l$. At the same time the sign $k_0(\omega)$ changes in the transition into the "transresonant" regime $(\omega > k(\omega)/m)$.

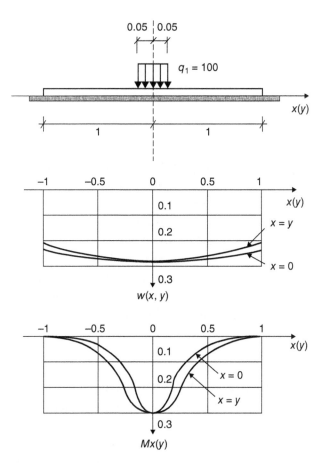

**Figure 7.14** Diagram of deflections and bending moments.

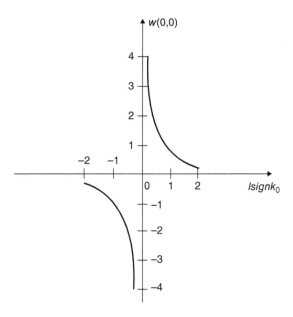

**Figure 7.15**   Dependence of the deflection of the center of the plate on frequencies of load.

Figure 7.15 shows the dependence of the deflection of the center of a plate on the parameter $\xi = lsignk_0$, which shows that this transition is accompanied by a change in the phase of oscillations. The calculations show that the use in the spectral method of the boundary elements expansions with the discrete spectrum leads to the very effective algorithms of other methods, connected with the eigenfunction expansions.

The specific advantage of SMBE in such problems is the generality of the procedure used and the opportunity to direct the use of the generalized functions. In the problems for the domains with the complex boundary, SMBE becomes particularly effective. For the first time in this work we give solutions of problems for the uninsulated beams and plates using spectral BEM (SMBM).

## 7.7   The Calculation of the Membrane of Arbitrary Shape on an Elastic Foundation

Let us examine the equilibrium of the membrane on an elastic foundation of Winklerian type (or similarly, that the membrane is located in the elastic discrete medium), as the simplest model for the illustration of the application of the spectral BEM. Such structures not only arise in the applications, but their design schema corresponds to the well-known two-parameter model of base. One of the versions of this model belonging M.M. Philonenko-Borodich is described exactly by such the design scheme – linear system of springs, connected together by the membrane. The equation of the equilibrium of the membrane can be written in the form:

$$-T\nabla^2 w(x,y) + kw(x,y) = q(x,y) \qquad (7.7.1)$$

where $\nabla^2$ – the operator of Laplace in the rectangular coordinates; $T$, $k$ – the tension of the membrane and the coefficient of the bed of base; and $q$ – external force. Equation (7.1.1) is the analog of the nonhomogeneous equation of Helmholtz; it is convenient because it is two-dimensional and has the second order, so that the calculation formulas acquire sufficient simplicity and transparency, which are necessary for the greatest clarity of an example. In addition the presence of the term $kw$ introduces additional simplifications during the introduction to the infinite extended domain, ensuring the boundedness of the solutions of Eq. (7.7.1) with any limited right part. Eq. (7.7.1) we will examine a certain bounded space $D^+$ of two-dimensional plane $R$ with the smooth boundary $\Gamma$. We will assign the boundary conditions on the border $\Gamma^+$:

$$\text{(i)} \qquad w^+(x,y) = f(x,y), \qquad (7.7.2)$$

or

$$\text{(ii)} \qquad \frac{\partial w^+}{\partial n}(x,y) = g(x,y), \qquad (7.7.3)$$

where the derivative of displacement along the normal to the boundary at point with the coordinates $x, y$ is equal

$$\frac{\partial w}{\partial n} = \frac{\partial w}{\partial x}(x,y)n_x(x,y) + \frac{\partial w}{\partial y}(x,y)n_y(x,y), \qquad (7.7.4)$$

$n_x, n_y$ – are the components of the unit vector, directed along the normal (see Figure 7.16).

Let us enlarge domain $D^+$ through the boundary $\Gamma$ to the entire plane $-\infty < x, y < \infty$ and continue $w(x, y)$ and $q(x, y)$ out of $D^+$. Using the

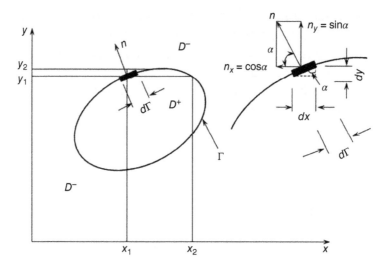

**Figure 7.16**   Schema of membrane and the element of the boundary.

transformation of Fourier in the extended domain, we obtain:

$$-\frac{1}{2\pi} \iint_{-\infty}^{\infty} \nabla^2 w(x,y) e^{i(\xi x + \eta y)} dxdy + \lambda^2 W(\xi, \eta) = Q(\xi, \eta), \qquad (7.7.5)$$

where $W(\xi, \eta)$ and $Q(\xi, \eta)$ – are transforms of Fourier of the desired function and load

$$W(\xi, \eta) = \frac{1}{2\pi} \iint_{-\infty}^{\infty} w(x,y) e^{i(\xi x + \eta y)} dxdy,$$

$$Q(\xi, \eta) = \frac{1}{2\pi T} \iint_{-\infty}^{\infty} q(x,y) e^{i(\xi x + \eta y)} dxdy.$$

After taking the integral in (7.7.5) by parts, we will take into account that on the boundary $\Gamma$ breakages of function $w$ and its derivative along the normal $n$ to the boundary are possible. Therefore let us assume:

$$\iint_{-\infty}^{\infty} \nabla^2 w(x,y) e^{i(\xi x + \eta y)} dxdy = \iint_{D^+} \nabla^2 w e^{i(\xi x + \eta y)} dxdy$$

$$+ \iint_{D^-} \nabla^2 w e^{i(\xi x + \eta y)} dxdy. \qquad (7.7.6)$$

Consider the integral:

$$\iint_{D^+} \frac{\partial^2 w}{\partial x^2} e^{i(\xi x + \eta y)} dx dy = \int_{y_1}^{y_2} \int_{x_1(y)}^{x_2(y)} \frac{\partial^2 w}{\partial x^2} e^{i(\xi x + \eta y)} dx dy$$

$$= \int_{y_1}^{y_2} \left\{ \left[ e^{i(\xi x + \eta y)} \frac{\partial w}{\partial x}(x,y) - i\xi e^{i(\xi x + \eta y)} w(x,y) \right]_{x_1(y)}^{x_2(y)} \right.$$

$$\left. - \xi^2 \int_{x_1(y)}^{x_2(y)} e^{i(\xi x + \eta y)} w(x,y) dx \right\} dy. \tag{7.7.7}$$

Under $x(y)$ a dependence between the Cartesian coordinates is understood. This dependence satisfies the equation of the curve, which describes the boundary $\Gamma$. The second integral in (7.7.7) with an accuracy to a multiplier is Fourier's transformation of function $w(x, y)$ on variable $\xi$ in the domain $D^+$.

$$-\xi^2 \int_{y_1}^{y_2} \left[ \int_{x_1(y)}^{x_2(y)} e^{i(\xi x + \eta y)} w(x,y) dx \right] dy = -\xi^2 \iint_{D^+} w(x,y) e^{i(\xi x + \eta y)} dx dy.$$

Taking into account that on the border $\Gamma$ the differential $dy = n_x d\Gamma$, the first integral in (7.7.7) can be written as the integral on the curve $\Gamma$ with the tendency of integrand function to $\Gamma$ from the domain $D^+$. Thus,

$$\iint_{D^+} \frac{\partial^2 w}{\partial x^2} e^{i(\xi x + \eta y)} dx dy = -\xi^2 \iint_{D^+} w(x,y) e^{i(\xi x + \eta y)} dx dy$$

$$+ \int_{\Gamma^+} e^{i(\xi x + \eta y)} \left( \frac{\partial w}{\partial x} - i\xi w \right) n_x d\Gamma. \tag{7.7.8}$$

Similarly, performing the same operations for the integral

$$\iint_{D^+} \frac{\partial^2 w}{\partial y^2} e^{i(\xi x + \eta y)} dx dy$$

and adding the obtained result with (7.7.8), we will come to the following expression

$$\iint_{D^+} \nabla^2 w e^{i(\xi x + \eta y)} dx dy = -(\xi^2 + \eta^2) \iint_{D^+} w(x,y) e^{i(\xi x + \eta y)} dx dy$$

$$+ \int_{\Gamma^+} e^{i(\xi x + \eta y)} \left[ n_x \left( \frac{\partial w}{\partial x} - i\xi w \right) + n_y \left( \frac{\partial w}{\partial y} - i\eta w \right) \right] d\Gamma. \tag{7.7.9}$$

It is not difficult to show that a similar expression with the opposite sign, as a result of a reversal (due to the change of the direction) of normal direction can be obtained for the integral on the domain $D^-$. Adding expressions for the integrals on $D^+$ and on $D^-$, and taking into account the expression for the derivative along the normal (see 7.101) we obtain:

$$\iint_{-\infty}^{\infty} \nabla^2 w e^{i(\xi x + \eta y)} dx dy = -2\pi(\xi^2 + \eta^2)W(\xi, \eta)$$

$$+ \int_{\Gamma^+} e^{i(\xi x + \eta y)} \left[ \Delta\frac{\partial w}{\partial n} - i(\xi n_x + \eta n_y)\Delta w \right] d\Gamma, \qquad (7.7.10)$$

where $W(\xi, \eta)$ – Fourier's transform of (from) displacement $w(x, y)$; and $\Delta$ – the jumps of the corresponding functions on $\Gamma$. Returning to (7.7.5), we find

$$(\xi^2 + \eta^2 + \lambda^2)W(\xi, \eta) = Q(\xi, \eta)$$

$$+ \frac{1}{2\pi} \int_{\Gamma^+} e^{i(\xi x + \eta y)} \left[ \Delta\frac{\partial w}{\partial n} - i(\xi n_x + \eta n_y)\Delta w \right] d\Gamma.$$
$$(7.7.11)$$

Using an inversion formula of Fourier, we come to the idea of solution of the problem through the densities of the force and kinematic actions on the border of the domain, represented by jumps of the function and its normal derivative

$$w(x_*, y_*) = \frac{1}{2\pi} \iint_{-\infty}^{\infty} \frac{Q(\xi, \eta) + V(\xi, \eta)}{\xi^2 + \eta^2 + \lambda^2} e^{i(\xi x + \eta y)} d\xi d\eta, \qquad (7.7.12)$$

where

$$V(\xi, \eta) = \frac{1}{2\pi} \int_{\Gamma} e^{i(\xi x + \eta y)} \left[ \Delta\frac{\partial w}{\partial n} - i(\xi n_x + \eta n_y)\Delta w \right] d\Gamma. \qquad (7.7.13)$$

In (7.7.14) the current coordinates $x$ and $y$ in (7.7.13) are designated through $x_*$ and $y_*$ to differentiate from (7.7.13). Let us designate also

$$\frac{1}{2\pi} \iint_{-\infty}^{\infty} \frac{Q(\xi, \eta)e^{-i(\xi x_* + \eta y_*)}}{\xi^2 + \eta^2 + \lambda^2} d\xi d\eta = w_q(x_*, y_*), \qquad (7.7.14)$$

$$\frac{1}{2\pi} \iint_{-\infty}^{\infty} \frac{V(\xi, \eta)e^{-i(\xi x_* + \eta y_*)}}{\xi^2 + \eta^2 + \lambda^2} d\xi d\eta = w_k(x_*, y_*), \qquad (7.7.15)$$

where $w_q(x_*, y_*)$ – the displacement in the infinite domain from the compensating load, applied on the contour. The last expression can be transformed (converted) as follows

$$w_k(x_*, y_*) = \int_\Gamma \left[ \Delta \frac{\partial w}{\partial n} G(x - x_*, y - y_*) - \Delta w \left( n_x \frac{\partial G}{\partial x} + n_y \frac{\partial G}{\partial y} \right) \right] d\Gamma,$$

where

$$G(x - x_*, y - y_*) = \frac{1}{4\pi^2 T} \iint_{-\infty}^{\infty} \frac{e^{-i[\xi(x_*-x)+\eta(y_*-y)]}}{\xi^2 + \eta^2 + \lambda^2} d\xi d\eta. \qquad (7.7.16)$$

Here, not accidentally, the integral (7.7.16) is designated through $G$, since it is the solution in the infinite domain due to the load $q(x, y) = \delta(x - x_*, y - y_*)$, i.e. Green's function. Actually, taking the load in the form $q(x, y) = \delta(x - x_*, y - y_*)$, we obtain

$$Q(\xi, \eta) = \frac{e^{i(\xi x_* + \eta y_*)}}{2\pi}.$$

Since for the infinite domain it should be assumed that $V(\xi, \eta) = 0$, then the formula (7.7.12) gives

$$w(x - x_*, y - y_*) = \frac{1}{4\pi^2 T} \iint_{-\infty}^{\infty} \frac{e^{-i[\xi(x_*-x)+\eta(y_*-y)]}}{\xi^2 + \eta^2 + \lambda^2} d\xi d\eta$$

$$= \frac{1}{\pi^2 T} \iint_0^{\infty} \frac{\cos \xi(x - x_*) \cos \eta(y - y_*)}{\xi^2 + \eta^2 + \lambda^2} d\xi d\eta$$

That also is the integral performance of the function of Green (7.7.16). Since Green's function is an even function [115], Green's function also is an even function of both arguments.

The Green function is known for the considered equation: $G(x - x_*; y - y_*) = G(x_* - x; y_* - y)$. However, since the considered problem is illustrative in nature, we will not focus our attention on the possibility of computing Green's function in the closed form. The application of spectral BEM is effective precisely in such cases when obtaining Green's function is difficult. Therefore, subsequently we will operate with the integral representation of the function of Green and its derivatives. Also taking into account that

$$n_x \frac{\partial G}{\partial x} + n_y \frac{\partial G}{\partial y} = \frac{\partial G}{\partial n},$$

let us finally write for $w(x_*, y_*)$ the following expression

$$w(x_*, y_*) = w_q(x_*, y_*) + \int_\Gamma \left( \Delta \frac{\partial w}{\partial n} G - \Delta w \frac{\partial G}{\partial n} \right) d\Gamma, \qquad (7.7.17)$$

which is characteristic for the method of potentials and for the method of delta-transformation. Thus, we are again convinced that the SMBE reduces to the same equations as the methods indicated, if it is possible to solve the integral which determines Green's function. The building of the boundary equations on the basis of formula (7.7.17) is achieved by means of the selection of various jumps on the boundary. With the indirect use of SMBE we assume continuity of the value of the unknown function ($\Delta w = 0$) and the unknown jumps of the normal derivatives ($\Delta \partial w / \partial n \neq 0$) on the border of the domain. In accordance with this, for the first problem of the theory of elasticity (7.7.1) and (7.7.2) we will have with $q \equiv 0$

$$\text{(i)} \qquad w_q(p) + \int_\Gamma \Delta \frac{\partial w}{\partial n}(q) G(p, q) d\Gamma_q = f(p) \qquad (7.7.18)$$

Here for simplicity and by analogy with the presentation of the previous chapters, we designated through point $p$ with the coordinates $x_*$, $y_*$ and through point $q$ with the coordinates $x$, $y$. For the solution of the second problem of the theory of elasticity (7.7.1), (7.7.3) it is necessary to differentiate the left and right side of (7.7.17) in the direction of the normal, and then to satisfy boundary conditions; at the same time at each point on the contour the direction of normal to its boundary must coincide with the direction in which the differentiation is conducted:

$$\frac{\partial w_q^+}{\partial n}(p) + \int_\Gamma \Delta \frac{\partial w}{\partial n}(q) \frac{\partial G^+}{\partial n}(p, q) = \frac{\partial w^+}{\partial n}(p). \qquad (7.7.19)$$

We will analogously have

$$\frac{\partial w_q^-}{\partial n}(p) + \int_\Gamma \Delta \frac{\partial w}{\partial n}(q) \frac{\partial G^-}{\partial n}(p, q) = \frac{\partial w^-}{\partial n}(p). \qquad (7.7.20)$$

Let us accumulate further the left and right parts of the Eq. (7.7.19) and (7.7.20); by analogy with the conclusion of the equations of BEM we will obtain:

$$\text{(ii)} \qquad \Delta \frac{\partial w}{\partial n}(p) + \int_\Gamma \Delta \frac{\partial w}{\partial n}(q) \frac{\partial G}{\partial n}(p, q) d\Gamma_q = f(p) - \frac{\partial w_q}{\partial w}(p) \qquad (7.7.21)$$

Similar operations reduce to the equations of the first and second kind, if instead of the jumps of normal derivative in the expression (7.7.17) the jumps of the unknown function $\Delta w \neq 0$ with $q \in \Gamma$ remain, or both types of jumps remain as this occurs in the method of delta-transformation. Thus, SMBE leads to the same results as the method of delta-transformation with the transition to Green's functions. As mentioned above it can be applied when it is not possible to obtain an analytical expression for function of Green. In this case it may be useful to consider the boundary equations relative to the Fourier transforms (or other spectral representations) unknown functions of entering in (7.7.18) and (7.7.21). Passing, for example, in (7.7.18) to the Fourier transforms, we obtain:

$$
W_q(\xi, \eta) + \int_\Gamma \Delta \frac{\partial w}{\partial n}(q)\overline{G}(q)d\Gamma_q = F(\xi, \eta), \qquad (7.7.22)
$$

where

$$
\overline{G}(q) = \frac{1}{2\pi T} \frac{e^{-i(\xi x + \eta y)}}{\xi^2 + \eta^2 + \lambda^2}.
$$

The solution of this equation does not cause difficulties; however, known difficulties are connected with the subsequent inversion of Fourier transformation and calculation of double improper integrals. This problem can be simplified for the semi-infinite domains, strips, etc. when it is possible to solve one of the integrals on the space coordinate, which is changed along the entire axis. SMBE ensues effective algorithms in such cases when the extended domain is taken for the finite domain. Using finite transformations with the kernel, which satisfy the differential equation of problem considered, we can reduce the problem to the solution of the system of algebraic equations on the border of the given domain. Let us examine the extended region in the form of the rectangle (Figure 7.17).

On the border of the extended area we can accept any boundary conditions at our discretion. Let us accept for certainty

$$
w(p) = 0, \quad (p \in \Gamma^*) \qquad (7.7.23)
$$

In this case for constructing the boundary equations it is convenient to use a finite sine-transformation. Passing in Eq. (7.7.1) to the sine transforms and accomplishing all operations in the same way as it was done using the Fourier transformation, we obtain Eq. (7.7.17), in which Green's function

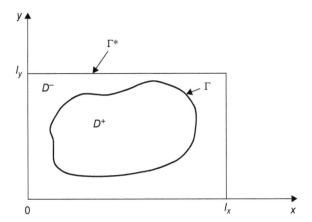

**Figure 7.17** Extended domain in the form of the rectangle.

will already take the form of the discrete expansion.

$$G(x, x_*; y, y_*) = \frac{4}{Tl_x l_y} \sum_{m,n=1}^{\infty} \frac{\sin \frac{m\pi x}{l_x} \sin \frac{n\pi y}{l_y} \sin \frac{m\pi x_*}{l_x} \sin \frac{n\pi y_*}{l_y}}{\left(\frac{m\pi}{l_x}\right)^2 + \left(\frac{n\pi}{l_y}\right)^2 + \lambda^2} \qquad (7.7.24)$$

The spectral expansions in the finite domain are also effective during the arrangement of the compensating actions on the border of the extended domain. As an example of the application of SMBE let us examine the solution of Eq. (7.7.1) in the triangular domain, shown in Figure 7.18.

Let us assign boundary conditions in the form of (7.7.2), we assume that the function $f(x, y)$ on the sides of triangle $x = 0$ and $y = 0$ becomes zero, and the diagonal is given according to the law of the triangle. As the extended domain we assume a rectangle with the sides $l_x$ and $l_y$, on which boundary conditions (7.7.23) are satisfied. The Green function will take the form (7.7.24). In this case the solution of equation will be written down in the form of:

$$w(x_*, y_*) = w_q(x_*, y_*) + \frac{4}{Tl_x l_y} \sum_{m=1}^{\infty} \sum_{n=1}^{\infty} g_{mn}(x_*, y_*)$$

$$\times \int_{\Gamma'} \varphi(x, y) \sin \frac{m\pi x}{l_x} \sin \frac{n\pi y}{l_y} d\Gamma' \qquad (7.7.25)$$

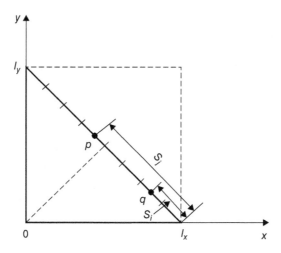

**Figure 7.18**　Schematic of triangular membrane.

where $\varphi(x, y)$–the unknown function, equal to the jump of the normal derivative of displacement $w(x_*, y_*)$ on the boundary of domain;

$$g_{mn}(x_*, y_*) = \frac{1}{\left(\dfrac{m\pi}{l_x}\right)^2 + \left(\dfrac{n\pi}{l_y}\right)^2 + \lambda^2} \sin\frac{m\pi x_*}{l_x} \sin\frac{n\pi y_*}{l_y}; \qquad (7.7.26)$$

$\Gamma'$ – the diagonal of rectangle, described by equation:

$$\frac{x}{l_x} + \frac{y}{l_y} = 1 \qquad (7.7.27)$$

The corresponding boundary equation relative to function $\varphi(x, y)$ will be:

$$w_q(p) + \frac{4}{Tl_x l_y} \sum_{m=1}^{\infty} \sum_{n=1}^{\infty} g_{mn}(p) \int_{\Gamma'} \varphi(q) \sin\frac{m\pi x_q}{l_x} \sin\frac{n\pi y_q}{l_y} d\Gamma_q' = f(p) \qquad (7.7.28)$$

The expression for $w_q$ in (7.7.25) and (7.7.27) can be obtained from the expression

$$w_q(x_*, y_*) = \int_0^{l_x l_y} \int q(x, y)G(x, y, x_*, y_*)dxdy. \qquad (7.7.29)$$

We will consider that the function inside the triangular domain is constant, and outside it is equal to zero. Then from (7.7.29) and taking into account (7.7.24) we obtain

$$w_q(x_*, y_*) = \frac{4q}{l_x l_y} \sum_{m=1}^{\infty} \sum_{n=1}^{\infty} g_{mn}(x_*, y_*) \int_0^{l_x} \sin \frac{m\pi x}{l_x} dx \int_0^{l_y\left(1-\frac{x}{l_x}\right)} \sin \frac{n\pi y}{l_y} dy$$

$$(7.7.30)$$

Calculating the integrals in the right part of (7.7.30), we find

$$w_q(x_*, y_*) = \frac{4}{\pi^2} \sum_{m=1}^{\infty} \sum_{n=1}^{\infty} q_{mn} \sin \frac{m\pi x_*}{l_x} \sin \frac{n\pi y_*}{l_y}, \qquad (7.7.31)$$

where

$$q_{mn} = \frac{\left[1 - \dfrac{(-1)^m n^2}{m^2 - n^2} + \dfrac{(-1)^n m^2}{m^2 - n^2}\right] \quad (m \neq n)}{mn\left[\lambda^2 + \left(\dfrac{m\pi}{l_x}\right)^2 + \left(\dfrac{n\pi}{l_y}\right)^2\right]^2}$$

$$\times [1 - (-1)^m], \qquad (m \neq n).$$

For the numerical solution of Eq. (7.7.28) let us introduce coordinate $s$ along diagonal $\Gamma'$, so that according to (7.7.27)

$$\frac{s}{l_s} + \frac{y}{l_y} = 1 - \frac{x}{l_x}, \qquad (7.7.32)$$

where

$$l_s = \sqrt{l_x^2 + l_y^2}.$$

So we divide the diagonal on the $N$ boundary element and we consider that the function $\varphi(s)$ is constant inside each element. Then Eq. (7.7.28) can be written in the form

$$\sum_{i=1}^{N} b_{ij}\varphi_j = c_j, \quad (j = 1, 2, \ldots N) \qquad (7.7.33)$$

where $\varphi_j = \varphi(s_j)$—the value of function $\varphi(s)$ on the $j$th boundary element;

$$b_{ij} = \frac{4}{Tl_x l_y} \sum_{m=1}^{\infty} \sum_{n=1}^{\infty} \frac{(-1)^{m+1}}{\lambda^2 + \left(\dfrac{m\pi}{l_x}\right)^2 + \left(\dfrac{n\pi}{l_y}\right)^2} \times a_{mni} \sin \frac{m\pi s_j}{l_s} \sin \frac{n\pi s_j}{l_s};$$

$$a_{mni} = \int_{S_i - \frac{\Delta s}{2}}^{S_i + \frac{\Delta s}{2}} \sin \frac{m\pi s}{l_s} \sin \frac{n\pi s}{l_s} ds;$$ (7.7.34)

$$c_j = f(s_j) - \sum_{m=1}^{\infty} \sum_{n=1}^{\infty} (-1)^{m+1} q_{mn} \sin \frac{m\pi s_j}{l_s} \sin \frac{n\pi s_j}{l_s};$$

$$s_j = \frac{2j-1}{2N} l_s; \qquad \Delta s = \frac{l_s}{N};$$

$f(s_j)$ – the assigned function, which is accepted (allowed) in the form:

$$f(s_j) = \begin{cases} \dfrac{2s_j}{l_s} & \left(0 \le s_j \le \dfrac{l_s}{2}\right) \\[2mm] 2 - \dfrac{2s_j}{l_s} & \left(\dfrac{l_s}{2} \le s_j \le l_s\right). \end{cases}$$ (7.7.35)

After solving the system of algebraic equations (7.7.33), the expression for the deflection can be written in the form

$$w(x_*, y_*) == \sum_{m=1}^{\infty} \sum_{n=1}^{\infty} \left[ q_{mn} + \frac{4}{Tl_x l_y} \frac{\sum\limits_{j=1}^{N} (-1)^{m+1} \varphi_j a_{mnj}}{\lambda^2 + \left(\dfrac{m\pi}{l_x}\right)^2 + \left(\dfrac{n\pi}{l_y}\right)^2} \right]$$
$$\times \sin \frac{m\pi x_*}{l_x} \sin \frac{n\pi y_*}{l_y}.$$ (7.7.36)

Figure 7.19 shows the distribution of the $w(x_*, y_*)$ on diagonal $y_* = x_*$ of the rectangle with

$$l_x = l_y = \frac{l_s}{\sqrt{2}}, \qquad \lambda = \frac{\sqrt{2}}{l_s}, \qquad q = \frac{2}{l_s^2}$$

obtained as a result of the partition of boundary into 20 boundary elements, taking into account 20 terms of series in formula (7.7.24)

Thus, let us summarize the results given in the current chapter: This chapter provides the description of the universal regular model of the elastic foundation, which possesses different degrees of coherency. The proposed model of elastic foundation ensures the limited displacements in the two-dimensional problem and under the concentrated loads, allowing jumps of displacements on the surface of the base, but it does not lead

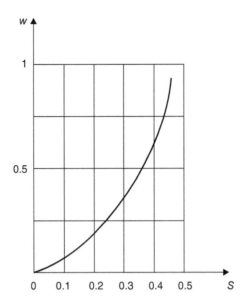

**Figure 7.19** Distribution of the displacements over the diagonal of the membrane.

to the infinite stresses on the boundaries of the contact zones. One of the parameters of model characterizes dependence between the load and displacement, and two others the coherence of base and its heterogeneity in the depth. The model can be used for describing the elastic properties of a wide circle of real ground bases, since with the limiting values of the parameters it turns into two extreme models with respect to the coherencies of the model – Winklerian base and elastic isotropic half-space.

The technique is developed on the basis of die soil tests utilizing the experimental determination of the characteristics of the base for the proposed model. Reviewing the results of experimental studies shows that real grounds (soils) meet the parameters that are considerably different from the limiting values that correspond to the Winklerian base and elastic half-space. The program is developed on the basis of the proposed model and obtains the numerical solution for the contact problems for the rigid die pressed by axial force into the elastic foundation. The analysis of the generated diagrams of contact pressures shows that they are close to those observed in the experiments.

The iteration technique of solution of contact problems for the structure on an elastic foundation is proposed. The convergence of which is ensured by using the proposed regular base model.

The proposed model makes it possible to obtain solutions for contact problems for calculating free uninsulated constructions on an elastic foundation (the flooring of industrial buildings, roads pavements, airfields, the surfaces of hydraulic structures, and others). In this work, solutions for such problems as the uninsulated beams and plates are given by the spectral method of boundary element (SMBM) for the first time.

# Appendix A
# Certificate of Essential Building Data [281]

*Static and Dynamic Analysis of Engineering Structures: Incorporating the Boundary Element Method,*
First Edition. Levon G. Petrosian and Vladimir A. Ambartsumian.
© 2020 John Wiley & Sons Ltd. Published 2020 by John Wiley & Sons Ltd.

**UNITED STATES DEPARTMENT OF COMMERCE**
**National Institute of Standards and Technology**
Gaithersburg, Maryland 20899-0001

Building 226, Room B158
Phone:   (301)-975-6078
FAX:     (301)-869-6275
e-mail:  ataylor@enh.nist.gov

PROPOSED FORMAT FOR A "CERTIFICATE OF ESSENTIAL BUILDING DATA"

Levon G. Petrosian
Visiting Scholar, Building and Fire Research Laboratory,
National Institute of Standards and Technology, Gaithersburg, Maryland
Formerly, Director, Armenian Earthquake Engineering Research Institute

Andrew W. Taylor
Research Engineer, Building and Fire Research Laboratory
National Institute of Standards and Technology, Gaithersburg, Maryland

Diana R. Todd
Research Engineer, Building and Fire Research Laboratory
National Institute of Standards and Technology, Gaithersburg, Maryland

In the aftermath of a natural or man-made disaster, rescue workers, engineers and public officials need quick access to accurate information about the essential structural features of damaged buildings. A minimal set of essential data is required to make judgments about the suitability of a structure for continued occupancy, to estimate the cost of damage and the feasibility of repairs, and to assess the vulnerability of the structure to further damage (e.g. from aftershocks). Experiences in past earthquakes, particularly the recent Northridge and Kobe events, have confirmed that even basic information about most damaged buildings is not readily available. Fundamental data such as the date of construction, the version of the building code under which the structure was designed, the type of structural framing system, and a history of significant modifications and repairs to the structure are almost never available on short notice. Such information, if recorded at all, usually resides in the files of government planning offices or consulting engineering firms. Therefore, it is proposed that a systematic method of recording, in a brief format, the fundamental information about a building be developed. This "Certificate of Essential Building Data" (CEBD) would be kept on file at the building site (much like an elevator certificate) and at a central municipal repository. Possible features of such a certificate and its proposed uses are described below.

The CEBD would be an abstract of the data essential to making informed decisions about the condition and use of structures following a natural disaster, such as an earthquake. The CEBD would be about two pages long, and would contain information in the following five categories:

1) Building Identification (name, address, owner, etc.)
2) General Description (occupancy class, number of stories, sketch of plan and elevation)
3) Structural System (structural frame type, dynamic and static structural characteristics)
4) Foundation System (foundation type, site soil conditions)
5) Building History (dates of design and constructions, record of significant modifications and repairs)

The information would be arranged in a format which corresponds to post-disaster reconnaissance survey forms, such as the one currently being developed by the Applied Technology Council in project ATC-38, and to other building data bases, such as the one currently being developed for Federal buildings by the Interagency Committee on Seismic Safety in Construction (ICSSC 1995).

Perhaps the main questions related to the development of a CEBD are not the content of the CEBD form itself, but rather how the CEBD would be implemented and the information used. Some of these issues are outlined below.

• Who would use the CEBD?

Although the main purpose of the CEBD is to assist rescue workers, engineers and government officials in post-disaster recovery efforts, there are many other uses for data contained in the CEBD. First, the information would be invaluable to researchers in earthquake engineering. The existing stock of structures, and their associated foundations and soil conditions, constitute the ultimate natural laboratory for the study of earthquake engineering. If CEBD's were available for a broad range of buildings in an area struck by an earthquake, much could be learned about the vulnerability of various structure types. Furthermore, if CEBD's are updated over the life of a building, the effects of previous weak, moderate and strong earthquakes, and structural modifications or repairs, could be monitored. Second, the CEBD could be used by disaster relief planners to estimate the extent and cost of damage resulting from hypothetical disaster scenarios. Third, the information could be used by owners, buyers and insurers of buildings as a capsule summary of the essential structural features of the building. The CEBD could not take the place of a detailed structural evaluation, but it would provide an overview of the essential characteristics of a structure.

• Which buildings would have a CEBD on file?

In a perfect world, nearly every public building would have a CEBD on file. However, practically speaking this would be impossible. The most likely candidates for CEBD's are new structures, existing structures which undergo significant modifications, and government buildings. The most qualified person to complete the CEBD is the structural engineer responsible for the design of a new building, or for renovation of an existing building. The completion of a CEBD could become part of the construction permit process. This would insure that major modifications to the building, which require a building permit, are recorded on the CEBD. Regarding government buildings, local jurisdictions which undertake a program of evaluating the vulnerability of their building stock to natural hazards could include completion of CEBD's as part of the evaluation process.

• Where would the CEBD be stored?

One copy of the CEBD would be kept on the building site with a building manager. Another copy would be kept at a central repository, such as a city building department or disaster response agency.

In short, the CEBD would fulfill two purposes. First, and foremost, the CEBD would be a readily available record of essential structural information which could be used in post-disaster relief and recovery efforts. Second, over time, as the CEBD data base grew, the CEBD would become an invaluable tool for studying the seismic vulnerability of various classes of structures, and for developing improved retrofit and repair measures for buildings.

Proposed Format for

# Certificate of Essential Building Data

| | |
|---|---|
| Levon G. Petrosian | Visiting Scholar, Ph. D., Professor, Building and Fire Research Laboratory, National Institute of Standards and Technology |
| Andrew W. Taylor | Research Structural Engineer, Ph.D., Building and Fire Research Laboratory, National Institute of Standards and Technology |
| Diana R. Todd | Research Structural Engineer, Ph.D., Building and Fire Research Laboratory, National Institute of Standards and Technology |

## What is it?

The Certificate of Essential Building Data (CEBD) is a systematized, concise record of the most important structural features of a building, including basic geometry, structural system type, foundation type, etc.

## Who would use it?

- Emergency workers (police, fire, rescue, building inspectors) immediately following a natural or man-made disaster, to make safety evaluations of damaged buildings.
- Disaster Planners , for community risk analysis and emergency response planning.
- Insurers, to estimate disaster risk exposure in a given area.
- Researchers, to learn about the actual performance of various types of buildings in natural disasters.

## How would it be implemented?

- For existing Federal buildings, the CEBD could be completed as part of the implementation of Executive Order 12941 (seismic safety of existing federal buildings)
- For new construction, the CEBD could be completed as part of the building permit process, since most of the information required for the CEBD can be supplied by the designer.
- For significant renovations, the CEBD could also be completed as part of the building permit process, since significant renovations require the designer to compile most of the information on the CEBD.
- For private buildings, the CEBD could be completed as part of local, state, or federal programs to evaluate significant buildings in a community.

## Where would the information be stored?

- At the building site
- In a central repository (e.g. city building department)
- On a publicly accessible electronic data base

## What information is in the CEBD?

### Examples of key data listed in the proposed CEBD:

- Identification of the structure
  - Name of structure
  - Address
  - Use and occupancy
  - Map location
- Structural configuration
  - Overall plan and elevation dimensions
  - Story heights and weights
  - Locations of seismic separation joints
- Structural system
  - Gravity load system description
  - Lateral load system description
  - Special structural conditions
  - Foundation system description
  - Experimentally obtained dynamic properties
  - Dates of design and construction
  - Dates of remodeling or reinforcement
  - Seismic design coefficients
  - Codes used in design
- Estimated or insured value of structure

NATIONAL INSTITUTE OF STANDARDS AND TECHNOLOGY

BUILDING AND FIRE RESEARCH LABORATORY

(Proposed format for)

# CERTIFICATE OF ESSENTIAL
# BUILDING DATA

### 1. IDENTIFICATION  OF STRUCTURE

1.1  Name of Structure (if any) _____

1.2  Address _____ City _____
     State _____ Zip Code _____ Building Name

1.3  County _____
1.4  Name of Building development, of which the structure is part
     (if any) _____
1.5  Building Owner/Manager Contact _____
     _____Phone _____
1.6  Jurisdiction Governing the Structure (name of city , county,
     state, or federal agency ) _____

1.7  Structure Use and Occupancy:
     1.7.1  Function  of the Structure (e.g. residential, commercial
            office, light manufacturing, heavy manufacturing, health
            care, auditorium, etc.) _____

     1.7.2  Maximum occupancy (number of persons) _____

     1.7.3  If hazardous materials are stored in the structure,
            type and location _____

1.8  Population setting ( urban , suburban, or rural) _____

1.9  Photograph (general view of structure)
1.10  Map and Plans:
      1.10.1  State Map Number, and coordinates in state plane
              coordinates system _____
      1.10.2  County or city map number, and coordinates _____

      1.10.3  Project - specific map number, and structure identify-
              cation number _____
      1.10.4  Structure plan number _____
      1.10.5  Latitude and longitude coordinates _____

### 2. STRUCTURE CONFIGURATION

2.1  Number of Stories (identify below-grade and above-grade

stories, including penthouse levels), and the height of-
each story _____

_____

2.2   Basic plan shape of the structure (sketch with dimensions
      sufficient to describe the overall plan shape of the
      structure_____
2.3   Basic elevations of the structure (sketches with dimensions
      sufficient to describe the overall elevation shape of the
      structure on all sides). Include the number of stories(below-
      grade and above-grade stories, including penthouse levels),
      and the height of each story _____

_____

_____

_____

2.4   Seismic separation joints (sketches showing the locations of
      joints)_____

_____

_____

### 3. STRUCTURAL SYSTEM

3.1   Structure system identification
      3.1.1   Gravity Load Structural System (using Applied
              Technology Council ATC-33 standard structural system
              types)_____

_____

      3.1.2   Lateral load structural system, if distinct from the
              gravity load structural system (using either Applied
              Technology Council ATC-33 or NEHRP Recommended
              Provisions standard structural system types)_____

_____

              3.1.2.1 Lateral load structural system along the longi
                      tudinal building axis_____

_____

              3.1.2.2 Lateral load structural system along the trans
                      verse building axis_____

_____

      3.1.3   Approximate weight of each floor, excluding contents
              of the structure_____

_____

      3.1.4 Secondary elements _____

3.1.4.1 Structural framing system of floors _____

3.1.4.2 Structural framing system of roof _____

3.1.4.3 Exterior cladding system _____

3.1.4.4 Description of interior,non-loadbearing parti
tions _____

3.1.5 Special structural conditions (e.g.mixed framing types
significant modifications, unusual geometry,unusual
mass distribution, unusual materials, etc.) _____

3.1.6 Is the structure essentially a standardized design?

3.1.7 Are there special seismic resistant elements in the
structure (e.g. base isolators, supplemental passive
dampers, semi-active, active dampers)? _____

3.1.8 Basic material grades _____

3.1.8.1 Predominant superstructure steel grade _____

3.1.8.1.1 Columns_____
3.1.8.1.2 Girders and floors _____

3.1.8.2 Predominant superstructure concrete strength

3.1.8.2.1 Columns _____
3.1.8.2.2 Girders and floors _____
3.1.8.2.3 Walls _____
3.1.8.3  Predominant superstructure concrete
          reinforcement grade_____

3.1.8.4  Other important superstructure materials and
          characteristics _____

## 3.2 Foundation system

3.2.1  Foundation types employed (spread footing ,pile and
        pile caps, mat, etc.) _____
        _____

3.2.2  Are there changes of surface elevation on the site
        which significantly affect the foundation design?
        _____
        _____

3.2.3 Soil conditions

        3.2.3.1 Soil category (NEHRP Recommended Provisions
                categories) _____

        3.2.3.2 Is a geotechnical engineering report available?
                _____

        3.2.3.3 Are special ground hazards present? (e.g. high
                potential for liquefaction ,landslide)_____
                _____

3.2.4 Have dynamic properties of the soil been obtained
      through field measurements? If so, what values were
      obtained? _____
                _____
                _____

## 3.3 Design

3.3.1 Dates of design (beginning to end) _____

3.3.2 Name of design organization (commercial firm, government
      agency) _____

3.3.3 Engineer of record _____
3.3.4 Architect of record _____
3.3.5 Building codes under which the structure was designed,
      and code version _____
3.3.6 Design seismic acceleration _____
3.3.7 Have dynamic properties of the structure been obtained

by field experiment ? If so, what values were obtained?
_____

3.4 Construction

   3.4.1 Dates of construction (beginning to end) _____

   3.4.2 Name of general contractor_____

3.5 Date when the structure was first approved for occupancy _____

### 4. STRUCTURE VALUE

4.1 Estimated or insured value of the structure, excluding
    contents. Include method of valuation _____
    _____

4.2 Estimated or insured value of structure contents. Include
    method of valuation_____
    _____
    _____

Notes:_____
    _____
    _____
    _____
    _____
    _____
    _____
    _____
    _____
    _____

Inspector: Name_____

           Title_____

           Address_____
                   _____
                   _____

Signatures_____

Date:_____  Time:_____

In addition the assessment of the structure of building must be provided.  Here we are using the same format:

5. Assessment of the structure

5.1    Assessment of the disaster resistance of the structure.  What criteria were used?

        Possible rating:      A.  Does not satisfy code requirements in existence at the time of design.

                                          B.  Does not satisfy the current code requirements, which are different from the code requirements in effect at the time of design.

                                          C.  Does not satisfy the current code requirements because of previous damage.

5.2    Assessment of fitness for occupancy.  What criteria were used?

        Possible rating:      A.  Fully fit for occupancy.

                                          B.  Fit for occupancy only after retrofit or repair.

                                          C.  Not fit for occupancy, and recommended for Demolition

5.3    Overall assessment of the structure

# Appendix B

# Contact Stresses on the Sole of the Circular Die and the Sole of the Plane Die

## B.1 Contact Stresses on the Sole of the Circular Die.

| $\varepsilon$ | $\nu$ | $r$ | | | | | |
|---|---|---|---|---|---|---|---|
| | | 0.0833 | 0.25 | 0.417 | 0.583 | 0.750 | 0.917 |
| $10^{-3}$ | 0 | 0.156 | 0.170 | 0.177 | 0.201 | 0.223 | 0.590 |
| | $10^{-3}$ | 0.156 | 0.170 | 0.178 | 0.201 | 0.223 | 0.590 |
| | $10^{-2}$ | 0.157 | 0.171 | 0.179 | 0.202 | 0.225 | 0.587 |
| | 0.1 | 0.165 | 0.184 | 0.191 | 0.213 | 0.238 | 0.560 |
| | 0.5 | 0.188 | 0.233 | 0.236 | 0.252 | 0.277 | 0.467 |
| | 1.0 | 0.182 | 0.273 | 0.275 | 0.284 | 0.301 | 0.399 |
| | 1.5 | 0.151 | 0.292 | 0.297 | 0.303 | 0.312 | 0.365 |

*(Continued)*

*Static and Dynamic Analysis of Engineering Structures: Incorporating the Boundary Element Method*,
First Edition. Levon G. Petrosian and Vladimir A. Ambartsumian.
© 2020 John Wiley & Sons Ltd. Published 2020 by John Wiley & Sons Ltd.

| | | | | $r$ | | | |
|---|---|---|---|---|---|---|---|
| $\varepsilon$ | $\nu$ | **0.0833** | **0.25** | **0.417** | **0.583** | **0.750** | **0.917** |
| $10^{-2}$ | 0 | 0.159 | 0.168 | 0.177 | 0.200 | 0.220 | 0.594 |
| | $10^{-3}$ | 0.159 | 0.168 | 0.177 | 0.200 | 0.220 | 0.593 |
| | $10^{-2}$ | 0.160 | 0.169 | 0.178 | 0.201 | 0.222 | 0.590 |
| | 0.1 | 0.169 | 0.182 | 0.190 | 0.212 | 0.235 | 0.563 |
| | 0.5 | 0.195 | 0.230 | 0.235 | 0.251 | 0.276 | 0.469 |
| | 1.0 | 0.196 | 0.271 | 0.274 | 0.283 | 0.299 | 0.401 |
| | 1.5 | 0.171 | 0.291 | 0.296 | 0.302 | 0.311 | 0.366 |
| 0.01 | 0 | 0.159 | 0.168 | 0.177 | 0.200 | 0.220 | 0.594 |
| | $10^{-3}$ | 0.159 | 0.168 | 0.177 | 0.200 | 0.220 | 0.593 |
| | $10^{-2}$ | 0.160 | 0.169 | 0.178 | 0.201 | 0.222 | 0.590 |
| | 0.1 | 0.169 | 0.182 | 0.190 | 0.212 | 0.235 | 0.563 |
| | 0.5 | 0.195 | 0.230 | 0.235 | 0.251 | 0.276 | 0.469 |
| | 1.0 | 0.196 | 0.271 | 0.274 | 0.283 | 0.299 | 0.401 |
| | 1.5 | 0.171 | 0.291 | 0.296 | 0.302 | 0.311 | 0.366 |
| 0.02 | 0 | 0.163 | 0.165 | 0.174 | 0.198 | 0.212 | 0.603 |
| | $10^{-3}$ | 0.164 | 0.165 | 0.174 | 0.198 | 0.212 | 0.603 |
| | $10^{-2}$ | 0.165 | 0.166 | 0.175 | 0.199 | 0.214 | 0.600 |
| | 0.1 | 0.175 | 0.178 | 0.187 | 0.210 | 0.228 | 0.572 |
| | 0.5 | 0.209 | 0.225 | 0.232 | 0.248 | 0.270 | 0.477 |
| | 1.0 | 0.224 | 0.267 | 0.270 | 0.280 | 0.296 | 0.406 |
| | 1.5 | 0.212 | 0.289 | 0.293 | 0.299 | 0.308 | 0.368 |
| 0.04 | 0 | 0.166 | 0.156 | 0.166 | 0.193 | 0.184 | 0.635 |
| | $10^{-3}$ | 0.166 | 0.156 | 0.167 | 0.193 | 0.184 | 0.635 |
| | $10^{-2}$ | 0.168 | 0.158 | 0.168 | 0.194 | 0.185 | 0.632 |
| | 0.1 | 0.179 | 0.169 | 0.179 | 0.204 | 0.202 | 0.603 |
| | 0.5 | 0.223 | 0.213 | 0.222 | 0.240 | 0.252 | 0.504 |
| 0.06 | 0 | 0.162 | 0.148 | 0.157 | 0.190 | 0.142 | 0.678 |
| | $10^{-3}$ | 0.162 | 0.148 | 0.157 | 0.190 | 0.142 | 0.678 |
| | $10^{-2}$ | 0.164 | 0.149 | 0.158 | 0.191 | 0.144 | 0.675 |
| | 0.1 | 0.176 | 0.160 | 0.169 | 0.199 | 0.163 | 0.645 |
| | 0.5 | 0.221 | 0.202 | 0.211 | 0.232 | 0.224 | 0.540 |
| 0.1 | 0 | 0.147 | 0.137 | 0.130 | 0.201 | 0.260 | 0.782 |
| | $10^{-3}$ | 0.147 | 0.137 | 0.130 | 0.201 | 0.260 | 0.782 |
| | $10^{-2}$ | 0.149 | 0.138 | 0.132 | 0.202 | 0.292 | 0.778 |
| | 0.1 | 0.160 | 0.147 | 0.143 | 0.206 | 0.562 | 0.745 |
| | 0.5 | 0.205 | 0.183 | 0.187 | 0.226 | 0.145 | 0.626 |
| | 1.0 | 0.249 | 0.220 | 0.226 | 0.249 | 0.209 | 0.527 |

## B.2    Contact Stresses on the Sole of the Plane Die.

| | | | | | | | $x$ | | | | |
|---|---|---|---|---|---|---|---|---|---|---|---|
| $\varepsilon$ | $\nu$ | 0.05 | 0.15 | 0.25 | 0.35 | 0.45 | 0.55 | 0.65 | 0.75 | 0.85 | 0.95 |
| $10^{-5}$ | $10^{-3}$ | 0.319 | 0.322 | 0.329 | 0.341 | 0.357 | 0.382 | 0.421 | 0.489 | 0.559 | 1.48 |
| | $10^{-2}$ | 0.321 | 0.324 | 0.331 | 0.342 | 0.359 | 0.384 | 0.422 | 0.490 | 0.561 | 1.47 |
| | 0.1 | 0.339 | 0.342 | 0.348 | 0.359 | 0.375 | 0.389 | 0.434 | 0.496 | 0.569 | 1.34 |
| | 0.5 | 0.403 | 0.405 | 0.410 | 0.417 | 0.428 | 0.445 | 0.469 | 0.508 | 0566 | 0.950 |
| | 1.0 | 0.455 | 0.456 | 0.458 | 0.461 | 0.467 | 0.474 | 0.486 | 0.504 | 0.535 | 0.705 |
| $10^{-3}$ | $10^{-3}$ | 0.319 | 0.322 | 0.329 | 0.340 | 0.357 | 0.382 | 0.420 | 0.489 | 0.559 | 1.48 |
| | $10^{-2}$ | 0.321 | 0.324 | 0.331 | 0.342 | 0.359 | 0.384 | 0.422 | 0.489 | 0.561 | 1.47 |
| | 0.1 | 0.339 | 0.342 | 0.348 | 0.359 | 0.375 | 0.398 | 0.434 | 0.496 | 0.569 | 1.34 |
| | 0.5 | 0.403 | 0.405 | 0.410 | 0.417 | 0.428 | 0.445 | 0.469 | 0.508 | 0.566 | 0.950 |
| | 1.0 | 0.455 | 0.456 | 0.458 | 0.461 | 0.467 | 0.474 | 0.486 | 0.504 | 0.535 | 0.705 |
| $10^{-2}$ | $10^{-3}$ | 0.318 | 0.321 | 0.328 | 0.339 | 0.355 | 0.381 | 0.418 | 0.486 | 0.545 | 1.51 |
| | $10^{-2}$ | 0.320 | 0.323 | 0.330 | 0.341 | 0.357 | 0.382 | 0.419 | 0.487 | 0.547 | 1.50 |
| | 0.1 | 0.337 | 0.340 | 0.347 | 0.357 | 0.373 | 0.396 | 0.432 | 0.494 | 0.558 | 1.37 |
| | 0.5 | 0.401 | 0.403 | 0.408 | 0.415 | 0.426 | 0.442 | 0.467 | 0.506 | 0.561 | 0.971 |
| | 1.0 | 0.452 | 0.453 | 0.456 | 0.459 | 0.465 | 0.473 | 0.484 | 0.503 | 0.534 | 0.720 |
| 0.02 | $10^{-3}$ | 0.314 | 0.318 | 0.324 | 0.335 | 0.351 | 0.375 | 0.411 | 0.483 | 0.504 | 1.58 |
| | $10^{-2}$ | 0.316 | 0.319 | 0.326 | 0.337 | 0.353 | 0.377 | 0.412 | 0.483 | 0.506 | 1.57 |
| | 0.1 | 0.333 | 0.336 | 0.343 | 0.353 | 0.368 | 0.391 | 0.425 | 0.490 | 0.532 | 1.44 |
| | 0.5 | 0.396 | 0.398 | 0.403 | 0.410 | 0.421 | 0.437 | 0.461 | 0.502 | 0.546 | 1.03 |
| | 1.0 | 0.447 | 0.448 | 0.450 | 0.160 | 0.460 | 0.468 | 0.481 | 0.501 | 0.530 | 0.760 |
| | 1.5 | 0.475 | 0.475 | 0.476 | 0.205 | 0.480 | 0.484 | 0.489 | 0.498 | 0.513 | 0.632 |
| | 2.0 | 0.488 | 0.489 | 0.489 | 0.249 | 0.490 | 0.491 | 0.493 | 0.497 | 0.504 | 0.569 |
| 0.04 | $10^{-3}$ | 0.306 | 0.308 | 0.315 | 0.325 | 0.340 | 0.363 | 0.389 | 0.486 | 0.358 | 1.81 |
| | $10^{-2}$ | 0.307 | 0.310 | 0.316 | 0.327 | 0.341 | 0.365 | 0.391 | 0.486 | 0.362 | 1.79 |
| | 0.1 | 0.324 | 0.326 | 0.332 | 0.342 | 0.356 | 0.379 | 0.405 | 0.489 | 0.398 | 1.65 |
| | 0.5 | 0.383 | 0.385 | 0.390 | 0.397 | 0.408 | 0.424 | 0.446 | 0.496 | 0.479 | 1.19 |
| | 1.0 | 0.433 | 0.435 | 0.437 | 0.441 | 0.448 | 0.457 | 0.470 | 0.496 | 0.501 | 0.882 |
| | 1.5 | 0.463 | 0.464 | 0.465 | 0.467 | 0.470 | 0.475 | 0.481 | 0.494 | 0.500 | 0.721 |
| | 2.0 | 0.480 | 0.480 | 0.480 | 0.481 | 0.482 | 0.484 | 0.478 | 0.493 | 0.497 | 0.634 |
| 0.06 | $10^{-3}$ | 0.301 | 0.304 | 0.310 | 0.319 | 0.334 | 0.358 | 0.374 | 0.498 | 0.255 | 1.95 |
| | $10^{-2}$ | 0.303 | 0.306 | 0.312 | 0.321 | 0.335 | 0.360 | 0.376 | 0.498 | 0.261 | 1.93 |
| | 0.1 | 0.319 | 0.322 | 0.327 | 0.337 | 0.350 | 0.373 | 0.391 | 0.498 | 0.308 | 1.77 |
| | 0.5 | 0.377 | 0.379 | 0.384 | 0.391 | 0.402 | 0.418 | 0.436 | 0.498 | 0.427 | 1.29 |
| | 1.0 | 0.427 | 0.428 | 0.431 | 0.435 | 0.441 | 0.451 | 0.463 | 0.496 | 0.474 | 0.954 |
| | 1.5 | 0.457 | 0.458 | 0.459 | 0.461 | 0.465 | 0.470 | 0.476 | 0.493 | 0.485 | 0.776 |
| | 2.0 | 0.475 | 0.475 | 0.476 | 0.477 | 0.478 | 0.480 | 0.484 | 0.492 | 0.488 | 0.676 |

*(Continued)*

| | | \multicolumn{10}{c}{$x$} | | | | | | | | |
|------|------------|-------|-------|-------|-------|-------|-------|-------|-------|-------|-------|
| $\varepsilon$ | $\nu$ | 0.05 | 0.15 | 0.25 | 0.35 | 0.45 | 0.55 | 0.65 | 0.75 | 0.85 | 0.95 |
| 0.08 | $10^{-3}$ | 0.288 | 0.289 | 0.297 | 0.305 | 0.311 | 0.356 | 0.299 | 0.607 | 0.177 | 2.42 |
|      | $10^{-2}$ | 0.290 | 0.292 | 0.297 | 0.308 | 0.313 | 0.356 | 0.303 | 0.604 | 0.166 | 2.40 |
|      | 0.1 | 0.305 | 0.307 | 0.312 | 0.322 | 0.328 | 0.367 | 0.325 | 0.588 | 0.718 | 2.22 |
|      | 0.5 | 0.360 | 0.362 | 0.366 | 0.374 | 0.381 | 0.405 | 0.393 | 0.542 | 0.192 | 1.62 |
|      | 1.0 | 0.409 | 0.410 | 0.413 | 0.417 | 0.423 | 0.436 | 0.436 | 0.514 | 0.338 | 1.20 |
|      | 1.5 | 0.440 | 0.441 | 0.442 | 0.445 | 0.448 | 0.456 | 0.457 | 0.501 | 0.401 | 0.969 |
|      | 2.0 | 0.460 | 0.460 | 0.461 | 0.462 | 0.464 | 0.468 | 0.469 | 0.494 | 0.432 | 0.829 |
| 0.1  | $10^{-3}$ | 0.279 | 0.283 | 0.284 | 0.303 | 0.286 | 0.377 | 0.207 | 0.762 | 0.584 | 2.80 |
|      | $10^{-2}$ | 0.281 | 0.284 | 0.286 | 0.303 | 0.290 | 0.374 | 0.214 | 0.756 | 0.569 | 2.78 |
|      | 0.1 | 0.296 | 0.299 | 0.301 | 0.316 | 0.307 | 0.380 | 0.246 | 0.720 | 0.434 | 2.57 |
|      | 0.5 | 0.350 | 0.352 | 0.355 | 0.365 | 0.365 | 0.407 | 0.343 | 0.616 | −0.409 | 1.89 |
|      | 1.0 | 0.398 | 0.399 | 0.402 | 0.407 | 0.410 | 0.432 | 0.405 | 0.554 | 0.194 | 1.40 |
|      | 1.5 | 0.430 | 0.430 | 0.432 | 0.435 | 0.437 | 0.450 | 0.436 | 0.524 | 0.306 | 1.12 |
|      | 2.0 | 0.450 | 0.451 | 0.452 | 0.453 | 0.455 | 0.461 | 0.454 | 0.508 | 0.365 | 0.951 |

# References

1. Abramyan, B.L. and Aleksandrov, A.Y. (1966). *The Axisymmetric Problems of the Theory of Elasticity. Trudy II Union Congress on Theoretical and Applied Mechanics*, 7–38. Moscow: Science.
2. Akour S.N. Dynamics of Nonlinear Beam on Elastic Foundation. Proceedings of the World Congress on Engineering, 2010, Vol II WCE 2010, June 30–July 2, 2010, London, U.K.
3. Aleksandrov, A.J. (1973). The solution of the basic three-dimensional problems of the theory of elasticity for the bodies of arbitrary shape by numerical realization of the method of integral equations. *DAN USSR* **208** (2): 290–294.
4. Aleksandrov, A.V., Lashchenikov, B.Y., Shaposhnikov, N.N., and Smirnov, V.A. (1976). *Methods of Calculation of Bars Systems, Plates, and Shells with Use of Computers*. Moscow: Stroyisdat Part I, 248p.
5. Aliabadi, M.H., Brebbia, C.A., and Makerle, J. (1996). *Boundary Element Reference Database*. Computational Mechanics Publications.
6. Aleynikov, S.M. *Spatial Contact Problems in Geotechnics. (Boundary-Element Method)*. Springer Originally published in Russian as "Boundary Element Method in Contact Problems for Elastic Spatial-and-Nonhomogeneous Bases" in 2000 by Publishing House of Civil Engineering Universities Association, Moscow, Russia, ISBN 5-93093-053-8, 2002. 601p.

---

*Static and Dynamic Analysis of Engineering Structures: Incorporating the Boundary Element Method*,
First Edition. Levon G. Petrosian and Vladimir A. Ambartsumian.
© 2020 John Wiley & Sons Ltd. Published 2020 by John Wiley & Sons Ltd.

7. Alexeyeva, L.A. (1998). Boundary element method of boundary value problems of elastodynamics by stationary running loads. *Int. J. Eng. Anal. Boundary Elem.* **11**: 37–44.

8. Ambartsumian, V.A. (1971). Periods of the free non-linear oscillations of frame buildings. *Izv. Acad. Sci. Arm. SSR, Ser. Tech. Sci.* **XXIV** (1): 16–22.

9. Ambartsumian, V.A. *To the Question of Calculation of Bar Bent Systems to the Seismic Impacts in Non-linear Elasticity Law*. Soobshenya ArmNIISA vip.19.

10. Ambartsumian, V.A. (1978). Design of frame structure of minimum volume with a given value of main frequency. *Izv. Acad. Sci. Arm. SSR, Ser. Tech. Sci.*

11. Ambartsumian, V.A. (1979). About one method of determination of frequencies and form of free non-linear oscillations of frame buildings. *Izv. Acad. Sci. Arm. SSR, Ser. Tech. Sci.* **XXXII** (5): 18–24.

12. Ambartsumian, V.A. (1980). A study of the interaction of structures with the ground under stationary seismic impacts. *J. Struct. Mech. Des. Struct.* Moscow (6): 63–66.

13. Ambartsumian, V.A. (1982). A free oscillations of hinge-supports beams with arbitrary varied cross-section. In: *The Design Structures to the Seismic Actions*, 14–20. Yerevan: Ayastan.

14. Ambartsumian, V.A. (1982). *Designing Underground Structures to the Seismic Actions*, 39–47. Yerevan: Ayastan.

15. Ambartsumian, V.A. (1982). Design of cantilever plates of minimum mass working to shift at a given main frequency. *Izv. Acad. Sci. Arm. SSR, Ser. Tech. Sci.* **XXXV** (1): 60–68.

16. Ambartsumian V.A. Diffraction of horizontal shear waves on the semi-cylindrical base of structures. Design of cantilever plate of minimum mass working to shift at a given main frequency. J. Intercollege Collection of scientific works "Engineering Problems in Structural Mechanics" Yerevan, 1987, p. 42–49.

17. Ambartsumian, V.A. and Petrosian, L.G. (1987). The propagation of seismic waves in a layered inhomogeneous medium. *Izv. Acad. Sci. Arm. SSR, Ser. Tech. Sci.* **XI** (1): 8–13.

18. Ambartsumian, V.A. and Khachiyan, E.E. (1972). *Analysis of Non-linear Reactions of kStructures with the Strong Earthquakes*, vol. **19**, 5–17. Yerevan: Soobshenya ArmNIISA.

19. Amiraslanov, N.A. and Barshtein, M.P. (1970). The oscillations of the structures extended in the plane with the earthquake. In: *Structural Mechanics and Calculation of Construction*, vol. **6**, 35–40. Moscow.

20. Anant, R.K. and Man-Gi, K. (1992). Analysis of rectangular plate on an elastic half space using an energy approach. *Appl. Math. Modell.* **16** (7): 338–356.

21. Anaugh, R.P. and Goldsmith, W. (1963). Diffraction of steady elastic waves by surfaces of arbitrary shape. *Trans. ASME: J. Appl. Mech.* **30**: 589–597.

22. Arman, J.L. (1977). *Applications of Optimal Control Theory for Distributed Parameter Systems to Problems of Structures Optimization*. Moscow: Mir 314p.

23. Banerjee, P.K. and Butterfield, R. (1979). *Developments in Boundary Element Methods*, vol. **I**, 28–39. London: Applied Science Publishers.

24. Banerjee, P.K. and Mukherjec, S. (1983). *Developments in Boundary Element Methods*, vol. **III**, 322–329. London: Applied Science Publishers.

25. Banerjee, P.K. and Shaw, R. (1982). *Developments in Boundary Element Methods*, vol. **II**, 214–223. London: Applied Science Publishers.

26. Banerjee, P.K. and Watson, J.O. (1984). *Developments in Boundary Element Methods*, vol. **IV**, 240–257. London: Applied Science Publishers.

27. Banerjee, P.K. and Butterfield, R. (1984). *Boundary Element Methods in Applied Science*. Moscow: Mir 494p.

28. Banerjee, P.K. and Mamoon, S.M. (1990). A fundamental solution due to a periodic point force in the interior of an elastic half space. *Earthquake Eng. Struct. Dyn.* **19**: 91–105.

29. Banerjee, P.K. (1994). *Boundary Elements Methods in Engineering*. London: McGraw-Hill.

30. Barshtein, M.P. (1968). The oscillations of the extended constructions under seismic action. In: *Structural Mechanics and Calculation of Construction*, vol. **6**, 30–36. Moscow.

31. Besuner, F.M. and Snow, D.U. (1985). Application of two-dimensional method of boundary integral equations for the solution of engineering problems. In: *Methods of Boundary Integral Equations. Computational Aspects and Applications in Mechanics*. Translation from English (ed. R.V. Goldstein). Moscow: Science 253p.

32. Belotserkovsky, S.M. and Lifanov, I.K. (1985). *Numerical Methods in the Singular Integral Equations*. Moscow: Science 256p.

33. Beskos, D.E. (1991). *Boundary Element Methods of Plates and Shells*. Berlin: Springer-Verlag.

34. Beskos, D.E. (1997). Boundary element methods in dynamic analysis: Part II (1986–1996). *Appl. Mech. Rev. ASME* **50**: 149–197.

35. Biot M.A. Linear Thermodynamics and the Mechanics of Solids. Proc. 3rd U.S. Nat. Congr. Appl. Mech, 1958, p. 1–18.

36. Bitsadze, A.V. (1966). *Boundary Value Problems for Elliptic Equations of the Second Order*. Moscow: Nauka 203p.

37. Bogolyubov, N.N. and Mitropolsky, Y.A. (1958). *Asymptotic Methods in the Theory of Nonlinear Oscillations*. Moscow: Fizmatgiz 472p.

38. Bode, T. (1948). *Circuit Theory and Design Amplifiers with Feedback*. Moscow: GITTL 642p.

39. Bolotin, V.V. (1965). *Statistical Methods in the Structural Mechanics*. Moscow: Stroyizdat 279p.

40. Bolotin, V.V. (1976). *The Dynamic Stability of Elastic Systems*. Moscow: Gostekhteoretizdat (State Publishing House of Technical and Theoretical Literature) 600p.

41. Bondar, N.G., Kazey, I.I., Lesokhin, B.F., and Kozmin, Y.G. (1965). *The Dynamics of Railroad Bridges*. Moscow: Transportizdat 410p.

42. Bormot, Y.L. (1977). Numerical analysis by the method of potential of spatial stress condition of the element of structures. *Izv. AS USSR, Ser. MTT* (4): 20–25.

43. Borodachev, N.M. (1976). Contact problem for a stamp with narrow rectangular base. *J. Appl. Math. Mech.* **40**: 505–512.

**44.** Bosakov, S.V. (2003, UDC539.3). Solving the contact problem for a rectangular die on an elastic foundation. *Int. App. Mech.* **39** (10): 1188–1189.

**45.** Brebbia, C.A. (ed.) (1978). *Recent Advances in Boundary Element Methods, Proc. 1st Int. Conference Boundary Element Methods, Southampton University, 1978.* London: Pentech Press.

**46.** Brebbia, C.A. (ed.) (1980). *New Developments in Boundary Element Methods, Proc. 2nd Int. Conference Boundary Element Methods, Southampton University, 1980.* Southampton/London: CML Publications/Butterworths.

**47.** Brebbia, C.A. (ed.) (1981). *Boundary Element Methods, Proc. 3rd Int. Conference Boundary Element Methods, Irvine. California, 1981.* Berlin: Springer-Verlag.

**48.** Brebbia, C.A. (ed.) (1982). *Boundary Element Methods in Engineering, Proc. 4th Int. Conference Boundary Element Methods, Southampton University, 1982.* Berlin: Springer-Verlag.

**49.** Brebbia, C.A., Tells, Z., and Wrobel, L. (1987). *Boundary Element Methods.* Moscow: Mir 524p.

**50.** Brebbia, C.A. and Walker, C. (1982). *The Application of Boundary Elements Methods in the Engineering.* Moscow: Mir 248p.

**51.** Brekhovskikh, L.M. (1973). *Waves in Layered Media.* Moscow: Nauka 343p.

**52.** Brychkov, Y.A. and Prudnilov, A.P. (1977). *Integral Transformations of Generalized Functions.* Moscow: Science 286p.

**53.** Burago Y.D., Mazya V.G., and Sapozhnikova V.D. To the theory of the potentials of the dual and single layer domains with irregular boundaries. // The problem of mathematical analysis. Boundary-value problems and integral equations / LGU. Leningrad, 1966, p. 3–35.

**54.** Burchuladze, T.V. and Gegelya, T.G. (1985). *Development of the Method of Potential in the Theory of Elasticity.* Tbilisi: Mecniereba 226p.

**55.** Burchuladze, T.V. and Rukhadze, R.V. (1974). About Geen's tensors in the elasticity theory. *Differ. Equations* **10** (6): 64–72.

**56.** Butkovsky, A.G. (1977). *The Structural Theory of the Continuous System.* Moscow: Nauka 320p.

**57.** Calderon, A.P. (1963). Boundary value problems for elliptic equations. In: *Outlines of the Joint Soviet-American Symposium on Partial Differential Equations,* 303–305. Novosibirsk: Academy Press.

**58.** Carrier, W.D. and Christian, J.T. (1973). Analysis of inhomogeneous elastic half-space. *J. Eng. Mech.* **99** (3): 301–306.

**59.** Case, K.M. and Hazeltine, R.D. (1970). Elastic radiation in a half space. *J. Math. Phys* **11** (8): 2546–2552. https://doi.org/10.1063/1.1665424.

**60.** Celep, Z., Turhan, D., and Al-Zaid, R.Z. (1988). Contact between a circular plate and a tensionless edge support. *Int. J. Mech. Sci.* **30**: 733–741.

**61.** Chander, S., Donaldson, B.K., and Negm, H.M. (1979). Improved extended field method numerical results. *J. Sound Vib.* **66** (1): 39–51.

**62.** Chang, R., Hang, E.J., and Rim, K. (1976). Analysis of unbounded contact problems by means of quadratic programming. *J. Optim. Theory Appl.* **20**: 171–189.

**63.** Chen, C. (1998). Solution of beam on elastic foundation by DQEM. *J. Eng. Mech.* **124**: 1381–1384.

**64.** Chen, L.H. and Schweikert, J. (1963). Sound radiation from an arbitrary body. *J. Acoust. Soc. Am.* **35**: 1626–1632.

**65.** Cheng, F. and Pantelides, C. (1988). Dynamic Timoshenko beam-columns on elastic media. *J. Struct. Eng.*, https://doi.org/10.1061/(ASCE)0733-9445(1988)114:7(1524): 1524–1550.

**66.** Chernov, Y.T. (1986). The calculation of structures on elastic foundations accounting for heterogeneity and physical nonlinearity. In: *Structural Mechanics and Calculation of Construction*, 27–32. Moscow.

**67.** Chernov, Y.T. and Osipova, M.V. (2015). General case plane vibrations of massive bodies on elastic supports. In: *Structural Mechanics and Calculation of Construction*, vol. **4**, 36–42. Moscow.

**68.** Cheung, Y.K. and Zienkiewicz, O.C. (1965). Plates and tanks on elastic foundation – An application of finite element method. *Int. J. Solids Struct.* **1**: 451–461.

**69.** Choi, S.W. and Jang, T.S. (2012). Existence and uniqueness of nonlinear deflections of an infinite beam resting on a non-linear and discontinuous elastic foundation. *Boundary Value Prob.* **5**.

**70.** Christensen, R. (1974). *Introduction to the Theory of Viscoelasticity*. Moscow: Mir 340p.

**71.** Clough, R.W. and Penzien, G. (1975). *Dynamics of Structures*. New York: 345p.

**72.** Cole, D.M., Kosloff, D.D., and Minster, J.B.A. (1978). Numerical boundary integral equation method for electrodynamics. *J. Bull. Seismol. Soc. Am.* **68**: 1331–1357.

**73.** Collar, A.R. (1953). The effect of shear flexibility and rotatory inertia of the bending vibrations of beams. *Q. J. Mech. Appl. Math.* **6**: 186–222. London.

**74.** Conry, T.F. and Seireg, A. (1971). A mathematical programming method for design of elastic bodies in contact. *ASME J. Appl. Mech.* **52**: 387–392.

**75.** Crandall, S.H. (1963). Dynamic response of system with structural damping. In: *Air Space and Instruments. Draper Anniversary Volume*, 197. McGrow-Hill.

**76.** Crandall, S.H. (1971). The role of dumping in the theory of vibration. *Mechanics*: (Translation from English), Moscow Mir **5** (129): 3–22.

**77.** Crandall S.H., Vildiz A. Random vibration of beams paper. Amer. Soc. Mech. Engineers, 1961.

**78.** Crouch, S.L. and Starfield, A.M. (1987). *Boundary Element Methods in Solid Mechanics: with Applications in Rock Mechanics and Geological Engineering.* (Translation from English). Moscow: Mir 328p.

**79.** Cruse, T.A. (1969). Numerical solution in three-dimensional electrostatics. *Int. J. Solids Struct.* **5** (12): 1259–1274.

**80.** Cruse T.A. Application of the Boundary Integral Equation Solution Method in Solid Mechanics. Int. Conference on Variational Methods in Engineering, 1972.

**81.** Cruse, T.A. (1973). Application of the boundary integral equation method to three-dimensional stress analysis. *Comput. Struct.* **3** (3): 509–527.

**82.** Cruse, T.A. and Rizzo, F.J. (1968). A direct formulation and numerical solution of the general transient elastodynamic problem. *Int. J. Math. Anal. Appl.* **22** (1): 244–259.

**83.** Cruse, T.A. and Rizzo, F.J. (1975). *Boundary Integral Equation Methods Computational Applications in Applied Mechanics*. New York: ASME.

**84.** Cruise, T.A. and Rizzo, F.J. (1979). Boundary integral equation method. In: *Applied the Integral Equation Method to Solve the Problems in Mechanics*. Moscow: Mir 220p.

**85.** Cruse T.A., Swedlov J.L. *Integral Program for Analysis and Design Problems in Advanced Composites Technology*. AFML-TR-71, 1971. 268p.

**86.** Cruse, T.A. and Van Buren, W. (1971). Three dimensional elastic stress analysis of a fracture specimen with an edge crack. *Int. J. Fract. Mech.* **7** (1): 1–15.

**87.** Darbinian, S.S. (1966). About the spectrum of the displacement during calculations of structures to the seismic resistance taking into account elasto-plastic deformations. *Izv. AN Arm. SSR, Ser. Tech. -Science* **XIX** (6): 7–21.

**88.** Dashevsky, M.A. (1967). Diffraction of elastic waves on the plane reinforced by a rigid ring. In: *Structural Mechanics and Calculation of Construction, Moscow*, vol. **2**, 23–28.

**89.** Di Paola, M., Marino, F., and Zingales, M. (2009). A generalized model of elastic foundation based on long-range interactions: integral and fractional model. *Int. J. Solids Struct.* ISSN: 0020-7683, **46** (17): 3124–3137.

**90.** Dinev, D. (2012). Analytical solution of beam on elastic foundation by singularity functions. *Eng. Mech.* **19** (6): 381–392.

**91.** Dinnik, A.N. (1952). *The Application of the Bessel Functions to the Problems of Theory of Elasticity*. Kiev: Izdatelstvo AH USSR.

**92.** Ditkin, V.A. and Prudnikov, A.P. (1974). *Integral Transformations and Operational Calculus*. Moscow: Gostekhizdat 542p.

**93.** Dorman, I.Y. (1996). *Seismic Resistance Transport Tunnels*. Moscow: Transport 175p.

**94.** Egorov, K.E. (1958). Concerning the question of the deformation of bases of finite thickness. *Mech. Gruntov, Sb. Tr.* **34**: 1–34, Gosstroiizdat, Moscow. (in Russian).

**95.** Egorov, K.E. (1965). Calculations of beds for foundations with ring footing. In: *Proceedings of the 6th International Conference on Soil Mechanics and Foundation Engineering*, vol. **2**, 41–45. Montreal.

**96.** Egorov, K.E., Barvashov, V.A., and Fedorovsky, V.G. (1975). Some applications of the elasticity theory to the design of foundations. In: *Soil Mechanics and Foundation Engineering*, vol. **2, 1**, 72–83. Moscow.

**97.** Egorov, K.E. and Simvulidi, I.A. (1969). Calculation of footings on compressible foundation beds. In: *Proceedings of the 7th International Conference on Soil Mechanics and Foundation Engineering*, vol. **2**, 77–84. Mexico.

**98.** Egupov, V.K. and Komandrina, T.A. (1969). *The Calculation of Buildings to Seismic Action*. Kiev, Ukraine: Budivelnik 208p.

**99.** Eisenberger, M. and Clastornik, J. (1986). Beams on variable two parameter elastic foundations. *Comput. Struct.* **23**: 351–356.

**100.** Erzhanov, J.S., Aytaliev, S.M., and Alekseeva, L.A. (1989). *The Dynamics of Tunnels and Underground Pipelines*. Alma-Ata: 240p.

**101.** Farrar C.R., Duffey P.J., Cornwell S.W. Dynamic Testing of Bridge Structures A Review, Los Alamos National Laboratory Report in Preparation, 1999.

**102.** Fikera, G. (1874). *The Theorems of Existence in the Theory of Elasticity*. Moscow: Mir 159p.

103. Filonenko-Borodich, M.M. (1940). Some approximate theories of elastic founda-tions (in Russian). *Uch. Zap. Mosk. Gos, Univ. Mekh.* **46**: 3–15.

104. Filonenko-Borodich, M.M. (1945). A very simple model of an elastic foundation capable of spreading the load. *Sb Tr. Mosk. Elektro. Inst. Inzh. Trans.* 53p.

105. Floris, C. and Lamacchia, F.P. (2006). Viscoelastic analysis of a Bernoulli-Navier beam resting on an elastic medium. In: *Proceedings of the Eighth International Con-ference on Computational Structures Technology* (eds. B.H.V. Topping, G. Montero and R. Montenegro). Stirlingshire, Scotland: Civil-Comp Press 286p.

106. Floris, C. and Lamacchia, F.P. (2011). Analytic solution for the interaction between a viscoelastic Bernoulli-Navier beam and a Winkler medium. *Struct. Eng. Mech.* **38** (5): 593–618.

107. Fotieva, N.N. (1980). *The Calculation of the Lining of Underground Structures in Seismically Active Regions*. Moscow: Nedra 222p.

108. Fox, C. (1961). The G and H functions as symmetrical Fourier kernels. *Trans. Am. Math. Soc.* **98**: 395–429.

109. Frangi, A. (1999). Elastodynamics by BEM: A new direct formulation. *Int. J. Numer. Methods Eng.* **45**: 721–740.

110. Frangi, A. (2000). Causal shape functions in the time domain boundary element method. *Comput. Mech.* **25**: 533–541.

111. Friedman, R. and Shaw, R.P. (1962). Diffraction of a plane shock wave by an arbi-trary rigid cylindrical obstacle. *Trans. ASME: J. Appl. Mech.* **29** (1): 40–46.

112. Forbes, D.J. and Robinson, A.R. (1969). *Numerical Analysis of Elastic Plates and Shallow Shells by on Integral Equation Method*. SPS 3345. University of Illinois.

113. Gagnon, P., Gosselin, C., and Cloutier, L. (1997). A finite strip element for the anal-ysis of variable thickness rectangular thick plates. *Comput. Struct.* **63** (2): 349–362.

114. Gamer, U. and Pao, Y.H. (1975). Wechselwirkung Swishen Halbraun und Hal-brzylinder Bei Erriegung Durch em Ebene Harmonishe. *ZAMM* **55** (4): 81–84.

115. Gelfand, I.M. and Shilov, G.E. (1959). *The Generalized Functions and Operations Over Them*. Moscow: Fizmatgiz 470p.

116. Gernet G., Cruz-Pascal J. Unsteady reaction of circular cylinder of arbitrary thick-ness locating in elastic medium to the action of the plane wave expansion. Works of ASME, E Series, Applied Mechanics, No. 3, 1966.

117. Gersevanov, N.M. (1948). *Collection of Scientific Articles*. Moscow: Stroyvoen-morizdat 373p.

118. Godunov, S.K. and Ryabenkiy, V.S. (1977). *The Difference Schemes. Introduction into Theory*. Moscow: Nauka 400p.

119. Goldenblat, I.I., Kartsivadze, G.N., Napetvaridze, S.G., and Nikolaenko, N.A. (1971). *The Design of Seismic Resistance Hydro-technical, Transportation and Some Special Structures*. Moscow: Stroyizdat 277p.

120. Goldstein, R.V., Ryskov, I.N., and Salganik, R.L. (1969). The central cross-section crack in an elastic medium. *Izv. AH CCCP, Ser. MTT* (4): 97–105.

121. Gorbunov-Pasadov, M.I. (1949). *Beams and Plates on Elastic Base*. Moscow: Stroiz-dat, Russia (former USSR) (in Russian).

122. Gorbunov-Pasadov, M.I. and Serebrjanyi, R.V. (1961). Design of structures on an elastic foundation. In: *Proceedings of the International Conference on Soil Mechan-ics and Foundation Engineering*, vol. **1**, 643–648.

**123.** Gradstein, I.S. and Ridjick, I.M. (1991). *Tables of Integrals, Sums, Series, and Products*. Moscow: Fizmatgiz 1100p.

**124.** Greenberg, M.D. (1971). *Application of Green's Function in Science and Engineering*. Englewoood Cliffs, NJ, USA: Prentice-Hall.

**125.** Greenberg, M.D. (1978). *Foundations of Applied Mathematics*, 655. NJ, USA: Prentice-Hall, Englewood Cliffs.

**126.** Grigolyuk, E.I. and Tolkachev, V.M. (1974). About the solution of integral equations for contact problems. In: *The Selected Problems of Applied Mechanics*. Moscow: Nayka 114p.

**127.** Grinev V.B., Filipov A.P. The optimal design of constructions, which have assigned natural frequencies. 1971. Applied Machanics, Vol. 7, Vipusk 10.

**128.** Grossman, V.A. (1967). Oscillations of extended systems. In: *The Seismic Stability of Buildings and Structures*, 26–45. Moscow: Stroyizdat.

**129.** Guenfoud, S., Bosakov, S.V., and Laefer, D.F. (2014). Dynamic analysis of a plate resting on elastic half-space with distributive properties. In: *The 2014 World Congress on: Advances in Civil, Environmental, and Materials Research (ACEM 14)*. Busan, Korea: BEXCO. 2014-08-28. Available at: http://hdl.handle.net/10197/7710.

**130.** Guz, A.N., Kubenko, V.D., and Cherevko, M.A. (1978). *Diffraction of Elastic Waves*. Kiev: Naukova Dumka 307p.

**131.** Gyugter, H.M. (1959). *The Theory of Potential and its Application to the Basic Problems of Mathematical Physics*. Moscow: Gostekhizdat 416p.

**132.** Hatzigeorgiou, G.D. and Beskos, D.E. (2000). Dynamic response of 3-D elastoplastic or damaged structures by BEM. In: *CD-Rom Proceedings of European Congress on Computational Methods in Applied Sciences and Engineering*. Barcelona: ECCOMAS, 11–14 September 2000. 9p.

**133.** Hatzigeorgiou, G.D. and Beskos, D.E. (2002). Dynamic response of 3-D damaged solids and structures by BEM. *Comput. Model. Eng. Sci.* **3**: 791–802.

**134.** Hatzigeorgiou, G.D. and Beskos, D.E. (2002). Dynamic elastoplastic analysis of 3-D structures by the D/BEM. *Comput. Struct.* **80**: 339–347.

**135.** Hausner, G.W. (1957). Interaction of building and ground during an earthquake. *Bull. Seismol. Soc. Am.* **47**: 179–186.

**136.** Heise, U. (1975). The calculations of Cauchy principle values in integral equations for boundary value problems of the plane and three-dimensional theory of elasticity. *J. Elast.* **5**: 99–110.

**137.** Heise, U. (1976). Non-integral terms in integral equations in the plane and three-dimensional theory of elasticity. *Mech. Res. Commun.* **3**: 119–124.

**138.** Heise, U. (1978). The spectra of same integral operators for plane elastostatical boundary value problems. *J. Elast.* **8**: 47–79.

**139.** Heise, U. (1978). Numerical properties of integral equations in the given boundary values and in which the given boundary values and the sought solution are defined on different curves. *Comput. Struct.* **8**: 199–205.

**140.** Herrera, I. and Sabina, F.J. (1978). Connectivity as an alternative to boundary integral equations. *Proc. Natl. Acad. Sci. USA.* **75**: 5–15.

**141.** Hess, J.L. and Smith, A.M.O. (1964). Calculations of non lifting potential flow about arbitrary three-dimensional bodies. *J. Ship Res.* **8** (2): 22–44.

**142.** Hetenyi, M. (1946). *Beams on Elastic Foundation.* Ann Arbor, MI: University of Michigan Press.

**143.** Horibe, T. (1987). An analysis for large deflection problems of beams on elastic foundations by boundary integral equation method. *Trans. Jpn. Soc. Mech. Eng. (JSME)* **53** (487): 622–629.

**144.** Horibe, T. (1996). Boundary integral equation method analysis for beam-columns on elastic foundation. *Trans. Jpn. Soc. Mech. Eng. (JSME)* **62** (601): 2067–2071.

**145.** Hu, C. and Hartley, G.A. (1994). Analysis of a thin plate on an elastic half-space. *Comput. Struct.* **52**: 277–235.

**146.** Huang, T.C. (1961). The effect of rotatory inertia and of shear deformation of the frequency and normal mode equations of uniform beams with simple and conditions. *Trans. ASME: J. Appl. Mech.* **28** (4): 579–584.

**147.** Hughes, W.F. and Gayford, W. (1964). *Basic Equations of Engineering Science.* New-York: McGraw-Hill.

**148.** Ilichev V.A. Questions of the calculation of bases and foundations to the dynamic impacts taking phenomena into account ground wave. The thesis of Doctor of Technical Sciences. Moscow, 1975. 315p.

**149.** Ilichev, V.A. and Shakhter, O.Y. (1976 No. 67,). *Determination of Dynamic Stresses and Displacements in the Elastic Half-Plane from the Internal Source, Which Imitates the Action of the Underground Metro Tunnel of Shallow Laying,* 42–64. Moscow: Trudi NII Institute of Ground and Underground Structures.

**150.** Ilichev, V.A. and Taranov, V.G. (1976). Experimental study of the interaction between a vertical vibration footing and its foundation. *Solid Mech. Found. Eng.* **13** (2): 111–120.

**151.** Ilyasevich, S.A. (1960). *Steel Bridges.* Moscow: Voenizdat 720p.

**152.** Ilyasevich, S.A. (1964). *About the Oscillations of Steel Bridges.* Moscow: Voenizdat 135p.

**153.** Inglis C.E. "Minutes of proceedings of the Institute of Civil Engineers," 234, Paper 4870, 1934.

**154.** Ishkova, A.G. (1957). On some generalizations of the solution of the problems concerning a circular plate and an infinite strip on a elastic half-space, (in Russian). *Prikl. Mat. Mech.* **21**: 287–290.

**155.** Jang, T.S. (2013). A new semi-analytical approach to large deflections of Bernoulli–Euler v. Karman beams on a linear elastic foundation: nonlinear analysis of infinite beams. *Int. J. Mech. Sci.* **66**: 22–32.

**156.** Jaswon, M.A. and Ponter, A.R. (1963). An integral equation method for a torsion problem. *Proc. R. Soc. London, Ser. A* **273**: 237–246.

**157.** Jaswon, M.A., Maiti, M., and Symm, M. (1967). Numerical Beharmonic analysis and some applications. *Int. J. Solids Struct.* **3** (3): 309–332.

**158.** Kadysh, F.S. (1962, ANLatv. SSR.). Comparison of the results of calculation and model tests with beams on elastic foundations, (in Russian). *Vopr. Dynamiki i Prochnosti, Riga.* **9**: 139–157.

**159.** Kalandia, A.I. (1973). *The Mathematical Methods of Two-Dimensional Theory of Elasticity*. Moscow: Nauka 303p.

**160.** Kamke, E. (1966). *Handbook of Differential Equations in Partial Derivatives of the First Order*. Moscow: Nauka 254p.

**161.** Kamke, E. (1966). *Handbook of Ordinary Differential Equations*. Moscow: Nauka 343p.

**162.** Kartsivadze, G.N. (1974). *The Seismic Stability Bar Structures*. Moscow: Transport 264p.

**163.** Kauder, G. (1961). *Nonlinear Mechanics*. Moscow: Inostranaya Literatura 776p.

**164.** Kazey, I.I. (1960). *Dynamic Calculation of Spans of the Railroad Bridges*. Moscow: Transzheldorizdat 466p.

**165.** Kellar, D.D. (1929). *Foundations of Potential Theory*, 1953. Berlin/New York: Springer/Dover.

**166.** Kerr, A.D. (1964). Elastic and viscoelastic foundation models. *J. Appl. Mech.* **31**: 491–498.

**167.** Kerr, A.D. (1965). A study of a new foundation model. *Acta Mech.* **1**: 135–147.

**168.** Kerr, A.D. and Coffin, D.W. (1991). Beams on a two-dimensional Pasternak base subjected to loads that cause lift-off. *Int. J. Solids Struct.* **28**: 413–422.

**169.** Ketch, B. and Teodoresku, P. (1978). *Introduction into the Theory of the Generalized Functions with the Applications in the Engineering*. Moscow: Mir 518p.

**170.** Khachiyan, E.E. (1963). *Some Applied Problems of the Theory of the Seismic Stability of Constructions*, vol. **3**. Yerevan: AICM 127p.

**171.** Khachiyan, E.E. (1973). *Seismic Influence on Tower Buildings and Structures*. Yerevan: Hayastan 328p.

**172.** Khachiyan, E.E. and Ambartsumyan, V.A. (1981). *Dynamic Models of Construction in the Theory of Seismic Stability*. Moscow: Hauka 204p.

**173.** Khachiyan, E.E., Ambartsumyan, V.A., and Arzumanyan, V.G. (1982). *The Study of Transverse and Torsional Oscillations of Structures at Seismic Impact*, 3–13. Yerevan: Hayastan.

**174.** Khachiyan, E.E., Ambartsumyan, V.A., and Arzumanyan, V.G. (1982). The study of the influence of shear deformation on the reactions of flexible structures under seismic impacts. In: *Calculation of Structures on Seismic Impact*, 3–13. Yerevan: Hayastan.

**175.** Khachiyan, E.E., Ambartsumyan, V.A., and Petrosian, L.G. (1977). The determination of the reactions of the extended buildings and structures taking into account the initial phase of the passage of seismic wave. *Yerevan: Izvestia. AN Arm. SSR* **30** (4): 37–47.

**176.** Khachiyan, E.E., Ambartsumyan, V.A., and Sarkisyan, A.G. (1987). The interaction of seismic waves with the structure. In: *Wave Propagation in Building Structures Under Seismic Actions*, 109–120. Moscow: Nauka.

**177.** Klein, G.K. (1954). Calculation of beams on a solid bases continuously non-union in depth. In: *Structural Mechanics and Constructions*, vol. **3**, 71–90. Moscow: Gosstroyizdat.

**178.** Klepikov V.P. The application of a method of the boundary integral equations in solving some two-dimensional problems of elasticity for the bodies with angular features. Thesis Ph.D. in technical sciences. Moscow, 1985.

179. Kobayaski S. Some Problems of the Boundary Integral Equation Method in Elastodynamics. Boundary Elm., Proc. 5th Int. Conf. Hiroshima, Nov., 1983. Berlin e.a. 1983, p. 775–784.

180. Kolar, V. and Nemic, I. (1989). *Modeling of Soil-Structure Interaction*. Elsevier.

181. Kolesnik, I.A. (1977). *The Oscillations of Combined Systems Under the Action of Moving Loads*. Kiev-Donetsk: Visha shkola 149p.

182. Konashenko, S.I. (1953). *The Free and Forced Oscillations of Flexible Arch with a Stiffening Beam*, vol. **23**, 173–197. Trudy: Dnepropetrovsk Institute of Railroads. Transzheldorizdat.

183. Kopeikin, Y.D. (1974). The direct solution of two- and three-dimensional boundary value problems of the theory of elasticity and plasticity with the aid of the singular integral equations of the method of potential. In: *Numerical Methods of Continuum Mechanics*, vol. **5**, 46–58. Novosibirsk.

184. Korchinsky, I.L. (1962). The influence of the length of the building on the value of the seismic load. In: *The Seismic Resistance of Industrial Buildings and Civil Constructions*, 42–69. Moscow: Gosstroyizdat.

185. Korchinsky, I.L. and Grill, A.A. (1969). The determination of seismic loads of the long-span cable-stayed bridges. In: *Design of Steel Structures*, vol. **VII**, 20–42. Moscow.

186. Korenev, B.G. (1940). The method of compensating loads in the application to the problems of equilibrium, vibration, and stability of plates and membranes. *Appl. Math. Mech.* Moscow **4** (5–6).

187. Korenev, B.G. (1960). *Some Tasks of the Theory of Elasticity and Thermal Conductivity Solving in Bessel Functions*. Moscow: Fizmatgiz 458p.

188. Korenev, B.G. (1962). *The Calculation of Beams and Plates on an Elastic Foundation*. Moscow: Goststroyizdat 335p.

189. Korenev, B.G. and Chernigovskaya, E.I. (1964). *The Calculation of Plates on an Elastic Foundation*. Moscow: Goststroyizdat 232p.

190. Kositsin, B.A. (1981). *The Static Calculation of Large Panel and Frame Buildings*. Moscow: Stroyizdat 215p.

191. Kovshov A.N., Neshcheretov N.I. About diffraction of non-stationary transverse waves to the cylindrical cavity. MTT 5, 1982.

192. Krylov, A.N. (1933). *On Analysis of Beams Resting on Elastic Foundations*. Leningrad: Science Academy of USSR.

193. Krylov, A.N. (1948). *Vibrations of Ships*, vol. **10**. Moscow: Doklady, USSR AN 403p.

194. Krynitsky, E. and Mazurkevich, Z. (1962). The natural oscillations of a hinge-supported beam with the linearly varying height of the cross section. *Trans. Am. Soc. Mech. Eng., Ser. E: Appl. Mech.* **28** (3).

195. Kubenko B.D. Dynamic stresses concentration for the elliptical paddle. Doklady AH USSR, #3, 1967.

196. Kubo K. A seismicity of suspension bridges forced to vibrate longitudinally. Proc. SWCEE. Tokyo-Kyoto, 7, 1960.

197. Kulakov, V.M. and Tolkatchev, B.M. (1976). Bending plates of arbitrary shape. *DAN USSR* 230p.

198. Kunnos, V.I. (1954). Free oscillations of combined systems. In: *Problems of Dynamics and Dynamics Strength*, vol. **2**, 21–58. Riga: Academy of Science of Latvian SSR.

**199.** Kupradze, V.D. (1963). *The Methods of Potential in the Theory of Elasticity.* Moscow: Fizmatgiz 472p.

**200.** Kupradze, V.D. and Burchuladze, T.B. (1975). Dynamic problems of the theory of elasticity and thermo-elasticity. In: *Modern Problems of Mechanics*, vol. **7**. Moscow: VINITI, USSR.

**201.** Kupradze, V.D., Hegelia, T.G., Basheleyshvili, M.O., and Burchuladze, T.B. (1976). *Three Dimensional Problems of Mathematical Theory of Elasticity and Thermo-Elasticity.* Moscow: Nauka 664p.

**202.** Lachat J.C. Further Developments of the Boundary Integral Techniques for Elastostatics. – Ph.D thes. – Southampton Univ., 1975.

**203.** Lachat, J.C. and Watson, J.O. (1977). Progress in the use of boundary integral equations, illustrated by examples. *Comput. Math. Appl. Mech. Eng.* **10**: 273–289.

**204.** Lardner, T. (1968). The solutions in the generalized hypergeometric functions of the problems about the lateral oscillations of one class of the bars with variable cross-section. *Trans. Am. Soc. Mech. Eng., Ser. E: Appl. Mech.* **35** (1).

**205.** Lashchenikov, B.Y. (1972). Semi-discrete analysis of the three dimensional systems, which consist of the discrete and continuous media. In: *Research Work - Theory of Construction*, vol. **XIX**. Moscow.

**206.** Lazarev, M.I. (1980). The solution of the basic problems of the theory of elasticity for incompressible media. *PMM* **5**: 867–874.

**207.** Lazarev, M.I. (1983). The solution of some exterior problems of the theory of elasticity. *Izv. AS USSR, PMM* **44** (5): 100–110.

**208.** Lazarev M.I. The method of boundary integral equations. Algorithms and their implementations. ONTI NTsBI of the AS USSR. 1984. 14p.

**209.** Lazarev, M.I. (1987). The potentials of the linear operators. *DAH USSR* **292** (5): 1045–1049.

**210.** Linkov, A.M. (1983). The plane problems about static uploading of the piecewise-homogeneous linear elastic media. *Appl. Math. Mech.* **47** (IV): 644–651.

**211.** Liu, Q. and Ma, J. (2013). Analytical model for beams on elastic foundations considering the coupling of horizontal and vertical displacements. *J. Eng. Mech.* **139** (12): 1757–1768.

**212.** Lopatinsky Ya., B. (1953). About one method of reduction of boundary problems to the regular integral equations for the system of differential equations elliptical type. *Ukr. Math. J.* **5** (2): 123–151.

**213.** Love, A.A. (1944). *Treatise of the Mathematical Theory of Elasticity*, 4e. New York, NY: Dover Publications.

**214.** Luzhin, O.V. (1964). The static and dynamic calculation of beams, frames, plates and shells by the method of the expansion of the given system. In: *Research of the Theory of Constructions*, vol. **13**, 63–74. Moscow: Stroyizdat.

**215.** Lurie, A.I. (1950). *Operational Calculus and Its Application to the Problems of Mechanics.* Moscow-Leningrad: Gostexisdat 432p.

**216.** Lurie, A.I. (1970). *The Theory of Elasticity.* Moscow: Nauka 939p.

**217.** Lurie, K.A. (1963). The Meyer–Boltz problem for multiple integrals and optimization of the behavior of the systems with distributed parameters. *Appl. Math. Mech.* **27** (5).

218. Lyapunov, A.M. (1949). *Works on the Theory of Potential.* Moscow: Gostexizdat 180p.

219. Makarov, B.P. and Kochetkov, B.E. (1987). *The Calculation of the Foundations of Construction on Non-uniform Base with Creep.* Moscow: Stroyizdat 255p.

220. Mansur W.J. Brebbia C.A. Transient Elastodynamic Using a Timestepping Technigue.// Boundary Elem.// Proc. 5th Int. Conf. Hiroshima, Nov., 1983. – Berline. 1983, p. 677–698.

221. Manvelov, L.I. and Bartoshevich, E.S. (1961). About selection of the design model of the elastic foundation. In: *Structural Mechanics and the Calculation of Construction,* vol. **4,** 18–22.

222. Massonnet, C. (1965). Numerical use of integral procedures. In: *Stress Analysis* (eds. O.S. Zienkiewicz and G.S. Holistern), 198–235. London.

223. Mau, S. and Mente, M. (1963). Dynamic stress and displacement near the cylindrical discontinuity surface due to the plane harmonic shear waves. *Trans. Am. Soc. Mech. Eng., Ser. E: Appl. Mech.* **4**: 135–140.

224. Mazya, V.G. and Plamenevsky, B.A. (1976). About coefficients in the asymptotic solutions of elliptical boundary value problems near the edge. *DAN USSR* **229** (1): 33–36.

225. Medvedev, S.V. (1962). *Engineering Seismology.* Moscow: Gosstroyizdat 284p.

226. Melerski, E.S. (2006). *Design of Beams, Circular Plates and Cylindrical Tanks on Elastic Foundation.* Taylor & Francis Group.

227. Melikyan, A.A. (1963). *Mountain Pressure Due to Seismic Stress State of Rocks,* vol. **X.** Tbilisi: ISMIS AN Gruz SSR.

228. Mikhaylov, S.E. (1983). Solution of the problems about the non-plane deformation of elastic bodies with angular points using integral equations method. *PMM* **47** (6): 981–987.

229. Mikhlin, S.G. (1948). Singular integral equations. *YMH* **3** (3): 29–119.

230. Mikhlin, S.G. (1957). *Variational Methods in Mathematical Physics.* Moscow: Gostexizdat 476p.

231. Mikhlin, S.G. (1959). *Lectures on Linear Integral Equations.* Moscow: Fizmatgiz 232p.

232. Mikhlin, S.G. (1962). *Multidimensional Singular Integrals and Integral Equations.* Moscow: Fizmatgiz 256p.

233. Mikhlin, S.G. (1967). *The Application of Integral Equations to the Some Problems of Mechanics, Mathematical Physics and Technology.* Moscow – Leningrad: Gostexizdat 304p.

234. Milovic, D. (1992). *Stresses and Displacements for Shallow Foundations.* B.V. Sara Burgerhartstraat 25, Amsterdam/London/New-York/Tokyo: Elsevier Science Publishers.

235. Morgaevsky, A.B. (1961). *About the Zones of the Dynamic Instability of a Structure, Loaded with the Concentrated Travelling Load,* vol. **42,** 54–78. Dnepropetrovsk, Ukraine: Dnipropetrovsk Metallurgic Institute.

236. Murusidze, R.X. (1965). Experimental study of the nonlinearity of deformation with the oscillations of reinforced concrete elements. In: *Seismic Resistance of Structures,* 129–139. Tbilisi.

**237.** Muskhelishvili, N.I. (1966). *Some Basic Problems of the Mathematical Theory of Elasticity*. Moscow: Nauka 708p.

**238.** Muskhelishvili, N.I. (1968). *Singular Integral Equations*. Moscow: Science 512p.

**239.** Nagdi, P. and Kalnins, A. (1962). About the oscillation of elastic spherical shells. *Trans. Am. Soc. Mech. Eng., Ser. E: Appl. Mech.* **1**: 75–83.

**240.** Naimark, M.A. (1969). *Linear Differential Operators*. Moscow: Nauka 526p.

**241.** Napetvaridze, S.G. (1959). *Seismic Resistance of Hydraulic Structures*. Moscow: Gosstroyizdat.

**242.** Napetvaridze, S.G. (1963). The influence of the length of a structure on the magnitude of the seismic force. *AS Gr. SSR* **IX**: 213–217.

**243.** Natroshvili D.G. The solution of the problem of Cauchy for the system of equations of the moment theory of elasticity. The annotations of reports, & seminars. IMP, TGU, #11, 1976, p. 45–50.

**244.** Nazarov, A.G. (1965). *About the Mechanical Similarity of the Solid Bodies Being Deformed*. Yerevan: Academy of Science Arm SSR 214p.

**245.** Newmark, N.M. and Rosenblueth, E. (1980). *Fundamentals of Earthquake Engineering*. Moscow: Stroyizdat 344p.

**246.** Neyfeh, A.H. and Nemat-Nasser, S. (1972). Elastic waves in inhomogeneous elastic media. *Trans. Am. Soc. Mech. Eng., Ser. E: Appl. Mech.* **39** (3): 58–65.

**247.** Nikiporets, G.L. (1977). The spectra of the reactions of the extensive in the plan constructions under the seismic actions. In: *Seismic Resistance*, vol. **2**, 5–10. Moscow.

**248.** Nikolaenko N.A. Oscillation of unlimited plate, lying on an elastic half-space and an elastic layer.// Abstract of dissertation thesis Ph.D., Moscow 1956. 13p.

**249.** Novatsky, V. (1963). *Dynamics of Structure*. Moscow: Gosstroyizdat 376p.

**250.** Obraztsov, I.P. and Onanov, G.G. (1973). *Structural Mechanics of the Oblique Thin-Walled Systems*. Moscow: Mashinostroenie 659c.

**251.** Ovsyannikov, A.S. (1981). Interaction of plane harmonics H wave with the reinforcement of semi-circular cylindrical groove in an elastic half-space. *Appl. Mech.* Moscow **XVII** (8): 101–104.

**252.** Okamoto, S. (1980). *Seismic Resistance of Engineering Structures*. Moscow: Stroyizdat 342p.

**253.** Palatnikov, E.A. (1964). *Rectangular Plate on an Elastic Foundation*. Moscow: Stroyizdat 236p.

**254.** Pao, I.S. (1962). Dynamic stress concentration in an elastic plate. *Trans. Am. Soc. Mech. Eng., Ser. E: Appl. Mech.* **29** (2): 147–154.

**255.** Park, J., Bai, H., and Jang, T.S. (2013). A numerical approach to static deflection analysis of an infinite beam on a nonlinear elastic foundation: one-way spring model. *J. Appl. Math.*: 136358. 10p.

**256.** Parton, V.Z. and Perlin, P.I. (1975). The integral equations of the basic spatial and plane problems of the elastic equilibrium. In: *Solid Mechanics of Deformable Bodies*, vol. **8**. Moscow: BINITI (All-Union Institute of Scientific and Technical Information) 14p.

**257.** Parton, V.Z. and Perlin, P.I. (1977). *The Integral Equations of the Theory of Elasticity*. Moscow: 311p.

**258.** Parton, V.Z. and Perlin, P.I. (1981). *The Methods of Mathematical Theory of Elasticity*. Moscow: Nauka 688p.

**259.** Parton, V.Z. and Kudryavtsev, B.A. (1969). Dynamic problems for the plane with the section. *Rep. AS USSR* **185** (3): 541–545.

**260.** Parton, V.Z. and Kudryavtsev, B.A. (1975). The dynamic problem of the mechanics of destruction for the plane with inclusion. In: *Mechanics of Deformable Elements and Structures*. Moscow: Mashinostroenie.

**261.** Pasternak P.L. On a new method of analysis of an elastic foundation by means of two foundation constants (in Russian). Gosuderevstvennae Izdatlesva Literaturi po Stroitelstvu i Arkihitekture, Moscow, USSR, 1954.

**262.** Petrofsky, I.G. (1951). *Lectures on the Theory of Integral Equations*. Moscow – Leningrad: 128p.

**263.** Petrosian L.G. About the vibration of combine system. Yerevan: YPTU. Series of Technical Sciences, Vol. XII, 5, 1978, p. 39–45.

**264.** Petrosian, L.G. (1979). The results of dynamic model tests of Dzhermuk's bridge. *Izv. AS Arm. CCP, Ser. Tech. Sci.* **XXXV** (3): 28–34.

**265.** Petrosian L.G., Research of some aspects of the stress-strain state of long-span structures. - Structure Mechanics - 01.02.03 code. Dissertation thesis. Technical Sciences (Ph. D). Moscow, 1980. 204p.

**266.** Petrosian, L.G. (1982). The selection of the rational primary systems for frame calculations and designs. Methodical instructions for the course and diploma research projects design. In: *Structural Mechanics and Calculation of Structures*. Yerevan: 58p.

**267.** Petrosian, L.G. (1984). *The Influence of the Length of Bridge Structures on the Formation of Seismic Loads*, 66–69. Yerevan: The Industry of Armenia.

**268.** Petrosian, L.G. (1984). *The Generalized Method of Finite Integral Transformations in the Calculation of Bar Systems*, 32–38. Yerevan: YPTU. Series of Technical Sciences.

**269.** Petrosian L.G. About reduction of boundary-value and initial problems to the "Standard" form with the aid of the delta-transformation. Izvestia AS Arm SSR, Structure Mechanics. BINITI, # 3965-B86. 1986. 12p.

**270.** Petrosian, L.G. (1986). The spectral method of boundary elements. In: *Structural Mechanics and Calculation of Construction*, vol. **4**, 45–50. Moscow.

**271.** Petrosian, L.G. (1986). The calculation of the internal friction in the one-dimensional dynamic systems. In: *Structure Mechanics*, 40–44. Yerevan: YPTU.

**272.** Petrosian, L.G. (1987). The equivalence of direct and indirect methods of the boundary elements. In: *Structural Mechanics*, 75–86. Yerevan: YPTU.

**273.** Petrosian, L.G. (1987). About one method of application of Hilbert transformation to the study of dynamic systems. *Izv. AS Arm. SSR, Ser. Mech.* (1): 17–24.

**274.** Petrosian, L.G. (1988). About one elastic foundation model. In: *Structural Mechanics and Calculation of Construction*, vol. **5**, 7–11. Moscow.

**275.** Petrosian, L.G. (1988). The contact problem for a rigid die on the generalized elastic foundation. *Izv. AS Arm. SSR, Mech.* (3): 59–64.

**276.** Petrosian, L.G. (1988). The calculation of plane foundation structures with the aid of the generalized base model. *Izv. AS Arm. SSR, Mech.* (6): 51–58.

**277.** Petrosian, L.G. (1988). *Static and Dynamic Design Problems for Structures on an Elastic Foundation.* Yerevan: Luis, BBK38.112 69p.

**278.** Petrosian L.G. The methods of boundary equations in the problems of calculating the structures on an elastic foundations. The dissertation on competition of a scientific degree of Doctor of Technical Sciences (Post Doctorate Research). The thesis of Doctor of Technical Sciences. Moscow: CSRICS, 1990. 310p.

**279.** Petrosian, L.G., Basilaya, V.M., and Khaselev, M.E. (1987). The application of the generalized finite integral transforms to the dynamic calculation of plates lying on an elastic foundations. In: *Structural Mechanics and Calculation of Construction*, vol. **5**, 51–56. Moscow.

**280.** Petrosian, L.G. and Rubanovich, S.G. (1979). The free oscillations of systems "Flexible Arch-Rigid Beam." *Izv. AS Arm. SSP, Ser. Tech. Sci.* **XXXII** (4): 22–28.

**281.** Petrosian, L.G., Taylor, A.W., and Todd, D.R. (1995). *Proposed Format for a Certificate of Essential Building Data United State Department of Commerce.* NIST/EERI 10p.

**282.** Petrov, A.A. and Bazilevsky, S.V. (1976). The non-stationary random oscillations of suspension bridges under the seismic influence. In: *Improving the Methods of Design and the Construction of Buildings and Structures in Seismic Regions*, 157–167. Moscow.

**283.** Petrov, A.A. and Duzinkevich, M.S. *The Determination of Seismic Loads on the Steel Arched Bridge Through the Arpa River (Armenia).* Moscow.

**284.** Philipov, A.P. (1971). *The Methods of Calculation of Construction to Oscillations.* Moscow – Leningrad: Stoyisdat 226p.

**285.** Plotnikov Yu. G. The application of Ritz's method to the solution of the stationary problems about the die on a linearly-deformed base. //Inform. Form #267 (24330)/ Turkmen INTI Gosplan of TSSR. Ashhabad, 1978. 4p.

**286.** Plotnikov Yu. G. The harmonic oscillation of circular dies on an viscose-elastic foundation.// Inform. Form #266 (24339)/ Turkmen INTI Gosplan of TSSR. Ashhabad, 1978. 7p.

**287.** Plotnikov Yu. G. The harmonic oscillations of dies on viscose-elastic foundations. // Dep. CINIS Gosstroy of USSR .Register # 1268, 1979.

**288.** Plotnikov Yu. G. The stationary oscillations of plane and axisymmetric dies on viscose-elastic foundations. // The dissertation thesis. Technical Sciences (Ph. D). Moscow, 1979. 200p.

**289.** Polozhiy, G.N. (1962). *The Numerical Solution of the Two-Dimensional and Three-Dimensional Boundary-Value Problems of Mathematical Physics and Functions of Discrete Argument.* Kiev: Izdatelstvo Kiev University 161p.

**290.** Ponomarev, S.D. and Biderman, V.L. (1952). *Strength Calculations in Mechanical Engineering.* Moscow: Mashgiz 947c.

**291.** Popov, G.J. (1982). *Concentration of Elastic Stresses Near the Dies, Sections, and Thin Inclusions.* Moscow: Nauka 342p.

**292.** Popov, G.J. (1982). *Contact Problems for Linear-Deformed Bases.* Kiev-Odessa: Bisha shkola 168p.

**293.** Preobrazhensky, I.N. (1981). *Stability, and the Vibration of Plates and Shells with Openings*. Moscow: Mashinostroenie 190p.

**294.** Providakis, C.P. and Beskos, D.E. (1999). Dynamic analysis of plates by boundary elements. *Appl. Mech. Rev. ASME* **52**: 213–236.

**295.** Prudnikov, A.P., Brychkov, Y.A., and Marichev, O.I. (1981). *Integrals and Series*. Moscow: Nauka 798p.

**296.** Pukhovsky, A.B. and Gordin, V.A. (1977). *Seismic Resisted Arch Structure Covering Spans of 62m., with Pre- Stressed Steel Panels*, vol. **3**, 3–10. Moscow: TsINIS/Earthquake Resistant Structures.

**297.** Rabotnov, Y.N. (1977). *The Elements of Hereditary Mechanics of Solid Bodies*. Moscow: Nauka 384p.

**298.** Rao, S. and Sundararajan, V. (1969). Plane bending oscillations of circular rings. *Trans. Am. Soc. Mech. Eng., Ser. E: Appl. Mech.* (3): 253–259.

**299.** Khachiyan, E.E. (1985). *Recommendations – Guidelines for Determining Dynamic Characteristics and Seismic Loads for Buildings and Structures Using Accelerograms of Earthquakes*. Yerevan: ArmNIISA 107p.

**300.** Reytman, M.I. and Shapiro, G.S. (1976). *The Methods of the Optimum Design of Deformable Bodies*. Moscow: Nauka.

**301.** Rizzo, F.J. (1967). An integral equation approach to boundary value problems of classical elasto-statics. *Q. Appl. Math.* **25**: 83–95.

**302.** Rizzo, F.J. and Shippy, D.J. (1977). An advanced boundary integral equation method for three dimensional thermoelasticity. *Int. J. Numer. Methods Eng.* **11**: 1753–1768.

**303.** Rizzo, F.J. and Shippy, D.J. (1968). A formulation and solution procedure for the general non-homogeneous elastic inclusion problem. *Int. J. Solids Struct.* **4**: 1161–1179.

**304.** Romanov V.G. The approximate solutions of the integral equations of basic two-dimensional static problems of the theory of elasticity for the domain with angles.// Computing systems VTSSO AH USSR. Novosibirsk. 1964, No. 6, p. 14–18.

**305.** Rosenberg, R.M. (1961). About the modes of oscillations of the normal type of nonlinear system with two degree of freedom. *Mechanics* (5): 3–15.

**306.** Rosenberg, R.M. (1962). The normal mode shapes of oscillation of nonlinear systems with *n* degree of freedom. *Trans. Am. Soc. Mech. Eng., Ser. E: Appl. Mech.* (5): 78–84.

**307.** Roytfarb, I.Z. and Kyung, T.V. (1976). Numerical method of the solution of the three-dimensional dynamic problems of the theory of elasticity on the basis of the method of potential. In: *Strength of Materials and the Theory of Construction*, vol. **29**, 54–60.

**308.** Ryabenkiy V.S. Some questions of the theory of difference boundary-value problems. // The dissertation thesis. Doctor of Phys.math.sciences. Moscow, 1969. 359c.

**309.** Ryabenkiy, V.S. (1983). The generalization of projectors and boundary equations of Kalderon on the basis of the concept of clear truck. *DAN USSR* **270** (2): 288–292.

**310.** Ryabenkiy, V.S. (1985). The boundary equations with projectors. *UMN* **40** (2): 221–249.

**311.** Ryabenkiy, V.S. (1987). *The Method of the Difference Potentials for Some Problems of Mechanics of Continuum Mediums.* Moscow: Nauka 320p.

**312.** Rzhanitsin, A.R. (1949). *Some Questions of the Mechanics of the System Which Are Deformed in Time.* Moscow – Leningrad: Gostexizdat 252p.

**313.** Sapountakis, E.J. and Katsikadelis, J.T. (1992). Unilaterally supported plates on elastic foundations by boundary element method. *Trans. Am. Soc. Mech. Eng. (ASME)* **59**: 580–586.

**314.** Sargsyan, A.E. (1986). The evaluation of the intensity of seismic impact on construction taking into account the pliability of its base. In: *Structural Mechanics and Calculation of Construction*, vol. **4**, 55–59. Moscow.

**315.** Sarkisian A.G. The application of the method of boundary elements to the solution of the plane dynamic problem of the theory of elasticity. //TsNIISK (Kucherenko) /. Dep.BHIIC Goststroya USSR, #6844, 1986.

**316.** Sarkisian, V.S. (1976). *Some Problems of the Mathematical Theory of Elasticity of Anisotropic Boby.* Yerevan: Arm SSP, State University 534p.

**317.** Savarensky, E.P. (1972). *Seismic Waves.* Moscow: Nedra 294p.

**318.** Schechter, O.I. (1964). The distribution of waves in a semi-infinite elastic bar with the impact on the top end taking into account the lateral damping reactions of the ground. In: *NII Institute of Foundations and Underground Structures*, vol. **54**, 66–79. Moscow: Gosstroyizdat.

**319.** Seeley, R.T. (1966). Singular integral and boundary value problems. *Am. J. Math.* **88** (4): 781–809.

**320.** Selvadurai, A.P.S. (1979). *Elastic Analysis of Soil – Foundation Interaction.* Amsterdam, Netherlands: Elsevier Scientific Publishing Co.

**321.** Serebryaniy, R.V. (1962). *The Calculation of Thin Hinged Joined Plates on an Elastic Foundation.* Moscow: Gostroyizdat 64p.

**322.** Seymov, V.M. (1986). *The Contact Problems of Dynamic.* Kiev: Haukova Dumka 233p.

**323.** Shapiro, G.S. (1973). Bending of a semi-infinite plate lying on an elastic foundation. *PMM* **7** (4): 316–320.

**324.** Shaposhnikov, N.N. and Myachenkov, B.I. (1981). *The Calculation of Machine-Building Constructions to Strength and Hardness.* Moscow: Machinostroenie 333p.

**325.** Sheynin I.S., Vilin G.V., and Kalitseva I. Ph. The method of solution of the plane problems about the oscillations of constructions on the border with the region of liquid of complex form. // Dynamic of Energy – Construction. Theses, Vol. 5 VNIIG (B.E. Vedeneeva), Leningrad, 1976, p. 3–17.

**326.** Sherman D.I. The method of integral equations in two dimensional and three-dimensional problems of the static theory of elasticity. // Theses of All Union Congress in Theoretical and Applied Mechanics. Moscow, 1962, p. 405–467.

**327.** Shippi, D.J. (1978). The application of the method of boundary integral equations to the study of nonstationary phenomena in solid bodies. In: *Methods of Boundary*

*Integral Equations. Computational Aspects and Application in Mechanics.* Translation from English (ed. R.V. Goldstein), 30–45. Mir.

328. Shtaerman, I.Y. (1949). *Contact Problem of the Theory of Elasticity.* Moscow: 270c.

329. Shulga, N.A. (1969). Wave diffraction on the circular obstacles in the half-plane. In: *Applied Mechanics*, vol. **5**, 115–119. Moscow.

330. Sidorov, V.N. and Zolotov, A.B. (1975). Algorithmization of the solution of boundary-value problems of structural mechanics on computers. In: *Structural Mechanics and Calculation of Construction*, vol. **5**, 36–42. Moscow.

331. Simvulidi, I.A. (1958). *Analysis of Beams on Elastic Foundations.* Moscow: Soviet Science.

332. Simvulidi, I.A. (1973). *Analysis of Engineering Structures on Elastic Foundations*, 3e. Moscow: Visshaya Shkola.

333. Sinha, S.N. (1963). Large deflection of plates on elastic foundations. *J. Eng. Mech. Div.* **89**: 1–24.

334. Sinitzin, A.P. (1961). *The Influence of a Running Seismic Wave on Massive Structures*, vol. **17**, 87–99. Moscow: The Institute of GeoPhysics.

335. Sinitsin, A.P. (1967). *The Method of Finite Elements in the Dynamic of Structures.* Moscow: Stroyizdat.

336. Sirosh, S.N., Ghali, A., and Razaqpur, A.G. (1989). A general finite element for beams or beam-columns with or without an elastic foundation. *Int. J. Numer. Methods Eng.* **28** (5): 1061–1076.

337. Skavutso, R., Bailey, J., and Raftopulos, D. (1971). Horizontal interaction of construction with seismic waves. *Trans. Am. Soc. Mech. Eng., Ser. E: Appl. Mech.* (1): 123–130.

338. Skott, F.R. (1981). *Foundation Analysis.* Englewood Cliffs, NJ: Prentice-Hall.

339. Skroll, C. (1972). Dynamic stresses near the elliptical inclusions. *Trans. Am. Soc. Mech. Eng., Ser. E: Appl. Mech.* (2).

340. Slepian, L.I. (1982). *Non-stationary Elastic Waves.* Leningrad: Sudostroenie 373p.

341. Smirnov, A.P. (1974). *Construction Stability and Oscillations.* Moscow: Transzheldorizdat 571p.

342. Smirnov, V.A. (1975). *Suspension Bridges of Large Spans.* Moscow: Vishaya Shkola 369p.

343. Sneddon, I. (1955). *Fourier Transform.* Moscow – IL: 667p.

344. Sobolev, D.N. (1986). The problem about the die, pressed into the statistically non-homogeneous elastic base. In: *Structural Mechanics and Calculation of Constructions*, vol. **2**, 15–18. Moscow.

345. Sorokin, E.S. (1960). *To the Theory of Internal Friction with the Oscillations of Elastic Systems.* Moscow: Gosstroyizdat 131p.

346. Straughan W.T. Analysis of Plates on Elastic Foundations. A Dissertation in Civil Engineering. Submitted to the Graduate Faculty of Texas Tech University in Partial Fulfillment of the Requirements for the Degree of Doctor of Philosophy May, 1990.

347. Strength, Stability, and Oscillations (1968). *Handbook of Mechanical Engineering*, vol. **I**. Moscow: Mashinostroenie 831p.

348. Strength, Stability, and Oscillations (1968). *Handbook of Mechanical Engineering*, vol. **II**. Moscow: Mashinostroenie 567p.

**349.** Svec, O.J. and Hardy, R.M. (1976). Thick plates on elastic foundations by finite elements. *J. Eng. Mech. Div.* **102** (3): 461–477.

**350.** Taylor, Z. (1967). The calculation of the bar of the smallest weight with the longitudinal vibrations with the given values of the natural frequency. *Rocket Eng. Cosmonaut.* **5** (10): 244–246.

**351.** Telles, D.K.F. (1987). *The Application of a Method of Boundary Elements for the Solution of Inelastic Problems*. Moscow: Stroyizdat 160p.

**352.** Teodoru I.B. Analysis of beams on elastic foundations: the finite differences approach, Proceedings of 'Juniorstav 2007' 9th Technical conference for Doctoral study, Brno, University of Technology, Czech Republic, 24th January, 2007

**353.** Teodoru, I.B. (2009). Beams on elastic foundations – the simplified continuum approach. *Bul. Inst. Polit. LV(LIX) Fasi. Sect. Construct. Arhitect.* **55** (4): 37.

**354.** Teodoru, I.B., Musat, V., and Vrabie, M. (2006). A finite element study of the bending behavior of beams resting on two-parameter elastic foundations. *Bul. Inst. Polit. Iasi, Tomul LII (LVI)*, Fasc: 3–4.

**355.** Thambiratnam, D. and Zhuge, Y. (1996). Dynamic analysis of beams on an elastic foundation subjected to moving loads. *J. Sound Vib.* **198** (2): 149–169.

**356.** Tikhonov, A.N. and Arsenin, V.J. (1986). *The Methods of Solving Incorrect Problems*. Moscow: Science 287p.

**357.** Timoshenko, S. (1956). *Strength of Materials, Part II, Advanced Theory and Problems*, 3e. Princeton, NJ: Van Nostrand.

**358.** Timoshenko, S.P. (1967). *Oscillations in Engineering*. Moscow: Nauka 444p.

**359.** Tomlin, G.R. and Butterfield, R. (1974). Elastic analysis of zoned orthotropic continua. *J. Eng. Mech. Div.* **100** (3): 511–529.

**360.** Travush, V.I. (1969). Bending uninsulated plates lying on linear-deformed bases. In: *Theory of Structures*, vol. **17**, 73–84. Moscow: Stroyizdat.

**361.** Travush, V.I. (1975). The rectangular uninsulated plate lying on linear-deformed base. In: *Structural Mechanics and Calculation of Construction*, vol. **3**, 17–22. Moscow.

**362.** Travush V.I. Calculation of building structures on the deformed base. The dissertation on competition of a scientific degree of Doctor of Technical Sciences (Post Doctorate Research). The thesis of Doctor of Technical Sciences. Moscow, 1976. 354p.

**363.** Travush, V.I. (1982). The method of generalized solutions in the problems of bending of plates on the linear-deformed base. In: *Structural Mechanics and Calculation of Construction*, vol. **1**, 24–29. Moscow.

**364.** Travush, V.I. and Sangadzhiev, V.K. (1984). Bending of rectangular plates with a free contour (outline) on an elastic base. In: *Structural Mechanics and Calculation of Construction*, vol. **6**, 37–40. Moscow.

**365.** Troitsky, V.A. (1976). *The Optimum Processes of the Vibrations of Mechanical Systems*. Leningrad: Engineering.

**366.** Tseitlin, A.I. (1966). About one class of paired integral equations. In: *Differential Equations*, vol. **8**, 1134–1139. Moscow.

**367.** Tseitlin, A.I. (1968). The calculation of hard round plates on a linear-deformable base by the method of paired integral equations. In: *Structural Mechanics and Calculation of Construction*, vol. **1**, 25–32. Moscow.

**368.** Tseitlin A.I. Some methods of calculation of structures lying on an elastic foundation. The dissertation author's abstract of a scientific degree of the Doctor Phys. Math. Science. Moscow: 1968. 31p.

**369.** Tseitlin, A.I. (1978). About the linear models of the frequency-independent internal friction. *Izv. AS USSR, MTT* (3): 18–28.

**370.** Tseitlin, A.I. (1980). The method of delta-transformation. In: *Engineering Problems of Structural Mechanics*, 139–147. Moscow: Stroyizdat.

**371.** Tseitlin, A.I. (1984). *The Applied Methods of Solving the Boundary-Value Problems of Structural Mechanics*. Moscow: Stroyizdat 334p.

**372.** Tseitlin, A.I., Atadzhanov, D.R., and Sarkisyan, A.G. (1986). *Grin's Function of Dynamic Problem for the Half-Plane.* TsNIISK named Kucherenko. DEP VHIIS, 9.01.86 # 6446 14p.

**373.** Tseitlin, A.I. and Guseva, N.I. (1979). *Statistical Methods of Calculation of Construction to the Group of Dynamic Impacts.* Moscow: Stroyisdat 175p.

**374.** Tseitlin, A.I., Kusainov, A.A., and Petrosian, L.G. (1986). *The Free Oscillations of Dissipative Systems*, vol. **9**, 50–56. Vestnick, Almaty: AS Kazakh SSR.

**375.** Tseitlin A.I., Neustroev E.A., and Petrosian L.G. Seismic interaction of structures with the ground. // Theses of reports of All – Union Conference and 3rd scientific conference of the Far-Eastern Section of MSSS. Vladivostok, 1982, p. 95–97.

**376.** Tseitlin, A.I. and Petrosian, L.G. (1984). About some generalizations of the method of integral transformations and their connection with the method of boundary equations. In: *Structural Mechanics and Calculation of Construction*, vol. **3**, 18–23. Moscow.

**377.** Tseitlin, A.I. and Petrosian, L.G. (1987). *The Methods of Boundary Elements in Structural Mechanics.* Yerevan: Luis 200p.

**378.** Tseitlin, A.I., Petrosian, L.G., and Atadzhanov, D.R. (1987). *About the Theory of Boundary Equations.* Moscow: TsNIISK named Kucherenko DEP VHIIS 9.01.86 #6445.

**379.** Tseitlin, A.I., Petrosian, L.G., and Atadzhanov, D.R. (1987). About the methods of calculation, based on the application of boundary equations. In: *Study in the Dynamics of Construction*, 4–18. Moscow: TsNIISK named Kucherenko.

**380.** Tseitlin A.I., Petrosian L.G., Basilaya V.M. The dynamic calculation of plates on an elastic foundation by the method of boundary equations. Theses of reports of All – Union Conference on the dynamic bases, foundations, and underground structures. Narva, Leningrad, 1985.

**381.** Tseitlin, A.I., Petrosian, L.G., Kusainov, A.A., and Mamaeva, G.V. (1988). Design structures on dynamic and seismic impacts, taking into account damping. In: *9 WCEE, Ninth World Conference on Earthquake Engineering*, vol. **1**, 182–189. Tokyo-Kyoto, Japan: Technical Sections.

**382.** Tseitlin, A.I. and Plotnikov, Y.G. (1979). The free oscillations of system with frequency-independent internal friction. In: *Structural Mechanics and Calculation of Construction*, vol. **1**, 29–35. Moscow.

**383.** Tsudik, E. (2006). *Analysis of Beams and Frames on Elastic Foundation.* Trafford Publishing 248p.

**384.** Tsudik, E. (2012). *Structures on Elastic Foundations.* J. Ross publishing 600p.

**385.** Turner, M. (1967). The design of the constructions of minimal volume having specified frequency. *Rocket Eng. Cosmonaut.* **5** (3): 27–35.

**386.** Tiwari, K. and Kuppa, R. (2014). Overview of methods of analysis of beams on elastic foundation. *IOSR J. Mech. Civil Eng. (IOSR-JMCE)* e-ISSN: 2278 – 1684, p-ISSN: 2320-334X **11** (5 Ver. VI): 22–29. www.iosrjournals.org.

**387.** Ugodchikov, A.G. and Khutoryansky, N.M. (1986). *The Method of Boundary Elements in the Mechanics of Deformable Solid Body*. Kazan: Izdatelstvo KGU 295p.

**388.** Uflyand, Y.S. (1948). Wave propagation with the transverse oscillations of bars and plates. *Appl. Math. Mech.* **12** (3).

**389.** Uflyand, Y.S. (1967). *Integral Transformations in the Problems of the Theory of Elasticity*. Leningrad: Nauka 420p.

**390.** Vaydiner, A.I. and Moskvitin, V.V. (1976). Singular integral equations of the three-dimensional elasticity problems. *Dokladi AH USSR* **228** (6): 1310–1314.

**391.** Vekua, N.P. (1970). *The System of Singular Integral Equations*. Moscow: Science 379p.

**392.** Veletsos, A.S. and Verbic, B. (1974). Basic response functions for elastic foundations. *ASCE J. Eng. Mech. Div.* **100** (2): 189–202.

**393.** Ventsel E.S. Construction of integral equations of thin plates with complex shape using the method of compensating loads. Zbornik Trudov Moscovskogo Instituta Ingenerov selskokoz proisvodstva. 1974, Vol. II.

**394.** Veryuzhsky, Y.V. (1975). *The Application of the Method of Potential for the Solution of the Problems of Theory of Elasticity*. Kiev: KICI 114p.

**395.** Veryuzhsky, Y.V. (1977). *The Method of Integral Equations in the Mechanics of Deformable Bodies*. Kiev: KICI 232p.

**396.** Veryuzhsky, Y.V. (1980). *The Numerical Methods of the Potential in Some Problems of Applied Mechanics*. Kiev: Bisha shkola 183p.

**397.** Veryuzhsky, Y.V., Petrenko, A.Y., Savitsky, V.V., and Sinegaliev, M.K. (1982). Application of integral representations of displacements and stresses for the solution of the boundary problems of the theory of elasticity by numerically-analytical methods of potential. In: *Strength of Materials and Theory of Structures*, vol. **40**, 65–70. Kiev: Budivelnik.

**398.** Vladimrov, V.S. (1978). *The Generalized Functions in Mathematical Physics*. Moscow: Nauka 312p.

**399.** Vladimirov, V.S. (1981). *The Equations of Mathematical Physics*. Moscow: Nauka 512p.

**400.** Vlasov, V.Z. and Leontev, N. (1960). *Beams Plates and Shells on an Elastic Foundation*. Moscow: Fizmatgiz 498p.

**401.** Vorovich, I.I. and Babeshko, V.A. (1979). *The Dynamic Mixed Problems of the Theory of Elasticity for Non-classical Domains*. Moscow: Nauka 320p.

**402.** Wang, T.M. and Stephens, J.E. (1977). Natural frequencies of Timoshenko beams on Pasternak foundations. *J. Sound Vib.* **39**: 149–155.

**403.** Wang, Y., Wang, Y., Zhang, B., and Shepard, S. (2011). Transient responses of beam with elastic foundation supports under moving wave load excitation. *Int. J. Eng. Technol.* **2**.

**404.** Watson J.O. Analysis of Thick Shells with Holes by Using Integral Equation Method. Ph.D. Thes. – Southamton Univ., 1973.

**405.** Watson, G.H. (1949). *The Theory of Bessel Functions. Part III*. Moscow: Izdatelstvo Inostranoi Literaturi 797p.

**406.** Weinberg, D.V. and Sinyavsky, A.L. (1959). *Calculation of Shells*. Kiev: Gosstroyizdat USSR 119p.

**407.** Weinberg, D.V. and Sinyavsky, A.L. (1961). Approximate calculation of shells with cutouts by methods of potential theory. In: *Problems of Continuum Mechanics*, 34–42. Moscow: Izdatelstvo AH USSR.

**408.** Werner, S.D., Lee, L.C., Wong, L.H., and Trifunac, M.D. (1979). Factors involved in the seismic design of bridge abutments. In: *Proceeding of a Workshop on the Earthquake Resistance of Highway Bridges*. Berkeley, CA: Applied Technology Council.

**409.** Yanke, E., Edme, P., and Lesh, P. (1968). *Special Functions. Formulas, Tables and Graphs*, 344. Moscow: Nauka.

**410.** Zakaryan, V.A. (1971). Experimental study of the dynamic characteristics of high-rise buildings. *Izv. AH Arm. SSR, Ser. Tech. Nauk.* **XXIV** (3): 33–44.

**411.** Zargaryan, S.S. and Enfiadzhyan, R.L. (1972). Two-dimensional problem of the theory of elasticity for the circle with a radial cut section. *DAN Arm. SSR* **54**: 138–147.

**412.** Zargaryan S.S. The solution of the problems of the theory of elasticity for the domains with irregular boundary by the method of integral equations. // Abstract of thesis of Doctor of Physmath. Sciences. Moscow, 1985. 34p.

**413.** Zelenev, V.M., Meshkov, S.I., and Rossikhin, Y.A. (1970). *Damped Oscillations of Elastic – Hereditary System with the Weakly Singular Kernels*, vol. **2**, 14–118. PMTF.

**414.** Zhemochkin, B.N. and Sinitsyn, A.P. (1962). *Practical Methods of the Calculation of Beams and Plates Resting on an Elastic Foundation*. Moscou: Stroyizdat Publishing Company, USSR (Russian Edition).

**415.** Zienkiewicz, O.C. (1975). *Finite Element Method in Techniques (in Russian)*. Moscow: Mir 541 p.

**416.** Zienkiewicz, O.C. and Taylor, R.L. (1991). The finite element method. In: *Soil and Fluid Mechanics, Dynamics and Non-Linearity*, vol. **2**. London: McGraw-Hill Book Co.

**417.** Zissimos, P., Mourelatos, Z.P., and Parsons, M.G. (1987). *A Finite Element Analysis of Beams on Elastic Foundation Including Shear and Axial Effects*, 2. Ann Arbor, MI 48109, U.S.A: Department of Naval Architecture and Maine Engineering, The Uuniversity of Michigan.

**418.** Zolotov A.B. The formulation of Neumann's task for the equation of Poisson and second boundary-value problem of the three-dimensional theory of elasticity in the terms of the generalized functions.// Nauchnie Trudi. TSNIISK. Numerical Methods and Algorithms. 1970 Release 9, p. 5–10.

**419.** Zygmund, A. and Calderon, A.P. (1957). Singular integral operators and differential equations. *Am. J. Math.* **79**: 289–369.

# Index

## A

Abel's kernel,  333

Absolutely rigid body,  179, 290

Acceleration,  132, 142, 170, 175, 176, 207, 208, 210, 212, 216, 223, 224, 226, 237, 239, 240, 242, 255, 292, 293

Acceleration of gravity,  176, 225

Acceleration of mass,  132

Accelerograms
  of ground motion,  213
  of a real earthquake,  209, 216, 237, 239, 240, 263, 297
  of strong earthquakes,  208, 210, 213, 231, 285–287, 293

Acoustic medium,  300

Additional conditions,  47, 48, 56, 316

Additional operator,  39

Adjoint differential forms,  46

Adjoint differential operation,  78

Adjustable single-span,  202

Algebraic equation(s),  25, 38, 40, 84, 89, 144, 179, 181, 266, 281, 310, 343, 344, 349, 357, 399, 421, 435, 439

Amplitudes of displacement,  140, 141, 212, 278, 300, 336, 351

Amplitudes of oscillations,  151, 157, 201, 220, 222, 223, 230, 236, 316, 325, 326, 350, 363, 374

Analytical functions,  187, 233, 435

*Static and Dynamic Analysis of Engineering Structures: Incorporating the Boundary Element Method,*
First Edition. Levon G. Petrosian and Vladimir A. Ambartsumian.
© 2020 John Wiley & Sons Ltd. Published 2020 by John Wiley & Sons Ltd.